THE LYING STONES
OF MARRAKECH

Stephen Jay Gould is the Alexander Agassiz Professor of Zoology and Professor of Geology at Harvard University, and the Vincent Astor Visiting Professor of Biology at New York University. His publications include *Ever Since Darwin, Eight Little Piggies, Life's Grandeur, Questioning the Millennium, Leonardo's Mountain of Clams and the Diet of Worms, Bully for Brontosaurus* and *Wonderful Life. Wonderful Life* won the Science Book Prize for 1991.

Stephen Jay Gould

THE LYING STONES
OF MARRAKECH

Penultimate Reflections
in Natural History

VINTAGE

Published by Vintage 2001

2 4 6 8 10 9 7 5 3 1

Copyright © Turbo, Inc. 2000

Stephen Jay Gould has asserted his right under the Copyright,
Designs and Patents Act 1988 to be identified as the author of
this work

All of the essays contained in this work were previously
published by *Natural History* magazine

First published in Great Britain by
Jonathan Cape 2000

Vintage
Random House, 20 Vauxhall Bridge Road,
London SW1V 2SA

Random House Australia (Pty) Limited
20 Alfred Street, Milsons Point, Sydney
New South Wales 2061, Australia

Random House New Zealand Limited
18 Poland Road, Glenfield, Auckland 10,
New Zealand

Random House (Pty) Limited
Endulini, 5A Jubilee Road, Parktown 2193,
South Africa

The Random House Group Limited Reg. No. 954009
www.randomhouse.co.uk

A CIP catalogue record for this book
is available from the British Library

ISBN 0 09 928583 5

Papers used by Random House are natural, recyclable
products made from wood grown in sustainable forests.
The manufacturing processes conform to the environmental regulations of the country of origin

Printed and bound in Great Britain by
Cox & Wyman Limited

For Jack Sepkoski (1948–1999),
who brought me one of the greatest possible joys
a teacher can ever earn or experience:
to be surpassed by his students.
Offspring should not predecease their parents,
and students should outlive their teachers.
The times may be out of joint,
but Jack was born to set the order of life's history right—
and he did!

CONTENTS

THE

Lying Stones

of

Marrakech

PREFACE

In the fall of 1973, I received a call from Alan Ternes, editor of *Natural History* magazine. He asked me if I would like to write columns on a monthly basis, and he told me that folks actually get paid for such activities. (Until that day, I had published only in technical journals.) The idea intrigued me, and I said that I'd try three or four. Now, 290 monthly essays later (with never a deadline missed), I look only a little way forward to the last item of this extended series—to be written, as number 300 exactly, for the millennial issue of January 2001. One really should follow the honorable principle of quitting while still ahead, a rare form of dignity chosen by such admirable men as Michael Jordan and Joe DiMaggio, my personal hero and mentor from childhood. (Joe died, as I put this book together, full of years and in maximal style and grace, after setting one last record—for number of times in receiving last rites and then rallying.) Our millennial transition may represent an arbitrary imposition of human decisions upon nature's true cycles, but what grander symbol for calling a halt and moving on could possibly cross the path of a man's lifetime? This ninth volume of essays will therefore be the penultimate book in a series that shall close by honoring the same decimal preference lying behind our millennial transition.

If this series has finally found a distinctive voice, I have learned this mode of speech in the most gradual, accumulating, and largely unconscious manner—against my deepest personal beliefs in punctuational change and the uniquely directive power (despite an entirely accidental origin) of human reason in evolution. I suppose I had read a bit of Montaigne in English 101, and I surely could spell the word, but I had no inkling about the definitions and traditions of the essay as a literary genre when Alan Ternes called me cold on that fine autumn day.

I began the series with quite conventional notions about writing science for general consumption. I believed, as almost all scientists do (by passively imbibing a professional ethos, not by active thought or decision), that nature speaks directly to unprejudiced observers, and that accessible writing for nonscientists therefore required clarity, suppression of professional jargon, and an ability to convey the excitement of fascinating facts and interesting theories. If I supposed that I might bring something distinctive to previous efforts in this vein, I managed to formulate only two vague personal precepts: first, I would try to portray all subjects at the same conceptual depth that I would utilize in professional articles (that is, no dumbing down of ideas to accompany necessary clarification of language); second, I would use my humanistic and historical interests as a "user friendly" bridge to bring readers into the accessible world of science.

Over the years, however, this mere device (the humanistic "bridge") became an explicit centrality, a feature that I permitted myself to accept (and regard as a source of comfort and pride rather than an idiosyncrasy to downplay or even to hide) only when I finally realized that I had been writing *essays,* not mere columns, all along—and that nearly five hundred years of tradition had established and validated (indeed, had explicitly defined) the essay as a genre dedicated to personal musing and experience, used as a gracious entrée, or at least an intriguing hook, for discussion of general and universal issues. (Scientists are subtly trained to define the personal as a maximally dangerous snare of subjectivity and therefore to eschew the first person singular in favor of the passive voice in all technical writing. Some scientific editors will automatically bluepencil the dreaded *I* at every raising of its ugly head. Therefore, "popular science writing" and "the literary essay" rank as an ultimately disparate, if not hostile, pairing of immiscible oil and water in our usual view—a convention that I now dream about fracturing as a preeminent goal for my literary *and* scientific life.)

I have tried, as these essays developed over the years, to expand my humanistic "take" upon science from a simple practical device (my original intention, insofar as I had any initial plan at all) into a genuine emulsifier that might fuse the literary essay and the popular scientific article into something distinctive, something that might transcend our parochial disciplinary divisions for the benefit of both domains (science, because honorable personal expression by competent writers can't ever hurt; and composition, because the thrill of nature's factuality should not be excluded from the realm of our literary efforts). At the very least, such an undertaking can augment the dimensionality of popular scientific articles—for we lose nothing of science's factual beauty and meaning, while we add the complexity of how we come to know (or fail to learn) to conventional accounts of what we think we know.

As this series developed, I experimented with many styles for adding this humanistic component about how we learned (or erred) to standard tales about what, in our best judgment, exists "out there" in the natural world—often only to demonstrate the indivisibility of these two accounts, and the necessary embeddedness of "objective" knowledge within worldviews shaped by social norms and psychological hopes. But so often, as both Dorothy and T. S. Eliot recognized in their different ways, traditional paths may work best and lead home (because they have truly withstood the test of time and have therefore been honed to our deep needs and best modes of learning, not because we fall under their sway for reasons of laziness or suppression).

Despite conscious efforts at avoidance, I find myself constantly drawn to biography—for absolutely nothing can match the richness and fascination of a person's life, in its wondrous mixture of pure gossip, miniaturized and personalized social history, psychological dynamics, and the development of central ideas that motivate careers and eventually move mountains. And try as I may to ground biography in various central themes, nothing can really substitute for the sweep and storytelling power of chronology. (I regard the Picasso Museum in Paris and the Turner Wing of the Tate Gallery in London as my two favorite art museums because each displays the work of a great creator in the strict chronological order of his life. I can then devise whatever alternative arrangement strikes my own fancy and sense of utility—but the arrow of time cannot be replaced or set aside; even our claims for invariance must seek constant features of style or subject *through* time's passage.)

So I have struggled, harder and more explicitly than for anything else in my life as a writer, to develop a distinctive and personal form of essay to treat great scientific issues in the context of biography—and to do so not by the factual chronology of a life's sorrows and accomplishments (a noble task requiring the amplitude of a full book), but rather by the intellectual synergy between a person and the controlling idea of his life. In this manner, when the conceit works, I can capture the essence of a scientist's greatest labor, including the major impediments and insights met and gathered along the way, while also laying bare (in the spare epitome demanded by strictures of the essay as a literary form of limited length) the heart of a key intellectual concept in the most interesting microcosm of a person's formulation and defense.

The first three parts of this book apply this strategy to three different times, places, subjects, and worldviews—an extended test of my claim for a distinctive voice based on applying biographical perspectives to the illumination of key scientific concepts and their history (following the basic strategy, in each essay, of linking a person's central operating idea, the focus of a professional life in development, to an important concept in human understanding of the natural world—in other words, to summarize the range and power of a principle by exemplifying its role in the intellectual development of a particularly interesting scientist). Thus I have tried to encapsulate, in the unforgiving form of an essay, the essence of both a person (as expressed in the controlling idea of his scientific life) and a concept (through the quintessentially human device of displaying its development in an individual life).

Part I treats the most fascinating period in my own subject of paleontology, the premodern struggle (sixteenth to early eighteenth centuries) to understand the origin of fossils while nascent science struggled with the deepest of all questions about the nature of both causality and reality themselves. Are fossils the remains of ancient organisms on an old earth, or manifestations of a stable and universal order, symbolically expressed by correspondences among nature's three kingdoms of animal, vegetable, and mineral, with fossils arising entirely within the mineral kingdom as analogs of living forms in the other two realms? No subject could be more crucial, and no alternative view more eerily unfamiliar, than this particular battleground for the nature of reality. I present three variations upon this theme, each biographically expressed: the early-eighteenth-century tale of paleontology's most famous hoax, combined with a weirdly similar story from modern Morocco; the linkage of the unknown Stelluti to the preeminent Galileo through their friendship, and through a common error that unites the master's original view of Saturn with Stelluti's erroneous belief that petrified wood arose in the mineral kingdom; and finally, a "reversed" biography expressed in terms of an organism under study (the brachiopod fossils that were once called "vulva stones" for their resemblance to female genitalia) rather than a person pursuing the investigation.

Part II then discusses the greatest conjunction of a time, a subject, and a group of amazing people in the history of natural history: late-eighteenth- and early-nineteenth-century France, when a group including some of the most remarkable intellects of the millennium invented the scientific study of natural history in an age of revolution. Georges Buffon establishes a discipline, by the grandest route of virtually defining a new and historically based way of knowing, in the forty-four volumes of his eminently literary *Histoire naturelle,* and then loses public recognition, for interesting and understandable reasons, in the midst of his ubiquity. Antoine Lavoisier, the most stunningly incisive intellect I have ever encountered, literally adds a new dimension to our understanding of nature in the geometry of geological mapping, his one foray (amidst intentions cut short by the guillotine) into my profession. Jean-Baptiste Lamarck belies his own unfairly imposed reputation for error and inflexibility with a heartrending reassessment of the foundations of his own deepest belief—in an odyssey that begins with a handwritten comment and drawing, inked by Lamarck into his own copy of his first evolutionary treatise, and here discovered and presented for the first time.

Part III then illustrates the greatest British challenge to this continental pre-eminence: the remarkable, and wonderfully literate, leading lights of Victorian science in Darwin's age of turmoil and reassessment: the heart of Lyell's uniformitarianism as seen (literally) by visiting the site of his most famous visual image, the pillars of Pozzuoli, used as a frontispiece to all editions of his *Principles of Geology;* Darwin's own intellectual development from such an unpromising temperament and early training to an ultimately understandable role as the most gentle, yet thorough revolutionary in the history of science; Richard Owen's invention of dinosaurs as an explicit device to subvert the evolutionary views of a generation before Darwin; and Alfred Russel Wallace on Victorian certainties and subsequent unpredictabilities.

The last three parts of this book do not invoke biography so explicitly, but they also use the same device of embodying an abstraction within a particular that can be addressed in sufficient detail and immediate focus to fit within an essay. The interlude of part IV presents some experiments in the different literary form of short takes (op-ed pieces, obituary notices, and even, in one case, an introductory statement for Penguin CD's series of famous classical compositions). Here I include six attempts (the literal meaning of *essay*) to capture the most elusive and important subject of all: the nature and meaning of excellence, expressed as a general statement about substrates (chapter 11) followed by five iterations on the greatness of individuals and their central passions across a full range of human activity—for excellence must be construed as a goal for all varieties of deeds and seasons, not only for mental categories—from bodily grace and dignity within domains debased by the confusion of celebrity with stature; to distinctive individuality within corporate blandness; to the intellectual innovations more commonly cited by scholars to exemplify this most precious (and uncommon) of human attributes.

Part V, on scientific subjects with more obvious and explicit social consequences (and often, unacknowledged social origins as well), also uses biography, but in a different way to link past stories with present realities—to convey the lesson that claims for objectivity based on pure discovery often replay episodes buried in history, and proving (upon exhumation and linkage) that our modern certainties flounder within the same complexities of social context and mental blockage: Spencer's social Darwinism, the Triangle Shirtwaist fire, and modern eugenics (chapter 17); contemporary boasts about the discovery of genes for specific behaviors, Davenport's heritability of wanderlust, and the old medical theory of humors (18); Dolly the cloned sheep, the nature of identical twins, and the decapitation of Louis XVI (19); J. B. S. Haldane on the "humaneness" of poison gas in warfare, and the role and status of unpredictability in science (20).

Finally, part VI abandons biography for another device of essayists: major themes (about evolution's different expression across scales of size and time) cast into the epitome of odd or intriguing particulars: fossil embryos nearly 600 million years old (21); three stories about measurable evolution in snails, lizards, and fishes (22), conventionally misinterpreted as modest enough to prove the efficacy of Darwin's mechanism extended across the immensity of geological time, but far too rapid and convulsive to convey any such meaning when properly read at this grand and unfamiliar scale; and avoidance in antipathy among several Christian groups (23) that "share" Jerusalem's Church of the Holy Sepulchre (the traditional site of Christ's crucifixion).

At this equipoise, with one more foray into the breach yet to come, I can only thank readers who have joined me on this rocky journey. For only the conjunction of growing fellowship and increasing knowledge—a loop of ethical and intellectual, emotional and rational feedback that positively rings with the optimism of potential survival, maybe even transcendence, in this endlessly fascinating world of woe—can validate the accident of our existence by our free decision to make maximal use of those simple gifts that nature and evolution have granted us.

I

Episodes

in

the Birth

of

Paleontology

The Nature of Fossils

and the

History of the Earth

1

The
Lying Stones
of Marrakech

WE TEND TO THINK OF FAKERY AS AN ACTIVITY DEDI-
cated to minor moments of forgivable fun (from the whoopie cush-
ion to the squirting lapel flower), or harmless embellishment (from
my grandfather's vivid eyewitness tales of the Dempsey-Firpo fight
he never attended, to the 250,000 people who swear they were there
when Bobby Thomson hit his home run in a stadium with a maximal capacity of some fifty thousand).

But fakery can also become a serious and truly tragic business,
warping (or even destroying) the lives of thousands, and misdirect-
ing entire professions into sterility for generations. Scoundrels may
find the matrix of temptation irresistible, for immediate gains in
money and power can be so great, while human gullibility grants
the skillful forger an apparently limitless field of operation. The Van
Gogh *Sunflowers*, bought in 1987 by a Japanese insurance company

for nearly 25 million pounds sterling—then a record price for a painting—may well be a forged copy made around 1900 by the stockbroker and artist manqué Emile Schuffenecker. The phony Piltdown Man, artlessly confected from the jaw of an orangutan and a modern human cranium, derailed the profession of paleoanthropology for forty years, until exposed as a fake in the early 1950s.

Earlier examples cast an even longer and broader net of disappointment. A large body of medieval and Renaissance scholarship depended upon the documents of Hermes Trismegistus (Thrice-Great Hermes), a body of work attributed to Thoth, the Egyptian god of wisdom, and once viewed as equal in insight (not to mention antiquity) to biblical and classical sources—until exposed as a set of forgeries compiled largely in the third century A.D. And how can we possibly measure the pain of so many thousands of pious Jews, who abandoned their possessions and towns to follow the false messiah Shabbetai Tzevi to Jerusalem in the apocalyptic year of 1666—only to learn that their leader, imprisoned by the sultan and threatened with torture, had converted to Islam, been renamed Mehmed Efendi, and made the sultan's personal doorkeeper.

The most famous story of fraud in my own field of paleontology may not qualify for this first rank in the genre, but has surely won both general fame and staying power by persistence for more than 250 years. Like all great legends, this story has a canonical form, replete with conventional moral messages, and told without any variation in content across the centuries. Moreover, this standard form bears little relationship to the actual course of events as best reconstructed from available evidence. Finally, to cite the third common property of such legends, a correction of the conventional tale wins added and general value in teaching us important lessons about how we use and abuse our own history. Thus, the old story merits yet another retelling—which I first provide in the canonical (and false) version known to so many generations of students (and no doubt remembered by many readers from their college courses in natural science).

In 1726, Dr. Johann Bartholomew Adam Beringer, an insufferably pompous and dilettantish professor and physician from the town of Würzburg, published a volume, the *Lithographiae Wirceburgensis* (Würzburg lithography), documenting in copious words and twenty-one plates a remarkable series of fossils that he had found on a mountain adjacent to the city. These fossils portrayed a large array of objects, all neatly exposed in three-dimensional relief on the surface of flattened stones. The great majority depicted organisms, nearly all complete, including remarkable features of behavior and soft anatomy that had never been noted in conventional fossils—lizards in their skin, birds complete with beaks and eyes, spiders with their webs, bees feeding on flowers, snails next to their eggs, and frogs copulating. But others showed heavenly objects—comets with tails, the crescent moon with rays, and the sun all effulgent with a glowing central face of human form. Still others depicted Hebrew letters, nearly all spelling out the tetragrammaton, the ineffable name of God—YHVH, usually transliterated by Christian Europe as "Jehovah."

Beringer did recognize the difference between his stones and conventional fossils, and he didn't state a dogmatic opinion about their nature. But he didn't doubt their authenticity either, and he did dismiss claims that they had been carved by human hands, either recently in an attempt to defraud, or long ago for pagan purposes. Alas, after publishing his book and trumpeting the contents, Beringer realized that he had indeed been duped, presumably by his students playing a prank. (Some sources say that he finally acknowledged the trickery when he found his own name written in Hebrew letters on one stone.) According to legend, the brokenhearted Beringer then impoverished himself by attempting to buy back all copies of his book—and died dispirited just a few years later. Beringer's false fossils have been known ever since as *Lügensteine,* or "lying stones."

To illustrate the pedigree of the canonical tale, I cite the version given in the most famous paleontological treatise of the early nineteenth century, Dr. James Parkinson's *Organic Remains of a Former World* (volume 1, 1804). Parkinson, a physician by training and a fine paleontologist by avocation, identified and gave his name to the degenerative disease that continues to puzzle and trouble us today. He wrote of his colleague Beringer:

> One work, published in 1726, deserves to be particularly noticed; since it plainly demonstrates, that learning may not be sufficient to prevent an unsuspecting man, from becoming the dupe of excessive credulity. It is worthy of being mentioned on another account: the quantity of censure and ridicule, to which its author was exposed, served, not only to render his cotemporaries [*sic*] less liable to imposition; but also more cautious in indulging in unsupported hypotheses. . . . We are here presented with the representation of stones said to bear petrifactions of birds; some with spread, others with closed, wings: bees and wasps, both resting in their curiously constructed cells, and in the act of sipping honey from expanded flowers . . . and, to complete the absurdity, petrifactions representing the sun, moon, stars, and comets: with many others too monstrous and ridiculous to deserve even mention. These stones, artfully prepared,

had been intentionally deposited in a mountain, which he was in the habit of exploring, purposely to dupe the enthusiastic collector. Unfortunately, the silly and cruel trick, succeeded in so far, as to occasion to him, who was the subject of it, so great a degree of mortification, as, it is said, shortened his days.

All components of the standard story line, complete with moral messages, have already fallen into place—the absurdity of the fossils, the gullibility of the professor, the personal tragedy of his undoing, and the two attendant lessons for aspiring young scientists: do not engage in speculation beyond available evidence, and do not stray from the empirical method of direct observation.

In this century's earlier and standard work on the history of geology (*The Birth and Development of the Geological Sciences,* published in 1934), Frank Dawson Adams provides some embellishments that had accumulated over the years, including the unforgettable story, for which not a shred of evidence has ever existed, that Beringer capitulated when he found his own name in Hebrew letters on one of his stones. Adams's verbatim "borrowing" of Parkinson's last line also illustrates another reason for invariance of the canonical tale: later retellings copy their material from earlier sources:

> Some sons of Belial among his students prepared a number of artificial fossils by moulding forms of various living or imaginary things in clay which was then baked hard and scattered in fragments about on the hillsides where Beringer was wont to search for fossils. . . . The distressing climax was reached, however, when later he one day found a fragment bearing his own name upon it. So great was his chagrin and mortification in discovering that he had been made the subject of a cruel and silly hoax, that he endeavored to buy up the whole edition of his work. In doing so he impoverished himself and it is said shortened his days.

Modern textbooks tend to present a caricatured "triumphalist" account in their "obligatory" introductory pages on the history of their discipline—the view that science marches inexorably forward from dark superstition toward the refining light of truth. Thus, Beringer's story tends to acquire the additional moral that his undoing at least had the good effect of destroying old nonsense about the inorganic or mysterious origin of fossils—as in this text for first-year students, published in 1961:

The idea that fossils were merely sports of nature was finally killed
by ridicule in the early part of the eighteenth century. Johann
Beringer, a professor at the University of Würzburg, enthusiastically
argued against the organic nature of fossils. In 1726, he published a
paleontological work . . . which included drawings of many true
fossils but also of objects that represented the sun, the moon, stars,
and Hebraic letters. It was not till later, when Beringer found a "fossil" with his own name on it, that he realized that his students, tired
of his teachings, had planted these "fossils" and carefully led him to
discover them for himself.

A recent trip to Morocco turned my thoughts to Beringer. For several years,
I have watched, with increasing fascination and puzzlement, the virtual
"takeover" of rock shops throughout the world by striking fossils from
Morocco—primarily straight-shelled nautiloids (much older relatives of the
coiled and modern chambered nautilus) preserved in black marbles and lime-
stones, and usually sold as large, beautifully polished slabs intended for table or
dresser tops. I wondered where these rocks occurred in such fantastic abundance; had the High Atlas Mountains been quarried away to sea level? I wanted
to make sure that Morocco itself still existed as a discrete entity and not only
as disaggregated fragments, fashioning the world's coffee tables.

I discovered that most of these fossils come from quarries in the rocky
deserts, well and due east of Marrakech, and not from the intervening mountains. I also learned something else that alleviated my fears about imminent dis-
persal of an entire patrimony. Moroccan rock salesmen dot the landscape in
limitless variety—from young boys hawking a specimen or two at every hairpin
turn on the mountain roads, to impromptu stands at every lookout point, to large
and formal shops in the cities and towns. The aggregate volume of rock must
be immense, but the majority of items offered for sale are either entirely phony
or at least strongly "enhanced." My focus of interest shifted dramatically: from
worrying about sources and limits to studying the ranges and differential exper-
tises of a major industry dedicated to the manufacture of fake fossils.

I must judge some "enhancements" as quite clever—as when the strong ribs
on the shell of a genuine ammonite are extended by carving into the smallest
and innermost whorls and then "improved" in regular expression on the outer
coil. But other "ammonites" have simply been carved from scratch on a
smoothed rock surface, or even cast in clay and then glued into a prepared hole in
the rock. Other fakes can only be deemed absurd—as in my favorite example

of a wormlike "thing" with circles on its back, grooves on both sides, eyes on a head shield, and a double projection, like a snake's forked tongue, extending out in front. (In this case the forger, too clever by half, at least recognized the correct principle of parts and counterparts—for the "complete" specimen includes two pieces that fit together, the projecting "fossil" on one slab, and the negative impression on the other, where the animal supposedly cast its form into the surrounding sediment. The forger even carved negative circles and grooves into the counterpart image, although these impressions do not match the projecting, and supposedly corresponding, embellishments on the "fossil" itself.)

But one style of fakery emerges as a kind of "industry standard," as defined by constant repetition and presence in all shops. (Whatever the unique and personal items offered for sale in any shop, this *vin ordinaire* of the genre always appears in abundance.) These "standards" feature small (up to four or six inches in length) flattened stones with a prominent creature spread out in three dimensions on the surface. The fossils span a full range from plausible "trilobites," to arthropods (crabs, lobsters, and scorpions, for example) with external hard parts that might conceivably fossilize (though never in such complete exactitude), to small vertebrates (mostly frogs and lizards) with a soft exterior, including such delicate features as fingers and eyes that cannot be preserved in the geological record.

After much scrutiny, I finally worked out the usual mode of manufacture. The fossil fakes are plaster casts, often remarkably well done. (The lizard that I bought, as seen in the accompanying photograph, must have been cast from life, for a magnifying glass reveals the individual pores and scales of the skin.) The forger cuts a flat surface on a real rock and then cements the plaster cast to this substrate. (If you look carefully from the side, you can always make out the junction of rock and plaster.) Some fakes have been crudely confected, but the best examples match the color and form of rock to overlying plaster so cleverly that distinctions become nearly invisible.

A fake fossil reptile from a Moroccan rock shop. Done in plaster from a live cast and then glued to the rock.

When I first set eyes on these fakes, I experienced the weirdest sense of déjà vu, an odd juxtaposition of old and new that sent shivers of fascination and discomfort up and down my spine—a feeling greatly enhanced by a day just spent in the medina of Fez, the ancient walled town that has scarcely been altered by a millennium of surrounding change, where only mules and donkeys carry the goods of commerce, and where high walls, labyrinthine streets, tiny open shops, and calls to prayer, enhanced during the fast of Ramadan, mark a world seemingly untouched by time, and conjuring up every stereotype held by an uninformed West about a "mysterious East." I looked at these standard fakes, and I saw Beringer's *Lügensteine* of 1726. The two styles are so uncannily similar that I wondered, at first, if the modern forgers had explicitly copied the plates of the *Lithographiae Wirceburgensis*—a silly idea that I dropped as soon as I returned and consulted my copy of Beringer's original. But the similarities remain overwhelming. I purchased two examples—a scorpion of sorts and a lizard—as virtual dead ringers for Beringer's *Lügensteine,* and I present a visual comparison of the two sets of fakes, separated by 250 years and a different process of manufacture (carved in Germany, cast in Morocco). I only wonder if the proprietor believed my assurances, rendered in my best commercial French, that I was a

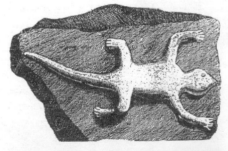

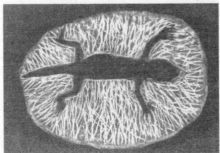

The striking similarity between the most famous fake in the history of paleontology (Beringer's Lügensteine, *or "lying stones," of 1726) and a modern Moroccan fabrication.*

professional paleontologist, and that his wares were *faux, absolument et sans doute*—or if he thought that I had just devised a bargaining tactic more clever than most.

But an odd similarity across disparate cultures and centuries doesn't provide a rich enough theme for an essay. I extracted sufficient generality only when I realized that this maximal likeness in appearance correlates with a difference in meaning that couldn't be more profound. A primary strategy of the experimental method in science works by a principle known since Roman times as *ceteris paribus* ("all other things being equal")—that is, if you wish to understand a controlling difference between two systems, keep all other features constant, for the difference may then be attributed to the only factor that you have allowed to vary. If, for example, you wish to test the effect of a new diet pill, try to establish two matched groups—folks of the same age, sex, weight, nutrition, health, habits, ethnicity, and so on. Then give the pill to one group and a placebo to the other (without telling the subjects what they have received, for such knowledge would, in itself, establish inequality based on differing psychological expectations). The technique, needless to say, does not work perfectly (for true *ceteris paribus* can never be obtained), but if the pill group loses a lot of weight, and the placebo group remains as obese as before, you may conclude that the pill probably works as hoped.

Ceteris paribus represents a far more distant pipe dream in trying to understand two different contexts in the developing history of a profession—for we cannot now manipulate a situation of our own design, but must study past circumstances in complex cultures not subject to regulation by our experimental ideals at all. But any constancy between the two contexts increases our hope of illustrating and understanding their variations in the following special way: if we examine the different treatment of the same object in two cultures, worlds apart, then at least we can attribute the observed variation to cultural distinctions, for the objects treated do not vary.

The effectively identical *Lügensteine* of early-eighteenth-century Würzburg and modern Marrakech embody such an interesting difference in proposed meaning and effective treatment by two cultures—and I am not sure that we should be happy about the contrast of then and now. But we must first correct the legend of Beringer and the original *Lügensteine* if we wish to grasp the essential difference.

As so often happens when canonical legends arise to impart moral lessons to later generations, the standard tale distorts nearly every important detail of Beringer's sad story. (I obtained my information primarily from an excellent

Note the exuberance and (by modern standards) whimsical nature of Beringer's fake fossils from 1726.

book published in 1963 by Melvin E. Jahn and Daniel J. Woolf, *The Lying Stones of Dr. Beringer*, University of California Press. Jahn and Woolf provide a complete translation of Beringer's volume, along with extensive commentary about the paleontology of Beringer's time. I used original sources from my own library for all quotations not from Beringer in this essay.)

First of all, on personal issues not directly relevant to the theme of this essay, Beringer wasn't tricked by a harmless student prank but rather purposely defrauded by two colleagues who hated his dismissive pomposity and wished to bring him down. These colleagues—J. Ignatz Roderick, professor of geography and algebra at the University of Würzburg, and Georg von Eckhart, librarian to the court and the university—"commissioned" the fake fossils (or, in Roderick's case, probably did much of the carving himself), and then hired a seventeen-year-old boy, Christian Zänger (who may also have helped with the carving), to plant them on the mountain. Zänger, a double agent of sorts, was then hired by Beringer (along with two other boys, both apparently innocent of the fraud) to excavate and collect the stones.

This information for revising the canonical tale lay hidden for two hundred years in the incomplete and somewhat contradictory records of hearings held in April 1726 before the Würzburg cathedral chapter and the city hall of Eivelstadt (the site of Beringer's mountain just outside Würzburg). The German scholar Heinrich Kirchner discovered these documents in 1934 in the town archives of Würzburg. These hearings focus on testimony of the three boys. Zänger, the "double agent," states that Roderick had devised the scheme because he "wished to accuse Dr. Beringer . . . because Beringer was so arrogant and despised them all." I was also impressed by the testimony of the two brothers hired by Beringer. Their innocence seems clear in the wonderfully ingenuous statement of Nicklaus Hahn that if he and his brother "could make such stones, they wouldn't be mere diggers."

The canonical tale may require Beringer's ruin to convey a desired moral, but the facts argue differently. I do not doubt that the doctor was painfully embarrassed, even mortified, by his exposed gullibility; but he evidently recovered, kept his job and titles, lived for another fourteen years, and published several more books (including, though probably not by his design or will, a posthumous second edition of his *Würzburg Lithography*!). Eckhart and Roderick, on the other hand, fell into well-earned disgrace. Eckhart died soon thereafter, and Roderick, having left Würzburg (voluntarily or not, we do not know), then wrote a humbling letter to the prince-bishop begging permission to return—which his grace allowed after due rebuke for Roderick's past deeds—and to regain access to the library and archives so that he could write a proper obituary for his deceased friend Eckhart.

But on the far more important intellectual theme of Beringer's significance in the history of paleontology, a different kind of correction inverts the conventional story in a particularly meaningful way. The usual cardboard tale of progressive science triumphant over past ignorance requires that benighted "bad guys," who upheld the old ways of theological superstition against objective evidence of observational science, be branded as both foolish and stubbornly unwilling to face nature's factuality. Since Beringer falls into this category of old and bad, we want to see him as hopelessly duped by preposterous fakes that any good observer should have recognized—hence the emphasis, in the canonical story, on Beringer's mortification and on the ridiculous character of the *Lügensteine* themselves.

The Würzburg carvings are, of course, absurd by modern definitions and understanding of fossils. We know that spiders' webs and lizards' eyes—not to mention solar rays and the Hebrew name of God—cannot fossilize, so the *Lügensteine* can only be human carvings. We laugh at Beringer for not making an identification that seems so obvious to us. But in so doing, we commit the greatest of all historical errors: arrogantly judging our forebears in the light of modern knowledge perforce unavailable to them. Of course the *Lügensteine* are preposterous, once we recognize fossils as preserved remains of ancient organisms. By this criterion, letters and solar emanations cannot be real fossils, and anyone who unites such objects with plausible images of organisms can only be a fool.

But when we enter Beringer's early-eighteenth-century world of geological understanding, his interpretations no longer seem so absurd. First of all, Beringer was puzzled by the unique character of his *Lügensteine,* and he adopted no dogmatic position about their meaning. He did regard them as natural and not carved (a portentous error, of course), but he demurred on further judgment and repeatedly stated that he had chosen to publish in order to provide information so that others might better debate the nature of fossils—a tactic that scientists supposedly value. We may regard the closing words of his penultimate chapter as a tad grandiose and self-serving, but shall we not praise the sentiment of openness?

> I have willingly submitted my plates to the scrutiny of wise men, desiring to learn their verdict, rather than to proclaim my own in this totally new and much mooted question. I address myself to scholars, hoping to be instructed by their most learned responses. . . . It is my fervent expectation that illustrious lithographers will shed light upon this dispute which is as obscure as it is unusual. I shall add thereto my own humble torch, nor shall I spare

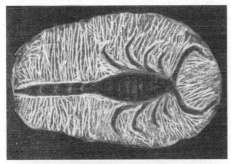

Another comparison between German fake fossils of 1726 and modern Moroccan fabrications.

any effort to reveal and declare whatever future yields may rise from the Würzburg field under the continuous labors of my workers, and whatever opinion my mind may embrace.

More importantly, Beringer's hoaxers had not crafted preposterous objects but had cleverly contrived—for their purposes, remember, were venomous, not humorous—a fraud that might fool a man of decent will and reasonable intelligence by standards of interpretation then current. Beringer wrote his treatise at the tail end of a debate that had engulfed seventeenth-century science and had not yet been fully resolved: what did fossils represent, and what did they teach us about the age of the earth, the nature of our planet's history, and the meaning and definition of life?

Beringer regarded the *Lügensteine* as "natural" but not necessarily as organic in origin. In the great debate that he knew and documented so well, many scientists viewed fossils as inorganic products of the mineral realm that somehow mimicked the forms of organisms but might also take the shapes of other objects, including planets and letters. Therefore, in Beringer's world, the *Lügensteine* could not be dismissed as preposterous prima facie. This debate could not have engaged broader or more crucial issues for the developing sciences of geology and biology—for if fossils represent the remains of organisms, then the earth must be ancient, life must enjoy a long history of consistent change, and rocks must form from the deposition and hardening of sediments. But if fossils can originate as inorganic results of a "plastic power" in the mineral kingdom (that can fashion other interesting shapes like crystals, stalactites, and banded agates in different circumstances), then the earth may be young and virtually unchanged (except for the ravages of Noah's flood), while rocks, with their enclosed fossils, may be products of the original creation, not historical results of altered sediments.

If pictures of planets and Hebrew letters could be "fossils" made in the same way as apparent organisms, then the inorganic theory gains strong support—for a fossilized aleph or moonbeam could not be construed as a natural object deposited in a streambed and then fossilized when the surrounding sediment became buried and petrified. The inorganic theory had been fading rapidly in Beringer's time, while the organic alternative gained continually in support. But the inorganic view remained plausible, and the *Lügensteine* therefore become clever and diabolical, not preposterous and comical.

In Beringer's day, many scientists believed that simple organisms arose continually by spontaneous generation. If a polyp can originate by the influence of sunshine upon waters, or a maggot by heat upon decaying flesh, why not con-

jecture that simple images of objects might form upon rocks by natural interactions of light or heat upon the inherent "lapidifying forces" of the mineral
kingdom? Consider, moreover, how puzzling the image of a fish *inside* a rock
must have appeared to people who viewed these rocks as products of an original creation, not as historical outcomes of sedimentation. How could an organism get inside; and how could fossils be organisms if they frequently occur
petrified, or made of the same stone as their surroundings? We now have simple and "obvious" answers to these questions, but Beringer and his colleagues
still struggled—and any sympathetic understanding of early-eighteenth-
century contexts should help us to grasp the centrality and excitement of these
debates and to understand the *Lügensteine* as legitimately puzzling.

I do not, however, wish to absolve Beringer of all blame under an indefen-
sibly pluralistic doctrine that all plausible explanations of past times may claim
the same weight of judicious argument. The *Lügensteine* may not have been
absurd, but Beringer had also encountered enough clues to uncover the hoax
and avoid embarrassment. However, for several reasons involving flaws in character and passable intelligence short of true brilliance, Beringer forged on,
finally trumping his judgment by his desire to be recognized and honored for
a great discovery that had consumed so much of his time and expense. How
could he relinquish the fame he could almost taste in writing:

> Behold these tablets, which I was inspired to edit, not only by my tire-
> less zeal for public service, and by your wishes and those of my many
> friends, and by my strong filial love for Franconia, to which, from
> these figured fruits of this previously obscure mountain, no less glory
> will accrue than from the delicious wines of its vine-covered hills.

I am no fan of Dr. Beringer. He strikes me, first of all, as an insufferable
pedant—so I can understand his colleagues' frustration, while not condoning
their solutions. (I pride myself on always quoting from original sources, and I
do own a copy of Beringer's treatise. I am no Latin scholar, but I can read and
translate most works in this universal scientific language of Beringer's time. But
I cannot make head or tail of the convoluted phrasings, the invented words, the
absurdly twisted sentences of Beringer's prose, and I have had to rely on Jahn
and Woolf's translation previously cited.)

Moreover, Beringer saw and reported more than enough evidence to
uncover the hoax, had he been inclined to greater judiciousness. He noted that
his *Lügensteine* bore no relationship to any other objects known to the bur-
geoning science of paleontology, not even to the numerous "real" fossils also

found on his mountain. But instead of alerting him to possible fraud, these differences only fueled Beringer's hopes for fame. He made many observations that should have clued him in (even by standards of his own time) to the artificial carving of his fossils: why were they nearly always complete, and not usually fragmentary like most other finds; why did each object seem to fit so snugly and firmly on the enclosing rock; why did only the top sides protrude, while the lower parts merged with the underlying rock; why had letters and sunbeams not been found before; why did nearly all fossils appear in the same orientation, splayed out and viewed from the top, never from the side or bottom? Beringer's own words almost shout out the obvious and correct conclusion that he couldn't abide or even discern: "The figures expressed on these stones, especially those of insects, are so exactly fitted to the dimensions of the stones, that one would swear that they are the work of a very meticulous sculptor."

Beringer's arrogance brought him down in a much more direct manner as well. When Eckhart and Roderick learned that Beringer planned to publish his work, they realized that they had gone too far and became frightened. They tried to warn Beringer, by hints at first but later quite directly as their anxiety increased. Roderick even delivered some stones to Beringer and later showed his rival how they had been carved—hoping that Beringer would then draw an obvious inference about the rest of his identical collection.

Beringer, however, was now committed and would not be derailed. He replied with the argument of all true believers, the unshakable faith that resists all reason and evidence: yes, you have proven that *these* psychics are frauds, but *my* psychics are the real McCoy, and I must defend them even more strongly now that you have heaped unfair calumnies upon the entire enterprise. Beringer never mentions Eckhart and Roderick by name (so their unveiling awaited the 1934 discovery in the Würzburg town archives), but he had been forewarned of their activities. Beringer wrote in chapter 12 of his book:

> Then, when I had all but completed my work, I caught the rumor circulating throughout the city . . . that every one of these stones . . . was recently sculpted by hand, made to look as though at different periods they had been resurrected from a very old burial, and sold to me as to one indifferent to fraud and caught up in the blind greed of curiosity.

Beringer then tells the tale of Roderick's warning but excoriates his rival as an oafish modern caricature of Praxiteles (the preeminent Greek sculptor), out to discredit a great discovery by artificial mimicry:

Our Praxiteles has issued, in an arrogant letter, a declaration of war. He has threatened to write a small treatise exposing my stones as supposititious [*sic*]—I should say, his stones, fashioned and fraudulently made by his hand. Thus does this man, virtually unknown among men of letters, still but a novice in the sciences, make a bid for the dawn of his fame in a shameful calumny and imposture.

If only Beringer had realized how truly and comprehensively he had spoken about "a shameful calumny and imposture." But Roderick succeeded because he had made his carvings sufficiently plausible to inspire belief by early-eighteenth-century standards. The undoing of all protagonists then followed because Beringer, in his overweening and stubborn arrogance, simply could not quench his ambition once a clever and plausible hoax had unleashed his ardor and vanity.

In summary, the *Lügensteine* of Würzburg played a notable role in the most important debate ever pursued in paleontology—a struggle that lasted for centuries and that placed the nature of reality itself up for grabs. By Beringer's time, this debate had largely been settled in favor of the organic nature of fossils, and this resolution would have occurred even if Beringer had never been born and the *Lügensteine* never carved. Beringer may have been a vain and arrogant man of limited talent, working in an academic backwater of his day, but at least he struggled with grand issues—and he fell because his hoaxers understood the great stakes and fashioned frauds that could be viewed as cleverly relevant to this intellectual battle, however preposterous they appear to us today with our additional knowledge and radically altered theories about the nature of reality and causation.

(One often needs a proper theory to set a context for the exposure of fraud. Piltdown Man fooled some of the world's best scientists for generations. I will never forget what W. E. le Gros Clark, one of the three scientists who exposed the fraud in the early 1950s, said to me when I asked him why the hoax had stood for forty years. Even an amateur in vertebrate anatomy— as this snail man can attest from personal experience—now has no trouble seeing the Piltdown bones for what they are. The staining is so crude, and the recent file marks on the orangutan teeth in the lower jaw so obvious—yet so necessary to make them look human in the forgers' plan, for the cusps of ape and human teeth differ so greatly. Le Gros Clark said to me: "One needed to approach the bones with the hypothesis of fraud already in mind. In such a context, the fakery immediately became obvious.")

The *Lügensteine* of Marrakech are, by contrast—and I don't know how else to say this—merely ludicrous and preposterous. No excuse save ignorance—and I do, of course, recognize the continued prevalence of this all-too-human trait—could possibly inspire a belief that the plaster blobs atop the Moroccan stones might be true fossils, the remains of ancient organisms. Beringer was grandly tricked in the pursuit of great truths, however inadequate his own skills. We are merely hoodwinked for a few dollars that mean little to most tourists but may make or break the lives of local carvers. *Caveat emptor.*

In contrasting the conflicting meanings of these identical fakes in such radically different historical contexts, I can only recall Karl Marx's famous opening line to *The Eighteenth Brumaire of Louis Napoleon,* his incisive essay on the rise to power of the vain and cynical Napoleon III after the revolution of 1848, in contrast with the elevated hopes and disappointments inspired by the original Napoleon. (The French revolutionary calendar had renamed the months and started time again at the establishment of the Republic. In this system, Napoleon's coup d'état occurred on the eighteenth of Brumaire, a foggy month in a renamed autumn, of year VIII—or November 9, 1799. Marx, now justly out of fashion for horrors later committed in his name, remains a brilliant analyst of historical patterns.) Marx opened his polemical treatise by noting that all great events in history occur twice—the first time as tragedy, and the second as farce.

Beringer was a pompous ass, and his florid and convoluted phrases represent a caricature of true scholarship. Still, he fell in the course of a great debate, using his limited talents to defend an inquiry that he loved and that even more pompous fools of his time despised—those who argued that refined people wouldn't dirty their hands in the muck of mountains but would solve the world's pressing issues under their wigs in their drawing rooms. Beringer characterized this opposition from the pseudo-elegant glitterati of his day:

> They pursue [paleontology] with an especially censorious rod, and condemn it to rejection from the world of erudition as one of the wanton futilities of intellectual idlers. To what purpose, they ask, do we stare fixedly with eye and mind at small stones and figured rocks, at little images of animals or plants, the rubbish of mountain and stream, found by chance amid the muck and sand of land and sea?

He then defended his profession with the greatest of geological metaphors:

> any [paleontologist], like David of old, would be able with one flaw-less stone picked from the bosom of Nature, to prostrate, by one blow

on the forehead, the gigantic mass of objections and satires and to vindicate the honor of this sublime science from all its calumniators.

Beringer, to his misfortune and largely as a result of his own limitations, did not pick a "flawless stone," but he properly defended the importance of paleontology and of empirical science in general. As a final irony, Beringer could not have been more wrong about the *Lügensteine,* but he couldn't have been more right about the power of paleontology. Science has so revolutionized our view of reality since 1726 that we, in our current style of arrogance, can only regard the Würzburg *Lügensteine* as preposterous, because we unfairly impose our modern context and fail to understand Beringer's world, including the deep issues that made his hoaxing a tragedy rather than a farce.

Our current reality features an unslayable Goliath of commercialism, and modern scientific Davids must make an honorable peace, for a slingshot cannot win this battle. I may be terribly old-fashioned (shades, I hope not, of poor Beringer)—but I continue to believe that such honor can only be sought in separation and mutual respect. Opportunities for increasing fusion with the world of commerce surround us with almost overwhelming temptation, for the immediate and palpable "rewards" are so great. So scientists go to work for competing pharmaceutical or computer companies, make monumental salaries, but cannot choose their topics of research or publish their work. And museums expand their gift shops to the size of their neglected exhibit halls, and purvey their dinosaurs largely for dollars in the form of images on coffee mugs and T-shirts, or by special exhibits, at fancy prices, of robotic models, built by commercial companies, hired for the show, and featuring, as their come-on, the very properties—mostly hideous growls and lurid colors—that leave no evidence in the fossil record and therefore remain a matter of pure conjecture to science.

I am relieved that Sue the *Tyrannosaurus,* sold at auction by Sotheby's for more than 8 million dollars, will go to Chicago's Field Museum and not to the anonymity of some corporate boardroom, to stand (perhaps) next to a phony Van Gogh. But I am not happy that no natural history museum in the world can pony up the funds for such a purpose—and that McDonald's had to provide the cash. McDonald's is not, after all, an eleemosynary institution, and they will legitimately want their piece for their price. Will it be the Happy Meal Hall of Paleontology at the Field Museum? (Will we ever again be able to view a public object with civic dignity, unencumbered by commercial messages? Must city buses be fully painted as movable ads, lampposts smothered, taxis festooned, even seats in concert halls sold one by one to donors and embellished in perpetuity with their names on silver plaques?) Or will we soon see Sue the

Robotic Tyrannosaur—the purchase of the name rather than the thing, for Sue's actual skeleton cannot improve the colors or sounds of the robots, and her value, in this context, lies only in the recognition of her name (and the memory of the dollars she attracted), not in her truly immense scientific worth.

I am neither an idealist nor a Luddite in this matter. I just feel that the world of commerce and the world of intellect, by their intrinsic natures, must pursue different values and priorities—while the commercial world looms so much larger than our domain that we can only be engulfed and destroyed if we make a devil's bargain of fusion for short-term gain. The worth of fossils simply cannot be measured in dollars. But the *Lügensteine* of Marrakech can only be assessed in this purely symbolic way—for the Moroccan fakes have no intellectual value and can bring only what the traffic (and human gullibility) will bear. We cannot possibly improve upon Shakespeare's famous words for this sorry situation—and this ray of hope for the honor and differences of intellect over cash:

> *Who steals my purse steals trash . . .*
> *But he that filches from me my good name*
> *Robs me of that which not enriches him,*
> *And makes me poor indeed.*

But we must also remember that these words are spoken by the villainous Iago, who will soon make Othello a victim, by exploiting the Moor's own intemperance, of the most poignant and tragic deception in all our literature. Any modern intellectual, to avoid Beringer's sad fate, must hold on to the dream—while keeping a cold eye on immediate realities. Follow your bliss, but remember that handkerchiefs can be planted for evil ends and fossils carved for ready cash.

2

The
Sharp-Eyed Lynx,
Outfoxed by Nature

I. GALILEO GALILEI AND THE THREE GLOBES
OF SATURN

IN 1603, FEDERICO CESI, THE DUKE OF ACQUASPARTA,
founded an organization that grew from uncertain beginnings to
become the first scientific society in modern European history. Cesi
(1585–1630), a teenaged nobleman, invited three slightly older
friends (all in their mid-twenties) to establish the Accademia dei
Lincei (Academy of the Lynxes), dedicated to scientific investigation
("reading this great, true, and universal book of the world," to cite
Cesi's own words), and named for a sleek and wily carnivore, then
still living in the forests of Italy and renowned in song and story for
unparalleled sight among mammals.

The legend of the sharp-eyed lynx had arisen in ancient times
and persisted to Cesi's day. Pliny's canonical compendium of

The official emblem of Europe's first scientific society, the Accademia dei Lincei (Academy of the Lynxes), founded in 1603 and including Galileo as an early member.

natural history had called the lynx "the most clear sighted of all quadrupeds." Plutarch had embellished the legend by speaking of "the lynx, who can penetrate through trees and rocks with its sharp sight." And Galen, ever the comparative anatomist, had written: "We would seem absurdly weak in our powers of vision if we compared our sight to the acuity of the lynx or the eagle." (I have translated these aphorisms directly from Conrad Gesner's 1551 compendium on mammals, the standard source for such information in Cesi's day.)

Still, despite Cesi's ambitious names and aims, the academy of four young men faltered at first. Cesi's father made a vigorous attempt to stop his son's foolishness, and the four Lynxes all dispersed to their native cities, keeping their organization alive only by the uncertain media of post and messages. But Cesi persevered and triumphed (for a time), thanks to several skills and circumstances. He acquired more power and prestige, both by growing up and by inheriting substantial wealth. Most importantly, he became a consummate diplomat and facilitator within the maximally suspicious and labyrinthine world of civil and ecclesiastical politics in Rome during the Counter-Reformation. The Lynxes flourished largely because Cesi managed to keep the suspicions of popes and cardinals at bay, while science prepared to fracture old views of the cosmos, and to develop radically new theories about the nature of matter and causation.

As a brilliant administrator, Cesi knew that he needed more clout among the membership of the Lynxes. He therefore recruited, as the fifth and sixth members of an organization that would eventually reach a roster of about thirty, two of the most prestigious thinkers and doers of early-seventeenth-century life. In 1610, he journeyed to Naples, where he persuaded the senior spokesman of the fading Neoplatonic school—the seventy-five-year-old Giambattista

Della Porta—to join a group of men young enough to be his grandsons. Then, in 1611, Cesi made his preeminent catch, when he recruited the hottest intellectual property in the Western world, Galileo Galilei (1564–1642), to become the sixth member of the Lynxes.

The year before, in 1610, Galileo had provided an ultimate proof for the cliché that good things come in small packages by publishing *Sidereus nuncius* (Starry messenger)—little more than a pamphlet really, but containing more oomph per paragraph than anything else ever achieved in the history of science or printing. Galileo shook the earth by turning his newly invented telescope upon the cosmos and seeing the moon as a planet with mountains and valleys, not as the perfect sphere required by conventional concepts of science and theology. Galileo also reported that thousands of previously invisible stars build the Milky Way, thus extending the cosmos beyond any previously conceivable limit; and that four moons orbit Jupiter, forming a miniature world analogous to the motion of planets around a central body. Moreover, Galileo pointed out, if satellites circle planets, then the crystalline spheres, supposedly marking the domain of each planet, and ordered as a set of concentric shells around the central earth, could not exist—for the revolution of moons would shatter these mystical structures of a geometrically perfect, unsullied, and unchanging cosmos, God's empyrean realm.

But Galileo also made some errors in his initial survey, and I have always been struck that standard books on the history of astronomy, written in the heroic or hagiographical mode, almost never mention (or relegate to an awkward footnote) the most prominent of Galileo's mistakes—for the story strikes me as fascinating and much more informative about the nature of science, and of creativity in general, than any of his valid observations.

Galileo also focused his telescope on Saturn, the most distant of the visible planets—and he saw the famous rings. But he could not properly visualize or interpret what he had observed, presumably because his conceptual world lacked the requisite "space" for such a peculiar object (while his telescope remained too crude to render the rings with enough clarity to force his mind, already benumbed by so many surprises, to the most peculiar and unanticipated conclusion of all).

The stymied Galileo looked and looked, and focused and focused, night after night. He finally interpreted Saturn as a threefold body, with a central sphere flanked by two smaller spheres of equal size, each touching the main planet. Following a common custom of the day—established to preserve claims of priority while not revealing preliminary conclusions ripe for theft by others—

Galileo encoded his interpretation as a Latin anagram, which he posted to his friend and leading compatriot in astronomical research, Johannes Kepler.

Kepler may have matched Galileo in brilliance, but he never resolved the anagram correctly, and he misinterpreted the message as a statement about the planet Mars. In frustration (and a bit of pique), he begged Galileo for the answer. His colleague replied with the intended solution:

> *Altissimum planetam tergeminum observavi.*
> [I have observed that the farthest planet is threefold.]

I regard the last word of Galileo's anagram as especially revealing. He does not advocate his solution by stating "I conjecture," "I hypothesize," "I infer," or "It seems to me that the best interpretation . . ." Instead, he boldly writes *"observavi"*—I have *observed.* No other word could capture, with such terseness and accuracy, the major change in concept and procedure (not to mention ethical valuation) that marked the transition to what we call "modern" science. An older style (as found, for example, in Gesner's compendium on mammals, cited above) would not have dishonored a claim for direct observation, but would have evaluated such an argument as a corroborative afterthought, surely secondary in weight to such criteria as the testimony of classical authors and logical consistency with a conception of the universe "known" to be both true and just—in other words, to authority and fixed "reasonableness."

But the new spirit of skepticism toward past certainty, coupled with respect for "pure" and personal observation—then being stressed by Francis Bacon in England, René Descartes in France, and the Lynxes in Italy—was sweeping through the intellectual world, upsetting all standard procedures of former times and giving birth to the modern form of an institution now called "science." Thus, Galileo supported his theory of Saturn with the strongest possible claim of the new order, the one argument that could sweep aside all opposition by claiming a direct, immediate, and unsullied message from nature. Galileo simply said: I have observed it; I have seen it with my own eyes. How could old Aristotle, or even the present pope himself, deny such evidence?

I do not intend, in this essay, to debunk the usual view that such a transition from old authority to direct observation marks a defining (and wonderfully salutary) event in the history of scientific methodology. But I do wish to note that all great mythologies include harmful simplicities amidst their genuine reforms—and that these negative features often induce the ironic consequence of saddling an original revolutionary doctrine with its own form of restrictive and unquestioned authority. The idea that observation can be pure and unsullied (and therefore beyond dispute)—and that great scientists are, by implication, people who can free their minds from the constraints of surrounding culture and reach conclusions strictly by untrammeled experiment and observation, joined with clear and universal logical reasoning—has often harmed science by turning the empiricist method into a shibboleth. The irony of this situation fills me with a mixture of pain for a derailed (if impossible) ideal and amusement for human foibles—as a method devised to undermine proof by authority becomes, in its turn, a species of dogma itself. Thus, if only to honor the truism that liberty requires eternal vigilance, we must also act as watchdogs to debunk the authoritarian form of the empiricist myth—and to reassert the quintessentially human theme that scientists can work only within their social and psychological contexts. Such an assertion does not debase the institution of science, but rather enriches our view of the greatest dialectic in human history: the transformation of society by scientific progress, which can only arise within a matrix set, constrained, and facilitated by society.

I know no better illustration of this central principle than the tale of Galileo's losing struggle with Saturn, for he insisted on validation by pure sight *(observavi)*, and he could never see his quarry correctly—presumably because his intellectual domain included no option for rings around a planet. Galileo did not just "see" Saturn; he had to interpret an object in his lens by classifying an ambiguous shape (the best that his poor optics could provide) within the structure of his mental space—and rings didn't inhabit this interior world.

The great Dutch astronomer Christiaan Huygens finally recognized the rings of Saturn in 1656, more than a decade after Galileo's death. Galileo, who had wrestled mightily with Saturn, never moved beyond his trigeminal claim, and finally gave up and turned to other pursuits. In his 1613 book on sunspots, published by the Lynxes (with the author designated on the title page as Galileo Galilei Linceo), he continued to insist that Saturn must be threefold because he had so observed the planet: "I have resolved not to put anything around Saturn except what I have already observed and revealed—that is, two small stars which touch it, one to the east and one to the west." Against a colleague who interpreted the planet as oblong, Galileo simply asserted his superior vision. The colleague, Galileo wrote, had viewed Saturn less often and with a much poorer telescope, "where perfection is lacking, [and] the shape and distinction of the three stars imperfectly seen. I, who have observed it a thousand times at different periods with an excellent instrument, can assure you that no change whatever is to be seen in it."

Yet just as Galileo prepared his book on sunspots for publication, he observed Saturn again after a hiatus of two years—and the two side planets had disappeared (a situation produced, we now know, when the planet's changing orientation presents the rings directly on edge—that is, as an invisible line in Galileo's poor telescope). The stunned Galileo, reduced to a most uncharacteristic modesty, had just enough time to make an addition to the last chapter of his book. He abjured nothing about his previous observations or about the righteousness of the empirical method in general. He merely confessed his puzzlement, making a lovely classical allusion to the primary myth about the planet's eponym:

> I had discovered Saturn to be three-bodied. . . . When I first saw them they seemed almost to touch, and they remained so for almost two years without the least change. It was reasonable to believe them to be fixed. . . . Hence I stopped observing Saturn for more than two years. But in the past few days I returned to it and found it to be solitary, without its customary supporting stars, and as perfectly round and sharply bounded as Jupiter. Now what can be said of this strange metamorphosis? That the two lesser stars have been consumed? . . . Has Saturn devoured his children? Or was it indeed an illusion and a fraud with which the lenses of my telescope deceived me for so long—and not only me, but many others who have observed it with me? . . . I need not say anything definite upon so strange and unexpected an event; it is too recent, too unparalleled, and I am restrained by my own inadequacy and the fear of error.

After this lengthy preamble on the maximally celebrated Galileo, let me now present the main subject of this essay: the virtually unknown Francesco Stelluti, one of the original four Lynxes, a loyal friend and supporter of Galileo, and the man who tried to maintain, and eventually disbanded with dignity (in 1652), the original Academy of the Lynxes, fatally weakened after Cesi's untimely death in 1630. The previously uncharted links between Stelluti and Galileo are rich and fascinating (I would have said "the links between these Lynxes," if the pun were not so egregious), and these connections provide a poignant illustration of this essay's central theme: the power and poverty of pure empiricism, and the need to scrutinize social and intellectual contexts, both for practicing scientists (so they will not be beguiled) and for all people who wish to understand the role and history of knowledge (so they will grasp the necessary and complex interdigitation of science and society).

The original Lynxes began with all the bravado and secrecy of a typical boys' club (Cesi, remember, was only eighteen years old, while his three compatriots were all twenty-six). They wrote complex rules and enunciated lofty ideals. (I do not know whether or not they developed a secret handshake!) Each adopted a special role, received a Latin moniker, and took a planet for his emblem. The leader Cesi commanded the botanical sciences as Coelivagus (the heavenly wanderer); the Dutchman Johannes van Heeck would read and interpret classical philosophy as Illuminatus; Anastasio de Filiis became the group's historian and secretary as Eclipsatus. Poor Francesco Stelluti, who published little and evidently saw himself as a systematic plodder, took up mathematics and geometry under the name of Tardigradus (the slow stepper). For his planet, Stelluti received the most distant and most slowly revolving body—Saturn, the subject of Galileo's error!

In their maturity, the Lynxes would provide powerful intellectual and institutional support for the open and empirical approach to science, as promoted by their most prominent member, Galileo. But at their beginnings, as a small club of young men, the Lynxes preferred the older tradition of science as an arcane and secret form of knowledge, vouchsafed only to initiates who learned the codes and formulae that could reveal the mysterious harmonies of universal order—the astrological links between planetary positions and human lives; the alchemical potions and philosopher's stones, heated in vats that could transmute base metals to gold (double, double toil and trouble; fire burn and cauldron bubble, to cite some famous witches); and the experiments in smoke, mirrors, and optical illusions that occupied an uncertain position between categories now labeled as magic and science, but then conflated. Giambattista Della Porta, the fifth Lynx, had survived as a living legend of this fading philosophy. Della Porta had made his reputation in 1558, long before the birth of any original Lynx, with a book entitled *Magia naturalis,* or *Natural Magic.* As a young man in Naples, Della Porta had founded his own arcane organization, the Accademia dei Segreti (Academy of Secrets), dedicated to alchemical and astrological knowledge, and later officially suppressed by the Inquisition.

By initiating the aged Della Porta into the Academy of Lynxes, Cesi and his compatriots showed the strength of their earlier intellectual allegiances. By inducting Galileo the next year, they displayed their ambivalence, and their growing attraction to a new view of knowledge and scientific procedure.

The election of both newcomers virtually guaranteed a period of definitional struggle within the Academy, for no love could unite Della Porta and Galileo, who not only differed maximally in their basic philosophical approaches to science, but also nearly came to blows for a much more specific reason vested in the eternally contentious issue of priority. Galileo never claimed that he had invented the telescope from scratch. He stated that he had heard reports about a crude version during a trip to Venice in 1609. He recognized the optical principles behind the device, and then built a more powerful machine that could survey the heavens. But Della Porta, who had used lenses and mirrors for many demonstrations and illusions in his *Natural Magic,* and who surely understood the rules of optics, then claimed that he had formulated all the principles for building a telescope (although he had not constructed the device) and therefore deserved primary credit for the invention. Although tensions remained high, the festering issue never erupted into overt battle because Galileo and Della Porta held each other in mutual respect, and Della Porta died in 1615 before any growing bitterness could bubble over.

Stelluti first encountered Galileo in the context of this struggle—and he initially took Della Porta's side! In 1610, with Della Porta inscribed as a Lynx but Galileo not yet a member, Stelluti wrote a gossipy letter to his brother about the furor generated by *Sidereus nuncius* and the dubious claims of the pamphlet's author:

> I believe that by now you must have seen Galileo, he of the *Siderius nuncius* . . . Giambattista Della Porta wrote about [the telescope] more than thirty years ago in his *Natural Magic* . . . so poor Galileo will be besmirched. But, nonetheless, the Grand Duke has given him 800 piastres.*

But when Galileo joined the Lynxes, and as his fame and success solidified and spread, Stelluti and his compatriots muted their suspicions and eventually became fervent Galileans. With Della Porta dead and *Starry Messenger* riding a truly cosmic crest of triumph, the Academy of the Lynxes grew to become Galileo's strongest intellectual (and practical) base, the primary institutional supporters of the new, open, empirical, and experimental view of scientific knowledge. Making the link between Galileo's error and Stelluti's emblem,

---

*The quotations from Galileo's *Letters on Sunspots* come from Stillman Drake's 1957 English translation, published by Anchor Books. I have translated all other quotes from the Italian of Stelluti's 1637 monograph on fossil wood, letters from several volumes of the *Edizione Nazionale* of Galileo's complete works, and three standard sources on the history of the Academy of the Lynxes: *Breve storia della Accademia dei Lincei* by D. Carutti (Rome: Salviucci, 1883); *Contributi alla storia della Accademia dei Lincei* by G. Gabrieli (Rome, 1989); and *L'Accademia dei Lincei e la cultura europea nel XVII secolo,* a catalog for a traveling exhibit about the Lynxes by A. M. Capecchi and several other authors, published in 1991.

Cesi wrote to Stelluti in 1611 about the wonders of the telescope, as revealed by Galileo himself, then paying a long visit to the duke of Acquasparta:

> Each evening we see new things in the heavens, the true office of the Lynxes. Jupiter and its four revolving satellites; the moon with its mountains, caverns, and rivers; the horns of Venus; and Saturn, your own triple-star [*il triplice suo Saturno*].

Such floods of reforming novelty tend to alienate reigning powers, to say the least—a generality greatly exacerbated in early-seventeenth-century Rome, where the papal government, besieged by wars and assaulted by the successes of the Reformation, felt especially unfriendly to unorthodoxy of any sort. Galileo had written a first note of cautious support for the Copernican system at the end of his *Letters on Sunspots* (published by the Lynxes in 1613). Soon afterward, in 1616, the Church officially declared the Copernican doctrine false and forbade Galileo to teach heliocentrism as a physical reality (though he could continue to discuss the Copernican system as a "mathematical hypothesis"). Galileo kept his nose clean for a while and moved on to other subjects. But then, in 1623, the Lynxes rejoiced in an unanticipated event that Galileo called a "great conjuncture" *(mirabel congiuntura):* the elevation of his friend and supporter Maffeo Barberini to the papacy as Urban VIII. (In an act of literal nepotism, Maffeo quickly named his nephew Francesco Barberini as his first new cardinal. In the same year of 1623, Francesco Barberini became the twenty-ninth member of the Lynxes.)

On August 12, 1623, Stelluti wrote from Rome to Galileo, then in Florence, expressing both his practical and intellectual joy in the outcome of local elections. Three members of the Lynxes would be serving in the new papal government, along with "many other friends." Stelluti then enthused about the new boss:

> The creation of the new pope has filled us all with rejoicing, for he is a man of such valor and goodness, as you yourself know so well. And he is a particular supporter of learned men, so we shall have a supreme patron. . . . We pray to the Lord God to preserve the life of this pope for a long time.

The Lynxes, suffused with hope that freedom of scientific inquiry would now be established, met for an extended convention and planning session at

Cesi's estate in 1624. Galileo had just built the first usable microscope for scientific investigation, after recognizing that lenses, properly arranged, could magnify truly tiny nearby objects, as well as enormous cosmic bodies rendered tiny in appearance by their great distance from human observers. Anticipating the forthcoming gathering of the Lynxes, Galileo sent one of his first microscopes to Cesi, along with a note describing his second great optical invention:

> I have examined a great many tiny animals with infinite admiration. Mosquitoes are the most horrible among them. . . . I have seen, with great contentment, how flies and other tiny animals can walk across mirrors, and even upside down. But you, my lord, will have a great opportunity to view thousands and thousands of details. . . . In short, you will be able to enjoy infinite contemplation of nature's grandness, and how subtly, and with what incredible diligence, she works.

Galileo's microscope entranced the Lynxes and became the hit of their meeting. Stelluti took a special interest and used the new device to observe and draw the anatomy of bees. In 1625, Stelluti published his results, including a large engraving of three bees drawn under Galileo's instrument. Historian of science Charles Singer cites these bees as "the earliest figures still extant drawn with the aid of the microscope." If the name of the sadly underrated Francesco Stelluti, the tardigrade among the Lynxes, has survived at all in conventional annals of the history of science, he perseveres only as an entry in the "list of firsts" for his microscopical drawing.

The Lynxes, always savvy as well as smart, did not choose to draw bees for abstract amusement. Not coincidentally, the family crest of Maffeo Barberini, the new pope and the Lynxes' anticipated patron, featured three bees. Stelluti dedicated his work to Urban VIII, writing in a banner placed above the three bees: "To Urban VIII Pontifex Optimus Maximus . . . from the Academy of the Lynxes, and in perpetual devotion, we offer you this symbol."

The emboldened Galileo now decided to come out of intellectual hiding, and to risk a discussion of the Copernican system. In 1632, he published his epochal masterpiece in the history of science and, from the resulting tragedy, the history of society as well: *Dialogo . . . sopra i due massimi sistemi nel mondo tolemaico e copernicano* (A dialogue on the two great systems of the world, Ptolemaic and Copernican). Galileo hoped that he could avoid any ecclesiastical trouble by framing the work as a dialogue—an argument between a sup-

The first published scientific figure based on observations under a microscope—Stelluti's 1625 image of bees, drawn to honor the new pope, Urban VIII, whose family crest featured three bees.

porter of the earth-centered Ptolemaic system and a partisan of Copernicus's sun-centered view.

We all know the tragic outcome of this decision only too well. The pope, Galileo's erstwhile friend, became furious and ordered the scientist to stand trial before the Roman Inquisition. This tribunal convicted Galileo and forced him to abjure, on his knees, his "false" and heretical Copernican beliefs. The Inquisition then placed him under a form of house arrest for the remainder of his life, on his small estate at Arcetri. Galileo's situation did not resemble solitary confinement at Alcatraz, and he remained fully active in scientific affairs by receiving visitors and engaging in voluminous correspondence to the moment of his death (even though blindness afflicted his last four years). In 1638, and partly by stealth, Galileo wrote his second great book in dialogue form (with the same protagonists) and had a copy smuggled to the Netherlands for publication: *Discourses and Mathematical Demonstrations Concerning Two New*

The famous frontispiece of Galileo's dialogue between Ptolemy and Copernicus (with Aristotle as their mediator).

Sciences. But he was not allowed to leave Arcetri either, as the vindictive pope, still feeling betrayed, refused Galileo's requests to attend Easter mass and to consult doctors in Florence when his sight began to fail.

The literature on the whys and wherefores of Galileo's ordeal could fill a large room in a scholarly library, and I shall not attempt even the barest summary here. (The most interesting and original of recent books include Mario Biagioli's *Galileo, Courtier,* University of Chicago Press, 1993; and Pietro Redondi's *Galileo Heretic,* Princeton University Press, 1987.) All agree that Galileo might have avoided his fate if any one of a hundred circumstances had unfolded in a slightly different manner. He was, in other words, a victim of bad luck and bad judgment (on both sides), not an inevitable sacrificial lamb in an eternal war between science and religion.

Until doing research for this essay, however, I had never appreciated the strength of one particularly relevant factor along the string of contingencies. From the vantage point of the Lynxes, Galileo would almost surely have managed to weave a subtle path around potential trouble, if the most final of all events had not intervened. In 1630, at age forty-five and the height of his influence, Federico Cesi, founder and perpetual leader of the Lynxes, died. Galileo learned the sad news in a letter from Stelluti: "My dear signor Galileo, with a trembling hand, and with eyes full of tears [*con man tremante, e con occhi pieni di lacrime*—such lamentation sounds so much better in Italian!], I must tell you the unhappy news of the loss of our leader, the duke of Acquasparta, as the result of an acute fever."

I feel confident that Cesi could have intervened to spare Galileo for two reasons. First, his caution and diplomacy, combined with his uncanny sense of the practical, would have suppressed Galileo's famous and fatal impetuosity. Galileo, ever testing the limits, ever pushing beyond into a realm of danger, did cast his work in the form of a dialogue between a Copernican and a supporter of Ptolemy's earth-centered universe. But he had scarcely devised a fair fight. The supporter of Ptolemy bore the name Simplicio, and the quality of his arguments matched his moniker. Moreover, Urban VIII developed a sneaking suspicion that Simplicio might represent a caricature of his own imperial self—hence his angry feeling that Galileo had betrayed an agreement to discuss Copernicanism as a coherent theory among equally viable alternatives. If Cesi had lived, he would, no doubt, have insisted that Galileo write his dialogue in a less partisan, or at least a more subtly veiled, form. And Cesi would have prevailed, both because Galileo respected his judgment so highly, and because the Lynxes intended to publish his book at Cesi's expense.

Second, Cesi operated as one of the most consummate politicians on the Roman scene. As a diplomat and nobleman (contrasted with Galileo's status as a commoner and something of a hothead), Cesi would have greased all the wheels and prepared a smooth way. Galileo recognized the dimensions of his personal misfortune only too well. He wrote to his friend G. B. Baliani in 1630, just before Cesi's death:

> I was in Rome last month to obtain a license to print the Dialogue that I am writing to examine the two great systems, Ptolemaic and Copernican. . . . Truly, I would have left all this in the hands of our most excellent prince Cesi, who would have accomplished it with much care, as he has done for my other works. But he is feeling indisposed, and now I hear that he is worse, and may be in danger.

Cesi's death produced two complex and intertwined results lying at the heart of this essay: the subsequent, and preventable, condemnation of Galileo; and the attrition and inevitable extinction of the Society of Lynxes. Stelluti tried valiantly to keep the Lynxes alive. He importuned Francesco Barberini, the cardinal nephew of the pope, and the only member of the Lynxes with enough clout to fill Cesi's shoes, to become the new leader. Barberini's refusal sealed the Lynxes' fate, for no other sufficiently rich and noble patron could be found. Cesi soldiered on for a while and, in a noble last hurrah, finally published, in 1651, the volume on the natural history of the New World that the Lynxes had been planning for decades: *Nova plantarum et mineralium mexicanorum historia*. In a final loving tribute, Stelluti included Cesi's unpublished work on botanical classification in an appendix. In 1652, Stelluti, the last original Lynx, died—and the organization that he had nourished for a lifetime, in his own slow and steady manner, ceased to exist.

II. FRANCESCO STELLUTI AND THE MINERAL WOOD OF ACQUASPARTA

Francesco Stelluti remained faithful to Galileo during his friend's final years of internal exile and arrest. On November 3, 1635, he wrote a long and interesting letter to Galileo at Arcetri, trying to cheer his friend with news from the world of science. Stelluti first expressed his sympathy for Galileo's plight: "God knows how grieved and pained I am by your ordeal" *(Dio sa quanto mi son doluto e doglio de' suoi travagli)*. Stelluti then attempted to raise Galileo's spirits with the latest report on an old project of the Lynxes—an analysis of some curious fossil wood found on Cesi's estate:

You should know that while I was in Rome, Signor Cioli visited the Duchess [Cesi's widow] several times, and that she gave him, at his departure, several pieces of the fossil wood that originates near Acquasparta. . . . He wanted to know where it was found, and how it was generated . . . for he noted that Prince Cesi, of blessed memory, had planned to write about it. The Duchess then asked me to write something about this, and I have done so, and sent it to Signor Cioli, together with a package of several pieces of the wood, some petrified, and some just beginning to be petrified.

This fossil wood had long vexed and fascinated the Lynxes. Stelluti had described the problem to Galileo in a letter of August 23, 1624, written just before the Lynxes' convention and the fateful series of events initiated by Stelluti's microscopical drawings of bees, intended to curry favor with the new pope.

Our lord prince [Cesi] kisses your hands and is eager to hear good news from you. He is doing very well, despite the enervating heat, which does not cause him to lose any time in his studies and most beautiful observations on this mineralized wood. He has discovered several very large pieces, up to eleven palms [of the human hand, not the tree of the same name] in diameter, and others filled with lines of iron, or a material similar to iron. . . . If you can stop by here on your return to Florence, you can see all this wood, and where it originates, and some of the nearby mouths of fire [steaming volcanic pits near Acquasparta that played a major role in Stelluti's interpretation of the wood]. You will observe all this with both surprise and enthusiasm.

We don't usually think of Galileo as a geologist or paleontologist, but his catholic (with a small c!) interests encompassed everything that we would now call science, including all of natural history. Galileo took his new telescope to his first meeting of 1611 with Cesi and the Lynxes, and the members all became enthralled with Galileo's reconstructed cosmos. But he also brought, to the same meeting, a curious stone recently discovered by some alchemists in Bologna, called the *lapis Bononensis* (the Bologna stone), or the "solar sponge"—for the rock seemed to absorb, and then reflect, the sun's light. The specimens have been lost, and we still cannot be certain about the composition or the nature of Galileo's stone (found in the earth or artificially made). But we do know that the Lynxes became entranced by this geological wonder. Cesi,

A comparison of title pages for Galileo's book on sunspots and Stelluti's treatise on fossil wood, with both authors identified as members of the Lynx society.

committed to a long stay at his estate in Acquasparta, begged Galileo for some specimens, which arrived in the spring of 1613. Cesi then wrote to Galileo: "I thank you in every way, for truly this is most precious, and soon I will enjoy the spectacle that, until now, absence from Rome has not permitted me" (I read this quotation and information about the Bologna stone in Paula Findlen's excellent book, *Processing Nature,* University of California Press, 1994).

Galileo then took a reciprocal interest in Cesi's own geological discovery—the fossil wood of Acquasparta; so Stelluti's letters reflect a clearly shared interest. Cesi did not live to publish his controversial theories on this fossil wood. Therefore, the ever-loyal Stelluti gathered the material together, wrote his own supporting text, engraved thirteen lovely plates, and published his most influential work (with the possible exception of those earlier bees) in 1637: *Trattato del legno fossile minerale nuovamente scoperto, nel quale brevemente si accenna la varia e mutabil natura di detto legno, rappresentatovi con alcune figure, che mostrano il luogo dove nasce, la diversita dell'onde, che in esso si vedono, e le sue cosi varie, e maravigliose forme*—a title almost as long as the following text (Treatise on newly discovered fossil mineralized wood, in which we point out the variable and mutable nature of this wood, as represented by several figures, which show the place where it originates, the diversity of waves [growth lines] that we see in it, and its highly varied and marvelous forms).

The title page illustrates several links with Galileo. Note the similar design and same publisher (Mascardi in Rome) for the two works. Both feature the official emblem of the Lynxes—the standard picture of the animal (copied from Gesner's 1551 compendium), surrounded by a laurel wreath and topped by the crown of Cesi's noble family. Both authors announce their affiliation by their name—the volume on sunspots by Galileo Galilei Linceo, the treatise on fossil wood by Francesco Stelluti Accad. Linceo. The ghosts of Galileo's tragedy also haunt Stelluti's title page, for the work bears a date of 1637 (lower right in Roman numerals), when Galileo lived in confinement at Arcetri, secretly writing his own last book. Moreover, Stelluti dedicates his treatise quite obsequiously "to the most eminent and most revered Signor Cardinal Francesco Barberini" (in type larger than the font used for Stelluti's own name), the nephew of the pope who had condemned Galileo, and the man who had refused Stelluti's invitation to lead (and save) the Lynxes after Cesi's death.

But the greatest and deepest similarity between Galileo's book on sunspots and Stelluti's treatise on fossil wood far transcends any visual likenesses, and resides instead in the nature of a conclusion, and a basic style of rhetoric and scientific procedure. Galileo presented his major discussion of Saturn in his book on sunspots (as quoted earlier in this essay)—where he stated baldly that an entirely false interpretation must be correct because he had observed the phenomenon with his own eyes. Stelluti's treatise on fossil wood presents a completely false (actually backward) interpretation of Cesi's discovery, and then uses exactly the same tactic of arguing for the necessary truth of his view because he had personally observed the phenomena he described!

Despite some practical inconveniences imposed by ruling powers committed neither to democracy nor to pluralism—one might, after all, end up burned like Bruno, or merely arrested, tried, convicted, and restricted like Galileo—the first half of the seventeenth century must rank as an apex of excitement for scientists. The most fundamental questions about the structure, meaning, and causes of natural phenomena all opened up anew, with no clear answers apparent, and the most radically different alternatives plausibly advocated by major intellects. By inventing a simple device for closer viewing, Galileo fractured the old conception of nature's grandest scale. Meanwhile, on earth, other scientists raised equally deep and disturbing questions about the very nature of matter and the basic modes of change and causality.

The nascent science of paleontology played a major role in this reconstruction of reality—primarily by providing crucial data to resolve the two debates that convulsed (and virtually defined) the profession in Stelluti and Galileo's time (see chapter 1 for more details on this subject):

1. What do fossils represent? Are they invariably the remains of organisms that lived in former times and became entombed in rocks, or can they be generated inorganically as products of formative forces within the mineral kingdom? (If such regular forms as crystals, and such complex shapes as stalactites, can arise inorganically, why should we deny that other petrified bodies, strongly resembling animals and plants, might also originate as products of the mineral kingdom?)

2. How shall we arrange and classify natural objects? Is nature built as a single continuum of complexity and vitality, a chain of being rising without a gap from dead and formless muds and clays to the pinnacle of humanity, perhaps even to God himself? Or can natural objects be placed into sharply separated, and immutably established, realms, each defined by such different principles of structure that no transitional forms between realms could even be imagined? Or in more concrete terms: does the old tripartite division of mineral, vegetable, and animal represent three loosely defined domains within a single continuum (with transitional forms between each pair), or a set of three utterly disparate modes, each serving as a distinct principle of organization for a unique category of natural objects?

Cesi had always argued, with force and eloquence, that the study of small objects on earth could yield as much reform and insight as Galileo's survey of the heavens. The microscope, in other words, would be as valuable as the telescope. Cesi wrote:

> If we do not know, collect, and master the smallest things, how will we ever succeed in grasping the large things, not to mention the biggest of all? We must invest our greatest zeal and diligence in the treatment and observation of the smallest objects. The largest of fires begins with a small spark; rivers are born from the tiniest drops, and grains of sand can build a great hill.

Therefore, when Cesi found a puzzling deposit of petrified wood near his estate, he used these small and humble fossils to address the two great questions outlined above—and he devised the wrong answer for each! Cesi argued that his fossil wood had arisen by transformation of earths and clays into forms resembling plants. His "wood" had therefore been generated from the mineral kingdom, proving that fossils could form inorganically. Cesi then claimed that his fossils stood midway between the mineral and vegetable kingdoms, providing a smooth bridge along a pure continuum. Nature must therefore be constructed as a chain of being. (Cesi had strongly advocated this position for a long time, so he can scarcely be regarded as a dispassionate or disinterested observer of fossils. His botanical classification, eventually published by Stelluti in 1651, arranged plants in a rising series from those he interpreted as most like minerals to forms that he viewed as most like animals.) Since Cesi could not classify his fossils into any conventional kingdom, he awarded them a separate name for a novel realm between minerals and plants—the Metallophytes.

Stelluti, playing his usual game of follow the leader, devoted his 1637 treatise to supporting Cesi's arguments for the transitional status of metallophytes and their origin from the mineral kingdom as transmuted earths and clays. The fossils may look like plants, but they originate from heated earths of the surrounding countryside (where subterranean magmas boil the local waters, thus abetting the conversion of loose earth to solid metallophyte). Stelluti concludes:

> The generation of this wood does not proceed from the seed or root of any plant, but only from a kind of earth, very much like clay, which little by little becomes transmuted to wood. Nature operates in such a manner until all this earth is converted into that kind of wood. And I believe that this occurs with the aid of heat from subterranean fires, which are found in this region.

To support this conclusion, Stelluti presented the following five basic arguments:

1. The fossil wood, generated from earth, only assumes the forms of tree trunks, never any other parts of true plants:

> It is clear that this wood is not born from seeds, roots or branches, like other plants, because we never find pieces of this wood with roots, or branches, or nerves [internal channels for fluids], as in other [truly vegetable] wood and trees, but only simple trunks of varied form.

Three figures of fossil wood from Stelluti's treatise of 1637.

2. The fossil trunks are not rounded, as in true trees, but rather compressed to oval shapes, because they grow *in situ* from earths flattened by the weight of overlying sediments (see the accompanying reproduction of Stelluti's figure):

> I believe that they adopt this oval shape because they must form under a great mass of earth, and cannot grow against the overlying weight to achieve the circular, or rather cylindrical, form assumed by the trunks of true trees. Thus, I can securely affirm that the original material of this wood must have been earth of a clayey composition.

3. Five of Stelluti's plates present detailed drawings of growth lines in the fossil wood (probably done, in part, with the aid of a microscope). Stelluti's argument for these inner details of structure follows his claim for the outward form of entire specimens: the growth lines form wandering patterns reflecting irregular pathways of generation from earth, following limits imposed by the weight of overlying sediments. These lines never form in regular concentric circles, as in true trees. Stelluti therefore calls them *onde,* or "waves," rather than growth lines:

> The waves and veins are not continuous, all following the same form, as in [vegetable] wood, but are shaped in a variety of ways—

some long and straight, others constricted, others thick, others contorted, others meandering. . . . This mineral wood takes its shape from the press of the surrounding earth, and thus it has waves of such varied form.

4. In the argument that he regarded as most decisive, Stelluti held that many specimens can be found in the process of transition, with some parts still made of formless earth, others petrified in the shapes of wood, and still others fully converted to wood. Stelluti views these stages as an actual sequence of transformation. He writes about a large specimen, exposed *in situ:*

In a ditch, we discovered a long layer of this wood . . . rather barrel-shaped, with one segment made of pure earth, another of mixed earth and wood, and another of pure wood. . . . We may therefore call it earth wood *(creta legno).*

Later, he draws a smaller specimen (reproduced here from Stelluti's figure) and states:

The interior part is made of wood and metal together, but the crust on the outside seems to be made of lateritic substance, that is, of terra cotta, as we find in bricks.

5. In a closing (and conclusive) flourish for the empirical method, Stelluti reports the results of a supposed experiment done several years before:

A piece of damp earth was taken from the interior of a specimen of this wood, and placed in a room of the palace of Acquasparta, belonging to Duke Cesi. After several months it was found to be completely converted into wood—as seen, not without astonishment, by the aforementioned Lord, and by others who viewed it. And not a single person doubted that earth was the seed and mother of this wood [*la terra è seme e madre di questo legno*].

With twentieth-century hindsight, we can easily understand how Stelluti fell into error and read his story backward. His specimens are ordinary fossil wood, the remains of ancient plants. The actual sequence of transformation runs from real wood, to replacement of wood by percolating minerals (petrifaction), to earth that may either represent weathered and degraded petrified wood, or may just be deposited around or inside the wood by flowing waters. In other words, Stelluti ran the sequence backward in his crucial fourth argument— from formless earth to metallophytes located somewhere between the mineral and vegetable kingdoms!

Moreover, Stelluti's criteria of shaping by overlying sediments (arguments 2 and 3) hold just as well for original wood later distorted and compressed, as for his reversed sequence of metallophytes actively growing within restricted spaces. Delicate parts fossilize only rarely, so the absence of leaves and stems, and the restriction of specimens to trunks, only records the usual pathways of preservation for ancient plants, not Stelluti's naive idea (argument 1) that the tree trunks cannot belong to the vegetable kingdom unless fossilized seeds or roots can also be found. As for the supposedly crucial experiment (argument 5)—well, what can we do with an undocumented three-hundred-year-old verbal report ranking only as hearsay even for Stelluti himself!

Nonetheless, Stelluti's treatise played an important role on the wrong side of the great debate about the nature of fossils—a major issue throughout seventeenth-century science, and not fully resolved until the mid-eighteenth century (see essay 1, about a late defense from 1726). Important authors throughout Europe, from Robert Plot in England (1677), to Olaus Worm in Denmark (1655), reported Stelluti's data as important support for the view that fossils can originate within the mineral kingdom and need not represent the remains of organisms. (Stelluti, by the way, did not confine his arguments to the wood of Acquasparta but made a general extrapolation to the nature and status of all fossils. In a closing argument, depicted on a fateful thirteenth plate of ammonites, Stelluti held that all fossils belong to the mineral kingdom and grow within rocks.)

When we evaluate the logic and rhetoric of Stelluti's arguments, one consistent strategy stands out. Stelluti had finally become a true disciple of Galileo and the primacy of direct empirical observation, viewed as inherently objective. Over and over again, Stelluti states that we must accept his conclusions because he has seen the phenomenon, often several times over many years, with his own eyes.

Stelluti had used this Galilean rhetoric to great advantage before. At the very bottom of his beautiful 1625 engraving of three bees for Pope Urban, Stelluti had added a little Latin note, just under his greatest enlargement of paired bee legs. In a phrase almost identical in form with Galileo's anagram about Saturn, Stelluti wrote: *Franciscus Stellutus Lynceus Fabr^{is} Microscopio Observavit*—"the Lynx Francesco Stelluti from [the town of] Fabriano observed [these objects] with a microscope." This time, at least, Stelluti had a leg up on Galileo—for the slow stepper among the Lynxes had made accurate observations, properly interpreted, while Galileo had failed for the much more difficult problem of Saturn. (This note, by the way, may represent the first appearance of the word *microscope* in print. Galileo had called his instrument an *occhiolino,* or "little eye," and his fellow Lynxes had then suggested the modern name.)

But Stelluti's luck had run out with Cesi's wood, when the same claim now buttressed his errors. Consider a sampling, following the order of his text, of Stelluti's appeals to the incontrovertible status of direct observation:

> The generation of this wood, which I have been able to see and observe so many times, does not proceed from seeds . . .

> The material of this wood is nothing other than earth, because I have seen pieces of it [*perche n'ho veduto io pezzi*] with one part made of hard earth and the other of wood.

> Figure 7 shows a drawing of a large oval specimen, which I excavated myself from the earth.

> The outer surface of the other piece appears to be entirely in wood, as is evident to the eye [in the drawing presented by Stelluti].

Stelluti ends his treatise with a flourish in the same mode: he need not write at great length to justify his arguments (and his text only runs to twelve pages), because he has based his work on personal observation:

> And this is all I need to say, with maximal brevity, about this material, which I have been able to see and observe so many times in those places where this new, rare, and marvelous phenomenon of nature originates.

But Stelluti had forgotten the old principle now embodied in a genre of jokes that begin by proclaiming: "I've got some good news, and some bad news." Galileo's empirical method can work wonders. But hardly any faith can be more misleading than an unquestioned personal conviction that the apparent testimony of one's own eyes must provide a purely objective account, scarcely requiring any validation beyond the claim itself. Utterly unbiased observation must rank as a primary myth and shibboleth of science, for we can only see what fits into our mental space, and all description includes interpretation as well as sensory reporting. Moreover, our mental spaces house a complex architecture built of social constraint, historical circumstance, and psychological hope—as well as nature's factuality, seen through a glass darkly.

We can be terribly fooled if we equate apparent sight with necessary physical reality. The great Galileo, the finest scientist of his or any other time, *knew* that Saturn—Stelluti's personal emblem—must be a triple star because he had so observed the farthest planet with good eyes and the best telescope of his day, but through a mind harboring no category for rings around a celestial sphere. Stelluti *knew* that fossil wood must grow from earths of the mineral kingdom because he had made good observations with his eyes and then ran an accurate sequence backward through his mind.

And thus, nature outfoxed the two Lynxes at a crucial claim in their careers—because both men concluded that sight alone should suffice, when genuine solutions demanded insight into mental structures and strictures as well.

As a final irony, Cesi had selected the emblem of Stelluti and Galileo's own society—the lynx—as an exemplar of this richer, dual pathway. The duke of Acquasparta had named his academy for a wild and wily cat, long honored in legend for possessing the sharpest sight among animals. Cesi chose well and subtly—and for a conscious and explicit reason. The maximal acuity of the lynx arose from two paired and complementary virtues—sharpness of vision *and* depth of insight, the outside and the inside, the eye and the mind.

Cesi had taken the emblem for his new society from the title page of Giambattista Della Porta's *Natural Magic* (1589 edition), where the same picture of a lynx stands below the motto: *aspicit et inspicit*—literally meaning "he looks at and he looks into," but metaphorically expressing the twinned ideals of observation and experimentation. Thus, the future fifth Lynx, the living vestige of the old way, had epitomized the richer path gained by combining insight with, if you will, "exsight," or observation. Cesi had stated the ideal in a document of 1616, written to codify the rules and goals of the Lynxes:

> In order to read this great, true and universal book of the world, it is necessary to visit all its parts, and to engage in both observation and experiment in order to reach, by these two good means, an acute and profound contemplation, by first representing things as they are and as they vary, and then by determining how we can change and vary them.

If we decide to embrace the entire universe as our potential domain of knowledge and insight—to use, in other words, the full range of scales revealed by Galileo's two great instruments, the telescope and the microscope (both, by the way, named by his fellow Lynxes)—we had better use all the tools of sensation *and* mentality that a few billion years of evolution have granted to our feeble bodies. The symbol of the lynx, who sees most acutely from the outside, but who also understands most deeply from the inside, remains our best guide. Stelluti himself expressed this richness, this duality, in a wonderfully poetic manner by extolling the lynx in his second major book, his translations of the poet Persius, published in 1630. Cesi had selected the lynx for its legendary acuity of vision, but Stelluti added:

> Not merely of the exterior eyes, but also of the mind, so necessary for the contemplation of nature, as we have taught, and as we practice, in our quest to penetrate into the interior of things, to know the causes and operations of nature . . . just as the lynx, with its superior vision, not only sees what lies outside, but also notes what arises from inside.

3

How the Vulva Stone Became a Brachiopod

WE USUALLY DEPICT THE RENAISSANCE (LITERALLY, THE "rebirth") as a clear, bubbling river of novelty that broke the medieval dam of rigidified scholasticism. But most participants in this great ferment cited the opposite of innovation as their motive. Renaissance thinkers and doers, as the name of their movement implied, looked backward, not forward, as they sought to rediscover and reinstitute the supposed perfection of intellect that Athens and Rome had achieved and a degraded Western culture had forgotten.

I doubt that anyone ever called Francis Bacon (1561–1626) a modest man. Nonetheless, even the muse of ambition must have smiled at such an audacious gesture when this most important British philosopher since the death of William of Ockham in 1347, his chancellor of England (until his fall for financial improprieties), declared "all knowledge" as his "province" and announced that he would write a Great Instauration (defined by *Webster's* as "a restoration after decay, lapse or dilapidation"), both to codify the fruitful rules of reason and to summarize all useful results. As a procedural starting point at the dawn of a movement that would become modern science, Bacon rejected both the scholastic view that equated knowledge with conservation, and the Renaissance reform that sought to recapture a long-lost perfection. Natural knowledge, he proclaimed, must be reconceptualized as a cumulative process of discovery, propelled by processing sensory data about the external world through the reasoning powers of the human brain.

Aristotle's writings on logic had been gathered into a compendium called the *Organon* (or "tool"). Bacon, with his usual flair, entitled the second book of his great instauration the *Novum Organum,* or new tool of reasoning—because the shift to such a different ideal of knowledge as cumulative, and rooted in an increasing understanding of external reality, also demanded that the logic of reasoning itself be reexamined. Bacon therefore began the *Novum Organum* by analyzing impediments to our acquisition of accurate knowledge about the empirical world. Recognizing the existence of such barriers required no novel insight. Aristotle himself had classified the common logical fallacies of human reasoning, while everyone acknowledged the external limits of missing data—stars too far away to study in detail (even with Galileo's newfangled telescope), or cities too long gone to leave any trace of their former existence.

But Bacon presented a brilliant and original analysis by concentrating instead on psychological barriers to knowledge about the natural world. He had, after all, envisioned the study of nature as a funneling of sensory data through mental processors, and he recognized that internal barriers of the second stage could stand as high as the external impediments of sensory limitations. He also understood that the realm of conceptual hangups extended far beyond the cool and abstract logic of Aristotelian reason into our interior world of fears, hopes, needs, feelings, and the structural limits of mental machinery. Bacon therefore developed an incisive metaphor to classify these psychological barriers. He designated such impediments as "idols" and recognized four major categories—*idola tribus* (idols of the tribe), *idola specus* (the cave), *idola fori* (the forum, or marketplace), and *idola theatri* (the theater).

Proceeding from the particular to the general, idols of the cave define the peculiarities of each individual. Some of us panic when we see a mathematical formula; others, for reasons of childhood suppression grafted upon basic temperament, dare not formulate thoughts that might challenge established orders.

Idols of the marketplace, perhaps Bacon's most original concept, designate limits imposed by language—for how can we express, or even formulate, a concept that no words in our language can specify? (In his brilliant story, "Averroës' Search," for example, Jorge Luis Borges—who loved Bacon's work and may well have written this tale to illustrate the idols—imagined the fruitless struggles of the greatest Arabic commentator on Aristotle to understand and translate the master's key concepts of "tragedy" and "comedy," for such notions could not be expressed, or even conceptualized, in Averroës's culture.)

Idols of the theater identify the most obvious category of impediments based on older systems of thought. We will have one hell of a time trying to grasp Darwinism if we maintain absolute and unquestioned fealty to the "old time religion" of Genesis literalism, with an earth no more than a few thousand years old and all organisms created by a deity, ex nihilo and in six days of twenty-four hours. Finally, idols of the tribe—that is, our tribe of *Homo sapiens*—specify those foibles and errors of thinking that transcend the peculiarities of our diverse cultures and reflect the inherited structures and operations of the human brain. Idols of the tribe, in other words, lie deep within the constitution of what we call "human nature" itself.

Bacon emphasized two tribal idols in his examples: our tendency to explain all phenomena throughout the spatial and temporal vastness of the universe by familiar patterns in the only realm we know by direct experience of our own bodies, the domain of objects that live for a few decades and stand a few feet tall; and our propensity to make universal inferences from limited and biased observations, ignoring evident sources of data that do not impact our senses. (Bacon cites the lovely example of a culture convinced that the Sea God saves shipwrecked men who pray for his aid because rescued sailors so testify. A skeptic, presented with this evidence and asked *"whether he did not now confess the divinity of Neptune?* returned this counter-question by way of answer; *yea, but where are they printed, that are drowned?* And there is the same reason of all such like superstitions, as in astrology, dreams, divinations, and the rest.")

In a 1674 translation of the *Great Instauration* (originally written in Latin), Bacon defines the idols in his characteristically pungent prose:

> Idols are the profoundest fallacies of the mind of man. Nor do they deceive in particulars [that is, objects in the external world] . . . but from a corrupt and crookedly-set predisposition of the mind; which doth, as it were, wrest and infect all the anticipations of the understanding. For the mind of man . . . is so far from being like a

smooth, equal and clear glass, which might sincerely take and reflect the beams of things, according to their true incidence; that it is rather like an enchanted glass, full of superstitions, apparitions, and impostures.

(Gilbert Wats, Bacon's translator, called his subject "a learned man, happily the learned'st that ever lived since the decay of the Grecian and Roman empires, when learning was at a high pitch." Wats also appreciated Bacon's distinctive approach to defining the embryonic field of modern science as accumulating knowledge about the empirical world, obtained by passing sensory data through the biased processing machinery of the brain. Wats described Bacon as "the first that ever joyn'd rational and experimental philosophy in a regular correspondence, which before was either a subtilty of words, or a confusion of matter." He then epitomized Bacon's view in a striking image: "For Truth, as it reflects on us, is a congruent conformity of the intellect to the object . . . when the intellectual globe, and the globe of the world, intermix their beams and irradiations, in a direct line of projection, to the generation of sciences.")

If our primary tribal idol resides in the ancient Greek proverb that "man [meaning all of us] is the measure of all things," then we should not be surprised to find our bodily fingerprints in nearly every assessment, even (or especially) in our words for abstractions—as in the strength of virility (from the Latin *vir,* "adult male"), the immaturity of puerility (from *puer,* "boy"), or the madness of hysteria (originally defined as an inherently feminine disease, from the Greek word for "womb"). Nevertheless, in our proper objection to such sexual stereotyping, we may at least take wry comfort in a general rule of most Indo-European languages (not including English) that assign genders to nouns for inanimate objects. Abstract concepts usually receive feminine designations—so the nobility of (manly) virtue presents herself as *la vertu* in France, while an even more distinctively manly virility also cross-dresses as *la virilité.*

We can, I believe, dig to an even deeper level in identifying tribal idols that probably lie in the evolved and inherited structures of neural wiring—the most basic and inherent substrate of "human nature" itself (if that ill-defined, overused, and much-abused term has any meaning at all). Some properties of human thinking seem so general, so common to all people, that such an evolutionary encoding seems reasonable, at least as a working hypothesis. For example, neurologists have identified areas of the brain apparently dedicated to the perception of faces. (One can easily speculate about the evolutionary value of such a propensity, but we must also recognize that these inherent biases of perception can strongly distort our judgment in other circumstances—Bacon's reason for designating such mental preferences as idols—as when we think we see a face in the random pitting of a large sandstone block on Mars, and then jump to conclusions about alien civilizations. I am not making this story up, by the way; the Martian face remains a staple of "proof" for the UFO and alien abduction crowd.) I suspect that the neural mechanism for facial recognition becomes activated by the abstract pattern of two equal and adjacent circles with a line below—a configuration encountered in many places besides real faces.

In this "deeper" category of tribal idols, I doubt that any rule enjoys wider application, or engenders greater trouble at the same time, than our tendency to order nature by making dichotomous divisions into two opposite groups. (Claude Lévi-Strauss and the French structuralists have based an entire theory of human nature and social history on this premise—and two bits from this corner says they're right, even if a bit overextended in their application.) Thus, we start with a few basic divisions of male versus female and night versus day—and then extend these concrete examples into greater generalities of nature versus culture ("the raw and the cooked" of Lévi-Strauss's book), spirit versus matter (of philosophical dualism), the beautiful versus the sublime (in Burke's theory of aesthetics); and thence (and now often tragically) to ethical belief, anathematization, and, sometimes, warfare and genocide (the good versus the bad, the godly who must prevail versus the diabolical, ripe for burning).

Again, one can speculate about the evolutionary basis of such a strong tendency. In this case, I rather suspect that dichotomization represents some "baggage" from an evolutionary past of much simpler brains built only to reach those quick decisions—fight or flight, sleep or wake, mate or wait—that make all the difference in a Darwinian world. Perhaps we have never been able to transcend the mechanics of a machinery built to generate simple twofold divisions and have had to construct our greater complexities upon such a biased and inadequate substrate—perhaps the most restrictive tribal idol of all.

I devoted the first part of this essay to a general discussion of our mental limitations because this framework, I believe, so well illuminates a particular problem in the history of paleontology that caught my fancy and attention both for unusual intrigue in itself, and for providing such an excellent "test case" of an important general pattern in the growth of scientific knowledge.

Classical authors, particularly Pliny in his *Natural History*, spoke in a limited way about fossils, usually (and correctly) attributing the shells found on mountaintops to a subsequent elevation of land from ancient seabeds. A few medieval authors (particularly Albert the Great in the thirteenth century) added a few comments, while Leonardo da Vinci, in the Leicester Codex (written in the early 1500s), made extensive and brilliant paleontological observations that were, however, not published until the nineteenth century, and therefore had no influence upon the field's later development. Essentially, then, the modern history of paleontology began in the mid-sixteenth century with the publication of two great works by two remarkable scholars: treatises on fossils published in 1546 by the German physician and mining engineer Georgius Agricola, and in 1565 (the year of his death in an epidemic of plague in Zurich) by the Swiss polymath Conrad Gesner.

In the compendium of Latinized folk names then used to identify fossils, most designations noted either a similarity in appearance to some natural or cultural phenomenon, or a presumed and legendary mode of origin. Thus, the flat and circular components of crinoid stems were called *trochites,* or wheel stones; the internal molds of rounded pairs of clamshells were *bucardites* (see accompanying figure, published in 1665), or bull's hearts; well-rounded concretions of the appropriate size were *enorchites,* or testicle stones (and if three were joined together, they became *triorchites,* or "three balls"); sea urchin tests became *brontia* (or thunder stones) because they supposedly fell from the sky in lightning storms.

A prominent group of fossils in this old taxonomy, and a puzzle (as we shall see) to early paleontologists, received the name of *hysteroliths,* also designated, in various vernaculars, as woman stones, womb stones, mother stones, or vulva stones (with the scholarly name derived from the same root as *hysteria,* an example cited earlier in this essay). The basis for this taxonomic consensus stands out in the first drawing of hysteroliths ever published—by the Danish natural historian Olaus Worm in 1665. A prominent median slit on one side

Olaus Worm's 1665 illustration of the internal mold of a clamshell pair. His generation did not know the source of these fossils and called them bucardites, *or "bull's hearts."*

Olaus Worm's original illustration of a hysterolith, *or "womb stone"—actually the internal mold of a brachiopod.*

(sometimes both) of a rounded and flattened object can hardly fail to suggest the anatomical comparison—or to cite Worm's own words, *"quod muliebre pudendum figura exprimat"* (because its form resembles the female genitalia).*
Interestingly, as Worm's second figure (to the right) shows, the opposite side of some (but not all) hysteroliths seems to portray a less obvious figure of the male counterparts! The men who wrote the founding treatises of modern paleontology could hardly fail to emphasize such a titillating object (especially in an age that provided few opportunities for approved and legitimate discussion, and illustration, of such intimate subjects).

This essay is not structured as a mystery yarn, so I spoil nothing, while (I hope) enhancing the intended intellectual theme, by providing the solution up front. Hysteroliths are the internal molds of certain brachiopod shells (just as bucardites, discussed and pictured above, are internal molds of certain clamshells). Brachiopods are not closely related to clams, but they also grow shells made of two convex valves that open along a hinge located at one end of the shell, and close by bringing the two valves together along their entire edges. Therefore, if you make an internal mold by pouring plaster of Paris into the closed shell, the resulting object will look roughly like a flattened sphere, with the degree of flattening specified by the convexity of the shell. Highly convex shells can produce nearly spherical molds (as in the fat clamshells that make bucardites). Shells of lower convexity—including most brachiopods and all the groups that make hysteroliths—yield more flattened molds.

Since molds are negative impressions of surrounding shapes, the suggestive parts of hysteroliths record features on the interior of a brachiopod shell in reverse. The slit that suggested a vulva and gave hysteroliths their name marks the negative impression of a raised and narrow linear ridge—called the median septum—that runs right down the middle of many brachiopod shell interiors, effectively dividing the valve in half. (For a clarifying analogy, think of the ridge as a knife and the slit as a cut.) The less pronounced "male" features on the

---

*Unless otherwise noted, all translations from the literature on hysteroliths are mine from Latin originals.

other side of some hysteroliths record, in positive relief, a cylindrical groove on the shell interior that houses part of the feeding skeleton (detached from the shell itself and rarely fossilized) in some groups of brachiopods.

By the mid-eighteenth century, paleontologists had reached a correct consensus. They knew that hysteroliths were internal molds of brachiopods, and they had even learned which kinds of brachiopods left such impressions upon their molds. They also recognized, of course, that the admittedly striking similarity with human genitalia recorded a sheer, if curious, accident with no causal meaning or connection whatsoever.

We therefore obtain, in the story of hysteroliths, a clean, clear, and lovely example of science operating at its best, by following the canonical definition of its very being and distinctiveness—a procedure dedicated to the sweetest of all goals: the construction of an accurate piece of natural knowledge. This odyssey through two centuries and several interesting stages progresses from the puzzled agnosticism of Agricola's first mention in 1546 toward Linnaeus's unchallenged consensus of 1753. I certainly do not deny the broad outline of this story. Agricola and Gesner possessed few clues for deciding among a wide range of alternatives—from the correct answer that eventually prevailed, to a hypothesis of inorganic origin by plastic forces circulating through rocks, to production by various ancient animals as a meaningful symbol that might even cure or alleviate human ailments of the genital organs. The correct answer may not have fulfilled all human hopes and uses, but hysteroliths really are brachiopod molds, and science supplied the tools for proper resolution.

I do, however, question the usual reading of such genuine scientific progress as a simple exercise in factual accumulation through accurate observation guided by objective principles of reasoning known as the scientific method. In this familiar model, the naïveté of Agricola and Gesner arises from their lack of accurate knowledge, not from any mental failures or barriers. In this sense, these sixteenth-century scholars might well be us in miniature, with the diminution established by what they couldn't know and we have since learned by living several centuries later and enjoying the fruits of advancing scientific understanding. But we should not so diminish these brilliant men and their interesting times. Gesner and Agricola do not rank below us; they only differed from us (while, no doubt, possessing more intrinsic "smarts" than the vast majority of us) in viewing the world from entirely divergent points of view that would be fascinating for us to comprehend.

I particularly appreciate Bacon's metaphor of the idols because this device can lead us toward a better appreciation for the complexities of creative thought, and the unifying similarities between the style we now call science and all other modes of human insight and discovery (while acknowledging, of course, that science presides over distinct subject matter and pursues particular goals in trying to understand the factual character of a "real" external world). Bacon argued that we must filter sensory data about this world through mental processors, and that these internal mechanisms always operate imperfectly because idols gum up the works. Discovery, therefore, arises from a complex intermeshing of these inside and outside components, and not by the accumulated input of facts from the outside world, processed through centuries by the universal and unchanging machinery of internal scientific logic.

Gesner did not use the same criteria for decisions that we employ today, so our differences cannot be attributed to his tiny molehill of reliable facts compared with our mountain. Rather, the idols conspired in him (as they still do in us, but with different resulting blockages) to construct a distinct kind of processing machine. Science prospers as much by retuning, or demolishing and then rebuilding, such mental machinery, as by accumulation of new factual information. Scientists don't simply observe and classify enough fossils until, one day, the status of hysteroliths as brachiopod molds becomes clear; rather, our theories about the nature of reality, and the meaning of explanation itself, must be decomposed and reconstructed before we can build a mental mansion to accommodate such information. And fruitful reconstruction requires, above all, that we acknowledge, examine, and challenge the Baconian idols of our own interior world.

I argued at the beginning of this essay that the Baconian idols could be ordered by degree of generality. In tracing the history of this internal component to solving the problem of hysteroliths, I noted an interesting progression in the release of blockages—from the most pervasive to the most particular idol, as paleontologists homed in upon a solution over two centuries. Perhaps we must first dig the right kind of mine before we can locate any particular nugget of great price.

1. *Idols of the Tribe in the Sixteenth Century.* Gesner and Agricola rediscover Pliny and the three dichotomies.

The hysterolith story begins as long ago as the recorded history of paleontology can venture, and as deeply as one can probe into the most pervasive and general of tribal idols: our propensity to dichotomize. Pliny the Elder, the great Roman statesman and natural historian who died with his boots on in the eruption of Mount Vesuvius in A.D. 79, wrote a compendium about the natural

world that survived as legions of hand copies made by monks and other scholars for more than a millennium before Gutenberg, and then became one of the
most widely published books in the first decades of printing. (In the trade,
books printed before 1500 are called *incunabulae,* or "from the cradle.")

Agricola and Gesner, as Renaissance scholars committed to the recovery of
ancient wisdom, sought above all to assign their specimens (and vernacular
names) to forms and categories mentioned by Pliny in his *Natural History.* In
an alphabetical list of rocks, minerals, and fossils, featured in the thirty-seventh
and last book of his great treatise, Pliny included a notable one-liner under letter D: *"Diphyes duplex, candida ac nigra, mas ac femina"*—having the character of
both sexes, white and black, male and female.

Pliny's treatise contained no pictures, so we can hardly know what object
he had meant to designate with this sparse description. But on the theme of
tribal idols, I am fascinated that the first mention of a possible hysterolith features two of the most general impediments in this category: our tendency to
read nature at all scales in terms of immediately familiar objects, particularly the
human body, and our propensity for classification by dichotomy. Pliny, in fact,
explicitly cites two of the most fundamental dichotomies in his single line: male
and female, and white and black. (Later commentators assumed that Pliny's
diphyes referred to stones that looked male on one side and female on the
other—hence their identification with hysteroliths.)

Moreover, we should also note the implicit inclusion of a third great
dichotomy—top and bottom—in Pliny's definition, for hysteroliths are composed of two distinct and opposite halves, a stunning representation, literally set
in stone, of our strongest mental idol, expressed geometrically. Moreover, all
three dichotomies carry great emotional weight both in their archetypal ideological status and in their embodiment of conventional rankings (by worth and
moral status) in a hierarchical and xenophobic society: male, white, and top versus female, black, and bottom. A modern perspective that we view as far more
valid, in both factual and moral terms, can only cause us to shiver when we
grasp the full implication of such a multiply dichotomized classification.

In his *De natura fossilium* of 1546, the first published treatise on paleontology (although the term *fossil* then designated any object found in the ground—
a broad usage consistent with its status as past participle of the Latin verb *fodere,*
"to dig up"—so this work treated all varieties of rocks, minerals, and the
remains of organisms now exclusively called fossils), Georgius Agricola
unearthed Pliny's one-liner, probably for the first time since antiquity, and
applied the name *diphyes* to some fossils found near the fortress of

Ehrenbreitstein. A generation later, in his *De rerum fossilium* (On fossil objects) of 1565, Conrad Gesner first connected Pliny's name and Agricola's objects with the folk designations and Latin moniker—*hysterolith*—that would then denote this group of fossils until their status as brachiopod molds became clear two hundred years later.

Sixteenth-century paleontology proceeded no further with hysteroliths, but we should not undervalue the achievements of Agricola and Gesner in terms of their own expressed aims. As men of the Renaissance, they wished to unite modern observations to classical wisdom—and the application of Pliny's forgotten and undocumented name to a clear category of appropriate objects seemed, to them, an achievement worth celebrating.

Moreover, when we note Gesner's placement of hysteroliths within his general taxonomy of fossils, we can peek through this window into the different intellectual domain of sixteenth-century explanation, and also begin to appreciate the general shifts in worldview that would have to occur before hysteroliths could be recognized as brachiopod molds. Gesner established fifteen categories, mostly based on presumed resemblances to more familiar parts of nature, and descending in a line of worth from the most heavenly, regular, and ethereal to the roughest and lowest. The first category included geometric forms (fossils of circular or spherical shape, for example); the second brought together all fossils that recalled heavenly bodies (including star-shaped elements of crinoid columns); while the third held stones that supposedly fell from the sky. At the other end, the disparaged fossils of class 15 resembled insects and serpents. Gesner placed hysteroliths into category 12, not at the bottom but not very near the honored pinnacle either, for "those that have some resemblance to men or quadrupedal animals, or are found within them." As his first illustration in category 12, Gesner drew a specimen of native silver that looked like a mat of human hair.

2. *Idols of the Theater in the Seventeenth Century.* Animal or mineral; useful symbol or meaningless accident?

If classic tribal idols played a founding role in setting the very name and definition of hysteroliths—their designation for some particularly salient features of female anatomy, and their description, by Pliny himself, in terms of three basic dichotomies that build the framework of our mental architecture—then some equally important theatrical idols (that is, constraints imposed by older, traditional systems of thought) underlay the major debate about the origin and meaning of hysteroliths that only began with seventeenth-century paleontology, but then pervaded the century: what are fossils?

The view of mechanism and causality that we call modern science answers this question without any ambiguity: fossils look like organisms in all their complex details, and we find them in rocks that formed in environments where modern relatives of these creatures now live. Therefore, fossils are remains of ancient organisms. This commonsense view had developed in ancient Greek times, and always held status as an available hypothesis. But the domain of seventeenth-century thought, the world that Bacon challenged and that modern science would eventually supplant, included other alternatives that may seem risible today, but that made eminent sense under other constructions of natural reality.

Bacon called these alternative worldviews idols of the theater, or impediments set by unfruitful systems of thought. Among the theatrical idols of seventeenth-century life, none held higher status among students of fossils than the Neoplatonic construction of nature as a static and eternal set of symbolic correspondences that reveal the wisdom and harmonious order of creating forces, and that humans might exploit for medical and spiritual benefit. A network of formal relationships (not direct causal connections, but symbolic resemblances in essential properties) pervaded the three kingdoms of nature—animal, vegetable, and mineral—placing any object of one kingdom into meaningful correspondence with counterparts in each of the other two. If we could specify and understand this network, we might hold the key to nature's construction, meaning, and utility.

Within this Neoplatonic framework, a close resemblance between a petrified "fish" enclosed within a rock and a trout swimming in a stream does not identify the stony version as a genuine former organism of flesh and blood, but suggests instead that plastic forces within the mineral kingdom can generate this archetypal form within a rock just as animate forces of another kingdom can grow a trout from an egg. Similarly, if various stones look like parts of the human body, then perhaps we can identify the mineral forces that resonate in maximal sympathy with the sources of our own animate being. Moreover, according to a theory of medicine now regarded as kooky and magical, but then perfectly respectable in a Neoplatonic framework, if we could identify the vegetable and mineral counterparts of human organs, then we might derive cures by potentiating our ailing animal versions with the proper sympathies of other realms, for every part in the microcosm of the human body must vibrate in harmony with a designated counterpart in the macrocosm of the earth, the central body of the universe. If the ingested powder of a pulverized "foot stone" could soothe the pains of gout, then hysteroliths might also alleviate sexual disorders.

The availability of this alternative view, based on the theatrical idol of Neoplatonism, set the primary context for seventeenth-century discussions about hysteroliths. Scholars could hardly ask: "what animal makes this shape as its mold?" when they remained stymied by the logically prior, and much more important, question: "are hysteroliths remains of organisms or products of the mineral kingdom?" This framework then implied another primary question— also posed as a dichotomy (and thus illustrating the continuing intrusion of tribal idols as well)—among supporters of an inorganic origin for hysteroliths: if vulva stones originate within the mineral kingdom, does their resemblance to female genitalia reveal a deep harmony in nature, or does the similarity arise by accident and therefore embody no meaning, a mode of origin that scholars of the time called *lusus naturae,* a game or sport of nature?

To cite examples of these two views from an unfamiliar age, Olaus Worm spoke of a meaningful correspondence in 1665, in the textual commentary to his first pictorial representation of hysteroliths—although he attributed the opinion to someone else, perhaps to allay any suspicion of partisanship:

> These specimens were sent to me by the most learned Dr. J. D. Horst, the archiater [chief physician] to the most illustrious Landgrave of Darmstadt. Dr. Horst states the following about the strength of these objects: these stones are, without doubt, useful in treating any loosening or constriction of the womb in females. And I think it not silly to believe, especially given the form of these objects [I assume that Dr. Horst refers here to hysteroliths that resemble female parts on one side and male features on the other] that, if worn suspended around the neck, they will give strength to people experiencing problems with virility, either through fear or weakness, thus promoting the interests of Venus in both sexes [*Venerem in utroque sexu promovere*].

But Worm's enthusiasm did not generate universal approbation among scholars who considered an origin for hysteroliths within the mineral kingdom. Anselm de Boot, in the 1644 French translation of his popular compendium on fossils (in the broad sense of "anything found underground"), writes laconically: *"Elles n'ont aucune usage que je sçache"* (they have no use that I know).

By the time that J. C. Kundmann—writing in vernacular German and living in Bratislava, relatively isolated from the "happening" centers of European intellectual life—presented the last serious defense for the inorganic theory of fossils in 1737, the comfortable rug of Neoplatonism had already been snatched away by time. (The great Jesuit scholar Athanasius Kircher had written the last major defense of Neoplatonism in paleontology in 1664, in his *Mundus subterraneus,* or *Underground World.*) Kundmann therefore enjoyed little intellectual maneuvering room beyond a statement that the resemblances to female genitalia could only be accidental—for, after all, he argued, a slit in a round rock can arise by many mechanical routes. In a long chapter devoted to hysteroliths, Kundmann allowed that hysteroliths might be internal molds of shells, and even admitted that some examples described by others might be so formed. But he defended an inorganic origin for his own specimens because he found no evidence of any surrounding shell material: "an excellent argument that these stones have nothing to do with clamshells, and must be considered as *Lapides sui generis"* (figured stones that arise by their own generation—a "signature phrase" used by supporters of an inorganic origin for fossils).

3. *Idols of the Marketplace in the Eighteenth Century.* Reordering the language of classification to potentiate the correct answer.

As stated above, the inorganic theory lost its best potential rationale when the late-seventeenth-century triumph of modern scientific styles of thinking (the movement of Newton's generation that historians of science call "*the* scientific revolution") doomed Neoplatonism as a mode of acceptable explanation. In this new eighteenth-century context, with the organic theory of fossils victorious by default, a clear path should have opened toward a proper interpretation of hysteroliths.

But Bacon, in his most insightful argument of all, had recognized that even when old theories (idols of the theater) die, and when deep biases of human nature (idols of the tribe) can be recognized and discounted, we may still be impeded by the language we use and the pictures we draw—idols of the marketplace, where people gather to converse. Indeed, in eighteenth-century paleontology, the accepted language of description, and the traditional schemes of classification (often passively passed on from a former Neoplatonic heritage without recognition of the biases thus imposed) established major and final barriers to solving the old problem of the nature of hysteroliths.

At the most fundamental level, remains of organisms had finally been separated as a category from other "things in rocks" that happened to look like parts or products of the animal and vegetable kingdom. But this newly restricted category commanded no name of its own, for the word *fossil* still covered everything found underground (and would continue to do so until the early nineteenth century). Scholars proposed various solutions—for example, calling organic remains "extraneous fossils" because they entered the mineral kingdom from other realms, while designating rocks and minerals as "intrinsic fossils"— but no consensus developed during the eighteenth century. In 1804, the British amateur paleontologist James Parkinson (a physician by day job, and the man who gave his name to Parkinson's disease), recognizing the power of Bacon's idols of the marketplace and deploring this linguistic impediment, argued that classes without names could not be properly explained or even conceptualized:

> But when the discovery was made, that most of these figured stones were remains of subjects of the vegetable and animal kingdoms, these modes of expression were found insufficient; and, whilst endeavoring to find appropriate terms, a considerable difficulty arose; language not possessing a sign to represent that idea, which the mind of man had not till now conceived.

The retention of older categories of classification for subgroups of fossils imposed an even greater linguistic restriction. For example, so long as some paleontologists continued to use such general categories as *lapides idiomorphoi* ("figured stones"), true organic remains would never be properly distinguished from accidental resemblances (a concretion recalling an owl's head, an agate displaying in its color banding a rough picture of Jesus dying on the cross, to cite two actual cases widely discussed by eighteenth-century scholars). And absent such a separation, and a clear assignment of hysteroliths to the animal kingdom, why should anyone favor the hypothesis of brachiopod molds, when the very name *vulva stone* suggested a primary residence in the category of accidents—

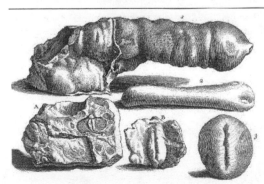

A 1755 illustration of hysteroliths on the same plate as a stalactite that accidentally resembles a penis.

for no one had ever argued that hysteroliths could be actual fossilized remains of detached parts of female bodies!

As a pictorial example, consider the taxonomic placement of hysteroliths in a 1755 treatise by the French natural historian Dezallier d'Argenville. He draws his true hysterolith (Figure A in the accompanying illustration) right next to slits in rocks that arose for other reasons (B and 3) and, more importantly, right under a stalactite that happens to look like a penis with two appended testicles. Now we know that the stalactite originated from dripping calcite in a cave, so we recognize this unusual resemblance as accidental. But if hysteroliths really belong in the same taxonomic category, why should we regard them as formed in any fundamentally different way?

When these idols of the marketplace finally receded, and hysteroliths joined other remains of plants and animals in an exclusive category of organic remains—and when the name *hysterolith* itself, as a vestige of a different view that emphasized accidental resemblance over actual mode of origin, finally faded from use—these objects could finally be seen and judged in a proper light for potential resolution.

Even then, the correct consensus did not burst forth all at once, but developed more slowly, and through several stages, as scientists, now and finally on the right track, moved toward a solution by answering a series of questions— all dichotomously framed once again—that eventually reached the correct solution by successive restriction and convergence. First, are hysteroliths molds of an organism, or actual petrified parts or wholes? Some proposals in the second category now seem far-fetched—for example, Lang in 1708 on hysteroliths as fossilized sea anemones of the coral phylum (colonies of some species do grow with a large slit on top), or Barrère in 1746 on *cunnulites* (as he called them, with an obvious etymology not requiring further explanation on my part) as the end pieces of long bones (femora and humeri) in juvenile vertebrates, before these termini fuse with the main shaft in adulthood. But at least paleontologists now operated within a consensus that recognized hysteroliths as remains of organisms.

Second, are hysteroliths the molds of plants or animals, with nuts and clams as major contenders in each kingdom—and with a quick and decisive victory for the animal kingdom in this case. Third, and finally, are hysteroliths the internal molds of clams or brachiopods—a debate that now, at the very end of the story, really could be solved by something close to pure observation, for consensus had finally been reached on what questions to ask and how they might be answered. Once enough interiors of brachiopod shells had been examined—

not so easy because almost all brachiopod fossils expose the outside of the shell, while few living brachiopods had been observed (for these animals live mostly in deep waters, or in dark crevices within shallower seas)—the answer could not be long delayed.

We may close this happy tale of virtue (for both sexes) and knowledge triumphant by citing words and pictures from two of the most celebrated intellectuals of the eighteenth century. In 1773, Elie Bertrand published a classification of fossils commissioned by Voltaire himself as a guide for arranging collections. His preface, addressed to Voltaire, defends the criterion of mode of origin as the basis for a proper classification—a good epitome for the central theme of this essay. Turning specifically to hysteroliths, Bertrand advises his patron:

> There is almost no shell, which does not form internal molds, sometimes with the shell still covering the mold, but often with only the mold preserved, though this mold will display all the interior marks of the shell that has been destroyed. This is the situation encountered in hysteroliths, for example, whose origin has been debated for so long. They are the internal molds of . . . terebratulids [a group of brachiopods]. (My translation from Bertrand's French.)

But if a good picture can balance thousands of words, consider the elegant statement made by Linnaeus himself in the catalog of Count Tessin's collection that he published in 1753. The hysteroliths (Figure 2, A–D), depicted with both their male and female resemblances, stand next to other brachiopod molds that do not resemble human genitalia (Figure 1, A and B)—thus establishing the overall category by zoological affinity rather than by external appearance. In Figures 3 to 7, following, Linnaeus seals his case by drawing the fossilized shells of related brachiopods. Two pictures to guide and establish a transition—from the lost and superseded world of Dezallier d'Argenville's theory of meaning by accidental resemblance to distant objects of other domains, to Linnaeus's modern classification by physical origin rather than superficial appearance.

Bacon's idols can help or harm us along these difficult and perilous paths to accurate factual knowledge of nature. Idols of the tribe may lie deep within the structure of human nature, but we should also thank our evolutionary constitution for another ineradicable trait of mind that will keep us going and questioning until we break through these constraining idols—our drive to ask and to know. We cannot look at the sky and not wonder why we see blue. We

In 1753, Linnaeus recognized hysteroliths as brachiopod molds and illustrated them with other brachiopods that do not mimic female genitalia.

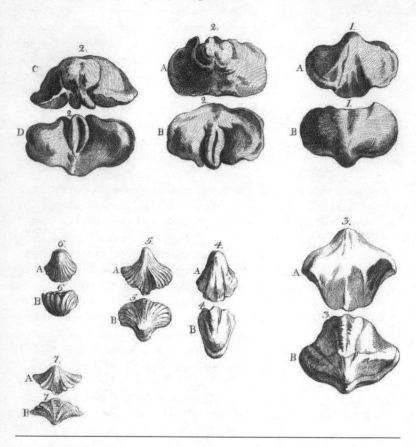

cannot observe that lightning kills good and bad people alike without demanding to know why. The first question has an answer; the second does not, at least in the terms that prompt our inquiry. But we cannot stop asking.

Let me close by tying the two parts of this essay together with a story that unites Bacon (the anchor of the first part) with Pliny (the progenitor of the second part) in their common commitment to this liberating compulsion to ask and know. Pliny died because he could not forgo a unique opportunity to learn something about the natural world—as he sailed too close to the noxious fumes of Vesuvius when he wished to observe a volcanic eruption more closely. Bacon died, albeit less dramatically, in the same noble cause and manner when he devised an experiment one cold day to determine whether snow could retard putrefaction. He stopped his carriage, bought a hen from a poultryman, and stuffed it with snow. The experiment worked, but the doctor died (not the patient this time, for the hen had expired before the procedure began!), as Bacon developed a cold that progressed to bronchitis, pneumonia, and death. He wrote a touching last letter (also quoted in a footnote to chapter 7) expressing a last wish for an explicit connection with Pliny: "I was likely to have the fortune of Caius Plinius the elder, who lost his life by trying an experiment about the burning of the mountain Vesuvius: for I was also desirous to try an experiment or two, touching on the conversion and induration of bodies. As for the experiment itself, it succeeded excellently well, but . . ."

Tribal idols may surround us, but our obsessively stubborn tribal need to ask and know can also push us through, as we follow Jesus' dictum that the truth will make us free. But we must also remember that Jesus then declined to answer Pilate's question: "what is truth?" Perhaps he understood that the idols conspire within us to convert this apparently simple inquiry into the most difficult question of all. But then, Jesus also knew, from the core of his being (in the conventional Christian interpretation), that human nature features an indivisible mixture of earthy constraint and (metaphorically at least) heavenly possibilities for liberation by knowledge—a paradox that virtually defines both the fascination and the frustration of human existence. We needed two hundred years of debate and discovery to turn a vulva stone into a brachiopod; but the same process has also stretched our understanding out to distant galaxies and back to the big bang.

II

Present

at the

Creation

How France's

Three Finest Scientists

Established

Natural History

in an

Age of Revolution

4

Inventing Natural History in Style

BUFFON'S STYLE AND SUBSTANCE

AN AVERAGE NOBLEMAN IN EIGHTEENTH-CENTURY France, including his wig, did not match the modern American mean in height. Nonetheless, at a shade under five feet five, Georges-Louis Leclerc, comte de Buffon, struck his own countrymen as short of stature. Yet he bestrode his world like a colossus. When he died, in 1788 at age eighty, his autopsy, performed by his own prior mandate, yielded fifty-seven bladder stones and a brain "of slightly larger size than that of ordinary [men]." Fourteen liveried horses, nineteen servants, sixty clerics, and a choir of thirty-six voices led his burial procession. The *Mercure* reported:

His funeral rites were of a splendor rarely accorded to power, opulence, dignity. . . . Such was the influence of this famous name that 20,000 spectators waited for this sad procession, in the streets, in the windows, and almost on the rooftops, with that curiosity that the people reserve for princes.

Buffon lived to see the first thirty-six volumes of his monumental *Histoire naturelle* (written with several collaborators, but under his firm and meticulous direction at all times); the remaining eight tomes were published after his death. No other eighteenth-century biologist enjoyed a wider readership or greater influence (with the possible exception of his archrival Linnaeus). Yet outside professional circles, we hardly recognize Buffon's name today. His one "standard" quotation— *"le style c'est l'homme même"* (style is the man himself)— comes from his inaugural address following his election as one of the "forty immortals" of the Académie Française, and not from his scientific publications. (See Jacques Roger's remarkable book, *Buffon's Life and Works,* translated by Sarah Lucille Bonnefoi, Cornell University Press, 1997.)

We must not equate the fading of a name through time with the extinction of a person's influence. In so doing, we propagate one of the many errors inspired by our generation's fundamental confusion of celebrity with stature. I will argue that, under certain definite circumstances—all exemplified in Buffon's life and career—a loss of personal recognition through time actually measures the spread of impact, as innovations become so "obvious" and "automatic" that we lose memory of sources and assign their status to elementary logic from time immemorial. (I do not, of course, challenge the truism that most fadings record the passage of a truly transient reason for celebrity; Linda Tripp and Tonya Harding come immediately to my mind, but surely not to the consciousness of any future grandchildren.)

Two prerequisites of intellectual fame have been well recognized: the gift of extraordinary intelligence, and the luck of unusual circumstances (time, social class, and so forth). I believe that a third factor—temperament—has not been given its equal due. At least in my limited observation of our currently depleted world, the temperamental factor seems least variable of all. Among people I have met, the few whom I would term "great" all share a kind of unquestioned, fierce dedication; an utter lack of doubt about the value of their activities (or at least an internal impulse that drives through any such angst); and above all, a capacity to work (or at least to be mentally alert for unexpected insights) at every available moment of every day in their lives. I have known other people of equal or greater intellectual talent who succumbed to mental illness, self-doubt, or plain old-fashioned laziness.

This maniacal single-mindedness, this fire in the belly, this stance that sets the literal meaning of *enthusiasm* ("the intake of God"), defines a small group of people who genuinely merit the cliché of "larger than life"—for they seem to live on another plane than we petty men who peep about under their huge legs. This mania bears no particular relationship to the external manifestation known as charisma. Some people in this category bring others along by exuding their zest; others may be glumly silent or actively dyspeptic toward the rest of the world. This temperament establishes an internal contract between you and your muse.

Buffon, all five feet and a bit of him, surely stood larger than life in this crucial sense. He established a rhythm of work in early adulthood and never deviated until his brief and final illness. Every spring, he traveled to his estate at Montbard in Burgundy, where he wrote *Histoire naturelle* and acted out the full life of a tough but fair seigneur and a restless entrepreneur (working to extend his agricultural projects, or building forges to smelt the local iron ore). Every fall he returned to Paris, where he dealt and cajoled to transform the Royal Botanical Garden (which he directed) into the finest general natural history museum in the world—a position certainly achieved by the following generation (and arguably still maintained today) when the successor to Buffon's expansion, the Muséum d'Histoire Naturelle, featured the world's three greatest naturalists as curators: Jean-Baptiste Lamarck, Georges Cuvier, and Etienne Geoffroy Saint-Hilaire.

Buffon worked at least fourteen hours every day. (He refused to alter any detail of this regimen, even in his last years when bladder stones, and various other maladies of old age, made travel so painful.) Jacques Roger describes the drill: "Those who worked with him or were under his orders had to adapt to his lifestyle. And everywhere, the same rule was in force: do not waste time." Buffon himself—in a passage that gives a good taste of the famous style (equal to the man himself!) of *Histoire naturelle*—attacked the Stoics with his personal formula for a life of continual enjoyment and action. If we accede to the stoical view, Buffon warned:

> Then let us say . . . that it is sweeter to vegetate than to live, to want nothing rather than satisfy one's appetite, to sleep a listless sleep rather than open one's eyes to see and to sense; let us consent to leave our soul in numbness, our mind in darkness, never to use either the one or the other, to put ourselves below the animals, and finally to be only masses of brute matter attached to the earth.

As for the other two prerequisites, the necessary brilliance shines forth in Buffon's work and needs no further comment. But Buffon's circumstances should have precluded his achievements (if temperament and brilliance had not pushed him through). As the son of a successful bourgeois family in Burgundy, he was not badly born (he received his later title of count from King Louis XV, and for his own efforts). But science, as a career, scarcely existed in his time— and non-Parisian nonnobility had little access to the few available opportunities. Buffon got a good education at a Jesuit *lycée* in Dijon, and he showed particular early talent for a field quite different from the later source of his triumph: mathematics. He wrote an important treatise on probability, translated Newton's *Fluxions* into French (from an English version of the Latin original), and applied his quantitative skills to important studies on the strength of timber grown on his own estate. He then worked through this botanical door to his eventual post as director of the king's gardens in Paris. The rest, as they say, is (natural) history.

Thirty-six volumes of *Natural History* appeared under Buffon's explicit authorship during his lifetime—one of the most comprehensive and monumental efforts ever made by one man (with a little help from his friends, of course) in science or literature. He intended to cover the entire range of natural objects in all three conventional kingdoms of animal, vegetable, and mineral. In truth, for he started at the traditional top and worked down, he never got to invertebrates or plants (or rather, he bypassed these "lower" manifestations of organic matter to write several volumes, late in life, on what he called "my dear minerals"). Moreover, despite plans and sketches, his own work on vertebrates didn't proceed "below" mammals and birds—and his colleague Lacépède published the last eight volumes (for a total of forty-four in the complete first edition) on reptiles and fishes (including whales) after Buffon's death.

Buffon treated all the great subjects of natural history in their full generality—from geology, to the origin of life, to embryology, physiology, biogeography, functional anatomy, and systematics. He regarded humans as a species of animal with unique properties, and therefore also covered most of anthropology, sociology, and cultural history as well. The general and theoretical articles of *Natural History* inspired endless and passionate debate—and made him a rarity in the history of literature: a man who became wealthy by his wits. (Inheritance and patronage didn't hurt either, but Buffon's volumes were bestsellers.) All sectors of French intellectual life, from the Encyclopedists to the Theological Faculty of the Sorbonne, took up his themes with gusto (agreeing with some and lambasting others, for Buffon's work was too multifarious, and too nuanced, for anyone's outright approbation or dismissal). He fought and made up with Voltaire, Rousseau, and nearly anyone who mattered in the closing years of the *ancien régime*.

But these general articles do not form the heart of *Natural History*. Rather, more than twenty volumes present long, beautifully crafted, descriptively detailed, and passionately opinionated treatises on mammals, birds, and minerals—with each species or kind granted its own chapter. These pieces, illustrated with engravings that became "standard," largely through endless pirating in later works by other authors, remain as charming (and often infuriating) as ever. As an example, consider Buffon's summary comments on his least favorite mammal, the sloth. (I imagine that Buffon, living at his own frenetic level, had even less patience with these slow creatures than those of us who operate at an ordinary human pace can muster):

> Whereas nature appears to us live, vibrant, and enthusiastic in producing monkeys; so is she slow, constrained and restricted in sloths. And we must speak more of wretchedness than laziness—more of default, deprivation, and defect in their constitution: no incisor or canine teeth, small and covered eyes, a thick and heavy jaw, flattened hair that looks like dried grass . . . legs too short, badly turned, and badly terminated . . . no separately movable digits, but two or three excessively long nails. . . . Slowness, stupidity, neglect of its own body, and even habitual sadness, result from this bizarre and neglected conformation. No weapons for attack or defense; no means of security; no resource of safety in escape; confined, not to a country, but to a tiny mote of earth—the tree under which it was born; a prisoner in the middle of great space . . . everything about them announces their misery; they are imperfect productions made by nature, which, scarcely having the ability to exist at all, can only persist for a while, and shall then be effaced from the list of beings. . . . These sloths are the lowest term of existence in the order of animals with flesh and blood; one more defect would have made their existence impossible. (My translation.)

I cannot begin to make a useful summary of the theoretical content of *Histoire naturelle,* if only because Buffon follows Bacon's lead in taking all (at least natural) knowledge for his province, and because Buffon's views do not always maintain full consistency either within or between sections. But short comments on three central subjects may provide some flavor of Buffon's approach to life, and his most important contributions to later research:

1. *Classification.* Carolus Linnaeus, Buffon's Swedish rival and close contemporary (both born in 1707, with Linnaeus dying ten years earlier than Buffon in 1778), developed the system of nomenclature that we continue to use today. Linnaeus prevailed because the formal rules of his system work well in practical terms, and also because his nested and hierarchical scheme of smaller-within-larger categories (species like dogs, within families like canids, within orders like carnivores, within classes like mammals, within phyla like vertebrates) could be slotted into a genealogical interpretation—the arborescent tree of life, with twigs on branches, on boughs, on trunks—that the discovery of evolution would soon impose upon any formal system of naming.

Buffon, on the other hand, sought to encompass all the overt complexity of organisms into a nonhierarchical system that recognized differing relationships for various properties (bats more like mammals in anatomy, more like birds in function). But this alternative model of a network with multiple linkages, rather than a strict hierarchy of inclusion, fails (in the admittedly retrospective light of evolution) to separate the superficial similarity of independent adaptation (wings of bats and birds) from the deep genealogical linkages of physical continuity through the ages (hair and live birth of bats and bears). Buffon's noble vision of equal treatment for all aspects of a species's life—placing ecology, function, and behavior at par with traditional anatomy—foundered on a false theory about the nature of relationships.

2. *Biogeography.* Previous naturalists, if they considered the question at all, generally envisaged a single center of creation for all animals, followed by spreading throughout the globe (a theory obviously consistent with the scenario of the biblical deluge, although not necessarily so inspired or defended). Buffon, on the other hand, recognizing that each species seemed to possess unique adaptations for its own region, argued for origination in appropriate places all over the globe, with only limited subsequent opportunities for migration—a more fruitful idea that founded the modern science of biogeography.

Buffon's notion of adaptation to local conditions directly inspired an important line of research in early American natural history. He argued that American mammals must be smaller than their Old World counterparts (rhino, giraffe, and tiger larger than tapir, llama, and jaguar, for example) because "the heat was in general much less in this part of the world, and the humidity much greater." American naturalists, Thomas Jefferson in particular, became annoyed at this charge of lesser stature for their New World, and sought vigorously to refute Buffon. This strong feeling led Jefferson to his most embarrassing error, when he misidentified the claw of a large fossil ground sloth (ironically, given Buffon's judgment of these creatures) as belonging to a giant lion that would have surpassed all European relatives in bulk. In correcting Jefferson's error, Georges Cuvier diplomatically named this new genus of sloths *Megalonyx jeffersoni*.

3. *The evolution and nature of species.* Most previous systems sought to define these basic units (for groups of organisms) in terms of unique structural features shared by all members and absent from all organisms in other species—an essentialist criterion doomed to failure in our actual world of shadings and exceptions. Buffon, on the other hand, sought a definition rooted in the status and behavior of groups in nature. He therefore held that the ability to interbreed with other members of the species, and to produce healthy and fertile offspring, must become the primary criterion for delimiting the boundaries of natural groups. In so doing, he laid the groundwork for modern notions of the interacting population as nature's basic entity, thus refuting the old Platonic alternative of searching for essential defining features to link any accidental configuration of actual matter (that is, a real organism) to the idealized *eidos* or archetype of its permanent species.

The venerable (and pernicious) tradition of defining past worthies by their supposed anticipation of modern views has misled many commentators into elevating this ecological definition of species, with its rejection of fixed Platonic archetypes, into a prototypical theory of evolution—thus making Buffon the worthy precursor of Darwin on a rectilinear path to truth. But such selective raiding parties from present knowledge into coherent, but fundamentally different, systems of past thought can only derail any effort to grasp the history of ideas as a fascinating panoply of changing worldviews, each fully developed in itself and worthy of our respect and understanding, despite the inevitability (if science has any value at all) of subsequent reformulations that will bring us closer to nature's modes and causes.

Buffon was not, and could not have been, an evolutionist in any modern sense (although *Histoire naturelle,* like the Bible, is so long and various that almost any position can be defended by partial quotation out of context). His system did allow for limited change within original species defined by their capacity for interbreeding. Buffon described these minor alterations as "degenerations" induced by changing climates. (In using the term *degeneration,* he did not invoke our modern meaning of "deterioration" for such changes usually improved a species's adaptation to local environments—but rather a departure from the "interior mold" or internal guardian of a species's identity in development.)

Buffon's complex and confusing notion of the *moule intérieur* (or "interior mold") underlay his basic theories both of embryology and of life's history through time. He accepted Aristotle's distinction between the controlling *form* of a species and the actual *matter* that builds any particular organism. He rejected Plato's notion of an external and eternal form, accepting Aristotle's alternative view of form as an attribute that shapes labile matter *from within*. For Buffon, the *moule intérieur* acts as the guardian of form and cannot be as labile as matter itself (or very plastic at all), lest general order disappear in nature (an unthinkable notion for an Enlightenment rationalist like Buffon), with each creature becoming no more than a glop of putty shaped only by the accidents of immediately surrounding conditions. For Buffon, a full theory of evolution would have destroyed the rational, albeit complex, order that he had pledged to define in his inimitable style.

BUFFON'S REPUTATION

If Buffon so shaped the science of his day, why did his name not survive as well as the imprint of his ideas? We can identify and distinguish several reasons, each relevant to the coordinating issue that I raised at the outset of this essay: the scaling of reputation with time, and the frequent failure of enduring fame to match continuing influence.

The sound bite does not just represent an invention of modern media in a restless age that has forgotten history. People have always needed simple labels to remember the reasons and meanings of events that shape our past. Unless such a distinctive label can be attached to a person's accomplishments, he will probably fade from sight. The major worthies and icons of the history of science all wear such labels (at least for popular recognition)— Copernicus for a new arrangement of the solar system, Newton for gravity, Darwin for evolution, Einstein for relativity (even if most of us can't define the concept very well). The principle extends beyond intellectual history; for everyone needs such a hook—Pandora her box, Lady Godiva her hair, Mark McGwire his bat. The generality also features a dark side, as good people with strong and consistent accomplishments become inevitably identified by an unforgettable and highly public moment of ultimate chagrin—Bill Buckner for a ball that bounced between his legs; another Bill for something else between his legs.*

---

*I wrote this essay in the summer of 1998, *in medio Monicae anni*, just before a presidential impeachment.

Buffon had a passion for order, but he developed no central theory that could be defined by a memorable phrase or concept. He wrote volumes of incomparable prose and propagated ideas, sometimes quite radical, about all major subjects in natural history. But no central thread unites his system. Moreover, Buffon may have been just a bit too worldly, just a tad too practical, ever to develop a transforming worldview clear and coherent enough (like Darwin's natural selection) both to attach distinctively to his person, and to apply consistently to a natural world strongly altered thereby.

In his uncomfortable duality of being both larger than life, and also so much *in* the life of his own society, Buffon often had to juggle and feint, to smooth over or to hide under, so that his readers or anyone in power, from priest to patron to Parisian pol, would not dismiss him as too far outside the sensibilities of his surrounding world. Buffon possessed a radical streak, the stubborn independence of all great thinkers. Mademoiselle Blesseau, his house manager and confidante, summed up Buffon's character in a letter written to his collaborator Faujas de Saint-Fond just after the master's death: "No one has ever been able to take credit for having controlled him." Jacques Roger comments:

> In the hierarchical society in which he lived, he knew how to carve out a place for himself, without excessive qualms or dishonoring servility. He used institutions as he found them and did not seek to change them because it was none of his business and because he did not have a great deal of confidence in human wisdom.

Buffon was just too engaged and too enmeshed to transform the world of thought with a consistent vision—too occupied with his seigneurial rights and funds (where he was fair but demanding, litigious if thwarted, and not particularly kind), and with wheeling and dealing to add land to his estates or to his (and the people's) Parisian gardens and museum. Too busy tending to his household after the early death of his wife, and worrying about his only and wastrel son, who suffered under his father's glory, bearing the diminutive nickname of Buffonet. (After his father's death, Buffonet ended up under the guillotine during the Reign of Terror.) Too involved also in pursuing his own tender, longstanding, and properly discreet relationship with Madame Necker, wife of the finance minister, who comforted and stood by him during his final illness and death. All this hubbub doesn't leave much time, or enough calm and extended space, for developing and propagating a consistent and radical reconstruction of nature.

Buffon's attitude toward religion and his relationships with France's Catholic hierarchy, best illustrate this defining (and ultimately constraining) feature of his personality. He was, without much doubt, a materialist at heart, and at least an agnostic in personal belief. A candid and private remark to Hérault de Séchelles epitomizes both his public stance and his personal attitude: "I have always named the Creator; but we need only remove this word and, of course, put in its place the power of Nature."

Buffon's publications play an extended cat and mouse game with religion. *Histoire naturelle* abounds with flowery and conventional hymns of praise to the omnipotent deity, creator of all things in heaven and earth. But Buffon's content often challenged traditional views and biblical texts. In fact, he began *Natural History* by forthrightly arguing, in volume 1 on the "Theory of the Earth" (published in 1749), that our planet had experienced an unlimited and cyclical history of gradual erosion and exposure of continents, uninterrupted by any catastrophe. (Buffon did not explicitly deny the Noachian deluge, but no one could have missed the implication.)

On January 15, 1751, the Theological Faculty of the Sorbonne attacked Buffon in a strong letter, demanding retraction or censorship. Buffon, in his usual worldly way, backed down in a note of apparent apology, stating that he believed "very firmly all that is told [in the book of Genesis] about Creation, both as to the order of time and the circumstances of the facts," and that he had presented his theory "only as a pure philosophical supposition." Buffon then published the Sorbonne's letter and his response at the beginning of the fourth volume of *Histoire naturelle* in 1753, and in all subsequent editions.

When I was younger, and beguiled by the false myth of warfare between science and religion as the path of progress in Western history, I viewed Buffon's retraction as a sad episode of martyrdom at an intermediary stage along an inevitable road. I now hold an entirely different view. Buffon surely prevailed in this incident. He reached a formal agreement with his enemies, staved off any future attacks, published a meaningless "apology" that no one would regard as sincere, and then never changed a single word of his original text. "It is better to be humble than hung," he wrote to a colleague in describing this contretemps; *Paris vaut bien une messe.*

Nonetheless, as Buffon lay dying, he clamored for last rites with a final ounce of passion that seemed ultimately and poignantly sincere. He had previously said to Hérault de Séchelles, in his usual and somewhat cynical manner, "When I become dangerously ill and feel my end approaching, I will not hesitate to send for the sacraments. One owes it to the public cult." Yet now, faced with the actuality, he seemed to plead only for himself. Madame Necker described his last moments (as a witness, not from secondhand reports): "He spoke to Father Ignace and said to him in a very anxious manner, 'Someone give me the good God quickly! Quickly! Quickly!' . . . Father Ignace gave him communion and M. de Buffon repeated during the ceremony, 'give it then! But give it then!'"

I do not know how to resolve this tangle of complexity, this mixture of practical posturing and sincere conviction. Perhaps we cannot go beyond Jacques Roger's insightful conclusions:

> That Buffon had a passion for order in everything—in his schedule, his accounts, his papers, and his life no less than in his study of nature—was such an obvious aspect of his temperament that his contemporaries noted it. He wanted an order, but not just any order; he wanted a true and legitimate order. Buffon wished there to be an order in society, and . . . he did define a few rules that should preside over such an order. Respect for the established religion is one of them, and he observed it all his life.

If we regard all these foregoing reasons for the eclipse of Buffon's name as primarily "negative" (his failure to construct and defend a transforming and distinctive view of life), another set of factors must be identified as the "positive" fate of all great reformers with such a broad palette and such an immediate impact. First of all, worldliness includes another side that promotes later invisibility. People who build institutions ("brick and mortar" folks, not mere dreamers and schemers), who lobby for educational reform, who write the textbooks that instruct generations of students, become widely known in their generation, if only because they demand explicit obeisance from anyone who wants to engage in the same business. But when they die and no longer hold the strings of power, their names fade quickly from view, even while their institutions and writings continue to mold the history of thought in profound and extensive ways.

Thus, we may note the irony of worldliness in the context of scales of time: one trades immediate recognition in life for the curious status of continuing, but anonymous, influence. How would French biology have developed without the Muséum, and without the forty-four volumes of *Histoire naturelle*? Can a great discovery by a recluse match the ultimately silent achievements of such worldliness? T. H. Huxley, with his tireless round of speeches, exhortations, popular books, politicking, and service on government committees, may have left a greater impact than Darwin upon British society. But Darwin, who, in the last decades of his life, almost never left his country house, even for trips to nearby London, persists (properly, I would claim, in another argument for another time) as the icon of our discoveries and our fears—while Huxley has become a faded memory.

How, similarly, can we measure Buffon's continuing presence? In the recent and brilliant reconstruction of the Grande Galérie of his Muséum into the world's finest modern exhibit on evolution? In *Histoire naturelle,* a treatise that has never been entirely out of print, and that taught students throughout the world as a primary text for more than a century—often in pirated editions that didn't acknowledge Buffon? (For example, few people know, I suspect, that the poet Oliver Goldsmith, to earn his bread, wrote a multivolumed *History of the Earth and Animated Nature* that amounts to little more than lightly annotated Buffon. My own collection of popular science books includes a volume, published in New York during the late nineteenth century, entitled *Buffon's Natural History*—a one-volume amalgam of bits and pieces from *Natural History,* and undoubtedly paying not a penny in royalties to Buffon's estate.)

Yet, and finally, the positive reasons for the paradoxical correlation of later anonymity with continuing impact also include a factor that should be judged as paramount, and that also distills the core of Buffon's greatest contribution to the history of ideas. Some of the grandest tools in the arsenal of our consciousness work so broadly and so generally that we can scarcely assign authorship to a single person. (Darwin can be identified as the discoverer of natural selection, even as the first comprehensive defender of biological evolution based on hard data drawn from all major subjects of natural history. But no one can be called the inventor of a developmental, rather than a static, view of nature.)

Buffon became the central figure in one of the great transformations of human thought—the discovery of history as a guiding principle for organizing the data of the natural world, including many aspects of human diversity (from language, to the arts, to social systems). As the earth's great age—its "deep time"—became apparent, and as revolutionary ideologies replaced monarchies in parts of Europe and America, such a reconstruction of knowledge hovered "in the air," and would have occurred if Buffon had never been born (see Paolo Rossi, *The Dark Abyss of Time,* University of Chicago Press, 1984). But, through a combination of the best subject matter to express such a change, an incomparable prose style, and a wide and dedicated readership, Buffon became the most influential focus of this transformation, with *Histoire naturelle* as the primary agent.

BUFFON'S DISCOVERY AND DEFINITION OF HISTORY

A truly historical account of nature demands deep time. But time can only provide a matrix for the unfolding of events. History requires the ordering of phe-

nomena in narrative form—that is, as a temporal series with direction given by
a sequence of complex and unrepeatable events, linked one to the next by sensible reasons for transition. In short, to qualify as history, such a sequence must
embody the last two syllables of its name: it must tell a story.

Most pre-Buffonian science included no history. Organisms had been created in primal perfection on a young earth, and none had become extinct
(except in the singularity of Noah's flood—and unique transforming events
don't constitute a history). The rocks of the earth represented either an original creation or the residues of Noah's flood. Even the influential cosmologies
of Newton, and of Buffon's younger and brilliant colleague Laplace, purposely
rejected history in positing exactly repeating cycles (perhaps with self-adjusting
variations) of "eternal return"—as Darwin recognized so well when he ended
his *Origin of Species* (1859) by contrasting the rich historicity of evolution with
the sterility of endless cosmological turning: "Whilst this planet has gone
cycling on according to the fixed law of gravity, from so simple a beginning
endless forms most beautiful and most wonderful have been, and are being,
evolved."

In the most important change of his own views between the inception of
Natural History in 1749 and the publication of its most important volume in
1778, Buffon became an advocate for historical thinking. His first volume of
1749, as discussed earlier, had been sufficiently radical in positing a long and
indeterminate age for the earth. Buffon did propose one historical item in this
initial *Theory of the Earth*—the first important hypothesis for the origin of planets by cometary impact into the sun, with ejection of masses to form the planetary spheres. But following this tempestuous origin, the earth of Buffon's first
volume experienced no further history—for geology had recorded only a series
of repeating cycles in erosion and exposure of continents.

But Buffon, confuting the cliché that scientists must develop their best ideas
in their youth or not at all, reversed his original belief, devised an intrinsically
and thoroughly historical theory of the earth, and published his results at the
age of seventy-one, in a volume that became by far his most popular, his most
influential, and his most controversial: *Des époques de la nature* (The epochs of
nature), which originally appeared in 1778 as supplementary volume five of
Histoire naturelle. This treatise became the most important scientific document
ever written in promoting the transition to a fully historical view of nature.
(Since Buffon's influence lay largely in his command of language, *The Epochs
of Nature* also illustrates the underappreciated principle that literary style may
not be irrelevant to the success of scientific ideas.) And yet, as argued above,
this shift to historical thinking raised too big an issue, involving too many

subjects and approaches, to lay in one man's lap—so Buffon's name never became firmly attached to his most important intellectual achievement.

The Epochs of Nature rose from complex roots in Buffon's psyche and activities. He did not simply devise this major transition from his armchair. Ever since developing his theory of planetary formation by cometary impact upon the sun, Buffon had searched for evidence that might indicate the time of origin, and the consequent age of the earth. ("Indeterminably long" could not satisfy a man of his restless energy.) After setting up his forges for smelting iron, a testable idea struck Buffon. If the earth had originated as a fireball, he could presumably calculate the length of time required for sufficient cooling to form a solid surface that could serve as a substrate for geological strata and life itself.

Buffon therefore began to experiment with the cooling of iron balls made in his forge. He then scaled his results upward to theoretical calculations for iron balls the size of the earth, and then to more realistic models for balls of various compositions more closely mimicking the earth's structure. Buffon pursued these experiments and calculations for years, obviously enjoying this return to his mathematical roots. He filled chapters of Histoire naturelle with his results, and finally decided that the earth must be at least 75,000 years old (and probably a good deal more ancient).

These experiments may have validated deep time in a quantitative manner, but they inspired an even more important change in Buffon's thinking, for they gave him the key to history. A continually cooling earth provided an arrow for time, a fundamental direction for the physical surface and for life as well. Since all organisms originate in perfect adaptation to surrounding environments— and since these environments have changed directionally through time to colder and colder conditions—the composition of faunas must also vary, as some species become extinct when climates alter beyond their power to cope, and new species, adapted to the changed circumstances, then appear.

(As one example of the radical nature of Buffon's historical view, the idea of extinction stuck in the craw of traditional naturalists, who remained committed to an earth made perfect in all ways at the outset, and who therefore could not abide the idea that species might disappear through failure of adaptation. Thomas Jefferson, Buffon's rival, could cite many good reasons for sending Lewis and Clark on their famous expedition, but one small factor lay in his hope that these explorers might find living mammoths in uncharted western lands, thus shaking Buffon's claim that species could die.)

Buffon constructed a rich history of seven successive epochs, all controlled by the continuous cooling of the earth from an original status as a solar fireball: first, the origin of the earth and planets by cometary impact; second, the formation of the solid earth and its mineral deposits; third, the covering of continents by oceans and the origin of marine life; fourth, the retreat of waters and the emergence of new continents; fifth, the appearance of animal life on land; sixth, the fragmentation of continents and the formation of current topography; seventh, the origin of humans and our accession to power.

Note that Buffon did not follow the most traditional arrow of history by arguing that life became continually more complex. In fact, he viewed the first marine creatures of epoch three (including ammonites and fishes) as already fully intricate. Buffon was not, after all, an evolutionist, and he built his arrow of time as a vector of decreasing warmth, not as a rising parade of organic progress. This arrow led him to a pessimistic conclusion, well constructed to fuel cosmic angst: the earth must eventually freeze over, leading to the extirpation of all life. This concept of a "heat death" for the earth became one of the most contentious and interesting ideas in late-eighteenth- and early-nineteenth-century thought, a theme of many poems, plays, and paintings.

Buffon's history also included a set of intriguing consequences, some internal to the theory, others inspired by the reactions of readers. To mention just two, Buffon might be cited by current ecoactivists (and I do say this facetiously) as an antihero—for he developed the notion of a greenhouse effect caused by human burning of forests, but welcomed such an imprint of advancing civilization as a device for postponing the earth's death by cold. Buffon wrote: "Cleansing, clearing, and populating a country gives it heat for several thousand years. . . . Paris and Quebec are at about the same latitude; Paris would therefore be as cold as Quebec if France and all the regions surrounding it were as lacking in men, and covered with woods . . . as are the neighboring lands of Canada."

Second, Buffon became the surprised recipient of several sumptuous gifts from Catherine II of Russia (a.k.a. Catherine the Great)—a collection of furs, all the medals of her reign (in gold), and her portrait on a gold snuff box encrusted with diamonds. Catherine had been delighted by Buffon's argument that since the earth becomes increasingly colder through time, new species originate in high latitudes and then migrate toward the tropics as temperatures continue to drop. Russia therefore became a cradle of life, rather than the frigid refugium envisioned by most previous writers. Buffon, ingratiating as always, thanked Catherine in a glowing letter that wished her well in campaigns against the Ottoman Empire ("that stagnating part of Europe"), and stated his hope to see "beautiful nature and the arts descend a second time from the North to the South under the standard of [her majesty's] powerful genius."

Moreover, and finally, the eminently orderly Buffon knew exactly what he had accomplished. He consciously promoted history as a novel and coordinating theme for all nature. He not only proposed a theory of origin, an arrow of time, and a narrative in seven epochs. He also knew that the triumph of history would require a fundamentally new way of thinking, and an explicit methodology, not yet familiar to scientists, for reconstructing the immensely long and poorly preserved record of the earth and life. He therefore suggested that natural scientists take their cue from procedures already worked out by students of human history. *The Epochs of Nature* begins with this call for an entirely new mode of thinking:

> In civil history, we consult titles, we research medals, we decipher ancient inscriptions in order to determine the time of human revolutions and to fix the dates of events in the moral order. Similarly, in natural history, it is necessary to excavate the archives of the world, to draw old monuments from the entrails of the earth, to collect their debris, and to reassemble into a single body of proof all the indices of physical changes which enable us to go back to the different ages of nature. This is the only way to fix points in the immensity of space, and to place a certain number of milestones on the eternal route of time. (My translation.)

No other person could possibly have provided better fuel for such a transformation in the history of human thought: this man of such restless energy; this man who operated forges and who developed the experimental and mathematical skill to infer the age of the earth from balls of iron; who composed thirty-six volumes of the greatest treatise ever written in natural history by working fourteen hours a day for more than forty years. And if all these skills and attributes could not turn the tide, Buffon also wrote in an elegant prose that placed him, a "mere" student of nature, among the leading men of letters in his interesting time. Buffon surely knew how to prevail—for style, after all, is the man himself.

5

The Proof of Lavoisier's Plates

I. WRITING IN THE MARGINS

I ONCE HAD A TEACHER WITH AN IDIOSYNCRATIC HABIT that distressed me forty years ago, but now—and finally, oh sweet revenge!—can work for me to symbolize the general process of human creativity. I never knew a stingier woman, and though she taught history in a junior high school in New York City, she might well have been the frugal New England farmer with a box marked "pieces of string not worth saving." Readers who attended New York public schools in the early 1950s will remember those small yellow slips of paper, three by six inches at most, that served all purposes from spot quizzes to "canvasses" for art class. Well, Mrs. Z. would give us one sheet—only one—for any classroom exam, no matter how elaborate the required answers. She would always reply to any plea for advice about containment or, God forbid, for an additional yellow sheet (comparable in her system of values to Oliver Twist's request for more soup) with a firm refusal followed by a cheery instruction expressed in her oddly lilting voice: ". . . and if you run out of room, just write in the margins!"

Margins play an interesting role in the history of scholarship, primarily for their schizophrenic housing of the two most contradictory forms of intellectual activity. Secondary commentaries upon printed texts (often followed by several layers of commentaries upon the commentaries) received their official designation as "marginalia" to note their necessary position at the edges. The usual status of such discourse as derivative and trivial, stating more and more about less and less at each iteration, leads to the dictionary definition of marginalia as "nonessential items" *(Webster's Third New International)* and inevitably recalls the famous, and literally biting, satire of Jonathan Swift:

> *So, naturalists observe, a flea*
> *Hath smaller fleas that on him prey;*
> *And these have smaller still to bite 'em;*
> *And so proceed* ad infinitum
> *Thus every poet, in his kind,*
> *Is bit by him that comes behind.*

But margins also serve the diametrically opposite purpose of receiving the first fruits and inklings of novel insights and radical revisions. When received wisdom has hogged all the central locations, where else can creative change begin? The curmudgeon and cynic in me regards Thoreau's *Walden* as the most overquoted (and underwhelming) American classic, but I happily succumb for the first time to cite his one-liner for a vibrant existence: "I love a broad margin to my life."

Literal margins, however, must usually be narrow—and some of the greatest insights in the history of human thought necessarily began in such ferociously cramped quarters. The famous story of Fermat's Last Theorem, no matter how familiar, cannot be resisted in this context: when the great mathematician died in 1665, his executors found the following comment in his copy of Diophantus' *Arithmetica,* next to a discussion of the claim that no natural numbers x, y, and z exist such that $x^n + y^n = z^n$, where n is a natural number greater than 2: "I have discovered a truly remarkable proof but this margin is too small to contain it." Mathematicians finally proved Fermat's Last Theorem just a few years ago, to great subsequent fanfare and an outpouring of popular books. But we shall never know if Fermat truly beat the best of the latest by three hundred years, or if (as my own betting money says, admittedly with no good evidence) he had a promising idea and never detected the disabling flaw in the midst of his excitement.

I devote this essay to the happier and opposite story of a great insight that a cramped margin did manage (just barely) to contain and nurture. This tale, for reasons that I do not fully understand, remains virtually unknown (and marginal in this frustrating sense) both to scientists and historians alike—although the protagonist ranks as one of the half dozen greatest scientists in Western history, and the subject stood at the forefront of innovation in his time. In any case, the movement of this insight from marginality in 1760 to centrality by 1810 marks the birth of modern geology, and gives us a rare and precious opportunity to eavesdrop on a preeminent thinker operating in the most exciting and instructive of all times: at a labile beginning in the codification of a major piece of natural knowledge—a unique moment featuring a landscape crossed by one hundred roads, each running in the right general direction toward a genuine truth. Each road, however, reaches a slightly different Rome, and our eventual reading of nature depends crucially upon the initial accidents and contingencies specifying the path actually taken.

In 1700, all major Western scholars believed that the earth had been created just a few thousand years ago. By 1800, nearly all scientists accepted a great antiquity of unknown duration, and a sequential history expressed in strata of the earth's crust. These strata, roughly speaking, form a vertical pile, with the oldest layers on the bottom and the youngest on top. By mapping the exposure of these layers on the earth's surface, this sequential history can be inferred. By 1820, detailed geological maps had been published for parts of England and France, and general patterns had been established for the entirety of both nations. This discovery of "deep time," and the subsequent resolution of historical sequences by geological mapping, must be ranked among the sweetest triumphs of human understanding.

Few readers will recognize the name of Jean-Etienne Guettard (1715–1786), a leading botanist and geologist of his time, and the initiator of the first "official" attempt to produce geological maps of an entire nation. In 1746, Guettard presented a preliminary "mineralogical map" of France to the Académie Royale des Sciences. In subsequent years, he published similar maps of other regions, including parts of North America. As a result, in 1766, the secretary of state in charge of mining commissioned Guettard to conduct a geological survey and to publish maps for all of France. The projected atlas would have included 230 maps, but everyone understood, I suspect, that such a task must be compared with the building of a medieval cathedral, and that no single career or lifetime could complete the job. In 1770, Guettard published the first sixteen maps. The project then became engulfed by political intrigue and finally by a revolution that (to say the least) tended to focus attention elsewhere. Only 45 of the 230 projected maps ever saw the published light of day, and control of the survey had passed to Guettard's opponents by this time.

Guettard's productions do not qualify as geological maps in the modern sense, for he made no effort to depict strata, or to interpret them as layers deposited in a temporal sequence—the revolutionary concepts that validated deep time and established the order of history. Rather, as his major cartographic device, Guettard established symbols for distinctive mineral deposits, rock types, and fossils—and then merely placed these symbols at appropriate locations on his map. We cannot even be sure that Guettard understood the principle of superposition—the key concept that time lies revealed in a vertical layering of strata, with younger layers above (superposed upon) older beds. Guettard did develop a concept of *"bandes,"* or roughly concentric zones of similar rocks, and he probably understood that a vertical sequence of strata might be expressed as such horizontal zones on a standard geographic map. But in any case, he purposely omitted these *bandes* on his maps, arguing that he wished only to depict facts and to avoid theories.

This focus on each factual tree, combined with his studious avoidance of any theoretical forest of generality or explanation, marked Guettard's limited philosophy of science, and also (however unfairly) restricted his future reputation, for no one could associate his name with any advance in general understanding. Rhoda Rappoport, a distinguished historian of science from Vassar College and the world's expert on late-eighteenth-century French geology, writes of Guettard (within a context of general admiration, not denigration): "The talent he most conspicuously lacked was that of generalization, or seeing the implications of his own observations. . . . Most of his work reveals . . . that he tried hard to avoid thinking of the earth as having a history."

But if Guettard lacked this kind of intellectual flair, he certainly showed optimal judgment in choosing a younger partner and collaborator for his geological mapping, for Guettard fully shared this great enterprise with Antoine-Laurent Lavoisier (1743–94), a mere fledgling of promise at the outset of their work in 1766, and the greatest chemist in human history when the guillotine literally cut his career short in 1794.

Guettard and Lavoisier took several field trips together, including a four-month journey in 1767 through eastern France and part of Switzerland. After completing their first sixteen maps in 1770, Lavoisier's interest shifted away from geology toward the sources of his enduring fame—a change made all the more irrevocable in 1777, when control of the geological survey passed to Antoine Monnet, inspector general of mines, and Lavoisier's enemy. (Later editions of the maps ignore Lavoisier's contributions and often don't even mention his name.)

Nonetheless, Lavoisier's geological interests persisted, buttressed from time to time by transient hope that he might regain control of the survey. In 1789, with his nation on the verge of revolution, Lavoisier presented his only major geological paper—a stunning and remarkable work that inspired this essay. Amidst his new duties as *régisseur des poudres* (director of gunpowder), and leading light of the commission that invented the meter as a new standard of measurement—and despite the increasing troubles that would lead to his arrest and execution (for his former role as a farmer-general, or commissioned tax collector)—Lavoisier continued to express his intention to pursue further geological studies and to publish his old results. But the most irrevocable of all changes fractured these plans on May 8, 1794, less than three months before the fall of Robespierre and the end of the Terror. The great mathematician Joseph-Louis Lagrange lamented the tragic fate of his dear friend by invoking the primary geological theme of contrasting time scales: "It took them only an instant to cut off his head, but France may not produce another like it in a century."

All the usual contrasts apply to the team of Guettard and Lavoisier: established conservative and radical beginner; mature professional and youthful enthusiast; meticulous tabulator and brilliant theorist; a counter of trees and an architect of forests. Lavoisier realized that geological maps could depict far more than the mere location of ores and quarries. He sensed the ferment accompanying the birth of a new science, and he understood that the earth had experienced a long history potentially revealed in the rocks of his maps. In 1749, Georges Buffon, the greatest of French naturalists, had begun his monumental treatise (*Histoire naturelle*, which would eventually run to forty-four volumes) with a long discourse on the history and theory of the earth (see chapter 4).

As he groped for a way to understand this history from the evidence of his field trips, and as he struggled to join the insights published by others with his own original observations, Lavoisier recognized that the principle of superposition could yield the required key: the vertical sequence of layered strata must record both time and the order of history. But vertical sequences differed in all conceivable features from place to place—in thickness, in rock types, in order of the layers. How could one take this confusing welter and infer a coherent history for a large region? Lavoisier appreciated the wisdom of his older colleague enough to know that he must first find a way to record and compile the facts of this variation before he could hope to present any general theory to organize his data.

A geological map by Guettard and Lavoisier, with Lavoisier's temporal sequence of strata in the right margin.

CARTE MINÉRALOGIQUE DES ENVIRONS DE PARIS.

Lavoisier therefore suggested that a drawing of the vertical sequence of sediments be included alongside the conventional maps festooned with Guettard's symbols. But where could the vertical sections be placed? In the margins, of course; for no other space existed in the completed design. Each sheet of Guettard and Lavoisier's *Atlas* therefore features a large map in the center with two marginal columns on the side: a tabular key for Guettard's symbols at the left, and Lavoisier's vertical sections on the right. If I wished to epitomize the birth of modern geology in a single phrase (admittedly oversimplified, as all such efforts must be), I would honor the passage—both conceptual and geometric—of Lavoisier's view of history, as revealed in sequences of strata, from a crowded margin to the central stage.

Many fundamental items in our shared conceptual world seem obvious and incontrovertible only because we learned them (so to speak) in our cradle and have never even considered that alternatives might exist. We often regard such notions—including the antiquity of the earth, the rise of mountains, and the deposition of sediments—as simple facts of observation, so plain to anyone with eyes to see that any other reading could only arise from the province of knaves or fools. But many of these "obvious" foundations arose as difficult and initially paradoxical conclusions born of long struggles to think and see in new ways.

If we can recapture the excitement of such innovation by temporarily suppressing our legitimate current certainties, and reentering the confusing transitional world of our intellectual forebears, then we can understand why all fundamental scientific innovation must marry new ways of thinking with better styles of seeing. Neither abstract theorizing nor meticulous observation can provoke a change of such magnitude all by itself. And when—as in this story of Lavoisier and the birth of geological mapping—we can link one of the greatest conceptual changes in the history of science with one of the most brilliant men who ever graced the profession, then we can only rejoice in the enlarged insight promised by such a rare conjunction.

Most of us, with minimal training, can easily learn to read the geological history of a region by studying the distribution of rock layers on an ordinary geographic map and then coordinating this information with vertical sections (as drawn in Lavoisier's margins) representing the sequence of strata that would be exposed by digging a deep hole in any one spot. But consider, for a moment, the intellectual stretching thus required, and the difficulty that such an effort would entail if we didn't already understand that mountains rise and erode, and that seas move in and out, over any given region of our ancient earth.

A map is a two-dimensional representation of a surface; a vertical section is a one-dimensional listing along a line drawn perpendicular to this surface and into the earth. To understand the history of a region, we must mentally integrate these two schemes into a three-dimensional understanding of time (expressed as vertical sequences of strata) across space (expressed as horizontal exposures of the same strata on the earth's surface). Such increases in dimensionality rank among the most difficult of intellectual problems—as anyone will grasp by reading the most instructive work of science fiction ever published, E. A. Abbott's *Flatland* (originally published in 1884 and still in print), a "romance" (his description) about the difficulties experienced by creatures who live in a two-dimensional world when a sphere enters the plane of their entire existence and forces them to confront the third dimension.

As for the second component of our linkage, I can only offer a personal testimony. My knowledge of chemistry remains rudimentary at best, and I can therefore claim no deep understanding of Lavoisier's greatest technical achievements. But I have read several of his works and have never failed to experience one of the rarest emotions in my own arsenal: sheer awe accompanied by spinal shivers. A kind of eerie, pellucid clarity pervades Lavoisier's writing (and simply makes me ashamed of the peregrinations in these essays).

Perhaps, indeed almost certainly, a few other scientists have combined equal brilliance with comparable achievement, but no one can touch Lavoisier in shining a light of logic into the most twisted corners of old conceptual prisons, into the most tangled masses of confusing observations—and extracting new truths expressed as linear arguments accessible to anyone. As an example of the experimental method in science (including the fundamental principle of double-blind testing), no one has ever bettered the document that Lavoisier wrote in 1784 as head of a royal commission (including Benjamin Franklin, then resident in Paris and, ironically, Dr. Guillotin, whose "humane" invention would end Lavoisier's life) to investigate (and, as results proved, refute) the claims of Dr. Mesmer about the role of animal magnetism in the cure of disease by entrancement (mesmerization).

Lavoisier did not compose his only geological paper until 1789, but Rhoda Rappoport has shown that he based this work upon conclusions reached during his mapping days with Guettard. Lavoisier did not invent the concept of vertical sections; nor did he originate the idea that sequences of strata record the history of regions on an earth of considerable antiquity. Instead, he resolved an issue that may seem small by comparison, but that couldn't be more fundamental to any hope for a workable science of geology (as opposed to the simpler pleasures of speculating about the history of the earth from an armchair): he showed how the geological history of a region can be read from variation in strata from place to place—or, in other words, how a set of one-dimensional lists of layered strata at single places can be integrated by that greatest of all scientific machines, the human mind, into a three-dimensional understanding of the history of geological changes over an entire region.

(I doubt that Lavoisier's work had much actual influence, for he published only one paper on this subject and did not live to realize his more extensive projects. Other investigators soon reached similar conclusions, for the nascent science of geology became the hottest intellectual property in late-eighteenth-century science. Lavoisier's paper has therefore been forgotten, despite several efforts by isolated historians of science through the years, with this essay as the latest attempt, to document the singularity of Lavoisier's vision and accomplishment.)

From my excellent sample of voluminous correspondence with lay readers during a quarter century of writing these essays, I have grasped the irony of the most fundamental misunderstanding about science among those who love the enterprise. (I am not discussing the different errors made by opponents of science.) Supporters assume that the greatness and importance of a work correlates directly with its stated breadth of achievement: minor papers solve local issues, while great works fathom the general and universal nature of things. But all practicing scientists know in their bones that successful studies require strict limitation: one must specify a particular problem with an accessible solution, and then find a sufficiently simple situation where attainable facts might point to a clear conclusion. Potential greatness then arises from cascading implications toward testable generalities. You don't reach the generality by direct assault without proper tools. One might as well dream about climbing Mount Everest in a T-shirt, wearing tennis shoes, and with a backpack containing only an apple and a bottle of water.

II. CAPTURING THE CENTER

When Lavoisier began his geological work with Guettard in 1766, he accepted a scenario, then conventional, for the history of the earth as revealed by the record of rocks: a simple directional scheme that envisaged a submergence of ancient landmasses (represented today by the crystalline rocks of mountains) under an ocean, with all later sediments formed in a single era of deposition from this stationary sea (on this topic, see Rhoda Rappoport's important article "Lavoisier's Theory of the Earth," *British Journal for the History of Science,* 1973). Since geologists then lacked techniques for unraveling the contorted masses of older crystalline rocks, they devoted their research to the later stratified deposits, and tried to read history as an uncomplicated tale of linear development. (No fossils had been found in the older crystalline rocks, so early geologists also assumed that the later stratified deposits contained the entire history of life.)

Lavoisier's key insight led him to reject this linear view (one period of deposition from a stationary sea) and to advocate the opposite idea that sea level had oscillated through time, and that oceans had therefore advanced and retreated through several cycles in any particular region—a notion now so commonplace that any geologist can intone the mantra of earth history: "the seas go in and the seas go out." Lavoisier reached this radical conclusion by combining the developing ideas of such writers as Buffon and De Maillet with his own observations on cyclical patterns of sedimentation in vertical sections.

Lavoisier christened his 1789 paper with a generous title characteristic of a time that did not separate literature and science: *Observations générales sur les couches modernes horizontales qui ont été déposées par la mer, et sur les conséquences qu'on peut tirer de leurs dispositions relativement à l'ancienneté du globe terrestre* (General observations on the recent horizontal beds that have been deposited by the sea, and on the consequences that one can infer, from their arrangement, about the antiquity of the earth). Lavoisier's title may be grand, general, and expansive, but his content remained precise, local, and particular—at first! Lavoisier begins his treatise by distinguishing the properties of sediments deposited in open oceans from those formed along shorelines—a procedure that he then followed to build the data for his central argument that seas advance and retreat in a cyclical pattern over any given region.

After two short introductory paragraphs, Lavoisier plunges right in by expressing puzzlement that two such opposite kinds of rock so often alternate to form multiple cycles in a single vertical section. Criteria of fossils and sediment indicate calm and gentle deposition for one kind: "Here one finds masses of shells, mostly thin and fragile, and most showing no sign of wear or abrasion. . . . All the features [of the rocks] that surround these shells indicate a completely tranquil environment" (my translations from Lavoisier's 1789 paper). But rocks deposited just above testify to completely different circumstances of formation: "A few feet above the place where I made these observations, I noted an entirely opposite situation. One now sees no trace of living creatures; instead, one finds rounded pebbles whose angles have been abraded by rapid and long-continued tumbling. This is the picture of an agitated sea, breaking against the shore, and violently churning a large quantity of pebbles." Lavoisier then poses his key question, already made rhetorical by his observations:

> How can we reconcile such opposite observations? How can such different effects arise from the same cause? How can movements that have abraded quartz, rock crystal, and the hardest stones into rounded pebbles, also have preserved light and fragile shells?

The simple answer to this specific and limited question may then lead to important generalities for the science of geology, and also to criteria for unraveling the particular history of the earth:

> At first glance, this contrast of tranquillity and movement, of organization and disorder, of separation and mixture, seemed inexplicable to me; nevertheless, after seeing the same phenomena again and again, at different times and in different places, and by combining these facts and observations, it seemed to me that one could explain these striking observations in a simple and natural manner that could then reveal the principal laws followed by nature in the generation of horizontal strata.

Lavoisier then presents his idealized model of a two-stage cycle as an evident solution to this conundrum:

Two kinds of very distinct beds must exist in the mineral kingdom: one kind formed in the open sea . . . which I shall call *pelagic* beds, and the other formed at the coast, which I shall call *littoral* beds.

Pelagic beds arise by construction, as "shells and other marine bodies accumulate slowly and peacefully during an immense span of years and centuries." But littoral beds, by contrast, arise by "destruction and tumult . . . as parasitic deposits formed at the expense of coastlines."

In a brilliant ploy of rhetoric and argument, Lavoisier then builds his entire treatise as a set of consequences from this simple model of two types of alternating sediments, representing the cycle of a rising and falling sea. This single key, Lavoisier claims, unlocks the great conceptual problem of moving from one-dimensional observations of vertical sequences in several localities to a three-dimensional reconstruction of history. (I call the solution three-dimensional for a literal reason, emphasized earlier in this essay in my discussion of geological maps: the two horizontal dimensions record geographic variation over the earth's surface, while the vertical dimension marks time in a sequence of strata):

This distinction between two kinds of beds . . . suddenly dispersed the chaos that I experienced when I first observed terranes made of horizontal beds. This same distinction then led me to a series of consequences that I shall try to convey, in sequence, to the reader.

The remainder of Lavoisier's treatise presents a brilliant fusion of general methodology and specific conclusions, a combination that makes the work such a wonderful exemplar of scientific procedure at its best. The methodological passages emphasize two themes: the nature of proof in natural history, and the proper interaction of theory and observation. Lavoisier roots the first theme in a paradox presented on pages 98–99: the need to simplify at first in order to generalize later. Science demands repetition for proper testing of observations—for how else could we learn that the same circumstances reliably generate the same results? But the conventional geologies of Lavoisier's time stymied such a goal—for one directional period of deposition from a single stationary sea offered no opportunity for testing by repetition. By contrast, Lavoisier's model of alternating pelagic and littoral beds provided a natural experiment in replication at each cycle.

But complex nature defies the needs of laboratory science for simple and well-controlled situations, where events can be replicated under identical conditions set by few variables. Lavoisier argues that we must therefore try to impose similar constraints upon the outside world by seeking "natural experiments" where simple models of our own construction might work adequately in natural conditions chosen for their unusual clarity and minimal number of controlling factors.

Consider three different principles, each exploited by Lavoisier in this paper, for finding or imposing a requisite simplicity upon nature's truly mind-boggling complexity.

1. Devise a straightforward and testable model. Lavoisier constructed the simplest possible scheme of seas moving in and out, and depositing only two basic (and strongly contrasting) types of sediment. He knew perfectly well that real strata do not arrange themselves in neat piles of exactly repeating pairs, and he emphasized two major reasons for nature's much greater actual complexity: first, seas don't rise and fall smoothly, but rather wiggle and jiggle in small oscillations superposed upon any general trend; second, the nature of any particular littoral deposit depends crucially upon the type of rock being eroded at any given coastline. But Lavoisier knew that he must first validate the possibility of a general enterprise—three-dimensional reconstruction of geological history—by devising a model that could be tested by replication. The pleasure of revealing unique details would have to wait for another time. He wrote:

> Beds formed along the coast by a rising sea will have unique characteristics in every different circumstance. Only by examining each case separately, and by discussing and explaining them in comparison to each other, will it be possible to grasp the full range of phenomena. . . . I will therefore treat [these details] in a separate memoir.

2. Choose a simple and informative circumstance. Nature's inherent complexity of irreducible uniqueness for each object must be kept within workable scientific bounds by intelligent choice of data with unusual and repeated simplicity. Here Lavoisier lucked out. He had noted the problem of confusing variation in littoral deposits based on erosion of differing rocks at varying coasts. Fortunately, in the areas he studied near Paris, the ancient cliffs that served as sources for littoral sediments might almost have been "made to order" for such a study. The cliffs had been formed in a widespread deposit of Cretaceous age called *La Craie* (or "the Chalk"—the same strata that build England's White

Cliffs of Dover). The Chalk consists primarily of fine white particles, swiftly washed out to sea as the cliffs erode. But the Chalk also includes interspersed beds of hard flint nodules, varying in size from golf balls to baseballs in most cases. These nodules provide an almost perfect experimental material (in uniform composition and limited range of size) for testing the effects of shoreline erosion. Lavoisier noted in particular (I shall show his engravings later in this essay) that the size and rounding of nodules should indicate distance of deposition from the shoreline—for pebbles should be large and angular when buried at the coast (before suffering much wear and erosion), but should then become smaller and rounder as they tumble farther away from the coastline in extensive erosion before deposition.

3. Ask a simple and resolvable question. You needn't (and can't) discover the deep nature of all reality in every (or any!) particular study. Better to pose smaller, but clearly answerable, questions with implications that then cascade outward toward a larger goal. Lavoisier had devised a simple, and potentially highly fruitful, model of oscillating sea levels in order to solve a fundamental question about the inference of a region's geological history from variation in vertical sections from place to place—the sections that he had placed in the margins of the maps he made with Guettard. But such a model could scarcely fail to raise, particularly for a man of Lavoisier's curiosity and brilliance, the more fundamental question—a key, perhaps, to even larger issues in physics and astronomy—of why oceans should rise and fall in repeated cycles. Lavoisier noted the challenge and wisely declined, recognizing that he was busy frying some tasty and sizable fish already, and couldn't, just at the moment, abandon such a bounty in pursuit of Moby Dick. So he praised his work in progress and then politely left the astronomical question to others (although he couldn't resist the temptation to drop a little hint that might help his colleagues in their forthcoming labors):

> It would be difficult, after such perfect agreement between theory and observation—an agreement supported at each step by proofs obtained from strata deposited by the sea—to claim that the rise and fall of the sea [through time] is only a hypothesis and not an established fact derived as a direct consequence of observation. It is up to the geometers, who have shown such wisdom and genius in different areas of physical astronomy, to enlighten us about the cause

of these oscillations [of the sea], and to teach us if they are still occurring, or if it is possible that the earth has now reached a state of equilibrium after such a long sequence of centuries. Even a small change in the position of the earth's axis of rotation, and a consequent shift in the position of the equator, would suffice to explain all these phenomena. But this great question belongs to the domain of physical astronomy, and is not my concern.

For the second methodological theme of interaction between observation and theory in science, Lavoisier remembered the negative lesson that he had learned from the failures of his mentor Guettard. A major, and harmful, myth of science—engendered by a false interpretation of the eminently worthy principle of objectivity—holds that a researcher should just gather facts in the first phase of study, and rigorously decline to speculate or theorize. Proper explanations will eventually emerge from the data in any case. In this way, the myth proclaims, we can avoid the pitfalls of succumbing to hope or expectation, and departing from the path of rigorous objectivity by "seeing" only what our cherished theory proclaims as righteous.

I do appreciate the sentiments behind such a recommendation, but the ideal of neutrally pure observation must be judged as not only impossible to accomplish, but actually harmful to science in at least two major ways. First, no one can make observations without questions in mind and suspicions about forthcoming results. Nature presents an infinity of potential observations; how can you possibly know what might be useful or important unless you are seeking an answer to a particular puzzle? You will surely waste a frightful amount of time when you don't have the foggiest idea about the potential outcomes of your search.

Second, the mind's curiosity cannot be suppressed. (Why would anyone ever want to approach a problem without this best and most distinctive tool of human uniqueness?) Therefore, you will have suspicions and preferences whether you acknowledge them or not. If you truly believe that you are making utterly objective observations, then you will easily tumble into trouble, for you will probably not recognize your own inevitable prejudices. But if you acknowledge a context by posing explicit questions to test (and, yes, by inevitably rooting for a favored outcome), then you will be able to specify— and diligently seek, however much you may hope to fail—the observations that can refute your preferences. Objectivity cannot be equated with mental blankness; rather, objectivity resides in recognizing your preferences and then subjecting them to especially harsh scrutiny—and also in a willingness to revise or abandon your theories when the tests fail (as they usually do).

Lavoisier had spent years watching Guettard fritter away time by an inchoate gathering of disparate bits of information, without any cohesive theory to guide and coordinate his efforts. As a result, Lavoisier pledged to proceed in an opposite manner, while acknowledging that the myth of objectivity had made his procedure both suspect and unpopular. Nonetheless, he would devise a simple and definite model, and then gather field observations in a focused effort to test his scheme. (Of course, theory and observation interact in subtle and mutually supporting ways. Lavoisier used his preliminary observations to build his model, and then went back to the field for extensive and systematic testing.) In an incisive contrast between naive empiricism and hypothesis testing as modes of science, Lavoisier epitomized his preference for the second method:

> There are two ways to present the objects and subject matter of science. The first consists in making observations and tracing them to the causes that have produced them. The second consists in hypothesizing a cause, and then seeing if the observed phenomena can validate the hypothesis. This second method is rarely used in the search for new truths, but it is often useful for teaching, for it spares students from difficulties and boredom. It is also the method that I have chosen to adopt for the sequence of geological memoirs that I shall present to the Academy of Sciences.

Lavoisier therefore approached the terranes of France with a definite model to test: seas move in and out over geographic regions in cycles of advancing and retreating waters. These oscillations produce two kinds of strata: pelagic deposits in deeper waters and littoral deposits fashioned from eroded coasts near the shoreline. Type of sediment should indicate both environment of deposition and geographic position with respect to the shoreline at any given time: pelagic deposits always imply a faraway shore. For the nearer littoral deposits, relative distance from shore can be inferred from the nature of any particular stratum. For littoral beds made mainly of flint nodules derived from Chalk, the bigger and more angular the nodules, the closer the shoreline.

From these simple patterns, all derived as consequences of an oscillating sea, we should be able to reconstruct the three-dimensional history of an entire region from variation in vertical sequences of sediments from place to place. (For example, if a continuous bed representing the same age contains large and angular flint nodules at point A, and smaller and more rounded nodules at point B, then A lay closer to the shoreline at the time of deposition.)

Lavoisier devotes most of his paper, including all of his seven beautifully drafted plates, to testing this model, but I can summarize this centerpiece of his treatise in three pictures and a few pages of text because the model makes such clear and definite predictions—and nature must either affirm or deny. Lavoisier's first six plates—in many ways, the most strikingly innovative feature of his entire work—show the expected geographic distribution of sediments under his model.

Lavoisier's first plate, for example, shows the predictable geographic variation in a littoral bed formed by a rising sea. The sea will mount from a beginning position (marked *"ligne de niveau de la basse mer,"* or line of low sea level, and indicated by the top of the illustrated waters) to a highstand marked *"ligne de niveau de la haute mer,"* or line of high sea level. The rising sea beats against a cliff, shown at the far left and marked *"falaise de Craye avec cailloux,"* or cliff of Chalk with pebbles. Note that, as discussed previously, this Chalk deposit contains several beds of flint nodules, drawn as thin horizontal bands of dark pebbles.

The rising sea erodes this cliff and deposits a layer of littoral beds underneath the waters, and on top of the eroded chalk. Lavoisier marks this bed with a sequence of letters BDFGHILMN, and shows how the character of the sediment varies systematically with distance from the shoreline. At B, D, and F, near the shore, large and angular pebbles (marked *"cailloux roulés,"* or rolled pebbles), formed from the eroded flint nodules, fill the stratum. The size of particles then decreases continually away from shore, as the pebbles break up and erode (changing from *"sable grossier,"* or coarse sand, to *"sable fin,"* or fine sand, to *"sable très fin ou argille,"* or very fine sand and clay). Meanwhile, far from shore (marked KK at the right of the figure), a pelagic bed begins to form (marked *"commencement des bancs calcaires,"* or beginning of calcareous beds).

From this model, Lavoisier must then predict that a vertical section at G, for example, would first show (as the uppermost stratum) a littoral bed made of large and angular pebbles, while a vertical section at M would feature a pelagic bed on top of a littoral bed, with the littoral bed now made of fine sand and clay. The two littoral beds at G and M would represent the same time, but their differences in composition would mark their varying distances from the shore. This simple principle of relating differences in beds of the same age to varying environments of deposition may seem straightforward, but geologists did not really develop a usable and consistent theory of such "facies" (as we call these variations) until this century. Lavoisier's clear vision of 1789, grossly simplified though his example may be, seems all the more remarkable in this context.

Lavoisier's first plate, showing spatial variation in sediments deposited in a rising sea.

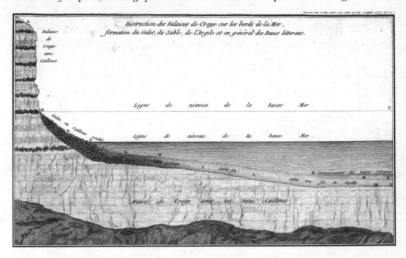

Lavoisier then presents a series of similar diagrams of growing complexity, culminating in plate 6, also reproduced here. This final plate shows the results of a full cycle—the sea, having advanced to its full height, has already retreated back to its starting point. The chalk cliff has been completely eroded away and now remains only as a bottom layer. (Note the distinctive bands of flint nodules for identification. I will discuss later the lowermost layer, marked *"ancienne terre,"* or ancient earth.) A lower littoral layer, formed by the rising sea, lies above the eroded chalk (marked HLMN as *"bancs littoraux inférieurs formés à la mer montante,"* or lower littoral beds formed by the rising sea). A pelagic bed, marked KKK (don't blame Lavoisier for a later and fortuitous American anachronism!), lies just above (labeled *"bancs pelagiens calcaires horizontaux supérieurs,"* or upper calcareous horizontal pelagic beds). Note how the pelagic bed pinches out toward shore because sediments of this type can be deposited only in deep water. This pelagic bed forms when sea level reaches its highest point. Then, as the sea begins to fall, another littoral bed will be deposited in progressively shallower water atop the pelagic bed (marked HIGG and labeled *"bancs littoraux supérieurs formés à la mer déscendante,"* or upper littoral beds formed by the falling sea).

Again, Lavoisier's insights are subtle and detailed—and several specific predictions can be derived from his model. For example, the upper and lower littoral beds will be confluent near the coast because the intervening pelagic bed didn't reach this far inland. Thus, a vertical section taken near the coast should show a single thick littoral bed made of large and angular pebbles. But farther away from shore, a vertical section should include a full array of alternating beds, illustrating the entire cycle and moving (top to bottom, as shown by the vertical line, located just left of center and marked 12345) from the upper littoral bed of the falling sea (1), to the intervening pelagic bed (2), to the lower littoral bed of the rising sea (3), to the underlying chalk (4), and finally to the foundation of the *ancienne terre* (5).

Thus, Lavoisier's model makes clear and specific predictions about how the sediments deposited in full cycles of rising and falling seas should be expressed in the vertical sections that had once adorned the margins of his maps with Guettard, and that represented his signal and original contribution to the developing science of geology. Moreover, the model specified predictions not only for vertical sequences in single places, but also for geographic variation in sequences from place to place. Therefore, in a last figure, Lavoisier presents some actual vertical sections measured in the field. The example presented here corresponds exactly to his prediction for section 12345 in the idealized model. Note the perfect correspondence between Lavoisier's *"Coupe des Montagnes des environ de St. Gobain"* (section through the mountains in the neighborhood of St. Gobain) and his model (except that the actual section doesn't extend below the chalk into the ancient basement). The measured section shows four layers labeled "upper littoral," "pelagic beds," "lower littoral," and "chalk" (note the layers of flint nodules in the lowermost chalk). Lavoisier had intended to write several more geological papers filled with similar empirical details to test his model. Thus, this pilot study presents only a few actual sections, but with impressive promise for continuing validation. Lavoisier had achieved a scientific innovation of the finest and most indubitable form: he had added a dimension (literally) to our knowledge of natural history.

As if he had not done enough already, Lavoisier then ended his treatise with two pages of admittedly hypothetical reasoning on the second great general theme in the study of time and history. His model of oscillating seas lies fully within the Newtonian tradition of complete and ahistorical generality. Lavoisier's oceanic cycles operate through time, but they do not express history because strings of events never occur in distinct and irreversibly directional sequences, and no single result ever denotes a uniquely definable moment. The cycles obey a timeless law of nature, and proceed in the same way no matter when they run. Cycle 100 will yield the same results as cycle 1; and the record

Lavoisier's final plate, showing the spatial and temporal complexity of sediments deposited in a full cycle of a rising and falling sea.

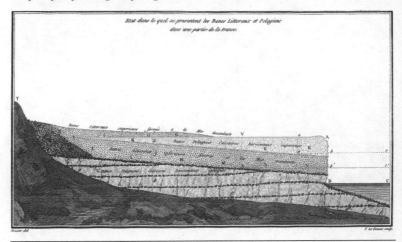

of rocks can never tell us where we stand in the flow of history. All variation reflects either a general environment (high or low sea) or a local circumstance (type of rock in the cliff being eroded), not the distinctive imprint of any unique and definite historical event.

Lavoisier, in other words, had worked brilliantly with the necessary concept of "time's cycle," so vital for any scientific account of the past because we need general laws to explain repeated physical events. But geology cannot render a full account of the earth's past without also invoking the fundamentally different, but intricately conjoined and equally necessary, concept of "time's arrow," so indispensable because geology embraces history as well—and historical accounts must tell stories defined by directional sequences of unique events.

As a prerequisite for interest and meaning, history must require a matrix of extensive time—which Lavoisier had already provided by combining his oscillating model of the oceans with empirical evidence for multiple cycles in vertical sections. If each cycle required considerable time (particularly for the formation of pelagic beds, so slowly built from the debris of organisms), then the evidence for numerous cycles implied an earth of great antiquity. By 1789 (and contrary to popular legend), few scientists still accepted a biblical chronology of just a few thousand years for the earth's history. But the true immensity of geological time still posed conceptual difficulties for many investigators, and Lavoisier's forthright claims mirrored the far more famous lines, published just

Lavoisier shows that an actual sequence of sediments conforms to his model of deposition in cycles of raising and lowering of sea level.

a year before in 1788, by the traditional "father" of modern geology, the Scotsman James Hutton: "time is, to nature, endless and as nothing." Lavoisier expressed his version of deep time in the more particular light of his model:

> The details that I have just discussed have no other object than to prove this proposition: if we suppose that the sea undergoes a very slow oscillatory movement, a kind of flux and reflux, that these movements occur during a period of several hundreds of thousands

of years, and that these movements have already occurred a certain number of times, then if we make a vertical section of rocks deposited between the sea and the high mountains, this section must present an alternation of littoral and pelagic beds.

Within such a matrix of deep time, the concept of a truly scientific history obtains new meaning and promise. At the end of his treatise, Lavoisier therefore touches upon this subject in his characteristically empirical way—by returning to the lowermost layer beneath the recorded sediments of his models and measured sections, a complex of rocks that he had bypassed with the simple label *"ancienne terre,"* or ancient earth. Lavoisier now states that he does not regard this foundation as part of the original earth at its time of formation, but rather as a probable series of sediments, much older than the Chalk, but also built as a sequence of littoral and pelagic beds (although now hard to identify because age has obliterated the characteristic features of such deposits):

> One will no doubt want to know about the rocks found underneath the Chalk, and what I mean by the expression *l'ancienne terre*. . . . This is almost surely not the original earth; on the contrary, it appears that what I have called *ancienne terre* is itself composed of littoral beds much older than those depicted in the figures.

In a remarkable passage, Lavoisier then invokes what would become the classic subject for juxtaposing the yin of history (or time's arrow) against the yang of constant features built by invariant laws (time's cycle) to form a complete science of geology: the directional character of life's pageant, the primary component of the earth's rip-roaring narrative story. (By the way, Lavoisier's particular claims turn out to be wrong in every detail, but I can hardly think of an observation more irrelevant to my present point. In 1789, no one knew much about paleontological particulars. I am stressing Lavoisier's keen and correct vision that life would provide the primary source of directional history, or time's arrow.)

Lavoisier bases his claim for history upon a clever argument. He believes that rocks of the *ancienne terre* contain no fossils. But if these rocks include (as he has just argued) the same alternation of pelagic and littoral beds found in younger sediments, then the invariant physical laws of time's cycle should lead us to expect fossils in these strata—for such sediments form in environments that now teem with life. Therefore, time's arrow of directional history must explain the difference. Physical conditions of the *ancienne terre* could not have differed from later circumstances that generated similar sediments, but the earth must then have housed no living creatures if these identical rocks contain no fossils.

Lavoisier then argues that sediments occasionally found below the Chalk (the oldest rocks with marine fossils), but above the *ancienne terre,* often contain fossils of plants. He therefore envisages a threefold directional history of life: an original earth devoid of organisms, followed by the origin of vegetation on land, and finally culminating in the development of animal life both in the sea and on land:

> It is very remarkable that the Chalk is usually the youngest rock to contain shells and the remains of other marine organisms. The beds of shale that we sometimes find below the Chalk often include vestiges of floating bodies, wood, and other vegetable matter thrown up along the coasts. . . . If we may be allowed to hazard a guess about this strange result, I believe we might be able to conclude, as Mr. Monge has proposed [the important French mathematician Gaspard Monge, who served with Lavoisier on the revolutionary commission to devise the metric system], that the earth was not always endowed with living creatures, that it was, for a long time, an inanimate desert in which nothing lived, that the existence of vegetables preceded that of most animals, or at least that the earth was covered by trees and plants before the seas were inhabited by shellfish.

And thus, hurriedly, at the very end of a paper intended only as a preliminary study, an introductory model to be filled in and fleshed out with extensive data based on field research, Lavoisier appended this little conjectural note—to show us, I suspect, that he grasped the full intellectual range of the problems set by geology, and that he recognized the power of combining a firm understanding of timeless and invariant laws with a confident narration of the rich directional history of an ancient earth. His last page bubbles with enthusiasm for future plans involving the whole earth, a project so soon cut off by the evil that only men can do. Consider the poignant paragraph just following his speculation about the history of life:

> In the next article, I will discuss in very great detail these opinions, which really belong more to Mr. Monge than to myself. But it is

indispensable that I first establish, in a solid way, the observations on which they are based.

I don't know why Lavoisier's execution affects me so deeply. We cannot assert with confidence that he would have completed his geological projects if he had lived (for all creative careers remain chock full of unrealized plans); and we know that he faced his end with a dignity and equanimity that can still provide comfort across the centuries. He wrote in a last letter:

> I have had a fairly long life, above all a very happy one, and I think that I shall be remembered with some regrets and perhaps leave some reputation behind me. What more could I ask? The events in which I am involved will probably save me from the troubles of old age. I shall die in full possession of my faculties.

Lavoisier needs no rescue, either from me or from any modern author. Yet speaking personally (a happy privilege granted to essayists ever since Montaigne invented the genre for this explicit purpose more than two hundred years before Lavoisier's time), I do long for some visceral sense of fellowship with this man who stands next to Darwin in my private pantheon of intellectual heroes. He died through human cruelty, and far too young. His works, of course, will live—and he needs no more. But, and I have no idea why, we also long for what I called visceral fellowship just above—some sense of *physical continuity,* some sign of an *actual presence* to transmit across the generations, so that we will not forget the person behind the glorious ideas. (Perhaps my dedication to such material continuity only marks a personal idiosyncrasy—but not, I think, a rare feeling, and certainly concentrated among those who choose paleontology for a profession because they thrill to the objective records of life's continuous history.)

So let me end with a personal testimony. Through the incredible good fortune of an odd coincidence in good timing and unfathomable pricing, I was able to buy a remarkable item at auction a while ago—the original set of proof plates, each personally signed by Lavoisier, of the seven figures (including the three reproduced here) that accompany his sole geological article of 1789. Two men signed each plate: first, in a thick and bold hand, Gabriel de Bory, vice-secretary of the Academy of Sciences (signed "Bory Vice-Secretaire"); and second, in a much more delicate flow composed of three flourishes surrounding the letters of his last name alone, Antoine-Laurent Lavoisier.

Lavoisier's signature (left) on one of his geological plates.

Lavoisier's own flourishes enhance the visual beauty of the plates that express the intellectual brilliance of his one foray into my field of geology—all signed in the year of the revolution that he greeted with such hope (and such willingness to work for its ideals); the revolution that eventually repaid his dedication in the most cruel of all possible ways. But now I hold a tiny little bit, only a symbol really, of Lavoisier's continuing physical presence in my professional world.

The skein of human continuity must often become this tenuous across the centuries (hanging by a thread, in the old cliché), but the circle remains unbroken if I can touch the ink of Lavoisier's own name, written by his own hand. A candle of light, nurtured by the oxygen of his greatest discovery, never burns out if we cherish the intellectual heritage of such unfractured filiation across the ages. We may also wish to contemplate the genuine physical thread of nucleic acid that ties each of us to the common bacterial ancestor of all living creatures, born on Lavoisier's *ancienne terre* more than 3.5 billion years ago— and never since disrupted, not for one moment, not for one generation. Such a legacy must be worth preserving from all the guillotines of our folly.

6

A Tree Grows in Paris: Lamarck's Division of Worms and Revision of Nature

I. THE MAKING AND BREAKING OF A REPUTATION

ON THE TWENTY-FIRST DAY OF THE AUSPICIOUSLY named month of *Floréal* (flowering), in the spring of year 8 on the French revolutionary calendar (1800 to the rest of the Western world), the former *chevalier* (knight) but now *citoyen* (citizen) Lamarck delivered the opening lecture for his annual course on zoology at the Muséum d'Histoire Naturelle in Paris—and changed the science of biology forever by presenting the first public account of his theory of evolution. Lamarck then published this short discourse in 1801 as the first part of his treatise on invertebrate animals (*Système des animaux sans vertèbres*).

Jean-Baptiste Lamarck (1744–1829) had enjoyed a distinguished career in botany when, just short of his fiftieth birthday, he became professor of insects and worms at the Muséum d'Histoire Naturelle, newly constituted by the revolutionary government in 1793. Lamarck would later coin the term *invertebrate* for his charges. (He also introduced the word *biology* for the entire discipline in 1802.) But his original title followed Linnaeus's designation of all nonvertebrated animals as either insects or worms, a Procrustean scheme that Lamarck would soon alter. Lamarck had been an avid shell collector and student of mollusks (then classified within Linnaeus's large and heterogeneous category of Vermes, or worms)—qualifications deemed sufficient for his change of subject.

Lamarck fully repaid the confidence invested in his general biological abilities by publishing distinguished works in the taxonomy of invertebrates throughout the remainder of his career, culminating in the seven volumes of his comprehensive *Histoire naturelle des animaux sans vertèbres* (Natural history of invertebrate animals), published between 1815 and 1822. At the same time, he constantly refined and expanded his evolutionary views, extending his introductory discourse of 1800 into a first full book in 1802 (*Recherches sur l'organisation des corps vivans,* or Researches on the organization of living beings), to his magnum opus and most famous work of 1809, the two-volume *Philosophie zoologique* (Zoological philosophy), to a final statement in the long opening section, published in 1815, of his great treatise on invertebrates.

The outlines of such a career might seem to imply continuing growth of prestige, from the initial flowering to a full bloom of celebrated seniority. But Lamarck's reputation suffered a spectacular collapse, even during his own lifetime, and he died lonely, blind, and impoverished. The complex reasons for his reverses include the usual panoply of changing fashions, powerful enemies, and self-inflicted wounds based on flaws of character (in his case, primarily an overweening self-assurance that led him to ignore or underestimate the weaknesses in some of his own arguments, or the skills of his adversaries). Most prominently, his favored style of science—the construction of grand and comprehensive theories following an approach that the French call *l'esprit de système* (the spirit of system building)—became notoriously unpopular following the rise of a hard-nosed empiricist ethos in early-nineteenth-century geology and natural history.

In one of the great injustices of our conventional history, Lamarck's disfavor has persisted to our times, and most people still know him only as the foil to Darwin's greatness—as the man who invented a silly theory about giraffes stretching their necks to reach the leaves on tall trees, and then passing the fruits of their efforts to their offspring by "inheritance of acquired characters," otherwise known as the hypothesis of "use and disuse," in contrast with Darwin's proper theory of natural selection and survival of the fittest.

Indeed, the usually genial Darwin had few kind words for his French predecessor. In letters to his friends, Darwin dismissed Lamarck as an idle speculator with a nonsensical theory. In 1844, he wrote to the botanist J. D. Hooker on the dearth of evolutionary thinking (before his own ideas about natural selection): "With respect to books on the subject, I do not know of any systematical ones except Lamarck's, which is veritable rubbish." To his guru, the geologist Charles Lyell (who had accurately described Lamarck's system for English readers in the second volume of his *Principles of Geology*, published in 1832), Darwin wrote in 1859, just after publishing *The Origin of Species:* "You often allude to Lamarck's work; I do not know what you think about it, but it appeared to me extremely poor; I got not a fact or idea from it."

But these later and private statements did Lamarck no practical ill. Far more harmfully, and virtually setting an "official" judgment from that time forward, his eminent colleague Georges Cuvier—the brilliant biologist, savvy statesman, distinguished man of letters, and Lamarck's younger and antievolutionary fellow professor at the Muséum—used his established role as writer of *éloges* (obituary notices) for deceased colleagues to compose a cruel masterpiece in the genre of "damning with faint praise"—a document that fixed and destroyed Lamarck's reputation. Cuvier began with cloying praise, and then described his need to criticize as a sad necessity:

> In sketching the life of one of our most celebrated naturalists, we have conceived it to be our duty, while bestowing the commendation they deserve on the great and useful works which science owes to him, likewise to give prominence to such of his productions in which too great indulgence of a lively imagination had led to results of a more questionable kind, and to indicate, as far as we can, the cause or, if it may be so expressed, the genealogy of his deviations.

Cuvier then proceeded to downplay Lamarck's considerable contributions to anatomy and taxonomy, and to excoriate his senior colleague for fatuous speculation about the comprehensive nature of reality. He especially ridiculed Lamarck's evolutionary ideas by contrasting a caricature of his own construction with the sober approach of proper empiricism:

> These [evolutionary] principles once admitted, it will easily be perceived that nothing is wanting but time and circumstances to enable

a monad or a polypus gradually and indifferently to transform themselves into a frog, a stork, or an elephant. . . . A system established on such foundations may amuse the imagination of a poet; a metaphysician may derive from it an entirely new series of systems; but it cannot for a moment bear the examination of anyone who has dissected a hand . . . or even a feather.

Cuvier's *éloge* reeks with exaggeration and unjust ridicule, especially toward a colleague ineluctably denied the right of response—the reason, after all, for our venerable motto, *de mortuis nil nisi bonum* ("say only good of the dead"). But Cuvier did base his disdain on a legitimate substrate, for Lamarck's writing certainly displays a tendency to grandiosity in comprehensive pronouncement, combined with frequent refusal to honor, or even to consider, alternative views with strong empirical support.

L'esprit de système, the propensity for constructing complete and overarching explanations based on general and exceptionless principles, may apply to some corners of reality, but this approach works especially poorly in the maximally complex world of natural history. Lamarck did feel drawn toward this style of system building, and he showed no eagerness to acknowledge exceptions or to change his guiding precepts. But the rigid and dogmatic Lamarck of Cuvier's caricature can only be regarded as a great injustice, for the man himself did maintain appropriate flexibility before nature's richness, and did eventually alter the central premises of his theory when his own data on the anatomy of invertebrate animals could no longer sustain his original view.

This fundamental change—from a linear to a branching system of classification for the basic groups, or phyla, of animals—has been well documented in standard sources of modern scholarship about Lamarck (principally in Richard W. Burkhardt, Jr.'s *The Spirit of System: Lamarck and Evolutionary Biology,* Harvard University Press, 1977; and Pietro Corsi's *The Age of Lamarck,* University of California Press, 1988). But the story of Lamarck's journey remains incomplete, for both the initiating incident and the final statement have been missing from the record—the beginning, because Lamarck noted his first insight as a handwritten insertion, heretofore unpublished, in his own copy of his first printed statement about evolution (the *Floréal* address of 1800, recycled as the preface to his 1801 book on invertebrate anatomy); and the ending, because his final book of 1820, *Système analytique des connaissances positives de l'homme* (Analytical system of positive knowledge about man), has been viewed only as an obscure swan song about psychology, a rare book even more rarely consulted (despite a fascinating section containing a crucial and novel wrinkle upon Lamarck's continually changing views about the classification of animals). Stories deprived of both beginnings and endings cannot satisfy our urges for fullness or completion—and I am grateful for this opportunity to supply these terminal anchors.

II. LAMARCK'S THEORY AND OUR MISREADINGS

Lamarck's original evolutionary system—the logical, pure, and exceptionless scheme that nature's intransigent complexity later forced him to abandon—featured a division of causes into two independent sets responsible for progress and diversity respectively. (Scholars generally refer to this model as the "two-factor theory.") On the one hand, a "force that tends incessantly to complicate organization" *(la force qui tend sans cesse à composer l'organisation)* leads evolution linearly upward, beginning with spontaneous generation of "infusorians" (single-celled animals) from chemical precursors, and moving on toward human intelligence.

But Lamarck recognized that the riotous diversity of living organisms could not be ordered into a neat and simple sequence of linear advance—for what could come directly before or after such marvels of adaptation as long-necked giraffes, moles without sight, flatfishes with both eyes on one side of the body, snakes with forked tongues, or birds with webbed feet? Lamarck therefore advocated linearity only for the "principal masses," or major anatomical designs of life's basic phyla. Thus, he envisioned a linear sequence mounting, in perfect progressive regularity, from infusorian to jellyfish to worm to insect to mollusk to vertebrate. He then depicted the special adaptations of particular lineages as lateral deviations from this main sequence.

These special adaptations originate by the second set of causes, labeled by Lamarck as "the influence of circumstances" *(l'influence des circonstances)*. Ironically, this second (and subsidiary) set has descended through later history as the exclusive "Lamarckism" of modern textbooks and anti-Darwinian iconoclasts (while the more important first set of linearizing forces has been forgotten). For this second set—based on change of habits as a spur to adaptation in new environmental circumstances—invokes the familiar (and false) doctrines now called "Lamarckism": the "inheritance of acquired characters" and the principle of "use and disuse."

Lamarck invented nothing original in citing these principles of inheritance, for both doctrines represented the "folk wisdom" of his time (despite their later disproof in the new world of Darwin and Mendel). Thus, the giraffe stretches its neck throughout life to reach higher leaves on acacia trees, and the shorebird extends its legs to remain above the rising waters. This sustained effort leads to longer necks or legs—and these rewards of hard work then descend to offspring in the form of altered heredity (the inheritance of acquired characters, either enhanced by use, as in these cases, or lost by disuse, as in eyeless moles or blind fishes living in perpetually dark caves).

As another irony and injustice (admittedly abetted, in part, by his own unclear statements), the ridicule that has surrounded Lamarck's theory since Cuvier's *éloge* and Darwin's dismissal has always centered upon the charge that Lamarck's views represent a sad throwback to the mystical vitalism of bad old times before modern science enshrined testable mechanical causes as the proper sources of explanation. What genuine understanding, the critics charge, can possibly arise from claims about vague and unknowable powers inherent in life itself, and propelling organisms either upward by an intrinsic complexifying force (recalling Molière's famous mock of vitalistic medicine, exemplified in the statement that morphine induces sleep *"quia est in eo virtus dormativa,"* because it contains a dormitive virtue) or sideward by some ineffable "willing" to build an adaptive branch by sheer organic effort or desire?

In a famous letter to J. D. Hooker, his closest confidant, Darwin first admitted his evolutionary beliefs in 1844 by contrasting his mechanistic account with a caricature of Lamarck's theory: "I am almost convinced . . . that species are not (it is like confessing a murder) immutable. Heaven forfend me from Lamarck nonsense of a 'tendency to progression,' 'adaptations from the slow willing of animals,' etc.! But the conclusions I am led to are not widely different from his; though the means of change are wholly so." And Cuvier, in a public forum, ridiculed the second set of adaptive forces in the same disdainful tone: "Wants and desires, produced by circumstances, will lead to other efforts, which will produce other organs. . . . It is the desire and the attempt to swim that produces membranes in the feet of aquatic birds. Wading in the water . . . has lengthened the legs of such as frequent the sides of rivers."

Lamarck hurt his own cause by careless statements easily misinterpreted in this mode. His talk about an "interior sentiment" *(sentiment intérieur)* to propel the upward force, or about organisms obeying "felt needs" *(besoins* in his terminology) to induce sideward branches of adaptation, led to suspicions about mysterious and unprovable vitalistic forces. But in fact, Lamarck remained a dedicated and vociferous materialist all his life—a credo that surely represents the most invariable and insistent claim in all his writings. He constantly sought to devise mechanical explanations, based on the physics and chemistry of matter in motion, to propel both sets of linear and lateral forces. I do not claim that his efforts were crowned with conspicuous success—particularly in his speculative attempts to explain the linear sequence of animal phyla by positing an ever more vigorous and ramifying flow of fluids, carving out spaces for organs and channels for blood in progressively more complex bodies. But one cannot deny his consistent conviction. *"La vie . . . n'est autre chose qu'un phénomène physique"* (life is nothing else than a physical phenomenon), he wrote in his last book of 1820. In a famous article, written to rehabilitate Lamarck at the Darwinian centennial celebrations (for *The Origin of Species*) in 1959, the eminent historian of science C. C. Gillispie wrote: "Life is a purely physical phenomenon in Lamarck, and it is only because science has (quite rightly) left behind his conception of the physical that he has been systematically misunderstood and assimilated to a theistic or vitalistic tradition which in fact he held in abhorrence."

Lamarck depicted his two sets of evolutionary forces as clearly distinct and destined to serve contrasting ends. The beauty of his theory—the embodiment of his *esprit de système*—lies in this clean contrast of both geometry and mechanism. The first set works upward to build progress in a strictly linear series of major anatomical designs (phyla) by recruiting a mechanism inherent in the nature of living matter. The second set works sideward to extract branches made of individual lineages (species and genera) that respond to the influence of external circumstances by precise adaptations to particular environments. (These side branches may be visualized as projecting at right angles, perpendicular to the main trunk of progress. Vectors at right angles are termed *orthogonal,* and are mathematically independent, or uncorrelated.)

Lamarck made this contrast explicit by stating that animals would form only a single line of progress if the pull of environmental adaptation did not interrupt, stymie, and divert the upward flow in particular circumstances:

> If the factor that is incessantly working toward complicating organization were the only one that had any influence on the shape and organs of animals, the growing complexity of organization would everywhere be very regular. But it is not; nature is forced to submit her works to the influence of their environment. . . . This is the special factor that occasionally produces . . . the often curious deviations that may be observed in the progression. (1809, *Philosophie zoologique;* my translations from Lamarck's original French text in all cases)

Thus, the complex order of life arises from the interplay of two forces in conflict, with progress driving lineages up the ladder, and adaptation forcing them aside into channels set by peculiarities of local environments:

The state in which we find any animal is, on the one hand, the result of the increasing complexity of organization tending to form a regular gradation; and, on the other hand, of the influence of a multitude of very various conditions ever tending to destroy the regularity in the gradation of the increasing complexity of organization. (1809, *Philosophie zoologique*)

Finally, in all his major evolutionary works, culminating in his multivolumed treatise on invertebrate anatomy (1815), Lamarck honored the first set of linear forces as primary, and identified the second set as superposed and contrary—as in this famous statement, marking the lateral pull of adaptation as foreign, accidental, interfering, and anomalous:

The plan followed by nature in producing animals clearly comprises a predominant prime cause. This endows animal life with the power to make organization gradually more complex. . . . Occasionally a foreign, accidental, and therefore variable cause has interfered with the execution of the plan, without, however, destroying it. This has created gaps in the series, in the form either of terminal branches that depart from the series in several points and alter its simplicity, or of anomalies observable in specific apparatuses of various organisms. (1815, *Histoire naturelle des animaux sans vertèbres*)

III. THE VALUES OF CHANGING THEORIES

Charles Darwin began the closing paragraph of his *Origin of Species* by noting the appeal of evolutionary explanation: "There is grandeur in this view of life." No thinking or feeling person can deny either nature's grandeur or the depth and dignity of our discovery that a history of evolution binds all living creatures together. But in our world of diverse passions and psychologies, primary definitions (and visceral feelings) about grandeur differ widely among students of natural history. Darwin emphasized the bounteous diversity itself, in all its buzzing and blooming variety—for the last sentence of this closing paragraph contrasts the "dullness" of repetitive planetary cycling with the endless expansion and novelty of evolution's good work: "whilst this planet has gone cycling on according to the fixed law of gravity, from so simple a beginning endless forms most beautiful and most wonderful have been, and are being, evolved."

But I suspect that Lamarck, following his own upbringing in the rigorous traditions of French rationalism during the Enlightenment, construed the primary definition of grandeur quite differently. As a devotee of *l'esprit de système,* Lamarck surely viewed the capacity of the human mind (his own in this case, for he was not a modest man) to apprehend the true and complete system of nature's rational order as a primary criterion (with the actual production of riotous diversity as a consequence requiring taxonomic arrangement, but only a product of the less important lateral set of adaptive forces that disturb the march of progress). Thus, the logical clarity of his two-factor theory—with a primary cause establishing a linear march of rational progress and an opposed and subsidiary cause generating a more chaotic forest of adaptive diversity—must have struck Lamarck as the defining ingredient of nature's grandeur and the power of evolution.

Our understanding of nature must always reflect a subtle interaction between messages from genuine phenomena truly "out there" in the real world and the necessary filtering of such data through all the foibles and ordering devices internal to the human mind and its evolved modes of action (see chapter 2). We cannot comprehend nature's complexity—particularly for such comprehensive subjects as evolution and the taxonomic structure of organic diversity—unless we impose our mental theories of order upon the overt chaos that greets our senses. The different styles followed by scientists to balance and reconcile these two interacting (but partly contradictory) sources of order virtually define the rich diversity of fruitful approaches pursued by a profession too often, and falsely, caricatured as a monolithic enterprise committed to a set of fixed procedures called "*the* scientific method."

Dangers and opportunities attend an overemphasis on either side. Rigid systematizers often misconstrue natural patterns by forcing their observations into rigidly preconceived structures of explanation. But colleagues who try to approach nature on her own terms, without preferred hypotheses to test, risk either being overwhelmed by a deluge of confusing information or falling prey to biases that become all the more controlling by their unconscious (and therefore unrecognized) status.

In this spectrum of useful approaches, Lamarck surely falls into the domain of scientists who place the logical beauty of a fully coherent theory above the messiness of nature's inevitable nuances and exceptions. In this context, I am all the more intrigued by Lamarck's later intellectual journey, so clearly contrary to his own inclinations, and inspired (in large part) by his inability to encompass new discoveries about the anatomy of invertebrates into the rigid confines of his beautiful system.

Nothing in the history of science can be more interesting or instructive than the intellectual drama of such a slow transformation in a fundamental view of life—from an initial recognition of trouble, to attempts at accommodation within a preferred system, to varying degrees of openness toward substantial change, and sometimes, among the most flexible and courageous, even to full conversion. I particularly like to contemplate the contributions of external and internal factors to such a change: new data mounting a challenge from the outside, coordinated with an internal willingness to follow the logic of an old system to its points of failure, and then to construct a revised theory imposing a different kind of consistency upon an altered world (with minimal changes for those who remain in love with their previous certainties and tend to follow conservative intellectual strategies, or with potentially revolutionary impact for people with temperaments that permit, or even favor, iconoclasm and adventure). Reward and risk go hand in hand, for the great majority of thoroughly radical revisions must fail, even though the sweetest fruits await the few victors in this chanciest and most difficult of all mental adventures.

When we can enjoy the privilege of watching a truly great intellect struggling with the most important of all biological concepts at a particularly interesting time in the history of science, then all factors coincide to produce a wonderful story offering unusual insight into the workings of science as well. When we can also experience the good fortune of locating a previously missing piece—in this case, the first record of a revision that would eventually alter the core of a central theory, although Lamarck, at this inception, surely had no inkling of how vigorously such a small seed could grow—then we gain the further blessing of an intriguing particular (the substrate of all good gossip) grafted onto a defining generality. The prospect of being an unknown witness—the "fly on the wall" of our usual metaphor—has always excited our fancy. And the opportunity to intrude upon a previously undocumented beginning—to be "present at the creation" in another common description—evokes an extra measure of intrigue. In this case, we begin with something almost inexpressibly humble: the classification of worms—and end with both a new geometry for animal life and a revised view of evolution itself.

IV. LAMARCK EMENDS HIS FIRST EVOLUTIONARY TREATISE

Once upon a time, in a faraway world before the electronic revolution, and even before the invention of typewriters, authors submitted literal manuscripts (from the Latin for "written by hand") to their publishers. When scholars revised a book for a second edition, they often worked from a specially prepared "inter-leaved copy" containing a blank sheet after each printed page. Corrections and additions could then be written on the blanks, enabling publishers to set a new edition from a coherent, bound document (rather than from a confusing mess of loose or pasted insertions).

Lamarck owned an interleaved copy of his first evolutionary treatise— *Système des animaux sans vertèbres* of 1801. Although he never published a second edition, he did write comments on the blank pages—and he incorporated some of these statements into later works, particularly *Philosophie zoologique* of 1809. This copy, which might have tempted me to a Faustian form of collusion with Mephistopheles, recently sold at auction for more money than even a tolerably solvent professor could ever dream of having at his disposition. But I was able to play the intellectual's usual role of voyeur during the few days of previewing before the sale—and I did recognize, in a crucial note in Lamarck's hand, a significance that had eluded previous observers. The eventual buyer (still unknown to me) expressed gratitude for the enhanced importance of his purchase, and kindly offered (through the bookseller who acted as his agent) to lend me the volume for a few days, and to allow me to publish the key note in this forum. I floated on cloud nine and happily rooted like a pig in . . . during those lovely days when I could hold and study my profession's closest approach to the holy grail.

Lamarck did not make copious additions, but several of his notes offer important insights, while their general tenor teaches us something important about the relative weighting of his concerns. The first forty-eight pages of the printed book contain the *Floréal* address, Lamarck's initial statement of his evolutionary theory. The final 350 pages present a systematic classification of invertebrates, including a discussion of principles and a list and description of all recognized genera, discussed phylum by phylum.

Of Lamarck's thirty-seven handwritten additions on the blank pages, twenty-nine offer only a word or two, and represent the ordinary activity of correcting small errors, inserting new information, or editing language. Lamarck makes fifteen comments about anatomy, mostly in his chapter on the genera of mollusks, the group he knew best. A further nine comments treat taxonomic issues of naming (adding a layman's moniker to the formal Latin designation, changing the name or affiliation of a genus); two add bibliographic data; and the final three edit some awkward language.

Taken as an ensemble, I regard these comments as informative in correcting a false impression that Lamarck, by this time of seniority in his career, cared only for general theory, and not for empirical detail. Clearly, Lamarck continued to cherish the minutiae of raw information, and to keep up with developing knowledge—the primary signs of an active scientific life.

Of the eight longer comments, four appear as additions to the *Floréal* address. They provide instructive insight into Lamarck's character and concerns by fulfilling the "conservative" function of making more explicit, and elaborating by hypothetical examples, the central feature of his original evolutionary theory: the sharp distinction between causes of "upward" progress and "sideward" adaptation to local circumstances.

In the two comments (among these four) that attracted most attention from potential buyers, Lamarck added examples of adaptation to local environments by inheritance of acquired characters (with both cases based on the evolutionary reorientation of eyes): first, flatfishes that live in shallow water, flatten their bodies to swim on their side, and then move both eyes to the upper surface of the head; and second, snakes that move their eyes to the top of their head because they live so close to the ground and must therefore be able to perceive a world of danger above them—and then need to develop a long and sensitive tongue to perceive trouble in front, and now invisible to the newly placed eyes. With these examples, Lamarck generalized his second set of forces by extending his stories to a variety of organisms: the *Floréal* address had confined all examples to the habits and anatomy of birds. (The purely speculative character of these cases also helps us to understand why more sober empiricists, like Darwin and Cuvier, felt so uncomfortable with Lamarck's supposed data for evolution.) In any case, Lamarck published both examples almost verbatim in his *Philosophie zoologique* of 1809.

A third comment then strengthens the other, and primary, set of linear causes by arguing that the newly discovered platypus of Australia could link the penultimate birds to the highest group of mammals. Finally, the fourth comment tries to explain the mechanisms of use and disuse by differential flow of fluids through bodies.

The other set of four longer comments adorns the second part of the book on taxonomic ordering of invertebrate animals. One insertion suggests that a small and enigmatic egg-shaped fossil should be classified within the phylum of corals and jellyfishes. A second statement, of particular interest to me, revises Lamarck's description of the clam genus *Trigonia*. This distinctive form had long been recognized as a prominent fossil in Mesozoic rocks, but no Tertiary fossils or living specimens had ever been found—and naturalists therefore supposed that this genus had become extinct. But two French naturalists then

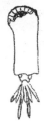

Lamarck's original illustration depicting the discovery of the squidlike animal that secretes the shell called Spirula.

found a living species of *Trigonia* in Australian waters, and Lamarck himself published the first description of this triumphant rediscovery in 1803. (As an undergraduate, I did my first technical research on dissection, and also wrote my first paper in the history of science, under the direction of Norman D. Newell at the Museum of Natural History in New York. He gave me a half dozen, still preciously rare, specimens of modern Australian trigonians. When I gulped and admitted that I had no experience with dissection and feared butchering such a valuable bounty, he said to me, in his laconic manner—so inspirational for self-motivated students, but so terrifying for the insecure—"Go down to the Fulton Fish Market and buy a bunch of quahogs. Practice on them first." I was far more terrified than inspired, but all's well that ends well.)

The final two comments provide the greatest visceral pleasure of all because Lamarck added drawings to his words (reproduced here with the kind permission of the book's new owner). The first sketch affirms Lamarck's continuing commitment to detail, and to following and recording new discoveries. A small, white, and delicate coiled shell of a cephalopod mollusk (the group including squid and octopuses) frequently washes up on beaches throughout the world. Lamarck himself had named this shell *Spirula* in 1799. But the animal that makes the shell had never been found. As a particular mystery, no one knew whether the animal lived inside the shell (as in a modern chambered nautilus) or grew the shell within its body (as in the "cuttlebone" of a modern squid). The delicacy of the object suggested a protected internal status, but the question remained open. Soon after Lamarck's book appeared, naturalists discovered the animal of *Spirula*, and affirmed an internal shell—a happy resolution that inspired Lamarck to a rare episode of artistic activity.

The last—and, as I here suggest, by far the most important—comment appears on the blank sheet following page 330, which contains the description of two remarkably different genera of "worms"—the medicinal leech *Hirudo* and the pond worm *Planaria*, known to nearly anyone who ever took a basic

Lamarck's original drawing and text, expressing his first and crucial recognition that annelid worms and parasitic internal worms represent very different kinds of animals.

laboratory course in biology. Here, Lamarck draws a simple sketch of the circulatory system of an annelid worm, and then writes the following portentous words:

> observation sur l'org^{on} des vers. dans les vers annelés et qui ont des organs externes, le sang est rouge et circule dans des vaisseaux arteriels et veineux. leur organisation les place avant les insectes. les vers intestins doivent seuls se trouver après les insectes. ils n'ont qu'un fluide blanc, libre, non contenue dans des vaisseaux. Cuvier. extrait d'un mém. lu a l'institut le 11 nivôse an 10.

> (Observation on the organization of worms. In annelid worms, which have external organs, the blood is red and circulates in arterial and venous vessels. Their organization places them before the insects. Only the internal worms come after the insects. They have only a white fluid, free, and not contained in vessels. Cuvier. Extract from a memoir read at the institute on the 11th day of [the month of] Nivôse year 10.)

Clearly, Lamarck now recognizes a vital distinction between two groups previously lumped together into the general category of "worms." He regards

one group—the annelids, including earthworms, leeches, and the marine poly-
chaetes—as highly advanced, even more so than insects. (Lamarck usually presented his scale of animal life from the top down, starting with humans and
ending with infusorians—and not from the bottom up, the system that became
conventional in later zoological writing. Thus, he states that annelids come
before insects because he views them as more advanced—that is, closer to the
mammalian top.) But another group, the internal worms* (mostly parasites living within the bodies of other animals), rank far lower on the scale—even after
(that is, anatomically simpler than) insects. These two distinct groups, previously
conflated, must now be widely separated in the taxonomic ordering of life.
Ironically, Lamarck acknowledges his colleague Cuvier (who would later turn
against him and virtually destroy his reputation) as the source for a key item of
information that changed his mind—Cuvier's report (presented at a meeting
during the winter of 1801–2, soon after the publication of Lamarck's book)
that annelids possessed a complicated circulatory system, with red blood running in arteries and veins, whereas internal worms grew no discrete blood vessels, and only moved a white fluid through their body cavity.

Obviously, Lamarck viewed this new information as especially important, for
no other anatomical note received nearly such prominence in his additions, while
only one other observation (a simple new bit of information, without much theoretical meaning) merits a drawing. But why did Lamarck view this division of
worms as so important? And how could such an apparently dull and technical
decision about naming act as a pivot and initiator for a new view of life?

V. An Odyssey of Worms

I have always considered it odd (and redolent either of arrogance or parochial-
ity) when a small minority divides the world into two wildly unbalanced categories of itself versus all others—and then defines the large category as an
absence of the small, as in my grandmother's taxonomy for *Homo sapiens:* Jews
and non-Jews. Yet our conventional classification of animals follows the same
strategy by drawing a basic distinction between vertebrates and invertebrates—
when only about forty thousand of more than a million named species belong
to the relatively small lineage of vertebrates.

---

*The standard English literature on this subject always translates Lamarck's phrase incorrectly as
"intestinal worms." These parasites dwell in several organs and places of vertebrate (and other) bodies, not only in the intestines. In French, the word *intestin* conveys the more general meaning of
"internal" or "inside."

On the venerable principle that bad situations can always be made worse, we can gain some solace by noting the even greater imbalance devised by the founder of modern taxonomy, Carolus Linnaeus. At least we now recognize vertebrates as only part of a single phylum, while most modern schemes divide invertebrates into some twenty to thirty separate phyla. But in his *Systema naturae* of 1758, the founding document of modern zoological nomenclature, Linnaeus identified only six basic animal groups: four among vertebrates (mammals, birds, reptiles, and fishes), and two for the entire realm of invertebrates (Insecta, for insects and their relatives, and Vermes, literally worms, for nearly everything else).

When Lamarck became professor of invertebrates at the Muséum in 1793 (with an official title in a Linnaean straitjacket as professor of insects and worms), he already recognized that reform demanded the dismemberment of Linnaeus's "wastebucket" category of Vermes, or worms. (*Wastebucket,* by the way, actually ranks as a semitechnical term among professional taxonomists, a description for inflated groups that become receptacles for heterogeneous bits and pieces that most folks would rather ignore—as in this relegation of all "primitive" bilaterally symmetrical animals to a category of "worms" ranking far beneath the notice of specialists on vertebrates.)

In his 1801 book, Lamarck identified the hodgepodge of Linnaeus's Vermes as the biggest headache and impediment in zoology:

> The celebrated Linnaeus, and almost all other naturalists up to now, have divided the entire series of invertebrate animals into only two classes: insects and worms. As a consequence, anything that could not be called an insect must belong, without exception, to the class of worms.

By the time Lamarck wrote his most famous book in 1809, his frustration had only increased, as he called Linnaeus's class of worms *"une espèce de chaos dans lequel les objets très-disparates se trouvent réunis"* (a kind of chaos where very disparate objects have been united together). He then blamed the great man himself for this sorry situation: "The authority of this scientist carried such great weight among naturalists that no one dared to change this monstrous class of worms." (I am confident that, in writing *"cette classe monstrueuse,"* Lamarck meant to attack the physical size based on number of included genera, not the moral status, of Linnaeus's Vermes.)

Lamarck therefore began his campaign of reform by raiding Vermes and gradually adding the extracted groups as novel phyla in his newly named category of invertebrates. In his first course of 1793, he had already expanded the Linnaean duality to a ladder of progress with five rungs—mollusks, insects, worms, echinoderms, and polyps (corals and jellyfish)—by liberating three new phyla from the wastebucket of Vermes.

This reform accelerated in 1795, when Georges Cuvier arrived and began to study invertebrates as well. The two men collaborated in friendship at first—and they surely operated as one mind on the key issue of dismembering Vermes. Thus, Lamarck continued to add phyla in almost every annual course of lectures, extracting most new groups from Vermes, but some from the overblown Linnaean Insecta as well. In year 7 (1799), he established the Crustacea (for marine arthropods, including crabs, shrimp, and lobsters), and in year 8 (1800) the Arachnida (for spiders and scorpions). Lamarck's invertebrate classification of 1801 therefore featured a growing ladder of progress, now bearing seven rungs. In 1809, he presented a purely linear sequence of progress for the last time in his most famous book, *Philosophie zoologique*. His tall and rigid ladder now included fourteen rungs, as he added the four traditional groups of vertebrates atop a list of invertebrate phyla that had just reached double digits (see accompanying chart directly reproduced from the 1809 edition).

So far, Lamarck had done nothing to inspire any reconsideration of the evolutionary views first presented in his *Floréal* address of 1800. His taxonomic reforms, in this sense, had been entirely conventional in adding weight and strength to his original views. The *Floréal* statement had contrasted a linear force leading to progress in major groups with a lateral force causing local adaptation in particular lineages. Lamarck's ladder included only seven groups in the *Floréal*

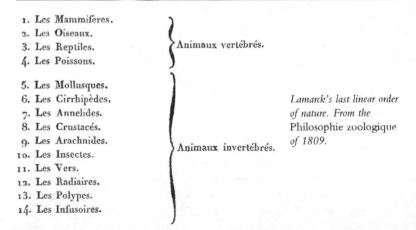

1. Les Mammifères.
2. Les Oiseaux.
3. Les Reptiles.
4. Les Poissons.

} Animaux vertébrés.

5. Les Mollusques.
6. Les Cirrhipèdes.
7. Les Annelides.
8. Les Crustacés.
9. Les Arachnides.
10. Les Insectes.
11. Les Vers.
12. Les Radiaires.
13. Les Polypes.
14. Les Infusoires.

} Animaux invertébrés.

Lamarck's last linear order of nature. From the Philosophie zoologique of 1809.

address. By 1809, he had doubled the length while preserving the same strictly linear form—thus strengthening his central contrast between two forces by granting the linear impetus a greatly expanded field for its inexorably exceptionless operation.

But if Lamarck's first reform of Linnaeus—the expansion of groups into a longer linear series—had conserved and strengthened his original concept of evolution, he now embarked upon a second reform, destined (though he surely had no inkling at the outset) to yield the opposite effect of forcing a fundamental change in his view of life. He had, heretofore, only extracted misaligned groups from Linnaeus's original Vermes. He now needed to consider the core of Vermes itself—and to determine whether waste and rot existed at the foundation as well.

"Worms," in our vernacular understanding, are defined both broadly and negatively (unfortunate criteria guaranteeing inevitable trouble down the road) as soft-bodied, bilaterally symmetrical animals, roughly cylindrical in shape and lacking appendages or prominent sense organs. By these criteria, both earthworms and tapeworms fill the bill. For nearly ten years, Lamarck did not seriously challenge this core definition.

But he could not permanently ignore the glaring problem, recognized but usually swept under the rug by naturalists, that this broad vernacular category seemed to include at least two kinds of organisms bearing little relationship beyond a superficial and overt similarity of external form. On the one hand, a prominent group of free-living creatures—earthworms and their allies—built bodies composed of rings or segments, and also developed internal organs of substantial complexity, including nerve tubes, blood vessels, and a digestive tract. But another assemblage of largely parasitic creatures—tapeworms and their allies—grew virtually no discretely recognizable internal organs at all, and therefore seemed much "lower" than earthworms and their kin under any concept of an organic scale of complexity. Would the heart of Vermes therefore need to be dismembered as well?

This problem had already been worrying Lamarck when he published the *Floréal* address in his 1801 compendium on invertebrate anatomy—but he was not yet ready to impose a formal divorce upon the two basic groups of "worms." Either standard of definition, taken by itself—different anatomies or disparate environments—might not offer sufficient impetus for thoughts about taxonomic separation. But the two criteria conspired perfectly together in the remaining Vermes: the earthworm group possessed complex anatomy *and* lived freely in the outside world; the tapeworm group maintained maximal simplicity among mobile animals *and* lived almost exclusively within the bodies of other creatures.

Lamarck therefore opted for an intermediary solution. He would not yet dismember Vermes, but he would establish two subdivisions *within* the class: *vers externes* (external worms) for earthworms and their allies, and *vers intestins* (internal worms) for tapeworms and their relatives. He stressed the simple anatomy of the parasitic subgroup, and defended their new name as a spur to further study, while arguing that knowledge remained insufficient to advocate a deeper separation:

> It is very important to know them [the internal worms], and this name will facilitate their study. But aside from this motive, I also believe that such a division is the most natural . . . because the internal worms are much more imperfect and simply organized than the other worms. Nevertheless, we know so little about their origin that we cannot yet make them a separate order.

At this point, the crucial incident occurred that sparked Lamarck to an irrevocable and cascading reassessment of his evolutionary views. He attended Cuvier's lecture during the winter of 1801–2 (year 10 of the revolutionary calendar), and became convinced, by his colleague's elegant data on the anatomy of external worms, that the extensive anatomical differences between his two subdivisions could not permit their continued residence in the same class. He would, after all, have to split the heart of Vermes. Therefore, in his next course, in the spring of 1802, Lamarck formally established the class Annelida for the external worms (retaining Vermes for the internal worms alone), and then separated the two classes widely by placing his new annelids above insects in linear complexity, while leaving the internal worms near the bottom of the ladder, well below insects.

Lamarck formally acknowledged Cuvier's spur when he wrote a history of his successive changes in classifying invertebrates for the *Philosophie zoologique* of 1809:

> Mr. Cuvier discovered the existence of arterial and venous vessels in distinct animals that had been confounded with other very differently organized animals under the name of worms. I soon used this new fact to perfect my classification; therefore, I established the class of annelids in my course for year 10 (1802).

The handwritten note and drawing in Lamarck's 1801 book, discussed and reproduced earlier in this essay, tells much the same story—but what a contrast, in both intellectual and emotional intrigue, between a sober memory written long after an inspiration, and the inky evidence of the moment of enlightenment itself!

But this tale should now be raising a puzzle in the minds of readers. Why am I making such a fuss about this particular taxonomic change—the final division of Vermes into a highly ranked group of annelids and a primitive class of internal worms? In what way does this alteration differ from any other previously discussed? In all cases, Lamarck subdivided Linnaeus's class Vermes and established new phyla in his favored linear series—thus reinforcing his view of evolution as built by contrasting forces of linear progress and lateral adaptation. Wasn't he just following the same procedure in extracting annelids and placing them on a new rung of his ladder? So it might seem—at first. But Lamarck was too smart, and too honorable, to ignore a logical problem directly and inevitably instigated by this particular division of worms—and the proper solution broke his system.

At first, Lamarck did treat the extraction of annelids as just another addition to his constantly improving linear series. But as the years passed, he became more and more bothered by an acute problem, evoked by an inherent conflict between this particular taxonomic decision and the precise logic of his overarching system. Lamarck had ranked the phylum Vermes, now restricted to the internal worms alone, just above a group that he named *radiaires*—actually (by modern understanding) a false amalgam of jellyfishes from the coelenterate phylum and sea urchins and their relatives from the echinoderm phylum. Worms had to rank above radiates because bilateral symmetry and directional motion trump radial symmetry and an attached (or not very mobile) lifestyle— at least in conventional views about ladders of progress (which, of course, use mobile and bilaterally symmetrical humans as an ultimate standard). But the parasitic internal worms also lack the two most important organ systems— nerve ganglia and cords, and circulatory vessels—that virtually define complexity on the traditional ladder. Yet echinoderms within the "lower" radiate phylum develop both nervous and circulatory systems. (These organisms circulate sea water rather than blood, but they do run their fluids through tubes.)

If the primary "force that tends incessantly to complicate organization" truly works in a universal and exceptionless manner, then how can such an inconsistent situation arise? If the force be general, then any given group must stand fully higher or lower than any other. A group cannot be higher for some features, but lower for others. Taxonomic experts cannot pick and choose. He who lives by the line must die by the line.

This problem did not arise so long as annelids remained in the class of worms. Lamarck, after all, had never argued that each genus of a higher group must rank above all members of a lower group in every bodily part. He only claimed that the "principal masses" of organic design must run in pure linear order. Individual genera may degenerate or adapt to less complex environments in various parts—but so long as some genera display the higher conformation in all features, then the entire group retains its status. In this case, so long as annelids remained in the group, then many worms possessed organ systems more complex than any comparable part in any lower group—and the entire class of worms could retain its unambiguous position above radiates and other primitive forms. But with the division of worms and the banishment of complex annelids, Lamarck now faced the logical dilemma of a coherent group (the internal parasitic worms alone) higher than radiates in some key features but lower in others. The pure march of nature's progress—the keystone of Lamarck's entire system—had been fractured.

Lamarck struggled with this problem for several years. He stuck to the line of progress in 1802, and again—for the last time, and in a particularly uncompromising manner that must, in retrospect, represent a last hurrah before the fall—in the first volume of his seminal work, *Philosophie zoologique,* of 1809. But honesty eventually trumped hope. Just before publication, Lamarck appended a short chapter of "additions" to volume two of *Philosophie zoologique.* He now, if only tentatively, floated a new scheme that would resolve his problem with worms, but would also unravel his precious linear system.

Lamarck had long argued that life began with the spontaneous generation of "infusorians" (single-celled animals) in ponds. But suppose that spontaneous generation occurs twice, and in two distinct environments—in the external world for a lineage beginning with infusorians, and inside the bodies of other creatures for a second lineage beginning with internal worms? Lamarck therefore wrote that "worms seem to form one initial lineage in the scale of animals, just as, evidently, the infusorians form the other branch."

Lamarck then faced the problem of allocating the higher groups. To which of the two great lines does each belong? He presented his preliminary thoughts in a chart—perhaps the first evolutionary branching diagram ever published in the history of biology—that directly contradicted his previous image of a single ladder. (Compare this figure with the version presented earlier in this essay, taken from volume one of the same 1809 work.) Lamarck begins (at the top, contrary to current conventions) with two lines, labeled *"infusoires"* (single-celled animals) and *"vers"* (worms). He then inserts light dots to suggest possible allocations of the higher phyla to the two lines. The logical problem that broke his system has now been solved—for the *radiaires* (radiate animals), standing below worms in some features, but above in others, now rank in an entirely separate series, directly following an infusorian beginning.

When mental floodgates open, the tide of reform must sweep to other locales. Once he had admitted branching and separation at all, Lamarck could hardly avoid the temptation to apply this new scheme to other old problems. Therefore, he also suggested some substantial branching at the end of his array. He had always been bothered by the conventional summit of reptiles to birds to mammals, for birds seemed just different from, rather than inferior to, mammals. Lamarck therefore proposed (and drew on his revolutionary chart) that reptiles branched at the end of the series, one line passing from turtles to birds *(oiseaux)* to *monotrèmes* (platypuses, which Lamarck now considers as separate from mammals), the other from crocodiles to marine mammals (labeled *m. amphibies*) to terrestrial mammals. Finally, and still in the new spirit, he even posited a threefold branching in the transition to terrestrial mammals, leading to separate lines for whales *(m. cétacés),* hoofed animals *(m. ongulés),* and mammals with nails *(m. onguiculés),* including carnivores, rodents, and primates (including humans).

Finally, Lamarck explicitly connected the two reforms: the admission of two sequences of spontaneous generation at the bottom, and a branching among higher vertebrates at the top: "The animal scale begins with at least two branches; in the course of its extent, several branches seem to end in different places."

After *Philosophie zoologique* of 1809, Lamarck wrote one additional major book on evolution, the introductory volume (1815) to his *Histoire naturelle des animaux sans vertèbres.* Here, he abandoned all the tentativeness of his 1809 revision, and announced his conversion to branching as the fundamental pattern of evolution. In direct contradiction to the linear model that had shaped all his previous work, Lamarck stated simply, and without ambiguity:

> *Dans sa production des differents animaux, la nature n'a pas executé une série unique et simple.*
> [In its production of the different animals, nature has not fashioned a single and simple series.]

He then emphasized the branching form of his new model, and explained how the division of worms, inspired by Cuvier's observations, had broken his former system and impelled his revision:

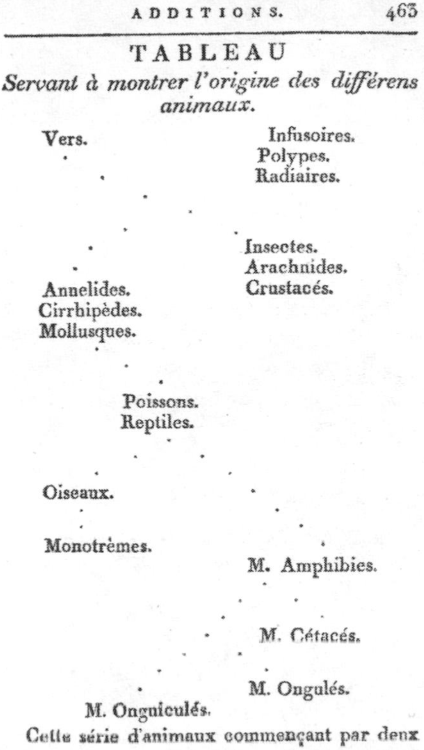

ADDITIONS.   463
_____

TABLEAU
Servant à montrer l'origine des différens
animaux.

Vers.   Infusoires.
  .   Polypes.
          .   Radiaires.

            .   Insectes.
            .   Arachnides.
Annelides.   Crustacés.
Cirrhipèdes.
Mollusques.   *Lamarck's first depiction of a*
          .   *branching model for the history of*
        .   *life. From the appendix to*
        Poissons.   Philosophie zoologique *(1809).*
        Reptiles.
            .   .

        .
Oiseaux.   .
          .
Monotrèmes.   .
                  M. Amphibies.

              .
          .   M. Cétacés.

          .   M. Ongulés.
    M. Onguiculés.
Cette série d'animaux commençant par deux

_____

The order is far from being simple; it is branching [*rameux*] and even appears to be constructed of several distinct series. . . . The animals that belonged to the class of worms display a great disparity of organization. . . . The most imperfect of these animals arise by spontaneous generation, and the worms [now restricted to *vers intestins*, with annelids removed] truly form their own series, later in origin than the one that began with infusorians.

Lamarck's third and last chart (reproduced here from his 1815 volume) shows how far he had progressed both in his own confidence, and in copious branching on his new tree of life. He titles the chart "presumed order of the formation of animals, showing two separate and subbranching series." Note how the two major lines of separate spontaneous generation—one beginning with infusorians, the other with internal worms—are now clearly marked and separated. Note also how each of the series also divides within itself, thus establishing the process of branching as a key theme at all scales of the system. The infusorian line branches at the level of polyps (corals and jellyfish) into a line of radiates and a line terminating in mollusks. The second line of worms also branches in two, leading to annelids on one side and insects on the other. But the insect line then splits again (a tertiary division) into a lineage of crustaceans and barnacles (labeled *cirrhipèdes*) and another of arachnids (spiders and scorpions).

Finally, we must recognize that these major changes do not only affect the overt geometry of animal organization. The conversion from linearity to branching also—perhaps even more importantly—marks a profound shift in Lamarck's underlying theory of nature. He had based his original system, defended explicitly and vociferously until 1809, on a fundamental division of two independent forces—a primary cause that builds basic anatomies in an unbroken line of progress, and a subsidiary lateral force that draws single lineages off the line into byways of immediate adaptation to local environments. A set of philosophical consequences then spring from this model: the predictable and lawlike character of evolution lies patent in the primary force and its ladder of progress reaching to man, while accidents of history (leading to local adaptations) can then be dismissed as secondary and truly independent from the overarching order.

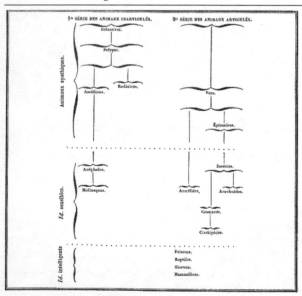

Lamarck's fully developed tree of life from 1815.

But the branching system destroys this neat and comforting scheme. First of all, the two forces become intermingled and conflated in the branching itself. We can no longer distinguish two independent and orthogonal powers working at right angles. Progress may occur along any branch to be sure, but the very act of division implies an environmental impetus to split the main line—and Lamarck had always advocated a complete and principled distinction between a single and inexorable main line and the numerous minor deviations that can draw off a long-necked giraffe or an eyeless mole, but can never disrupt or ramify the major designs of animal life. In the new model, however, environment intrudes at the first construction of basic order—as one group arises spontaneously in ponds, and another inside the bodies of other creatures! Moreover, each of the two resulting lines then branches further, and unpredictably, under environmental impetuses that were not supposed to derail the force of progress among major groups—as when insects split into a terrestrial line of arachnids and a marine line leading to crustaceans and barnacles.

Second, the forces of history and natural complexity have now triumphed over the scientific ideal of a predictable and lawlike system. The taxonomy of animals could no longer embody an overarching plan of progress, illustrating the fundamental order, harmony, and predictable good sense of the natural world (perhaps even the explicit care of a loving deity, whose plans we may hope to understand because he thinks as we do). Now, the confusing, particular, local, and unpredictable forces of complex environments hold sway, ready at any time to impose a deviation upon any group with enough hubris to suppose that Emerson's forthcoming words could describe their inevitable progress:

> *And striving to be man, the worm*
> *Mounts through all the spires of form.*

VI. Lamarck's Epilogue and My Own

Following his last and greatest treatise on the anatomy of invertebrate organisms, Lamarck published only one other major work—*Analytic System of Positive Knowledge About Man* (1820). This rare book has not been consulted by previous historians who traced the development of Lamarck's changing views on the classification of animals. Thus, traditional accounts stop at Lamarck's 1815 revision, with its fundamental distinction between two separate lineages of spontaneous generation. The impression therefore persists that Lamarck never fully embraced the branching model, later exemplified by Darwin as the "tree of life"—with a common trunk of origin for all creatures and no main line of growth thereafter. Lamarck had compromised his original ladder of progress by advocating two separate origins, but he could continue to stress linearity in each of the resulting series.

But his 1820 book, although primarily a treatise on psychology, does include a chapter on the classification of animals—and I discovered, in reading these pages, that Lamarck did pursue his revisionary path further, and did finally arrive at a truly branching model for a tree of life. Moreover, in a remarkable passage, Lamarck also recognizes the philosophical implications of his full switch by acknowledging a reversal in his ranking of natural forces in one of the most interesting (and honorable) intellectual conversions that I have ever read.

Lamarck still talks about forces of progress and forces of branching, and he does argue that progress will proceed along each branch. But branching has triumphed as a primary and controlling theme, and Lamarck now frames his entire discussion of animal taxonomy by emphasizing successive points of division. For example, consider this epitome of vertebrate evolution:

> Reptiles come necessarily after fishes. They build a branching sequence, with one branch leading from turtles to platypuses to the diverse group of birds, while the other seems to direct itself, via lizards, toward the mammals. The birds then . . . build a richly varied branching series, with one branch ending in birds of prey.

(In previous models, Lamarck had viewed birds of prey as the top rung of a single avian ladder.)

But much more radically, his 1815 model based on two lines of spontaneous generation has now disappeared. In its place, Lamarck advocates the same tree of life that would later become conventional through the influence of Darwin and other early evolutionists. Lamarck now proposes a single common ancestor for all animals, called a monad. From this beginning, infusorians evolve, followed by polyps, arising "directly and almost without a gap." But polyps then branch to build the rest of life's tree: "instead of continuing as a single series, the polyps appear to divide themselves into three branches"—the radiates, which end without evolving any further; the worms, which continue to branch into all phyla of segmented animals, including annelids, insects, arachnids, crustaceans, and barnacles, each by a separate event of division; and the tunicates (now regarded as marine organisms closely related to vertebrates), which later split to form several lines of mollusks and vertebrates.

Lamarck then acknowledges the profound philosophical revision implied by a branching model for nature's fundamental order. He had always viewed the linear force of progress as primary. As late as 1815, even after he had changed his model to permit extensive branching and two environmentally induced sequences of spontaneous generation, Lamarck continued to emphasize the primary power of the linear force, compared with disturbing and anomalous exceptions produced by lateral environmental causes, called *l'influence des circonstances*. To restate the key passage quoted earlier in this essay:

> The plan followed by nature in producing animals clearly comprises a predominant prime cause. This endows animal life with the power to make organization gradually more complex. . . . Occasionally a foreign, accidental, and therefore variable cause has interfered with the execution of the plan . . . [producing] branches that depart from the series in several points and alter its simplicity.

But Lamarck, five years later in his final book of 1820, now abandons this controlling concept of his career, and embraces the opposite conclusion. The influence of circumstances (leading to a branching model of animal taxonomy) rules the paths of evolution. All general laws, of progress or anything else, must be regarded as subservient to the immediate singularities of environments and histories. The influence of circumstances has risen from a disturbing and peripheral joker to the true lord of all (with an empire to boot):

> Let us consider the most influential cause for everything done by nature, the only cause that can lead to an understanding of everything that nature produces. . . . This is, in effect, a cause whose power is absolute, superior even to nature, since it regulates all nature's acts, a cause whose empire embraces all parts of nature's domain. . . . This cause resides in the power that circumstances have to modify all operations of nature, to force nature to change continually the laws that she would have followed without [the intervention of] these circumstances, and to determine the character of each of her products. The extreme diversity of nature's productions must also be attributed to this cause.

Lamarck's great intellectual journey began with a public address about evolution, delivered in 1800 during a month that the revolutionary government had auspiciously named *Floréal,* or flowering. He then developed the first comprehensive theory of evolution in modern science—an achievement that won him a secure place in any scientific hall of fame or list of immortals—despite the vicissitudes of his reputation during his own lifetime and immediately thereafter.

But Lamarck's original system failed—and not for the reasons that we usually specify today in false hindsight (the triumph of Mendelism over Lamarck's erroneous belief in inheritance of acquired characters), but by inconsistencies that new information imposed upon the central logic of Lamarck's system during his own lifetime. We can identify a fulcrum, a key moment, in the unraveling of Lamarck's original theory—when he attended a lecture by Cuvier on the anatomy of annelids, and recognized that he would have to split his taxonomic class of worms into two distinct groups. This recognition—which Lamarck recorded with excitement (and original art) as a handwritten insertion into his first published book on evolution—unleashed a growing cascade of consequences that, by Lamarck's last book of 1820, had destroyed his original theory of primary ladders of progress versus subsidiary lateral deviations, and led him to embrace the opposite model (in both geometry of animal classification, and basic philosophy of nature) of a branching tree of life.

A conventional interpretation would view this tale as fundamentally sad, if not tragic, and would surely note a remarkable symbol and irony for a literary conclusion. Lamarck began his adventure in the springtime month of flowering. But he heard Cuvier's lecture, and his system began to crumble, on the eleventh day of Nivôse—the winter month of snow. How fitting—to begin with springtime joy and promise, and to end in the cold and darkness of winter.

How fitting in one distorted sense—but how very, very wrong. I do not deny or belittle Lamarck's personal distress, but how can we view his slow acknowledgment of logical error, and his willingness to construct an entirely new and contrary explanation, as anything other than a heroic act, worthy of our greatest admiration and identifying Lamarck as one of the finest intellects in the history of biology (the name that he invented for his discipline). Two major reasons lead me to view Lamarck's intellectual odyssey in this eminently positive light. First, what can be more salutary in science than the flexibility that allows a person to change his mind—and to do so not for a minor point under the compulsion of irrefutable data, but to rethink and reverse the most fundamental concept underlying a basic philosophy of nature?

I would argue, secondly, that Lamarck's journey teaches us something vitally important about the interaction between nature and our attempts to understand her ways. The fallacies and foibles of human thinking generate systematic and predictable trouble when we try to grasp the complexities of external reality. Among these foibles, our persistent attempts to build abstractly beautiful, logically impeccable, and comprehensively simplified systems always lead us astray. Lamarck far exceeded most colleagues in his attraction to this perilous style of theorizing—this *esprit de système*—and he therefore fell further and harder because he also possessed the honesty and intellectual power to probe his mistakes.

Nature, to cite a modern cliché, always bats last. She will not succumb to the simplicities of our hopes or mental foibles, but she remains eminently comprehensible. Evolution follows the syncopated drumbeats of complex and contingent histories, shaped by the vagaries and uniquenesses of time, place, and environment. Simple laws with predictable outcomes cannot fully describe the pageant and pathways of life. A linear march of progress must fail as a model for evolution, but a luxuriantly branching tree does express the basic geometry of history.

When Lamarck snatched victory from the jaws of his defeat (by abandoning his beloved ladder of life and embracing the tree), he stood in proper humility before nature's complexity—a lesson for us all. But he also continued to wrestle with nature, to struggle to understand and even to tame her ways, not simply to bow down and acknowledge sovereignty. Only the most heroic people can follow Job's great example in owning error while continuing to hurl defiance and to shout "I am here." Lamarck greeted nature (traditionally construed as female) with Job's ultimate challenge to God (construed as male, in equally dubious tradition): "Though he slay me, yet will I trust in him; but I will maintain mine own ways before him" (Job 13:15).

I therefore propose that we reinterpret the symbolic meaning of Lamarck's undoing in the month of Nivôse. Cuvier's challenge elicited a cascade of discovery and reform, not the battering of bitter defeat. And snow also suggests metaphors of softness, whiteness, and purification—not only of frost, darkness, and destruction. God, in a much kinder mood than he showed to poor Job, promised his people in the first chapter of Isaiah: "though your sins be as scarlet, they shall be as white as snow; though they be red like crimson, they shall be as wool." We should also remember that this biblical verse begins with an even more famous statement—a watchword for an intellectual life, and a testimony to Lamarck's brilliance and flexibility: "Come now, and let us reason together."

III

Darwin's Century
— and Ours

Lessons from

Britain's

Four Greatest

Victorian Naturalists

7

Lyell's Pillars of Wisdom

I. Controlling the Fires of Vesuvius

The two classical scenarios for a catastrophic end of all things—destruction by heat and flames or by cold and darkness—offer little fodder for extended discussion about preferences, a point embedded, with all the beauty of brevity, in Robert Frost's poem "Fire and Ice," written in 1923:

> Some say the world will end in fire,
> Some say in ice.
> From what I've tasted of desire
> I hold with those who favor fire.
> But if it had to perish twice,
> I think I know enough of hate
> To say that for destruction ice
> Is also great
> And would suffice.

Among natural phenomena that poets and scholars have regarded as heralds or harbingers of the final consummation, volcanic eruptions hold pride of place. Mount Vesuvius may represent a mere pimple of activity compared with the Indonesian explosion of Tambora in 1815 or Kratakau in 1883, but a prime location on the Bay of Naples, combined with numerous eruptions at interesting times, has promoted this relatively small volcano into a primary symbol of natural terror.

Given our traditional dichotomy for unpleasant finality, I note with some amusement that the two most famous encounters of celebrated scientists with this archetypal volcano—one in each millennium of modern history—have elicited contrasting comparisons of Vesuvian eruptions with the end of time: "lights out" for the first, "up in flames" for the second.

Pliny the Elder (A.D. 23–79) wrote a massive compendium, *Natural History,* divided into thirty-seven *libri* (books) treating all aspects, both factual and folkloric, of subjects now gathered under the rubric of science. Pliny's encyclopedia exerted enormous influence upon the history of Western thought, particularly during the Renaissance (literally "rebirth"), when rediscovery of classical knowledge became the primary goal of scholarship (see chapter 3). Several editions of Pliny's great work appeared during the first few decades of printing, following the publication of Gutenberg's Bible in 1455.

In August of A.D. 79, while serving as commander of the fleet in the Bay of Naples, Pliny noted a great cloud arising from Mount Vesuvius. Following the unbeatable combination of a scientist's curiosity and a commander's duty, Pliny sailed toward the volcano, both to observe more closely and to render aid. He went ashore at a friend's villa, made a fateful decision to abandon the shaking houses for the open fields, and died by asphyxiation in the same eruption that buried the cities of Pompeii and Herculaneum.

Pliny the Younger, his nephew and adopted son, remained at their villa, a few miles farther west of the volcano, to continue (as he stated) his studies of Livy's historical texts. After the dust had settled—sorry, but I couldn't resist this opportunity to use a cliché literally—he wrote two famous letters to the historian Tacitus, describing what he had heard of his uncle's fate and what he had experienced on his own. Pliny the Younger recounted all the horrors of shaking houses, falling rocks, and noxious fumes, but he emphasized the intense darkness produced by the spreading volcanic cloud, a pall that he could compare only with one scenario for the end of time.*

---

*In one of those odd coincidences that make writing, and intellectual life in general, such a joy, I happened to be reading, just two days after completing this essay, a volume of Francis Bacon's complete works. I knew the old story about his death in 1626. Bacon, who loved to perform and report simple experiments of almost random import (his last and posthumous work, *Sylva sylvarum* [The forest of forests], lists exactly one thousand such observations and anecdotes), wanted to learn if

A darkness overspread us, not like that of a cloudy night, or when there is no moon, but of a room when it is shut up and all the lights are extinguished. Nothing then was to be heard but the shrieks of women, the screams of children, and the cries of men . . . some wishing to die from the very fear of dying, some lifting up their hands to the gods; but the greater part imagining that the last and eternal night had come, which was to destroy both the gods and the world together.

Athanasius Kircher (1602–80), a German Jesuit who lived in Rome, where he served as an unofficial "chief scientist" for the Vatican, cannot be regarded as a household name today (although he served as a primary character and inspiration for Umberto Eco's novel *The Island of the Day Before*). Nonetheless, Kircher ranked among the most formidable intellects of the seventeenth century. He wrote, for example, the most famous works of his time on magnetism, music, China (where the Jesuit order had already established a major presence), and the interpretation of Egyptian hieroglyphics (his system ultimately failed, but did offer important clues and inspiration for later scholars). Kircher tumbled into intellectual limbo largely because his Neoplatonic worldview became so completely eclipsed by an alternative concept of causality that we call modern science—a reform that Galileo (whom Kircher had more or less replaced as a

---

snow could retard putrefaction. He therefore stopped his carriage on a cold winter day, bought a hen from a poultryman, and stuffed it with snow. He was then overtaken with a sudden chill that led to bronchitis. Too ill to reach London, Bacon sought refuge instead at the home of a friend, the earl of Arundel, where he died a few days later.

But I had never read Bacon's last and poignant letter, with its touching reference to Pliny the Elder's similar demise in his boots—and, in the context of this essay, the ironic likeness of icy scenarios for endings: Pliny the Younger's primary invocation of darkness, and Bacon's literal encounter with cold:

My very good lord,

I was likely to have had the fortune of Caius Plinius the elder, who lost his life by trying an experiment about the burning of the mountain Vesuvius: for I was also desirous to try an experiment or two, touching on the conversion and induration of bodies. As for the experiment itself, it succeeded excellently well; but in the journey (between London and Highgate) I was taken with such a fit of casting [an old term for vomiting, from *casting* in the sense of "throwing out or up," as in dice or a fishing line] as I know not whether it were the stone, or some surfeit [that is, kidney or gall stones, or overeating], or cold, or indeed a touch of them all three. But when I came to your lordship's house, I was not able to go back, and therefore was forced to take up my lodging here. . . . I kiss your noble hands for the welcome. . . . I know how unfit it is for me to write to your lordship with any other hand than my own, but by my troth my fingers are so disjointed with this fit of sickness, that I cannot steadily hold a pen.

leading scientist in Vatican eyes) had espoused in the generation just before, and that Newton would carry to triumph in the generation to follow.

Kircher published his masterpiece in 1664, an immense and amazing work entitled *Mundus subterraneus* (Underground world), and covering all aspects of anything that dwelled or occurred within the earth's interior—from lizards in caves, to fossils in rocks, to mountain springs, earthquakes, and volcanoes. Kircher had been inspired to write this work in 1637–38 when he witnessed the major eruptions of Etna and Stromboli. Mount Vesuvius, after centuries of quiescence, had also erupted in 1631, and Kircher eagerly awaited the opportunity to visit this most famous volcano on his return route to Rome.

He climbed the mountain at night, guided by flames still issuing from the active crater, and then lowered himself as far as he dared into the fuming and bubbling vent the next morning. When he published his great treatise twenty-five years later, the memories of his fear and wonder remained so strong that he prefaced his entire volume with a vivid personal tale of his encounter with a primary symbol for the end of time. But Kircher favored the alternative scenario of fire:

> In the middle of the night, I climbed the mount with great diffi-
> culty, moving upward along steep and rugged paths, toward the
> crater, which, horrible to say, I saw before me, lit entirely by fire and
> flowing pitch, and enveloped by noxious fumes of sulphur. . . . Oh,
> the immensity of divine power and God's wisdom! How incom-
> prehensible are thy ways! If, in thy power, such fearful portents of
> nature now punish the duplicity and maliciousness of men, how
> shall it be in that last day when the earth, subjected to thy divine
> anger, is dissolved by heat into its elements. (My translation from
> Kircher's Latin)

I like to imagine that, as he wrote these lines, this greatest of priestly scientists hummed, sotto voce, the haunting Gregorian tune of the *Dies irae,* the most famous prayer about the last judgment:

> *Dies irae, dies illa*
> *Solvet saeclum in favilla*
> [On this day of anger,
> the world will dissolve into ashes]

Vesuvius looms over modern Naples even more ominously than Mount Rainier over Seattle, for Vesuvius lies much closer to the city center and sports a record of much more recent and frequent activity—though neither city could ever claim a medal from the global commission on safe geological siting. (My father, as a GI in World War II, observed the aftermath of the last eruption of Vesuvius in 1944.) In the light of this historical testimony, combined with a continuing and pervasive presence for any modern visitor (from a majestic mountain standing tall on the horizon, to the petrified bread and bathroom graffiti of ordinary life suddenly extinguished one fine day in Pompeii), how could anyone fail to draw from Vesuvius the same geological lesson that led Pliny and Kircher to extrapolate from a raging local volcano to a globally catastrophic end of time: the history of our planet must be ruled by sudden cataclysms that rupture episodes of quiescence and mark the dawn of a new order.

And yet the most famous geological invocation of volcanism in the Bay of Naples, bolstered by the most celebrated visual image in the profession's entire history, led scientific views of the earth in the opposite direction—toward a theory that currently observable processes, operating at characteristically gradual rates, could explain the full pageant of planetary history without any invocation of episodic global paroxysms or early periods of tumultuous planetary change, superseded later by staid global maturity.

Charles Lyell (1797–1875), the primary architect of this "uniformitarian" view, and the most famous name in the history of anglophonic geology, visited Naples on the "grand tour" of European cultural centers that nearly all Britons of good breeding undertook as an essential part of a gentleman's education. He made all the customary stops, from the steaming vents and bubbling pools of the Phlegraean Fields, to the early excavations of Pompeii, to the obligatory ascent of Vesuvius (still putting on a good show after erupting throughout the late eighteenth century, during the long tenure in Naples of British diplomat, and aficionado of volcanoes, Sir William Hamilton—a level of ardent activity matched only by the torrid, and rather public, affair between Hamilton's wife Emma and Lord Nelson himself).

How, then, could Lyell redefine Naples as a source of support for a theory so contrary both to traditional interpretations and to the plain meaning of the grandest local sights? This question occupied the forefront of my mind as I prepared for my first trip to Naples. In contemplating this geological mecca, I could hardly wait to visit the palpable signs of Pliny's misfortune (the excavations of Pompeii and Herculaneum), and to follow Kircher's path to their immediate source. But most of all, I wanted to stand upon the site of Lyell's visual epiphany, the source of his frontispiece for *Principles of Geology* (1830–33)—perhaps the most important scientific textbook ever written—and the primary icon for transforming the Vesuvian landscape from a poster for catastrophism into a paradoxical proof of triumphant gradualism: the three Roman columns of the so-called Temple of Serapis (actually a marketplace) at Pozzuoli. (I shall document, in the second part of this essay, how Lyell used these three pillars as a "tide gauge" to record extensive and gradual changes of land and sea levels during the past two thousand years—a uniformitarian antidote to the image of fiery Vesuvius as a symbol for catastrophic global endings.)

The clichés of travel literature require an arduous journey sparked with tales of adventure and danger. But I have never managed to strike up a friendship with this stylistic convention, and I remain a city boy at heart (and therefore quite unafraid of rather different kinds of dangers). In truth, I never got to the top of Vesuvius. My rented car carried no tire chains, and a sheet of January ice had closed the road. As for Pozzuoli, I can't claim any more adventure than a trip to South Ferry or Ozone Park would provide. Pozzuoli is the last stop on the Neapolitan subway.

But then, why should intellectual content correlate with difficulty of physical access—a common supposition that must rank among the silliest of romantic myths? Some of the greatest discoveries in the history of science have occurred in libraries or resided, unsuspected for decades, in museum drawers. By all means, take that dogsled across the frozen wastes if no alternative exists, but if the A train also goes to the same destination, why not join Duke Ellington for a smoother ride?

To reach the specifics of Pozzuoli on a literary journey, we must follow the path of Lyell's general theory. Lyell, a barrister by original profession, sought to reform the science of geology on both substantive and methodological grounds. He based his system—one might say his brief—on two fundamental propositions. *First,* the doctrine of gradualism: modern causes, operating entirely within the range of rates now observable, can explain the full spectrum of geological history. Apparently grandiose or catastrophic events truly arise by a summation of small changes through the immensity of geological time—the deep canyon carved grain by grain, the high mountain raised in numerous increments of earthquake and eruption over millions of years.

Second, the claim for a nondirectional or steady-state earth. Standard geological causes (erosion, deposition, uplift, and so on) show no trend either to increase or decrease in general intensity through time. Moreover, even the physical state of the earth (relative temperatures, positions of climatic belts, percentages of land and sea) tends to remain roughly the same, or to cycle around and around, through time. Change never slows or ceases; mountains rise and erode; seas move in and out. But the average state of the earth experiences no systematic trend in any sustained direction. Lyell even believed at first, though he changed his mind by the 1850s, when he finally concluded that mammals would not be found in the oldest strata, that the average complexity of life had remained constant. Old species die, and new species originate (by creation or by some unknown natural mechanism). But clams remain clams, and mammals mammals, from the earliest history of life until now.

When a scientist proposes such a comprehensive system, we often gain our best insights into the sources and rationale for his reforms by explicating the alternative worldview of his opponents. New theories rarely enter a previous conceptual void; rather, they arise as putative improvements or replacements for previous conventionalities. In this case, Lyell's perceived adversaries advocated an approach to geology often labeled either as catastrophism or directionalism (in opposition to Lyell's two chief tenets of gradualistic change on an earth in steady state).

Catastrophists argued that most geological change occurred in rare episodes of truly global paroxysm, marked by the "usual suspects" of volcanism, mountain building, earthquakes, flooding, and the like. Most catastrophists also held that the frequency and intensity of such episodes had decreased markedly through time, thus contrasting a feisty young earth with a much calmer planet in its current maturity.

For most catastrophists, these two essential postulates flowed logically from a single theory about the earth's history—the origin of the planet as a molten fireball spun off from the sun (according to the hypothesis, then favored, of Kant and Laplace), followed by progressive cooling. As this cooling proceeded, the outer crust solidified while the molten interior contracted continuously. The resulting instability—caused, almost literally, by an enlarging gap between the solidified crust and the contracting molten interior—eventually induced a sudden global readjustment, as the crust fractured and collapsed upon the contracted molten core. Thus, directionalism based on continuous cooling linked the catastrophism of occasional readjustment by crustal collapse with the hypothesis of a pervasive "arrow of time" leading from a fiery beginning, replete with more frequent and more intense paroxysms, to our current era of relative calm and rarer disruption.

Incidentally, this account of catastrophism as a genuine and interesting scientific alternative to Lyellian uniformity disproves the conventional canard, originally floated as a rhetorical device by Lyell and his partisans, but then incorporated uncritically as the conventional wisdom of the profession. In this Manichaean account, catastrophism represented the last stronghold for enemies of modern science: theologically tainted dogmatists who wanted to preserve both the literal time scale of Genesis and the miraculous hand of God as a prime mover by invoking a doctrine of global paroxysm to compress the grand panoply of geological change into a mere few thousand years. In fact, by the 1830s, all scientists, catastrophists and uniformitarians alike, had accepted the immensity of geological time as a central and proven fact of their emerging profession (see chapter 5). Catastrophists upheld a different theory of change on an equally ancient earth—and their views cannot be judged less "scientific," or more theologically influenced, than anything touted by Lyell and his school.

The personal, social, and scientific reasons behind Lyell's chosen commitments represent a complex and fascinating subject well beyond the scope of this essay. But we may at least note the overt strategy, chosen by this master of persuasive rhetoric, this barrister manqué, to promulgate his uniformitarian doctrine as the centerpiece of his textbook, *Principles of Geology*. In part, he chose the substantive route of arguing that the world, as revealed by geological evidence, just happens to operate by gradual and nondirectional change. But Lyell awarded primacy of place to a methodological claim: only such a uniformitarian approach, he urged, could free the emerging science of geology from previous fetters and fanciful, largely armchair, speculation.

If global paroxysms forge most of history, Lyell argued, then how can we ever develop a workable science of geology—for we have not witnessed such events in the admittedly limited duration of human history, and we can therefore identify no observational basis for empirical study. And if a tumultuous past operated so differently from a calmer present, then how can we use modern processes—the only mechanisms subject to direct observation and experiment, after all—to resolve the past? But on an earth in steady state, built entirely by modern causes acting at current intensities, the present becomes, in an old pedagogical cliché, "the key to the past," and the earth's entire history opens to scientific study. Thus, in a famous statement of advocacy, Lyell condemned catastrophism as a doctrine of despair, while labeling his uniformitarian reform as the path to scientific salvation:

> Never was there a dogma more calculated to foster indolence, and
> to blunt the keen edge of curiosity, than this assumption of the dis-
> cordance between the former and the existing causes of change. It

produced a state of mind unfavourable in the highest conceivable degree to the candid reception of the evidence of those minute, but incessant mutations, which every part of the earth's surface is undergoing. . . . The student, instead of being encouraged with the hope of interpreting the enigmas presented to him in the earth's structure,—instead of being prompted to undertake laborious inquiries into . . . causes now in operation, was taught to despond from the first. Geology, it was affirmed, could never rise to the rank of an exact science,—the greater number of phenomena must for ever remain inexplicable. . . .

In our attempt to unravel these difficult questions, we shall adopt a different course, restricting ourselves to the known or possible operations of existing causes. . . . We shall adhere to this plan . . . because . . . history informs us that this method has always put geologists on the road that leads to truth,—suggesting views which, although imperfect at first, have been found capable of improvement, until at last adopted by universal consent. (From the introductory chapter to the third and final volume of Lyell's *Principles*, 1833)

Major intellectual struggles cannot be won by success in easy and rudimentary skirmishes. Adversaries must also be outflanked on their home ground, where superior knowledge and forces should have rendered them invincible. A new theory must meet and encompass the hardest and most apparently contradictory cases head-on. Lyell understood this principle and recognized that he would have to bring the Vesuvius of Pliny and Kircher, of Pompeii and Emma Hamilton's fire, into his uniformitarian camp—not as a prisoner, but as a proud example. No other place or subject receives even half so much attention throughout the three volumes of *Principles of Geology*.

Lyell centered his uniformitarian case for Naples and Vesuvius upon two procedural themes that embodied all his logical and literary brilliance as geology's greatest master of argument. He first invoked the cardinal geological principle of appropriate scale by pointing out that a Vesuvian eruption, while ultimately catastrophic for the baker or blacksmith of Pompeii, not only causes no planetary disruption at its own moment of maximal intensity, but then falls even further into insignificance when several hundred years of subsequent quiescence erase its memory from the populace and erode its products from the landscape.

Why, then, should such a local catastrophe serve as an unquestioned model for extrapolation to sudden global doom? Perhaps we should draw an opposite lesson from the same event: local means local—and just as the canyon deepens grain by grain, so does the mountain chain rise gradually, eruption by eruption over extended time. At most, Vesuvius teaches us that increments of gradualism may be large at human scale—the lava field versus the eroded sand grain—but still small by global standards. In 1830, Lyell summarized a long chapter, "History of the volcanic eruptions in the district around Naples," by writing:

> The vast scale and violence of the volcanic operations in Campania, in the olden time, has been a theme of declamation. . . . Instead of inferring, from analogy that . . . each cone rose in succession,—and that many years and often centuries of repose intervened between each eruption—geologists seem to have conjectured that the whole group sprung up from the ground at once, like the soldiers of Cadmus when he sowed the dragon's teeth.

Moreover, Lyell continued in closing the first volume of his tenth edition (1867), even by purely local standards, natural catastrophes usually impose only a fleeting influence upon history. Most inhabitants view Campania as a land of salubrious tranquillity. As for Vesuvius itself, even the worst natural convulsion cannot match the destructive power of human violence and venality. In a striking literary passage, Lyell reminds us that Vesuvius posed maximal danger to the Roman empire when Spartacus housed the troops of his slave revolt in the volcano's quiescent crater in 73 B.C., not when lavas and poisonous gases poured out in A.D. 79:

> Yet what was the real condition of Campania during those years of dire convulsion? "A climate," says Forsyth, "where heaven's breath smells sweet and wooingly—a vigorous and luxuriant nature unparalleled in its productions—a coast which was once the fairy-land of poets, and the favourite retreat of great men." . . . The inhabitants, indeed, have enjoyed no immunity from the calamities which are the lot of mankind; but the principal evils which they have suffered must be attributed to moral, not to physical, causes—to disastrous events over which man might have exercised a control, rather than to inevitable catastrophes which result from subterranean agency. When Spartacus encamped his army of ten thousand gladiators in

the old extinct crater of Vesuvius, the volcano was more justly a subject of terror to Campania than it has ever been since the rekindling
of its fires.

For his second theme, Lyell emphasized the importance of interpreting evidence critically, but not necessarily literally. The geological record, like most
archives of human history, features more gaps than documents. (In a famous
metaphor, later borrowed by Darwin for a crucial argument in *The Origin of
Species,* Lyell compared the geological record to a book with very few pages
preserved, of these pages few lines, of the lines few words, and of the words few
letters.) Moreover, the sources of imperfection often operate in a treacherous
way because information does not disappear at random, but rather in a strongly
biased fashion—thus tempting us to regard some causes as dominant merely
because the evidence of their action tends to be preserved, while signs of truly
more important factors may differentially disappear from the record.

Lyell recognized that catastrophes usually leave their signatures, for extensive outpourings of lava, or widespread fracturing of strata by earthquakes, resist
erasure from the geological record. But the publishers of time often print
equally important evidence for gradual change—the few inches of sediment
that may accumulate during millions of years in clear calm seas, or the steady
erosion of a riverbed grain by grain—upon missing pages of the geological
book. This bias not only overemphasizes the role of catastrophes in general, but
may also plant the false impression that intensity of geological change has
diminished through time—for if the past favors the preservation of catastrophes, while the present yields more balanced data for all modes of change, then
a literal and uncritical reading of geological evidence may inspire erroneous
inferences about a more tumultuous past.

Lyell summarized this crucial argument about biases of preservation in a
brilliant metaphor for Mount Vesuvius. "Suppose," he writes, "we had discovered two buried cities at the foot of Vesuvius, immediately superimposed upon
each other, with a great mass of tuff and lava intervening, just as Portici and
Resina, if now covered with ashes, would overlie Herculaneum." (When Lyell
visited the area in 1828, excavations at Herculaneum had proceeded further
than those at Pompeii—hence Lyell's primary citation of a town that now ranks
second to Pompeii for memorializing the destructive powers of Vesuvian eruptions.) If we read such a sequence literally, we would have to infer a history
built by sudden and catastrophic changes. The remains of an Italian city, littered
with modern debris of beer cans and bicycles, would overlie the strata of a

The frontispiece to Lyell's Principles of Geology, *showing the three pillars of Pozzuoli, with evidence for a substantial rise and fall of sea level in historic times.*

Roman town replete with fragments of amphoras and chariots—with only a layer of volcanic rocks between. We would then conclude that a violent catastrophe had triggered a sudden mutation from Latin to Italian, and from chariot wheels to automobile tires (for we would note the genuine relationships while missing all the intermediary stages)—simply because the evidence for nearly two thousand years of gradual transitions failed to enter a historical record strongly biased toward the preservation of catastrophic events.

A successful campaign for substantial intellectual reform also requires a new and positive symbol or icon, not just a set of arguments (as presented so far) to refute previous interpretations. Vesuvius in flames, the icon of Pliny or Kircher, must be given a counterweight—some Neapolitan image, also a consequence of Vesuvian volcanism, to illustrate the efficacy of modern causes and the extensive results produced by accumulating a series of small and gradual changes through substantial time. Lyell therefore chose the Roman pillars of Pozzuoli— an image that he used as the frontispiece for all editions of *Principles of Geology* (also as an embossed golden figure on the front cover of later editions). By assuming this status as an introductory image in the most famous geological book ever written, the pillars of Pozzuoli became icon numero uno for the earth sciences. I cannot remember ever encountering a modern textbook that does not discuss Lyell's interpretation of these three columns, invariably accompanied by a reproduction of Lyell's original figure, or by an author's snapshot from his own pilgrimage.

II. RAISING (AND LOWERING) THE COLUMNS OF POZZUOLI

In exchanging the pillars of Pozzuoli for the fires of Vesuvius as a Neapolitan symbol for the essence of geological change, Lyell made a brilliant choice and a legitimate interpretation. The three tall columns—originally interpreted as remains of a temple to Serapis (an Egyptian deity much favored by the Romans as well) but now recognized as the entranceway to a marketplace—had been buried in later sediment and excavated in 1750. The marble columns, some forty feet tall, are "smooth and uninjured to the height of about twelve feet above their pedestals." Lyell then made his key observation, clearly illustrated in his frontispiece: "Above this is a zone, about nine feet in height, where the marble has been pierced by a species of marine perforating bivalve—*Lithodomus.*"

From this simple configuration, a wealth of consequences follow—all congenial to Lyell's uniformitarian view, and all produced by the same geological agents that shaped the previously reigning icon of Vesuvius in flames. The columns, obviously, were built above sea level in the first or second century A.D. But the entire structure then became partially filled by volcanic debris, and subsequently covered by sea water to a height of twenty feet above the bases of the columns. The nine feet of marine clam holes (the same animals that, as misnamed "shipworms," burrow into piers, moorings, and hulls of vessels throughout the world) prove that the columns then stood entirely underwater to this level—for these clams cannot live above the low-tide line, and the Mediterranean Sea experiences little measurable tide in any case. The nine feet of clam borings, underlain by twelve feet of uninjured column, implies that an infill of volcanic sediments had protected the lower parts of the columns—for these clams live only in clear water.

But the bases of the columns now stand at sea level—so this twenty-foot immersion must have been reversed by a subsequent raising of land nearly to the level of original construction. Thus, in a geological moment of fewer than two thousand years, the "temple of Serapis" experienced at least two major movements of the surrounding countryside (without ever toppling the columns)—more than twenty feet down, followed by a rise of comparable magnitude. If such geological activity can mark so short a time, how could anyone deny the efficacy of modern causes to render the full panoply of geological history in the hundreds of millions of years actually available? And how could anyone argue that the earth has now become quiescent, after a more fiery youth, if the mere geological moment of historical time can witness so much mobility? Thus, Lyell presented the three pillars of Pozzuoli as a triumphant icon for both key postulates of his uniformitarian system—the efficacy of modern causes, and the relative constancy of their magnitude through time.

The notion of a geologist touring Naples, but omitting nearby Pozzuoli, makes about as much sense as a tale of a pilgrim to Mecca who visited the casbah but skipped the Kaaba. Now I admire Lyell enormously as a great thinker and writer, but I have never been a partisan of his uniformitarian views. (My very first scientific paper, published in 1965, identified a logical confusion among Lyell's various definitions of uniformity.) But my own observations of the pillars of Pozzuoli seemed only to strengthen and extend his conclusions on the extent and gradual character of geological change during historical times.

I had brought only the first edition (1830–33) of Lyell's *Principles* with me to Naples. In this original text, Lyell attributed (tentatively, to be sure) all changes in level to just two discrete and rapid events. He correlated the initial subsidence (to a level where marine clams could bore into the marble pillars) to "earthquakes which preceded the eruption of the Solfatara" (a volcanic field on the outskirts of Pozzuoli) in 1198. "The pumice and other matters ejected from that volcano might have fallen in heavy showers into the sea, and would thus immediately have covered up the lower part of the columns." Lyell then ascribed the subsequent rise of the pillars to a general swelling and uplift of land that culminated in the formation of Monte Nuovo, a volcanic mound on the outskirts of Pozzuoli, in 1538.

But at the site, I observed, with some surprise, that the evidence for changing levels of land seemed more extended and complex. I noticed the high zone of clam borings on the three columns, but evidence—not mentioned by Lyell—for another discrete episode of marine incursion struck me as even more obvious and prominent, and I wondered why I had never read anything about this event. Not only on the three major columns, but on every part of the complex (see the accompanying figure)—the minor columns at the corners of the quadrangular market area, the series of still smaller columns surrounding a circular area in the middle of the market, and even the brick walls and sides of structures surrounding the quadrangle—I noted a zone, extending two to three feet up from the marble floor of the complex and terminated by a sharp line of demarcation. Within this zone, barnacles and oyster shells remain cemented to the bricks and columns—so the distinct line on top must represent a previous high-water mark. Thus, the still higher zone of clam borings does not mark the only episode of marine incursion. This lower, but more prominent, zone of shells must signify a later depression of land. But when?

Lyell's original frontispiece (redrafted from an Italian publication of 1820), which includes the bases of the large columns, depicts no evidence for this zone. Did he just fail to see the barnacles and oysters, or did this period of marine flooding occur after 1830? I scoured some antiquarian bookstores in Naples and found several early-nineteenth-century prints of the columns (from travel books about landscapes and antiquities, not from scientific publications). None showed the lower zone of barnacles and oysters. But I did learn something interesting from these prints. None depicted the minor columns now standing both in the circular area at the center, and around the edge of the quadrangle—although these locations appear as flat areas strewn with bric-a-brac in some prints. But a later print of 1848 shows columns in the central circular area. I must therefore assume that the excavators of Pozzuoli reerected the smaller columns of the quadrangle and central circle sometime near the middle of the nineteenth century—while we know that Lyell's three major columns stood upright from their first discovery in 1749. (A fourth major column still lies in several pieces on the marble floor of the complex.)

All these facts point to a coherent conclusion. The minor columns of the central circle and quadrangle include the lower zone of barnacles and oysters. These small columns were not reerected before the mid-nineteenth century. Lyell's frontispiece, and other prints from the early nineteenth century, show the three large columns without encrusting barnacles and oysters at the base. Therefore, this later subsidence of land (or rise of sea to a few feet above modern levels) must have culminated sometime after the 1840s—thus adding further evidence for Lyell's claim of substantial and complex movements of the earth *within* the geological eye-blink of historic times.

For a few days, I thought that I had made at least a minor discovery at Pozzuoli—until I returned home (and to reality), and consulted some later editions of Lyell's *Principles,* a book that became his growing and changing child (and his lifelong source of income), reaching a twelfth edition by the time of his death. In fact, Lyell documented, in two major stages, how increasing knowledge about the pillars of Pozzuoli had enriched his uniformitarian view from his initial hypothesis of two quick and discrete changes toward a scenario of more gradual and more frequent alterations of level.

1. In the early 1830s, Charles Babbage, Lyell's colleague and one of the most interesting intellectuals of Victorian Britain (more about him later), made an extensive study of the Pozzuoli columns and concluded that both the major fall of land (to the level of the clam borings) and the subsequent rise had occurred in a complex and protracted manner through several substages, and not all at once as Lyell had originally believed. Lyell wrote in the sixth edition of 1840:

> Mr. Babbage, after carefully examining several incrustations . . . as also the distinct marks of ancient lines of water-level, visible below the zone of lithophagous perforations [holes of boring clams, in plain English], has come to the conclusion, and I think, proved, that the subsidence of the building was not sudden, or at one period only, but gradual, and by successive movements. As to the re-elevation of the depressed tract, that may also have occurred at different periods.

2. When Lyell first visited Pozzuoli in 1828, the high-water level virtually matched the marble pavement. (Most early prints, including Lyell's frontispiece, show minor puddling and flooding of the complex. Later prints, including an 1836 version from Babbage that Lyell adopted as a replacement for his original frontispiece in later editions of *Principles,* tend to depict deeper water.) In 1838, Lyell read a precise account of this modern episode of renewed subsidence—and

he then monitored this most recent change in subsequent editions of *Principles*. Niccolini, "a learned architect [who] visited the ruins frequently for the sake of making drawings," found that the complex had sunk about two feet from his first observations in 1807 until 1838, when "fish were caught every day on that part of the pavement where in 1807, there was never a drop of water in calm weather."

Lyell continued to inquire about this active subsidence—from a British colleague named Smith in 1847, from an Italian named Scacchi in 1852, and from his own observations on a last trip in 1858. Lyell acknowledged several feet of recent sinking and decided to blame the old icon of Vesuvius! The volcano had been active for nearly a hundred years, including some spectacular eruptions during Hamilton's tenure as British ambassador—after several centuries of quiescence. Lyell assumed that this current subsidence of surrounding land must represent an adjustment to the loss of so much underground material from the volcano's crater. He wrote: "Vesuvius once more became a most active vent, and has been ever since, and during the same lapse of time the area of the temple, so far as we know anything of its history, has been subsiding."

In any case, I assume that the prominent layer of encrustation by marine barnacles and oysters, unmentioned by Lyell and undepicted in all my early-nineteenth-century sources—but (to my eyes at least) the most obvious sign of former geological activity at Pozzuoli today, and far more striking, in a purely visual sense, than the higher zone of clam borings—occurred during a more recent episode of higher seas. Again, we can only vindicate Lyell's conviction about the continuing efficacy of current geological processes.

A conventional essay in the hagiographical mode would end here, with Lyell triumphant even beyond the grave and his own observations. But strict uniformity, like its old alternative of uncompromising catastrophism, cannot capture all the complexity of a rich and flexible world that says yes to at least part of most honorable extremes in human system building.

Uniformity provided an important alternative and corrective to strict catastrophism, but not the complete truth about a complex earth. Much of nature does proceed in Lyell's slow and nondirectional manner, but genuine global catastrophes have also shaped our planet's history—an idea once again in vogue, given virtual proof for triggering of the late Cretaceous mass extinction, an event that removed dinosaurs along with some 50 percent of all marine species, by the impact of an extraterrestrial body. Our city of intellectual possibilities includes many mansions, and restriction to one great house will keep us walled off from much of nature's truth.

As a closing example, therefore, let us return to Lyell's fascinating colleague, Charles Babbage (1792–1871), Lucasian professor of mathematics at Cambridge, and inventor of early calculating machines that presaged the modern digital computer. The *Encyclopaedia Britannica* ends an article on this versatile genius by writing: "He assisted in establishing the modern postal system in England and compiled the first reliable actuarial tables. He also invented a type of speedometer and the locomotive cowcatcher." So why not geology as well!

Babbage presented his studies of Pozzuoli to the Geological Society of London in 1834, but didn't publish his results until 1847 because, as he stated in a preface written in the third person, "other evocations obliged him to lay it aside"—primarily that cowcatcher, no doubt! Babbage had pursued his studies to affirm Lyell's key uniformitarian postulate, as clearly indicated in the ample subtitle of his publication: "Observations on the Temple of Serapis at Pozzuoli near Naples, with an attempt to explain the causes of the frequent elevation and depression of large portions of the earth's surface in remote periods, and to prove that those causes continue in action at the present time."

By delaying publication until 1847, Babbage needed to add an appendix to describe the recent subsidence also noted by Lyell in later editions of *Principles of Geology.* Babbage discussed the observations of Niccolini and, especially, of Smith as reported to the Geological Society of London: "Mr. Smith found the floor of the temple dry at high water in 1819, and 18 inches on it at high water in 1845." But Babbage then integrated these latest data with his previous observations on earlier changes in historical times to reach his general uniformitarian conclusions:

> The joint action of certain existing and admitted causes must necessarily produce on the earth's surface a continual but usually slow change in the relative levels of the land and water. Large tracts of its surface must be slowly subsiding through the ages, whilst other portions must be rising irregularly at various rates.

To generalize this Neapolitan conclusion, Babbage then cited the ongoing work of a young naturalist, based on entirely different phenomena from the other side of the globe: coral atolls of the tropical Pacific Ocean. This young man had not yet become the Charles Darwin whom we revere today. (Publication of *The Origin of Species* still lay twelve years in the future, and Darwin had revealed his evolutionary suspicions only to a few closest confidants, not including Babbage.) Therefore, Babbage and the scientific community of Britain knew Darwin only as a promising young naturalist who had undertaken a five-year voyage around the world, published a charming book on his adventures and three scientific volumes on the geology of South

America and the formation of coral atolls, and now labored in the midst of a comprehensive treatise, which would eventually run to four volumes, on the taxonomy of barnacles.

Darwin's theory on the origin of coral atolls surely struck his colleagues as the most important and original contribution of his early work. Darwin, labeling his explanation as the "subsidence theory" of coral reefs, explained the circular form of atolls as a consequence of subsidence of the surrounding sea floor. Reefs begin by growing around the periphery of oceanic islands. If the islands then subsided, the corals might continue to grow upward, eventually forming a ring as the central island finally disappeared below the waves.

This brilliant—and largely correct—explanation included two implications particularly favorable to Lyell and his fellow uniformitarians, hence their warm embrace for this younger colleague. First, the subsidence theory provided an excellent illustration for the efficacy and continuity of gradual change—for corals could not maintain their upward growth unless the central islands sank slowly. (Reef corals, filled with symbiotic photosynthetic algae, cannot live below the level of penetration for sunlight into oceanic waters—so any rapid subsidence would extinguish the living reefs.)

Second—and more crucial to the work of Babbage and Lyell at Pozzuoli— the large geographic range of atolls proves that major regions of the earth's crust must be subsiding, thus also implying that other regions of comparable extent must rise at the same time. Therefore, the fluctuations recorded on Pozzuoli's pillars need not represent only a local phenomenon, but may also illustrate one of the most fundamental principles of the gradualist, nondirectionalist, and uniformitarian mechanics of basic planetary behavior. In fact, and above all other implications, Darwin had emphasized his discovery that coral atolls do not form in regions with active volcanoes, while no atolls exist where volcanoes flourish in eruption. This mutual avoidance indicates that large tracts of the earth's crust, not merely local pinpoints, must be subsiding or rising in concert—with atolls as primary expressions of subsidence, and volcanoes as signs of uplift.

Babbage wrote to praise the young Darwin, but also to assert that he had reached the same uniformitarian conclusions independently, during his own studies of Pozzuoli:

> Mr. Darwin, whose voyages and travels extended from 1826 to 1836 [*sic;* the *Beagle* voyage lasted from 1831 to 1836], was gradually accumulating and arranging an immense collection of facts relating to the formation of coral and lagoon islands, as well as to the relative changes of level of land and water. In 1838 Mr. Darwin published

his views on those subjects, from which, amongst several other very important inferences, it resulted, that he had, from a large induction of facts, arrived at exactly the same conclusion as that which it has been the chief object of this paper to account for, from the action of known and existing causes.

So far, so good—and so fair, and so just. But Babbage then proceeded further—into one of the most ludicrously overextended hypotheses ever advanced in the name of uniformitarian geology. He appended a "supplement" to his 1847 publication on the pillars of Pozzuoli entitled "Conjectures concerning the physical condition of the surface of the moon." In Babbage's day, most scientists interpreted lunar craters as volcanic cones—a catastrophic explanation that Babbage wished to challenge. He noted that a region of lunar craters would look very much like a field of earthly coral atolls standing in the bed of a vanished sea:

> The perusal of Mr. Darwin's explanation of the formation of coral reefs and of lagoon island led me to compare these islands with those conical crater-shaped mountains which cover the moon's surface; and it appears to me that no more suitable place could be found for throwing out the following conjectures, than the close of a paper in which I have endeavoured to show, that known and existing causes lead necessarily to results analogous to those which Mr. Darwin has so well observed and recorded. . . .
>
> If we imagine a sea containing a multitude of such lagoon islands to be laid dry, the appearance it would present to a spectator at the moon would strongly resemble that of a country thickly studded with volcanic mountains, having craters of various sizes. May not therefore much of the apparently volcanic aspect of the moon arise from some cause which has laid dry the bottom of a former ocean on its surface?

Babbage became bolder near the end of his commentary, as he explicitly wondered "if those craters are indeed the remains of coral lagoon islands." To be fair, Babbage recognized the highly speculative nature of his hypothesis:

> The proceeding remarks are proposed entirely as speculations, whose chief use is to show that we are not entirely without principles from which we may reason on the physical structure of the

moon, and that the volcanic theory is not the only one by which the phenomena could be explained.

But later discoveries only underscore the irony of what may be the greatest overextension of uniformitarian preferences ever proposed by a major scientist. Babbage suggested that lunar craters might be coral atolls because he wished to confute their catastrophic interpretation as volcanic vents and mountains. Indeed, lunar craters are not volcanoes. They are formed by the even more sudden and catastrophic mechanism of meteoritic impact.

Comprehensive worldviews like uniformitarianism or catastrophism provide both joys and sorrows to their scientific supporters—the great benefits of a guide to reasoning and observation, a potential beacon through the tangled complexities and fragmentary character of nature's historical records; but also and ineluctably combined with the inevitable, ever-present danger of false assurances that can blind us to contrary phenomena right before our unseeing eyes. Lyell himself emphasized this crucial point, with his characteristic literary flair, in the closing paragraph to his discussion about the pillars of Pozzuoli— in this case, to combat the prejudice that landmasses must be rock stable, with all changes of level ascribed to movements of the sea:

> A false theory it is well known may render us blind to facts, which are opposed to our prepossessions, or may conceal from us their true import when we behold them. But it is time that the geologist should in some degree overcome those first and natural impressions which induced the poets of old to select the rock as the emblem of firmness—the sea as the image of inconstancy.

But we also know that no good deed goes unpunished and that any fine principle can turn around and bite you in the ass. Lyell had invoked this maxim about the power of false theories to emphasize that conventional preferences for catastrophism had been erroneously nurtured by the differential preservation of such evidence in our imperfect geological records. But Georges Cuvier, Lyell's French colleague, leading catastrophist, and perhaps the only contemporary who could match Lyell's literary and persuasive flair, had issued the ultimate *touché* in a central passage of the most celebrated defense for geological catastrophism—his *Discours préliminaire* of 1812.

In this manifesto, Cuvier reaches an opposite conclusion from the same valid argument about the blinding force of ordinary presuppositions. We are misled, Lyell had remarked, by the differential preservation of catastrophes in the geological record. Cuvier held, *au contraire,* that we become equally blinded by the humdrum character of daily experience. Most moments, Cuvier argues, feature no local wars or deaths, and certainly no global cataclysms. So we do not properly credit these potential forces as agents of history, even though one global paroxysm every few million years (and therefore rarely, if ever, observable in a human lifetime) can shape the pageant of life on earth. Cuvier writes:

> When the traveller voyages over fertile plains and tranquil waters that, in their courses, flow by abundant vegetation, and where the land, inhabited by many people, is dotted with flourishing villages and rich cities filled with proud monuments, he is never troubled by the ravages of war or by the oppression of powerful men. He is therefore not tempted to believe that nature has her internal wars, and that the surface of the globe has been overturned by successive revolutions and diverse catastrophes.

I must now leave these two great geological gladiators, each using the same excellent tool of reason, to battle for his own particular theory about the earth's behavior. I return then to the pillars of Pozzuoli, just down the road from the third-largest preserved amphitheater of the Roman world (where we may site those warriors for a closing image). When I visited in early January of the premillennial year of 1999, I noticed a small, modern monument at one end of the Pozzuoli complex, a chipped and neglected slab of marble festooned with graffiti scrawled over a quotation with no identifying author. But I did copy the text as a good summary, less literary to be sure than the warring flourishes of Lyell or Cuvier, but equally eloquent in support of their common principle—a good guide to any scientist, and to any person who wishes to use the greatest human gift of independent reason against the presuppositions that bind us to columns of priestly or patriotic certainty, or to mountains of cultural stolidity:

> *Cio che piu importa e che i popolo, gli uomini tutti, perdano gli istinti e le abitudini pecorili che la millenaria schiavitu ha loro ispirato ed apprendano a pensare ed agire liberamente.*
> [What is most important, is that the populace, all people, lose the instincts and habits of the flock, which millennia of slavery have inspired in them, and learn to think and act in freedom.]

8

A Sly Dullard Named Darwin: Recognizing the Multiple Facets of Genius

I.

MOST YOUNG MEN OF HIS TIME COULD ONLY FANTA-
size; but Charles Darwin experienced the overt drama of his century's archetypal episode in a genre of personal stories that we now
call "coming of age": a five-year voyage of pure adventure (and
much science), circumnavigating the globe on H.M.S. *Beagle*.
Returning to England at age twenty-seven, Darwin became a
homebody and never again left his native land, not even to cross the
English Channel. Nonetheless, his subsequent life included two
internal dramas far more intense, far more portentous, and (for anyone who can move beyond the equation of swashbuckling with
excitement), far more thrilling than anything he had experienced as

a world traveler: first, the intellectual drama of discovering both the factuality and mechanism of evolution; and second, the emotional drama of recognizing (and relishing) the revolutionary implications of his theory of natural selection, while learning the pain that revelation would impose upon both immediate family and surrounding society.

What could possibly be more exciting than this story, set in London in 1837. The *Beagle* had docked a few months before, and Darwin now lived in town, where he courted the right contacts and worked on his specimens. He learned that his small Galápagos birds all belong to the family of finches, and not to several disparate groups, as he had thought. He never suspected this result, and had therefore not recorded the separate islands where he had collected these birds. (Theory always influences our style of collecting facts. As a creationist on the voyage itself, Darwin never imagined that the birds could have originated from a common source and then differentiated locally. According to the creationist view, all species must have been created "for" the Galápagos, and the particular island of discovery therefore held no importance. But in any evolutionary reading, and with all the birds so closely related, precise locality now mattered intensely.) He therefore tried to reconstruct the data from memory. Ironically (in view of the depth of their later enmity over evolution), he even wrote to Captain FitzRoy of the *Beagle* in order to get birds that his old boss had collected—and labeled much more carefully!

On March 14, his ornithological consultant John Gould (no relation) presented a paper at the Zoological Society, showing that the small rhea, a large flightless bird, collected by Darwin in southern Patagonia, represented a new species, not merely a geographical variant as Darwin had thought. Gould heightened Darwin's interest enormously by naming the bird *Rhea darwinii*. Janet Browne writes in her fine biography of Darwin:*

> This moment more than any other in Darwin's life . . . deserves to be called a turning point. Darwin was tantalized by the week's results. Why should two closely similar rheas agree to split the country between them? Why should different finches inhabit identical islets? The Galapagos iguana, he was further told by Thomas Bell, similarly divided themselves among the islands, and the heavily built tortoises with their individualized shells again came to mind.

---

*The original version of this essay appeared in the *New York Review of Books* as a review of Janet Browne's *Voyaging*.

Darwin now made a key analogy. (Has any truly brilliant insight ever been won by pure deduction, and not by metaphor or analogy?) Darwin realized that the different species of finches and rheas each inhabited specific territories, each adjacent to the domain of another species. If both finches and rheas replaced each other geographically, then shouldn't temporal succession also occur in continuity—that is, by evolution rather than successive creation? Darwin had collected important and entirely novel fossils of large mammals. He thought, and his expert consultant Richard Owen had affirmed, that the fossils of one creature, later named *Macrauchenia* by Owen, stood close to the modern guanaco, a modern South American mammal closely related to the llama. Darwin experienced a key flash of insight and wrote in a small private notebook: "The same kind of relation that common ostrich [rhea] bears to Petisse [the new species *Rhea darwinii*], extinct guanaco to recent; in former case position, in latter time."

Darwin had not become an evolutionist during the *Beagle* voyage, but he had fallen under the spell of gradualism and uniformity in the earth's development, a view identified with his intellectual hero, the English geologist Charles Lyell (see preceding chapter). Darwin, at this stage of his career, worked primarily as geologist, not a biologist. He wrote three books on geological subjects inspired by the *Beagle* voyage—on coral reefs, volcanic islands, and the geology of South America—but none strictly on zoology.

Lyell, well apprised of Darwin's beliefs and accomplishments, rejoiced at first in the prospect of a potential disciple, schooled in the field of nature. "How I long for the return of Darwin!" he wrote to Adam Sedgwick, Darwin's old Cambridge geology teacher. Darwin and Lyell quickly became inseparable—in part as guru and disciple, in part simply as friends. Janet Browne writes of Lyell:

> Darwin was the first naturalist to use his "Principles" effectively: Lyell's first, and in many ways his only fully committed disciple. "The idea of the Pampas going up at the rate of an inch in a century, while the western coast and Andes rise many feet and unequally, has long been a dream of mine," Lyell excitedly scrawled to him in October. "What a field you have to write on! If you cannot get here for dinner, you must if possible join the evening party."

In other words, in these crucial weeks after the return of the *Beagle*, Darwin had reached evolution by a double analogy: between geographic and temporal variation, and between geological and biological gradualism. He began to fill notebook after notebook with cascading implications. He numbered these private volumes, starting with A for more factual matters of zoology, but describing a second set, M and N, as "full of metaphysics on morals and speculations on expression." He drew a tree of life on one of the pages, and then experienced an attack of caution, writing with a linguistic touch from *Beagle* days: "Heaven knows whether this agrees with Nature—Cuidado [watch out]."

I tell this story at some length both for its intrinsic excitement, and to present an interesting tidbit that has eluded previous historians, but that professional paleontologists must recognize and relish: for those who still cherish the myth that fact alone drives any good theory, I must point out that Darwin, at his key moment of insight—making his analogy from geography to time and evolution—chose an entirely false example to illustrate his correct principle! *Macrauchenia* is not, after all, an ancestor (or even a close relative) of guanacos, but a member of a unique and extinct South American mammalian group, the Litopterna. South America was an island continent—a kind of "Superaustralia" with a fauna even richer and more bizarre than Australia's until the Isthmus of Panama rose just a few million years ago and joined the continent with North America. Several orders of large mammals, now extinct, had evolved there, including the litopterns, with lineages that converged by independent adaptation upon horses and camels of other continents.

One may not wish to become as cavalier as Charles's brother Erasmus, who considered *The Origin of Species* "the most interesting book" he had ever read, and who wrote of any factual discrepancy: "The *a priori* reasoning is so entirely satisfactory to me that if the facts won't fit in, why so much for the facts is my feeling." Still, beautiful (and powerful) theories can rarely be killed by "a nasty, ugly little fact" of T. H. Huxley's famous statement—nor should major ideas be so destroyed in a recalcitrant world where reported "facts" so often turn out to be wrong. Fact and theory interact in wondrously complex, and often mutually reinforcing, ways. Theories unsupported by fact may be empty (and, if unsupportable in principle, meaningless in science): but we cannot even know where to look without some theory to test. As Darwin wrote in my favorite quotation: "How can anyone not see that all observation must be for or against some view if it is to be of any service." Whatever the historical interest in this tale, and despite the irony of the situation, we do not denigrate Darwin's achievement, or evolution's truthful power, by noting that Darwin's crucial analogy, at his moment of eureka, rested upon a factual error.

This issue of interaction between fact and theory brings us to the core of fascination with Darwin's biography. Darwin worked as an accumulator of facts nonpareil—in part because he had found the right theory and therefore knew where to look; in part as a consequence of his obsessive thoroughness; in part as a benefit of his personal wealth and connections. But he also developed one of the most powerful and integrative theoretical constructions—and surely the most disturbing to traditional views about the meaning of human life—in Western history: natural selection. How could Darwin accomplish so much? He seems so unlikely a candidate.

II.

In addition to general benefits conferred by wealth and access to influential circles, Darwin enjoyed specific predisposing advantages for becoming the midwife of evolution. His grandfather Erasmus had been a famous writer, physician, and freethinker. (In the first sentence of his preface to Mary Shelley's *Frankenstein,* P. B. Shelley had, in order to justify Dr. Frankenstein's experiment, alluded to Erasmus Darwin's atheistical view on the possibility of quickening matter by electricity.) Erasmus died before Charles's birth, but the grandson studiously read and greatly admired his grandfather's writing—and Erasmus Darwin had been a thoroughgoing evolutionist. Charles studied medicine in Edinburgh, where he became close to his teacher Robert Grant, a committed Lamarckian evolutionist delighted to have Erasmus's grandson as a student. And then, of course, Darwin enjoyed the grandest privilege of five years' exposure to nature's diversity aboard the *Beagle.* Still, he remained a creationist, if suffused with nascent doubt, when he returned to London in 1836.

Some people display their brilliance in their cradles—as with Mill learning classics and Mozart writing symphonies almost before either could walk. We are not surprised when such men become "geniuses"; in fact, we expect such an eventual status, unless illness or idiosyncrasy conquers innate promise. But descriptions of Darwin's early years could lead only to a prediction of a worthy, but undistinguished life. Absolutely nothing in any record documents the usual accoutrements of intellectual brilliance. Geniality and fecklessness emerge as Darwin's most visible and distinctive traits. "He was so quiet," Janet Browne writes, "that relatives found it difficult to say anything about his character beyond an appreciative nod towards an exceedingly placid temperament. Geniality was what was most often remembered by Darwin's schoolfriends: the good-humored acquiescence of an inward-looking boy who did not appear much to mind whatever happened in life. . . . Some could barely remember Darwin when asked for anecdotes at the close of his life."

Darwin did develop a passion for natural history, expressed most keenly in his beetle collection—but so many children, then and now, become total devotees to such a hobby for a transient moment in a life leading elsewhere. No one could have predicted *The Origin of Species* from a childhood insect collection. Darwin performed as an indifferent student in every phase of his formal education. Sickened by the sight of blood, he abandoned medical studies in Edinburgh. His father became so frustrated when Charles quit Edinburgh that he admonished his son: "You care for nothing but shooting, dogs, and rat-catching, and you will be a disgrace to yourself and all your family." Charles recounted the episode in his *Autobiography*, written late in life with characteristic Victorian distance and emotional restraint: "He was very properly vehement against my turning an idle sporting man, which then seemed my probable destination."

Robert Waring Darwin therefore sent his unpromising boy to Cambridge, where he could follow the usual course for unambitious later-born sons and train for the sinecure of a local parsonage. Charles showed the same interest in religion that he manifested at the time for all other academic subjects save natural history—none at all. He went along, *faute de mieux*, in his usual genial and feckless way. He obtained the Victorian equivalent of a "gentleman's C" degree, spending most of his time gambling, drinking, and hunting with his upper-class pals. He still planned to become a minister during the entire *Beagle* voyage—though I am quite sure that his thoughts always focused upon the possibilities for amateur work in natural history that such a job provided, and not at all upon the salvation of souls, or even the weekly sermon.

The *Beagle* worked its alchemy in many ways, mostly perhaps in the simple ontogenetic fact that five years represents a lot of living during one's mid-twenties and tends to mark a passage to maturity. Robert Waring Darwin, apprised by scientific colleagues of his son's remarkable collections and insights, surrendered to the inevitable change from religion to science. Charles's sister Susan wrote as the *Beagle* sailed home: "Papa and we often cogitate over what you will do when you return, as I fear there are but small hopes of your still going in the church—I think you must turn professor at Cambridge."

But the mystery remains. Why Darwin? No one thought him dull, but no one marked him as brilliant either. And no one discerned in him that primary emotional correlate of greatness that our modern age calls "fire in the belly." Thomas Carlyle, a good judge, who knew both Darwin brothers Charles and Erasmus well, considered Erasmus as far superior in intelligence.

I believe that any solution to this key puzzle in Darwinian biography must begin with a proper exegesis of intelligence—one that rejects Charles Spearman's old notion of a single scalar quantity recording overall mental might (called *g* or general intelligence, and recently revived by Murray and Herrnstein as the central fallacy of their specious book, *The Bell Curve*—see the second edition of my book *The Mismeasure of Man*). Instead, we need a concept of intelligence defined as a substantial set of largely independent attributes. This primary alternative to *g* has its own long and complex history, from an extreme in misuse by the old phrenologists, to modern tenable versions initiated by Louis L. Thurstone and J. P. Guilford, and best represented today by the work of Howard Gardner.

I do not know what *g*-score might be given to Darwin by a modern Spearmanian. I do know, however, that we must follow the alternative view of independent multiplicity to grasp Darwin's triumph in the light of such unpromising beginnings (unpromising in the apparently hopeless sense of little talent with maximal opportunity, rather than the more tractable Horatio Alger mode of great promise in difficult circumstances).

Moreover, the theory of multiplicity has an important historical and philosophical consequence for understanding human achievement. If Spearman had been correct, and if intelligence could be construed as a single, innately provided, and largely invariant scalar quantity that could be plotted as a single linear order, then we might frame a predictive and largely biological model of achievement with a predominant basis in bloodlines and a substrate in neurology. But the theory of multiplicity demands an entirely different style of biographical analysis that builds a totality from narrative details.

If the sum of a person's achievement must be sought in a subtle combination of differing attributes, each affected in marvelously varying ways by complexities of external circumstances and the interplay of psyche and society, then no account of particular accomplishment can be drawn simply by prediction based on an overall inherited mental rank. Achieved brilliance must arise from (1) a happy combination of fortunate strength in several independent attributes, joined with (2) an equally fortuitous combination of external circumstances (personal, familial, societal, and historical), so that (3) such a unique mental convergence can lead one mind and personality to solve a major puzzle about the construction of natural reality. Explanations of this kind can only be achieved in the mode of dense narrative. No shortcuts exist; the answer lies in a particular concatenation of details—and these must be elucidated and integrated descriptively.

I used to think that the last section of Darwin's autobiography (on "mental qualities") represented little more than a lie, enforced by conventions of Victorian public modesty, since Darwin could not speak openly about his strengths. The very last line may indeed be regarded as a tad disingenuous:

"With such moderate abilities as I possess, it is truly surprising that I should have influenced to a considerable extent the belief of scientific men on some important points."

In rereading this section while writing this essay, I have changed my mind. I now view Darwin's assessment of his strengths and weaknesses as probably quite accurate, but set in the false context of his own belief in something close to a Spearmanian definition of brilliance. He had internalized a fairly stereotypical notion of an acme in scientific reasoning (based largely upon mathematical ability and lightning-quick powers of deduction), and he recognized that he possessed no great strength in these areas. He understood what he could do well, but granted these powers only secondary rank and importance. If Darwin had embraced the notion of intelligence as a plethora of largely independent attributes, and had also recognized (as he did for the evolution of organisms) that great achievement also requires a happy concatenation of uncontrollable external circumstances, then he might not have been so surprised by his own success.

Darwin begins the last section of his autobiography with deep regret for his negatives:

> I have no great quickness of apprehension or wit which is so remarkable in some clever men, for instance, Huxley. . . . My power to follow a long and purely abstract train of thought is very limited; and therefore I could never have succeeded with metaphysics or mathematics.

He then almost apologizes for his much humbler positive qualities:

> Some of my critics have said, "Oh, he is a good observer, but he has no power of reasoning!" I do not think that this can be true, for the 'Origin of Species' is one long argument from the beginning to the end, and it has convinced not a few able men. No one could have written it without having some power of reasoning. I think that I am superior to the common run of men in noticing things which easily escape attention, and in observing them carefully. My industry has been nearly as great as it could have been in the observation and collection of facts. What is more important, my love of natural science has been steady and ardent. . . . From my early youth I have had the strongest desire to understand or explain whatever I

observed. . . . These causes combined have given me the patience
to reflect or ponder for any number of years over any unexplained
problem.

The beginning of the final paragraph beautifully summarizes the argument
that I advocate here—sublime achievement as a unique joining, at a favorable
social and historical moment, of a synergistic set of disparate mental attributes.
But Darwin does not accept this definition and therefore views his achievement—which he does not deny, for he was not, internally at least, a modest
man—as something of a puzzle:

Therefore my success as a man of science, whatever this may have
amounted to, has been determined, as far as I can judge, by complex and diversified mental qualities and conditions.

Janet Browne did not write her book with such a theory of intelligence and
accomplishment explicitly in mind, but her biography of Darwin explains his
achievements better than any previous work because she has provided the requisite thick description of both the various attributes of mind (multiple intelligences) that motivated Darwin's work and powered his conclusions, and the
conjunction of numerous external factors that fed his triumph.

Darwin's multiple intelligences: As the greatest veil-coverers of recent times,
the Victorians did not only hide their sexual habits. More generally, they concealed most displays of passion about anything. Since passion may be the common ground for Darwin's diverse strengths, and since he so carefully
constructed an external persona of dispassionate gentility, this wellspring of his
greatness can easily be missed. But the sheer accumulative density of Browne's
documentation eventually prevails.

We come to understand, first of all, Darwin's enormous energy—whether
overt and physical during his active youth on the *Beagle,* or cerebral when he
became an invalid for most of his adult life. (Some people just seem to live at
a higher plane of intensity, and must see most of us as we view the languorous
world of a sloth—see chapter 4 on Buffon.) We often miss this theme with
Darwin because he led such a quiet life as an adult and spent so much time
prostrated by illness. But I am, of course, speaking about an internal drive. Our
minds are blank or unproductive most of the time (filled with so much Joycean
buzz that we can't sort out a useful theme). Darwin must have been thinking
with useful focus all the time, even on his sickbed. I don't quite understand how

this intense energy meshes with Darwin's placidity of personality (as expressed so strongly from earliest days), the geniality that makes him so immensely likable among history's geniuses—usually a far more prickly lot. Perhaps he just kept the prickly parts under wrap because he had been schooled as such an eminent Victorian. Perhaps (a more likely alternative, I think) emotional placidity and level of intrinsic energy just represent different and separable aspects of human personalities.

In any case, this "energy"—expressed as passion, range, thoroughness, zeal, even ruthlessness at times—drove Darwin's achievements. He expressed the most overt form in roving over South America, trekking for weeks across mountains and deserts because he heard some rumor about fossil bones at the other end—and then, with equal restlessness, thinking and thinking about his results until he could encompass them in a broad theoretical conception. (For example, Darwin developed a correct theory for the origin of coral reefs—his first great contribution to science—by reading and pondering before he ever reached Pacific atolls for direct observation.)

Back in London, Darwin virtually moved into the Athenaeum Club by day, using its excellent library as his private preserve, reading the best books on all subjects from cover to cover. Browne writes:

> Something of Darwin's mettle also showed through in the way he set off on a lifetime's program of reading in areas formerly holding only faint attractions. He tackled David Hume, Adam Smith, and John Locke in turn; Herbert Mayo's *Philosophy of Living* (1837), Sir Thomas Browne's *Religio Medici,* and John Abercrombie's *Inquiries Concerning the Intellectual Powers and the Investigation of Truth* (1838) came between Gibbon and Sir Walter Scott.

As Darwin read theory and philosophy in all fields, he also began an almost obsessive querying and recording of anyone in any station who might supply information about natural history:

> He asked Mark, Dr. Darwin's [his father's] coachman, for his opinion on dogs, and Thomas Eyton for his views on owls and pigs. He made Fox [his cousin] struggle with a deluge of farmyard questions of all shapes and sizes. He struck up a correspondence with his Uncle Jos about Staffordshire worms. . . . Darwin elaborated this way of proceeding into one of the most distinctive aspects of his

life's work. When seeking information on any new topic, he learned to go straight to the breeders and gardeners, the zookeepers, Highland ghillies, and pigeon fanciers of Victorian Britain, who possessed great practical expertise and, as Darwin fondly imagined, hardly any interest in pursuing larger theoretical explanations. . . . Being a gentleman—being able to use his social position to draw out material from people rarely considered scientific authorities in their own right—was important. His notebooks began bulging with details methodically appropriated from a world of expertise normally kept separate from high science.

Darwin pushed himself through these intense cycles of reading, pondering, noting, asking, corresponding, and experimenting over and over again in his career. He proceeded in this way when he wrote four volumes on the taxonomy of barnacles in the late 1840s and mid-1850s, when he experimented on the biogeography of floating seeds in the 1850s, on fertilization of orchids by insects in the early 1860s, and when he bred pigeons, studied insectivorous and climbing plants, and measured rates of formation of soil by worms.

He coordinated all the multiple intelligences that can seek, obtain, and order such information with his great weapon, a secret to all but a few within his innermost circle until he published *The Origin of Species* in 1859—the truth of evolution explained by the mechanism of natural selection. I don't think that he could have made sense of so much, or been able to keep going with such concentration and intensity, without a master key of this kind. He must have used yet another extraordinary intelligence to wrest this great truth from nature. But once he did so, he could then bring all his other mentalities to bear upon a quest never matched for expansiveness and import: to reformulate all understanding of nature, from bacterial physiology to human psychology, as a history of physical continuity, "descent with modification" in his words. Had he not noted, with justified youthful hubris in an early notebook, "He who understands baboon would do more towards metaphysics than Locke"?

Darwin's fortunate circumstances: All the world's brilliance, and all the soul's energy, cannot combine to produce historical impact without a happy coincidence of external factors that cannot be fully controlled: health and peace to live into adulthood, sufficient social acceptability to gain a hearing, and life in a century able to understand (though not necessarily, at least at first, to believe). George Eliot, in the preface to *Middlemarch,* wrote about the pain of brilliant women without opportunity:

Here and there a cygnet is reared uneasily among the ducklings in the brown pond, and never finds the living stream in fellowship with its own oary-footed kind. Here and there is born a Saint Theresa, foundress of nothing, whose loving heartbeats and sobs after an unattained goodness tremble off and are dispersed among hindrances instead of centering in some long-recognizable deed.

Darwin experienced the good fortune of membership in that currently politically incorrect but ever-so-blessed group—upper-class white males of substantial wealth and maximal opportunity. This subject, a staple of recent Darwinian biographies, has been particularly well treated by Adrian Desmond and James Moore in their biography, *Darwin* (Warner Books, 1992). Though the theme is now familiar to me, I never cease to be amazed by the pervasive, silent, and apparently frictionless functioning (in smoothness of operation, not lack of emotionality) of the Victorian gentleman's world—the clubs, the networks, the mutual favors, the exclusions of some people, with never a word mentioned. Darwin just slid into this world and stuck there. He used his wealth, his illnesses, his country residence, his protective wife for one overarching purpose: to shield himself from ordinary responsibility and to acquire precious time for intellectual work. Darwin knew what he was doing and wrote in his autobiography: "I have had ample leisure from not having to earn my own bread. Even ill-health, though it has annihilated several years of my life, has saved me from the distractions of society and amusement."

Janet Browne's greatest contribution lies in her new emphasis upon a theme that has always been recognized but, strangely, never exploited as a major focus for a Darwin biography—the dynamics of immediate family. I had never realized, for example, just how wealthy and powerful Darwin's father had been. I had known that he was a famous physician, but I hadn't appreciated his role as the most prominent moneylender in the county. He was a fair and patient man, but nearly everyone who was anyone owed him something. And I feel even more enlightened about the warm and enabling (in the good sense) relationship of Charles and his remarkable wife, Emma, whose unwritten biography remains, in my view, one of the strangest absences in our scholarship about nineteenth-century life. (Sources exist in abundance for more than a few Ph.D.s. We even know the cumulative results of thirty years of nightly backgammon games between Charles and Emma; for Charles, as I have said, was an ardent recorder of details!) Much has been written about Emma and other family ties as sources of Charles's hesitancies and cautions (and I accept, as substantially true, the old cliché that Charles delayed publication because he feared the impact of his freethinking ideas upon the psyche of his devout wife). But we need to give more attention and study to Darwin's family as promoters of his astonishing achievement.

If I had to summarize the paradoxes of Darwin's complex persona in a phrase, I would say that he was a philosophical and scientific radical, a political liberal, and a social conservative (in the sense of lifestyle, not of belief)—and that he was equally passionate about all three contradictory tendencies. Many biographers have argued that the intellectual radical must be construed as the "real" Darwin, with the social conservative as a superficial aspect of character, serving to hide an inner self and intent. To me, this heroically Platonic view can only be labeled as nonsense. If a serial killer has love in his heart, is he not a murderer nonetheless? And if a man with evil thoughts works consistently for the good of his fellows, do we not properly honor his overt deeds? All the Darwins build parts of a complex whole; all are equally him. We must acknowledge all facets to fully understand a person, and not try to peel away layers toward a nonexistent archetypal core. Darwin hid many of his selves consummately well, and we shall have to excavate if we wish to comprehend. I, for one, am not fazed. Paleontologists know about digging.

9

An Awful Terrible
Dinosaurian Irony

STRONG AND SUBLIME WORDS OFTEN LOSE THEIR
sharp meanings by slipping into slangy cuteness or insipidity. Julia
Ward Howe may not win History's accolades as a great poet, but
the stirring first verse of her "Battle Hymn" will always symbolize
both the pain and might of America's crucial hour:

> Mine eyes have seen the glory of the coming of the Lord;
> He is trampling out the vintage where the grapes of wrath are
>   stored;
> He hath loosed the fateful lightning of his terrible, swift sword;
> His truth is marching on.

The second line, borrowed from Isaiah 63:3, provided John
Steinbeck with a title for the major literary marker of another troubled time in American history. But the third line packs no punch
today, because *terrible* now means "sorta scary" or "kinda sad"—as
in "Gee, it's terrible your team lost today." But *terrible,* to Ms. Howe
and her more serious age, embodied the very opposite of merely

mild lament. *Terrible*—one of the harshest words available to Victorian writers—invoked the highest form of fear, or "terror," and still maintains a primary definition in *Webster's* as "exciting extreme alarm," or "overwhelmingly tragic."

Rudyard Kipling probably was a great poet, but "Recessional" may disappear from the educational canon for its smug assumption of British superiority:

> God of our fathers, known of old,
> Lord of our far-flung battle line,
> Beneath whose awful Hand we hold
> Dominion over palm and pine.

For Kipling, "awful Hand" evokes a powerful image of fearsome greatness, an assertion of majesty that can only inspire awe, or stunned wonder. Today, an awful hand is only unwelcome, as in "keep your awful hand off me"—or impoverishing, as in your pair of aces versus his three queens.

Unfortunately, the most famous of all fossils also suffer from such a demotion of meaning. Just about every aficionado knows that *dinosaur* means "terrible lizard"—a name first applied to these prototypes of prehistoric power by the great British anatomist Richard Owen in 1842. In our culture, reptiles serve as a prime symbol of slimy evil, and scaly, duplicitous, beady-eyed disgust—from the serpent that tempted Eve in the Garden, to the dragons killed by Saint George or Siegfried. Therefore, we assume that Owen combined the Greek *deinos* (terrible) with *sauros* (lizard) to express the presumed nastiness and ugliness of such a reprehensible form scaled up to such huge dimensions. The current debasement of *terrible* from "truly fearsome" to "sorta yucky" only adds to the negative image already implied by Owen's original name.

In fact, Owen coined his famous moniker for a precisely opposite reason. He wished to emphasize the awesome and fearful majesty of such astonishingly large, yet so intricate and well-adapted creatures, living so long ago. He therefore chose a word that would evoke maximal awe and respect—*terrible,* used in exactly the same sense as Julia Ward Howe's "terrible swift sword" of the Lord's martial glory. (I am, by the way, not drawing an inference in making this unconventional claim, but merely reporting what Owen actually said in his etymological definition of dinosaurs.)*

Owen (1804–92), then a professor at the Royal College of Surgeons and at the Royal Institution, and later the founding director of the newly independent

---

*A longstanding scholarly muddle surrounding the where and when of Owen's christening has been admirably resolved in two recent articles by my colleague Hugh Torrens, a geologist and historian of science at the University of Keele in England—"Where did the dinosaur get its name?" *New*

natural history division of the British Museum, had already achieved high status as England's best comparative anatomist. (He had, for example, named and described the fossil mammals collected by Darwin on the *Beagle* voyage.) Owen was a complex and mercurial figure—beloved for his wit and charm by the power brokers, but despised for alleged hypocrisy, and unbounded capacity for ingratiation, by a rising generation of young naturalists, who threw their support behind Darwin, and then virtually read Owen out of history when they gained power themselves. A recent biography by Nicolaas A. Rupke, *Richard Owen: Victorian Naturalist* (Yale University Press, 1994), has redressed the balance and restored Owen's rightful place as brilliantly skilled (in both anatomy and diplomacy), if not always at the forefront of intellectual innovation.

Owen had been commissioned, and paid a substantial sum, by the British Association for the Advancement of Science to prepare and publish a report on British fossil reptiles. (The association had, with favorable outcome, previously engaged the Swiss scientist Louis Agassiz for an account of fossil fishes. They apparently took special pleasure in finding a native son with sufficient skills to tackle these "higher" creatures.) Owen published the first volume of his reptile report in 1839. In the summer of 1841, he then presented a verbal account of his second volume at the association's annual meeting in Plymouth. Owen published the report in April 1842, with an official christening of the term *Dinosauria* on page 103:

> The combination of such characters . . . all manifested by creatures far surpassing in size the largest of existing reptiles will, it is presumed, be deemed sufficient ground for establishing a distinct tribe or suborder of Saurian Reptiles, for which I would propose the name of Dinosauria.

(From Richard Owen, *Report on British Fossil Reptiles, Part II*, London, Richard and John E. Taylor, published as *Report of the British Association for the Advancement of Science for 1841*, pages 60–204.)

Many historians have assumed that Owen coined the name in his oral presentation of 1841, and have cited this date as the origin of dinosaurs. But as Torrens shows, extensive press coverage of Owen's speech proves that he then included all dinosaur genera in his overall discussion of lizards, and had not yet chosen either to separate them as a special group, or to award them a definite

---

Scientist, volume 134, April 4, 1992, pages 40–44; and "Politics and paleontology: Richard Owen and the invention of dinosaurs," in J. O. Farlow and M. K. Brett-Surman, editors, *The Complete Dinosaur,* Indiana University Press, 1997, pages 175–90.

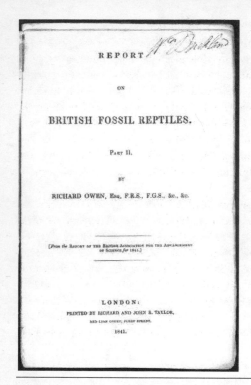

William Buckland's personal copy of Richard Owen's initial report on dinosaurs. Note Buckland's signature on the title page (left) and his note (right) showing his concern with the subject of evolution ("transmutation").

name. ("Golden age" myths are usually false, but how I yearn for a time when *local* newspapers—and Torrens got his evidence from the equivalent of the *Plymouth Gazette* and the *Penzance Peeper,* not from major London journals— reported scientific talks in sufficient detail to resolve such historical questions!) Owen therefore must have coined his famous name as he prepared the report for printing—and the resulting publication of April 1842 marks the first public appearance of the term *dinosaur.* As an additional problem, a small initial run of the publication (printed for Owen's own use and distribution) bears the incorrect date of 1841 (perhaps in confusion with the time of the meeting itself, perhaps to "backdate" the name against any future debate about priority, perhaps just as a plain old mistake with no nefarious intent)—thus confounding matters even further.

In any case, Owen appended an etymological footnote to his defining words cited just above—the proof that he intended the *dino* in *dinosaur* as a mark of awe and respect, not of derision, fear, or negativity. Owen wrote: "Gr. [Greek] *deinos* [Owen's text uses Greek letters here], fearfully great; *sauros,* a lizard." Dinosaurs, in other words, are awesomely large ("fearfully great"), thus inspiring our admiration and respect, not terrible in any sense of disgust or rejection.

I do love the minutiae of natural history, but I am not so self-indulgent that I would impose an entire essay upon readers just to clear up a little historical matter about etymological intent—even for the most celebrated of all prehistoric critters. On the contrary: a deep and important story lies behind Owen's conscious and explicit decision to describe his new group with a maximally positive name marking their glory and excellence—a story, moreover, that cannot be grasped under the conventional view that dinosaurs owe their name to supposedly negative attributes.

Owen chose his strongly positive label for an excellent reason—one that could not possibly rank as more ironic today, given our current invocation of dinosaurs as a primary example of the wondrous change and variety that evolution has imparted to the history of life on our planet. In short, Owen selected his positive name in order to use dinosaurs as a focal argument *against* the most popular version of evolutionary theory in the 1840s. Owen's refutation of evolution—and his invocation of newly minted dinosaurs as a primary example— forms the climax and central point in his concluding section of a two-volume report on British fossil reptiles (entitled "Summary," and occupying pages 191–204 of the 1842 publication).

This ironic tale about the origin of dinosaurs as a weapon against evolution holds sufficient interest for its own immediate cast of characters (involving, in equal measure, the most important scientists of the day, and the fossils deemed most fascinating by posterity). But the story gains even more significance by illustrating a key principle in the history of science. All major discoveries suffer from simplistic "creation myths" or "eureka stories"—that is, tales about momentary flashes of brilliantly blinding insight by great thinkers. Such stories fuel one of the primal legends of our culture—the lonely persecuted hero, armed with a sword of truth and eventually prevailing against seemingly insuperable odds. These sagas presumably originate (and stubbornly persist against contrary evidence) because we so strongly want them to be true.

Well, sudden conversions and scales falling from eyes may work for religious epiphanies, as in the defining tale about Saul of Tarsus (subsequently renamed Paul the Apostle) on the Damascus Road: "And as he journeyed, he came near Damascus: and suddenly there shined round about him a light from heaven: And he fell to the earth, and heard a voice saying unto him, Saul, Saul, why persecutest thou me? . . . And immediately there fell from his eyes as it had been scales; and he received sight forthwith, and arose, and was baptized"* (Acts 9:4). But scientific discoveries are deep, difficult, and complex. They require a rejection of one view of reality (never an easy task, either conceptually or psychologically), in favor of a radically new order, teeming with consequences for everything held precious. One doesn't discard the comfort and foundation of a lifetime so lightly or suddenly. Moreover, even if one thinker experiences an emotional and transforming eureka, he must still work out an elaborate argument, and gather extensive empirical support, to persuade a community of colleagues often stubbornly committed to opposite views. Science, after all, operates both as a social enterprise and an intellectual adventure.

A prominent eureka myth holds that Charles Darwin invented evolution within the lonely genius of his own mind, abetted by personal observations made while he lived on a tiny ship circumnavigating the globe. He then, as the legend continues, dropped the concept like a bombshell on a stunned and shocked world in 1859. Darwin remains my personal hero, and *The Origin of Species* will always be my favorite book—but Darwin didn't invent evolution and would never have persuaded an entire intellectual community without substantial priming from generations of earlier evolutionists (including his own grandfather). These forebears prepared the ground, but never devised a plausible mechanism (as Darwin achieved with the principle of natural selection), and they never recorded, or even knew how to recognize, enough supporting documentation.

We can make a general case against such eureka myths as Darwin's epiphany, but such statements carry no credibility without historical counterexamples. If we can show that evolution inspired substantial debate among biologists long before Darwin's publication, then we obtain a primary case for interesting and extended complexity in the anatomy of an intellectual revolution. Historians have developed many such examples (and pre-Darwinian evolutionism has long been a popular subject among scholars), but the eureka myth persists, perhaps because we so yearn to place a name and a date upon defining episodes in our history. I know of no better example, however little known and poorly documented, than Owen's invention of the name *dinosaur* as an explicit weapon—ironically for the wrong side, in our current and irrelevant judgment—in an intense and public debate about the status of evolution.

---

*Since this essay focuses on the changing meaning of words, I just can't resist citing the next line (Acts 9:5) after the ellipsis in my quotation—surely the unintentionally funniest biblical verse based on the passage of a word from high culture into slang between the seventeenth-century King James Bible and our current vernacular: "And he said, Who art thou, Lord? And the Lord said, I am Jesus, whom thou persecutest: it is hard for thee to kick against the pricks" (then meaning "obstacles," as symbolized by sharp thorns on bushes).

Scattered observations of dinosaur bones, usually misinterpreted as human giants, pervade the earlier history of paleontology, but the first recognition of giant terrestrial reptiles from a distant age before mammalian dominance (the marine ichthyosaurs and plesiosaurs had been defined a few years earlier) did not long predate Owen's christening. In 1824, the Reverend William Buckland, an Anglican divine by title, but a leading geologist by weight of daily practice and expertise, named the first genus that Owen would eventually incorporate as a dinosaur—the carnivorous *Megalosaurus*.

Buckland devoted his professional life to promoting paleontology and religion with equal zeal. He became the first officially appointed geologist at Oxford University, and presented his inaugural lecture in 1819 under the title "Vindiciae geologicae; or the connexion of geology with religion explained." Later, in 1836, he wrote one of the eight Bridgewater Treatises, a series generously endowed by the earl of Bridgewater upon his death in 1829, and dedicated to proving "the power, wisdom and goodness of God as manifested in the creation." Darwin's circle referred to this series as the "bilgewater" treatises, and the books did represent a last serious gasp for the venerable, but fading, doctrine of "natural theology" based on the so-called "argument from design"—the proposition that God's existence, and his attributes of benevolence and perfection, could both be inferred from the good design of material objects and the harmonious interaction among nature's parts (read, in biological terms, as the excellent adaptations of organisms, and the harmony of ecosystems expressed as a "balance of nature").

Buckland (1784–1856) provided crucial patronage for several key episodes in Owen's advance, and Owen, as a consummate diplomat and astute academic politician, certainly knew and honored the sources of his favors. When Owen named dinosaurs in 1842, the theoretical views of the two men could only evoke one's favorite metaphor for indistinction, from peas in a pod, to Tweedledum and Tweedledee. Owen later became an evolutionist, though never a supporter of Darwinian natural selection. One might be cynical, and correlate Owen's philosophical shift with the death of Buckland and other powerful men of the old guard, but Owen was too intelligent (and at least sufficiently honorable) to permit such a simple interpretation, and his later evolutionary views show considerable subtlety and originality. Nonetheless—in a point crucial to this essay—Owen remained an unreconstructed Bucklandian creationist, committed to the functionalist approach of the argument from design, when he christened dinosaurs in 1842.

Gideon Mantell, a British surgeon from Sussex, and one of Europe's most skilled and powerful amateur naturalists, named a second genus (that Owen would later include among the dinosaurs) in 1825—the herbivorous *Iguanodon*, now classified as a duckbill. Later, in 1833, Mantell also named *Hylaeosaurus*, now viewed as an armored herbivorous dinosaur ranked among the ankylosaurs.

Owen united these three genera to initiate his order Dinosauria in 1842. But why link such disparate creatures—a carnivore with two herbivores, one now viewed as a tall, upright, bipedal duckbill, the other as a low, squat, four-footed, armored ankylosaur? In part, Owen didn't appreciate the extent of the differences (though we continue to regard dinosaurs as a discrete evolutionary group, thus confirming Owen's basic conclusion). For example, he didn't recognize the bipedality of some dinosaurs, and therefore reconstructed all three genera as four-footed creatures.

Owen presented three basic reasons for proposing his new group. First, the three genera share the most obvious feature that has always set our primal fascination with dinosaurs: gigantic size. But Owen knew perfectly well that similarity in bulk denotes little or nothing about taxonomic affinity. Several marine reptiles of the same age were just as big, or even bigger, but Owen did not include them among dinosaurs (and neither do we today).

Moreover, Owen's anatomical analysis had greatly reduced the size estimates for dinosaurs (though these creatures remained impressively large). Mantell had estimated up to one hundred feet in length for *Iguanodon,* a figure reduced to twenty-eight feet by Owen. In the 1844 edition of his *Medals of Creation* (only two years after Owen's shortening, thus illustrating the intensity of public interest in the subject), Mantell capitulated and excused himself for his former over-estimate in the maximally exculpatory passive voice:

> In my earliest notices of the *Iguanodon* . . . an attempt was made to estimate the probable magnitude of the original by instituting a comparison between the fossil bones and those of the Iguana.

But the modern iguana grows short legs, splayed out to the side, and a long tail. The unrelated dinosaur *Iguanodon* had very long legs by comparison (for we now recognize the creature as bipedal), and a relatively shorter tail. Thus, when Mantell originally estimated the length of *Iguanodon* from very incomplete material consisting mostly of leg bones and teeth, he erred by assuming the same proportions of legs to body as in modern iguanas, and by then appending an unknown tail of greatly extended length.

Second, and most importantly, Owen recognized that all three genera share a set of distinct characters found in no other fossil reptiles. He cited many technical details, but focused on the fusion of several sacral vertebrae to form an unusually strong pelvis—an excellent adaptation for terrestrial life, and a feature long known in *Megalosaurus*, but then only recently affirmed for *Iguanodon*, thus suggesting affinity. Owen's first defining sentence in his section on dinosaurs (pages 102–3 of his 1842 report) emphasizes this shared feature: "This group, which includes at least three well-established genera of Saurians, is characterized by a large sacrum composed of five anchylosed [fused] vertebrae of unusual construction."

Third, Owen noted, based on admittedly limited evidence, that dinosaurs might constitute a complete terrestrial community in themselves, not just a few oddball creatures living in ecological corners or backwaters. The three known genera included a fierce carnivore, an agile herbivore, and a stocky armored herbivore—surely a maximal spread of diversity and ecological range for so small a sample. Perhaps the Mesozoic world had been an Age of Dinosaurs (or rather, a more inclusive Age of Reptiles, with dinosaurs on land, pterodactyls in the air, and ichthyosaurs, plesiosaurs, and mosasaurs in the sea). In this view, dinosaurs became the terrestrial component of a coherent former world dominated by reptiles.

Owen then united all his arguments to characterize dinosaurs in a particularly flattering way that justified his etymological choice of "fearfully great." In short, Owen depicted dinosaurs not as primitive and anatomically incompetent denizens of an antediluvian world, but rather as uniquely sleek, powerful, and well-designed creatures—mean and lean fighting and eating machines for a distinctive and distinguished former world. Owen emphasized this central point with a striking rhetorical device, guaranteed to attract notice and controversy: he compared the design and efficiency of dinosaurs with modern (read superior) mammals, not with slithery and inferior reptiles of either past or present worlds.

Owen first mentioned this argument right up front, at the end of his opening paragraph (quoted earlier in this essay) on the definition of dinosaurs:

> The bones of the extremities are of large proportional size, for Saurians. . . . [They] more or less resemble those of the heavy pachydermal Mammals, and attest . . . the terrestrial habits of the species.

Owen then pursues this theme of structural and functional (not genealogical) similarity with advanced mammals throughout his report—as, for example, when he reduces Mantell's estimate of dinosaurian body size by comparing their leg bones with the strong limbs of mammals, attached under the body for max-imal efficiency in locomotion, and not with the weaker limbs of reptiles, splayed out to the side and imposing a more awkward and waddling gait. Owen writes:

> The same observations on the general form and proportions of the animal [*Iguanodon*] and its approximation in this respect to the Mammalia, especially to the great extinct Megatherioid [giant ground sloth] or Pachydermal [elephant] species, apply as well to the *Iguanodon* as to the *Megalosaurus*.

Owen stresses this comparison again in the concluding paragraphs of his report (and for a definite theoretical purpose embodying the theme of this essay). Here Owen speaks of "the Dinosaurian order, where we know that the Reptilian type of structure made the nearest approach to Mammals." In a final footnote (and the very last words of his publication), Owen even speculates—thus anticipating an unresolved modern debate of great intensity—that the effi-cient physiology of dinosaurs invites closer comparison with warm-blooded mammals than with conventional cold-blooded modern reptiles:

> The Dinosaurs, having the same thoracic structure as the Croco-diles, may be concluded to have possessed a four-chambered heart; and, from their superior adaptation to terrestrial life, to have enjoyed the function of such a highly-organized center of circulation in a degree more nearly approaching that which now characterizes the warm-blooded Vertebrata.

When we contrast artistic reconstructions made before and after Owen's report, we immediately recognize the dramatic promotion of dinosaurs from ungainly, torpid, primeval reptilian beasts to efficient and well-adapted creatures more comparable with modern mammals. We may gauge this change by com-paring reconstructions of *Iguanodon* and *Megalosaurus* before and after Owen's report. George Richardson's *Iguanodon* of 1838 depicts a squat, elongated crea-ture, presumably relegated to a dragging, waddling gait while moving on short legs splayed out to the side (thus recalling God's curse upon the serpent after a portentous encounter with Eve: "upon thy belly shalt thou go, and dust shalt thou eat all the days of thy life"). But Owen, while wrongly interpreting all dinosaurs as four-footed, reconstructed them as competent and efficient run-ners, with legs held under the body in mammalian fashion. Just compare

Richardson's torpid dinosaur of 1838 with an 1867 scene of *Megalosaurus* fighting with *Iguanodon,* as published in the decade's greatest work in popular science, Louis Figuier's *The World Before the Deluge.*

Owen also enjoyed a prime opportunity for embodying his ideas in (literally) concrete form, for he supervised Waterhouse Hawkins's construction of the first full-sized, three-dimensional dinosaur models—built to adorn the reopening of the Crystal Palace at Sydenham in the early 1850s. (The great exhibition hall burned long ago, but Hawkins's dinosaurs, recently repainted, may still be seen in all their glory—less than an hour's train ride from central London.) In a famous incident in the history of paleontology, Owen hosted a New Year's Eve dinner on December 31, 1853, *within* the partially completed

An 1838 (top) compared with an 1867 (bottom) reconstruction of dinosaurs, to show the spread of Owen's view of dinosaurs as agile and active.

model of *Iguanodon*. Owen sat at the head of the table, located within the head of the beast; eleven colleagues won coveted places with Owen inside the model, while another ten guests (the Victorian version of a B-list, I suppose) occupied an adjoining side table placed outside the charmed interior.

I do not doubt that Owen believed his favorable interpretation of dinosaurs with all his heart, and that he regarded his conclusions as the best reading of available evidence. (Indeed, modern understanding places his arguments for dinosaurian complexity and competence in a quite favorable light.) But scientific conclusions—particularly when they involve complex inferences about general worldviews, rather than simple records of overtly visible facts—always rest upon motivations far more intricate and tangled than the dictates of rigorous logic and accurate observation. We must also pose social and political questions if we wish to understand why Owen chose to name dinosaurs as fearfully great: who were his enemies, and what views did he regard as harmful, or even dangerous, in a larger than purely scientific sense?

When we expand our inquiry in these directions, the ironic answer that motivated this essay rises to prominence as the organizing theme of Owen's deeper motivations: he delighted in the efficiency and complexity of dinosaurs—and chose to embody these conclusions in a majestic name—because dinosaurian competence provided Owen with a crucial argument against the major evolutionary theory of his day, a doctrine that he then opposed with all the zeal of his scientific principles, his conservative political beliefs, and his excellent nose for practical pathways of professional advance.

I am not speaking here of the evolutionary account that would later bear the name of Darwinism (and would prove quite compatible with Owen's observations about dinosaurs)—but rather of a distinctively different and earlier version of "transmutation" (the term then generally used to denote theories of genealogical descent), best described as the doctrine of "progressionism." The evolutionary progressionists of the 1840s, rooting their beliefs in a pseudo-Lamarckian notion of inherent organic striving for perfection, looked upon the fossil record as a tale of uninterrupted progress within each continuous lineage of organisms. Owen, on the contrary, viewed the complexity of dinosaurs as a smoking gun for annihilating such a sinister and simplistic view.

Owen opposed progressionistic evolution for a complex set of reasons. First of all, this opinion endeared him to his patrons, and gave him leverage against his enemies. William Buckland, Owen's chief supporter, had used his Bridgewater Treatise of 1836 to argue against evolutionary progressionism by citing the excellent design of ancient beasts that should have been crude and primitive by virtue of their primeval age. Buckland invoked both *Megalosaurus* and *Iguanodon* to advance this argument (though these genera had not yet been designated as dinosaurs).

In his preface, Buckland announced an intention to show that "the phenomena of Geology are decidedly opposed" to "the derivation of existing systems of organic life . . . by gradual transmutation of one species into another." He then argued that the superb design of ancient organisms proved the constant superintendence of a loving deity, rather than a natural process of increasing excellence from initial crudity to current complexity. According to Buckland, the superb design of giant Mesozoic reptiles "shows that even in those distant eras, the same care of the common Creator, which we witness in the mechanism of our own bodies . . . was extended to the structure of creatures, that at first sight seem made up only of monstrosities." He then inferred God's direct benevolence from the excellent adaptation of *Iguanodon* teeth to a herbivorous lifestyle: we cannot "view such examples of mechanical contrivance, united with so much economy of expenditure . . . without feeling a profound conviction that all this adjustment has resulted from design and high intelligence." In his 1842 report, Owen dutifully quoted all these key statements about dinosaurian excellence as proof of God's commanding love and wisdom, while copiously citing and praising Buckland as his source of insight.

This defense of natural theology and attack upon evolutionary progressionism also positioned Owen well against his enemies. Owen denigrated the amateur and bucolic Mantell as much as he revered his urbane professional patron Buckland, but this irrelevant spat centered on social rather than ideological issues. In London, Owen faced one principal enemy at this early stage of his career, when advance to domination seemed most precarious—the newly appointed professor of zoology at University College London: Robert E. Grant.

Grant (1793–1874) ended up in disgrace and poverty for reasons that remain unresolved and more than a little mysterious.* But, in the late 1830s, Grant (who had just moved south from Edinburgh) enjoyed prominence as a newly leading light in London's zoological circles. He also became Owen's obvious and only rival for primacy. Grant had published an excellent and highly respected series of papers on the biology and classification of "lower" invertebrates, and he held advantages of age and experience over Owen.

But Grant was also a political radical, a man of few social graces, and—more relevantly—the most prominent public supporter of evolutionary progressionism in Great Britain. (In a wonderful tale for another time   and a primary illustration

---

*See the excellent intellectual and "forensic" work of historian Adrian Desmond on this vexatious question. Incidentally, my own inspiration for this essay began with an invitation to speak at

for the sociological principle of six degrees of separation, and the general doctrine of "what goes 'round comes 'round"—Charles Darwin had spent an unhappy student year at Edinburgh before matriculating at Cambridge. As the only light in this dark time, Darwin became very close to Grant, who must be regarded as his first important academic mentor. Of course, Darwin knew about evolution from general readings (including the works of his own grandfather Erasmus, whom Grant also much admired), and Grant's Lamarckism stood in virtual antithesis to the principle of natural selection that Darwin would later develop. But the fact remains that Darwin first learned about evolution in a formal academic setting from Grant. The mystery of Grant only deepens when we learn that the impeccably generous and genial Darwin later gave Grant such short shrift. He apparently never visited his old mentor when the two men lived in London, literally at a stone's throw of separation, after Darwin returned from the *Beagle* voyage. Moreover, Darwin's autobiography, written late in his life, contains only one short and begrudging paragraph about Grant, culminating in a single statement about Grant's evolutionism: "He one day, when we were walking together, burst forth in high admiration of Lamarck and his views on evolution. I listened in silent astonishment, and as far as I can judge without any effect on my mind."

Grant represented a threat and a power in 1842, and Owen used his antievolutionary argument about dinosaurs as an explicit weapon against this archrival. The nastiest statement in Owen's report on fossil reptiles records his unsubtle skewering of Grant's evolutionary views:

> Does the hypothesis of the transmutation of species, by a march of progressive development occasioning a progressive ascent in the organic scale, afford any explanation of these surprising phenomena? . . . A slight survey of organic remains may, indeed, appear to support such views of the origin of animated species; but of no stream of science is it more necessary, than of Paleontology, to "drink deep or taste not."

To illustrate the supposed superficiality and ignorance behind such false arguments for evolution, Owen then cites only an 1835 paper by Grant—thus clearly identifying the target of his jibes. Could anyone, moreover, make a more dismissive and scurrilous statement about a colleague than Owen's rejection of Grant by Alexander Pope's famous criterion (the subject of chapter 11):

University College in a celebration to honor the reopening of Grant's zoological museum. In reading about Grant, and developing considerable sympathy for his plight, I naturally extended my research to his enemy Owen, to dinosaurs, and ultimately to this essay.

A little learning is a dangerous thing;
Drink deep, or taste not the Pierian spring.

Owen's concluding section cites several arguments to buttress his antitrans-mutationist message—but dinosaurs take the stand as his star and culminating witnesses. Owen, following a standard formulation of his generation, not a unique insight or an idiosyncratic ordering, makes a primary distinction between an evolutionary version of progressionism—where each lineage moves gradually, inexorably, and unidirectionally toward greater complexity and increasing excellence of design—and the creationist style of progressionism espoused by Buckland and other natural theologians—where God creates more complex organisms for each new geological age, but the highest forms of one period do not evolve into dominant creatures of the next age.

(To rebut the obvious objection that God could then be viewed as a bum-bler who couldn't get things right at the outset, and then had to use all of geo-logical time for practice runs, Buckland and company—including Owen in the final paragraphs of his 1842 report—argued that God always creates organisms with optimal adaptations for the environments of each geological period. But these environments change in a directional manner, requiring a progressive advance of organic architecture to maintain a level of proper adaptation. Specifically, Buckland argued that climates had worsened through time as the earth cooled from an initially molten state. Cold-blooded torpor worked best on a hot and primitive earth, but a colder and tougher world required the cre-ation of warm-blooded successors.)

Buckland and Owen held that the fossil record could act as an arbiter for this vital zoological debate; indeed, they both believed that paleontology might win primary importance as a science for this capacity to decide between evo-lutionary and creationist versions of progressionism. The two theories differed starkly in their predictions on a crucial matter: for transmutationists, each sep-arate lineage should progress gradually and continuously through time, while the newly dominant form of each geological age should descend directly from the rulers of the last period. But for progressive creationists, an opposite pattern should prevail in the fossil record: individual lineages should show no definite pattern through time, and might even retrogress; while the dominant form of each age should arise by special creation, without ancestors and with no ties to the rulers of past ages.

With this background, we can finally grasp the central significance of Owen's decision to reconstruct dinosaurs as uniquely complicated beasts, more comparable in excellence of design with later (and advanced) mammals, than with lowly reptiles of their own lineage. Such dinosaurian excellence refuted transmutation and supported progressive creationism on both crucial points. First, the high status of dinosaurs proved that reptiles had degenerated through time, as the old and best—the grand and glorious *Megalosaurus* and *Iguanodon*— gave way to the later and lowlier snakes, turtles, and lizards. Second, these highest reptiles did not evolve into the next dominant group of mammals, for small and primitive mammals had already been discovered in Mesozoic rocks that housed the remains of dinosaurs as well.

Owen clearly rejoiced (and gave thanks for the support thereby rendered to his patron Buckland) in the fearful greatness of his newly christened dinosaurs as a primary argument against the demonizing doctrine of progressive transmutation. He wrote in the concluding passages of his 1842 report:

> If the present species of animals had resulted from progressive devel-
> opment and transmutation of former species, each class ought now
> to present its typical characters under their highest recognized con-
> ditions of organization: but the review of the characters of fossil
> Reptiles, taken in the present Report, proves that this is not the case.
> No reptile now exists which combines a complicated . . . dentition
> with limbs so proportionally large and strong, having . . . so long
> and complicated a sacrum as in the order *Dinosauria*. The
> Megalosaurs and Iguanodons, rejoicing in these undeniably most
> perfect modifications of the Reptilian type, attained the greatest
> bulk, and must have played the most conspicuous parts, in their
> respective characters as devourers of animals and feeders upon veg-
> etables, that this earth has ever witnessed in oviparious [egg-laying]
> and cold-blooded creatures.

Owen then closed his argument, and this entire section of his report, by using dinosaurs to support the second antitransmutationist principle as well: not only do dinosaurs illustrate a lack of progress within reptilian lineages, but they also demonstrate that higher mammals could not have evolved from dominant reptiles:

> Thus, though a general progression may be discerned, the interrup-
> tions and faults, to use a geological phrase, negative [*sic*] the notion
> that the progression had been the result of self-developing energies
> adequate to a transmutation of specific characters; but, on the con-

trary, support the conclusion that the modifications of osteological structure which characterize the extinct Reptiles, were originally impressed upon them at their creation, and have been neither derived from improvement of a lower, nor lost by progressive development into a higher type.

As a closing fillip and small (but pretty) footnote to this argument from the public record, I can also add a previously unknown affirmation for the centrality of antievolutionism as a primary motivation in Owen's designation of dinosaurs as "fearfully great." Owen certainly stressed an antitransmutationist message in discoursing on the significance of dinosaurs. But how do we know that transmutation represented a live and general debate in the zoology of Owen's day—and not just a funny little side issue acting as a bee in Owen's own idiosyncratic bonnet? The public record does provide support for the generality. For example, the first full-scale defense of evolution written in English, the anonymously printed *Vestiges of the Natural History of Creation* (by the Scottish publisher Robert Chambers), became the literary sensation and hottest press item of 1844.

But I can add a small testimony from a personal source. Several years ago, I had the great good fortune to purchase, at a modest price before the hype of *Jurassic Park* sent dinosaur memorabilia through the financial roof, Buckland's personal copy of Owen's 1842 report. (This copy, inscribed to Buckland by Owen, and signed by Buckland in two places, bears the incorrect date of 1841, and must therefore belong to the original lot of twenty-five, printed for Owen's private distribution.) Buckland obviously read the document with some care, for he underlined many passages and wrote several marginal annotations. But the annotations follow a clear pattern: Buckland only highlights factual claims in his marginal notes; he never comments on theoretical or controverted points. Mostly, he just lists the taxonomic names of species under discussion, or the anatomical terms associated with Owen's immediate descriptions. For example, in Owen's section on dinosaurs, Buckland writes "sacrum" next to Owen's identification of fused sacral vertebrae as a defining character of dinosaurs. And he writes "28 feet" next to Owen's defense of this smaller length for *Iguanodon*.

Buckland breaks this pattern only once—to mark and emphasize Owen's discussion of dinosaurs as an argument against transmutation of species. Here, on page 196, Buckland writes the word "transmutation" in the margin—his only annotation for a theoretical point in the entire publication. Moreover, and in a manner that I can only call charming in our modern age of the 3M

Post-it, Buckland cut out a square of white paper and fastened it to page 197 by a single glob of marginal glue, thus marking Owen's section on evolution by slight projection of the square above the printed page when the book lies closed. Finally, Buckland again wrote the single word "Transmutation" on a loose, rectangular slip of paper that he must have inserted as a bookmark into Owen's report. Buckland, one of the leading English geologists of his time, evidently regarded Owen's discussion of evolution as the most important theoretical issue addressed by the giant reptiles that Buckland had first recognized and that Owen had just named as dinosaurs.

Evolution must be a genuinely awful and absolutely terrible truth if Owen felt compelled to employ the most fearfully great of all fossil creatures in an ultimately vain attempt to refute this central principle of life's history, the source of all organic diversity from *Megalosaurus* to Moses, from *Iguanodon* to the "lowly" infusorians residing inside those zoological oddballs who learn to name dinosaurs and strive to contemplate the great I Am.

10

Second-Guessing
the Future

FROM ANONYMOUS VICE-PRESIDENTS TO NAMELESS PALOOKAS, a special kind of opprobrium seems to haunt those who finish second—close but no cigar, in an old cliché. I once met "Two Ton Tony" Galento in a bar in upstate New York, a pitiful figure as an old man, still cadging drinks in exchange for the true story of his moment of glory: when he knocked Joe Louis down before losing their fight for the heavyweight championship. And just consider the stereotype of the sidekick—old, fat, foolish, and in servitude—from Gabby Hayes and Andy Devine in the quintessential epic of our pop culture, to Leporello and Sancho Panza in the literary world. (Strong and noble sidekicks like Tonto get cast as "ethnics" to advertise their secondary rank by another route, now happily—or at least hopefully—fading from the collective consciousness of white America.)

Second in time fares no better than second in status. I was, at first, surprised by a statement that made perfect sense once I punctured the apparent paradox. A composer friend told me that he could easily obtain funding for a premiere performance of any new work—as special grants and scholarships abound for such a noble purpose. A philanthropist who truly loved music, he told me, would endow the most unprofitable and unfashionable of all genres: *second* performances of new works.

I recently had the privilege of speaking with Larry Doby, one of the toughest, most courageous, and most admirable men I have ever met. But how many readers recognize his name? We all know Jackie Robinson, who came first; Larry Doby was the second black player in Major League Baseball (and first in the American League). We all recognize the tune when Rodolfo grasps Mimi's cold little hand in Puccini's *La Bohème,* first performed in 1896. But how many people know that Leoncavallo (who had scored the hit of 1892 with *I Pagliacci*) also wrote an opera with the same title (and tale) in 1897?

I can think of only one second finisher who became more famous (at least among Anglophones) than the victor—but only for special circumstances of unusual heroism in death, mingled with a dose of overextended British patriotism: Robert Scott, who reached the South Pole on January 18, 1912, only to find that Roald Amundsen had beat him to the bottom by a full month. Confined to a tent by a blizzard, and just eleven miles from his depot, Scott froze to death, leaving a last journal entry that has never been matched in all the annals of British understatement, and that, I confess, still brings tears to my eyes: "It seems a pity, but I do not think I can write more."

In my parish, the dubious (and admittedly somewhat contradictory) status of most famous second-place finisher goes without contest to Alfred Russel Wallace, who, in 1858 during a malarial fit on the Indonesian island of Ternate, devised virtually the same theory of natural selection that Darwin had developed (but never published) in 1838. In a familiar story, Wallace sent his short paper to Darwin, a naturalist he greatly admired and who, as Wallace knew, maintained a strong interest in "the species question" (though Wallace had no inkling of Darwin's particular and nearly identical theory, and probably didn't even realize that Darwin had a theory at all). Darwin, in understandable panic, turned to his best friends, Charles Lyell and Joseph Hooker, for advice. In a resolution known to later history as the "delicate arrangement," Darwin's friends made a joint presentation to the Linnaean Society of London in July 1858. At this meeting, they read both Wallace's paper and some unpublished letters and manuscripts by Darwin, illustrating his earlier authorship of the same idea.

Conspiracy theorists always stand at the ready, and several salvos have been launched for this particular episode, but to no avail or validity, in my judgment. Yes, Wallace was never asked (being quite incommunicado, half a world away, while time did press). Yes, Darwin was wealthy and well established; Wallace, poor, younger, and struggling for livelihood and reputation. (But why, then, grant him equal billing with Darwin for a joint presentation of unpublished results?) No, I think that, as usual (and unfortunately for the cause of a good tale), the more boring resolution of ordinary decency applies.

"Delicate arrangement" describes the result quite accurately: a fair solution to a tough problem. Darwin held legitimate priority, and he had not been shilly-shallying, or resting on old claims and laurels. He had been diligently working on his evolutionary views and had already, when he received Wallace's paper, finished nearly half of a much longer book on natural selection that he then abandoned (spurred no doubt by fears of further anticipations) to write the shorter "abstract" known to the world as *The Origin of Species* (a pretty hefty book of 490 pages), published in 1859.

Wallace, at least, never complained, and seemed to feel honored that his exercise of an evening had been thus linked with Darwin's long effort. (I do not, of course, base this claim on Wallace's public pronouncements, where his secondary status to Darwin would have precluded any overt expression of bitterness. Rather, in his truly voluminous private jottings, letters, and conversations, Wallace never expressed anything but pleasure at Darwin's willingness to share at least partial credit.)

I do not, however, deny the usual assessment of Wallace as a man trammeled by meager circumstances and dogged by hard luck. He spent several youthful years of difficult and dangerous fieldwork in the Amazon, only to lose all his specimens in a shipwreck that nearly ended his own life as well. Wallace did not despair, but quickly set sail in the other direction, and spent several years engaged in similar work around the Malay Archipelago, where he took second place in the greatest biological discovery in history. He grew up in poverty (in a family of middle-class social status but much lower means), and while comfortable enough during most of his adult life, he never accumulated adequate resources to reach his true goal: doing science without impediment, and without needing to live by his own wits as a writer and lecturer. (A government pension, secured for Wallace by Darwin and his friends—perhaps partly to assuage a tinge of guilt—didn't hurt, but didn't guarantee solvency either.)

Because Wallace lived a long time (1823–1913), wrote copiously both for his bread and from his convictions, and held a variety of passionate and quirky views, he left us a vast legacy of varied content and quality. He campaigned ardently for the right and the just according to his idiosyncratic standards, and he fought valiantly for a set of causes usually deemed "cranky," both in his own

The illustrations for this chapter were published in France in the late nineteenth century and represent predictions for twentieth-century achievements.

EN L'AN 2000

time and today—including phrenology and spiritualism (where he nearly came to blows with skeptics like Darwin and Huxley)—and against vaccination, which he called "one of the foulest blots on the civilization of the 19th century." His politics defy simple characterization, but generally fall into a camp that might be labeled as democratic socialism of a Fabian bent, but spiced by utter devotion to a few favored causes that did not rank high on most people's list of indispensable reforms.

I have often called upon Wallace's large body of work for essays in these books, both for his wisdom (in debunking Percival Lowell's ideas on Martian canal builders), and for his crankiness (in claiming virtual proof for the proposition that, throughout the entire universe, no planet but the earth could house intelligent life). But now, for the first time, I invoke Wallace proactively, and after considerable patience in waiting for the appropriate moment.

An impassioned author, approaching a public turning point at the height of his own supposed wisdom and maturity, could scarcely resist such a temptation for proclamation. The turnings of our centuries may bear no relationship to any natural cycle in the cosmos. (I label such passages as "precisely arbitrary" in the subtitle to my previous book, *Questioning the Millennium*.) But we construe such artificial transitions as occasions for taking stock, especially at the centurial boundaries that have even generated their own eponymous concept of cyclical *Angst*—the *fin de siècle* (end of century) phenomenon. (The forthcoming millennium might provoke an even greater burst, but we have too little experience for any prediction. I am at least amused by a diminution in the quality of anxiety for the two documented transitions: last time around, Europe feared all the gory prophecies of Armageddon, as recorded in Revelation 20. For this second experience in Western history, we focus our worries on what might happen if computers misread the great turning as a recursion to the year 1900.)

Thus, Alfred Russel Wallace could not let the nineteenth century expire without presenting his summation, his evaluation, and his own predictions to the world. He published *The Wonderful Century: Its Successes and Failures* in 1898, and I have been waiting for several years to celebrate the hundredth anniversary of this book near the dawn of our own new millennium.* I saved my remarks for this forum of evolutionary essays, both because Wallace plays a major role on this particular stage, and because the genre of fin-de-siècle summations includes two linked and distinctive themes that have served as linchpins for these essays: the relationship between science and society (an unavoidable centerpiece in assessing the nineteenth century, with its technologically inspired industrial and colonial expansions), and the unpredictability of evolutionary and social futures (ironically, the theme that ultimately undermines this entire genre of summing up the past in hopes of securing a better future).

Wallace presents a simple thesis as the foundation for his epitome of the nineteenth century—a standard view about the relation of science to society, stated in the context of a particular time. Science, Wallace argues, has made unprecedented gains, largely expressed as technological advance (at least in terms of impacts upon everyday life), but this progress has been blunted, if not perverted, by our failure to make any moral improvements, especially as expressed in the alleviation of social inequities. Thus, and ironically, the progress of science, however bursting with potential for social improvement as well, has actually operated to increase the sum total of human misery.

Wallace opens with a statement of his thesis:

> The present work is not in any sense a history, even on the most limited scale. It may perhaps be termed an appreciation of the century—of what it has done, and what it has left undone. . . . A comparative estimate of the number and importance of these [material and intellectual] achievements leads to the conclusion that not only is our century superior to any that have gone before it, but that it

---

* I wrote and first published this essay in 1998.

may be best compared with the whole preceding historical period. It must therefore be held to constitute the beginning of a new era in human progress. But this is only one side of the shield. Along with these marvelous Successes—perhaps in consequence of them—there have been equally striking Failures, some intellectual, but for the most part moral and social. No impartial appreciation of the century can omit a reference to them; and it is not improbable that, to the historian of the future, they will be considered to be its most striking characteristic.

In his first, and shorter, section on scientific and technological progress, Wallace even tries to quantify the relative value of nineteenth-century achievements, reaching the conclusion that this single century had surpassed the summation of all previous human history in weight of accumulated progress:

> In order to estimate its [the nineteenth century's] full importance and grandeur—more especially as regards man's increased power over nature, and the application of that power to the needs of his life today, with unlimited possibilities in the future—we must compare it, not with any preceding century, or even with the last millennium, but with the whole historical period—perhaps even with the whole period that has elapsed since the stone age.

The chapters of this first part then detail the major inventions, spurred by advancing science, that brought such great potential improvement to nineteenth-century life: control of fire (with wide-ranging implications from steam engines to generating plants), labor-saving machinery, transportation, communication, and lighting (culminating in the incandescent bulb). Wallace's examples often combine charm with insight (as we recall, from yet a century further on, the different lives of not-so-distant forebears). For example, Wallace writes of his own childhood:

> The younger generation, which has grown up in the era of railways and of ocean-going steamships, hardly realize the vast change which we elders have seen. . . . Even in my own boyhood the wagon for the poor, the stage coach for the middle class, and the post-chaise for the wealthy, were the universal means of communication, there being only two short railways then in existence. . . . Hundreds of

four-horse mail and stage coaches, the guards carrying horns or bugles which were played while passing through every town or village, gave a stir and liveliness and picturesqueness to rural life which is now almost forgotten.

I confess to a personal reason for intrigue with Wallace's best example for regarding the nineteenth century as exceeding all previous history in magnitude of technological improvement: the trip from London to York, he states, took less time during the Roman occupation than in 1800, just before the advent of railroads—for the Romans built and maintained better roads, and horses moved no faster in 1800 than in A.D. 300. (I am amused by the analogous observation that rail travel on my frequent route between New York and Boston has slowed during the last century. A nineteenth-century steam engine could make the journey faster than Amtrak's quickest train, which now runs by electricity from New York to New Haven, but must then lose substantial time in switching engines for the diesel run on a nonelectrified route from New Haven to Boston. Yes, they tell us, vast improvement and full electrification lie just around the temporal corner. But how long, oh Lord, how long!)

In reading Wallace's examples, I also appreciated the numerous reminders of the central principle that all truly creative invention must be tentative and flexible, for many workable and elegant ideas will be quickly superseded—as in this temporary triumph for news over the newly invented telephone:

> Few persons are aware that a somewhat similar use of the telephone is actually in operation at Buda Pesth [*sic* for Budapest, a city then recently amalgamated from two adjoining towns with Wallace's separate names] in the form of a telephonic newspaper. At certain fixed hours throughout the day a good reader is employed to send definite classes of news along the wires which are laid to subscribers' houses and offices, so that each person is able to hear the particular items he desires, without the delay of its being printed and circulated in successive editions of a newspaper. It is stated that the news is supplied to subscribers in this way at little more than the cost of a daily newspaper, and that it is a complete success.

But Wallace's second and longer section then details the failures of the nineteenth century, all based on the premise that moral stagnation has perverted the application of unprecedented scientific progress:

EN L'AN 2000.

We of the 19th century were morally and socially unfit to possess and use the enormous powers for good or evil which the rapid advance of scientific discovery had given us. Our boasted civilization was in many respects a mere surface veneer; and our methods of government were not in accordance with either Christianity or civilization. This view is enforced by the consideration that all the European wars of the century have been due to dynastic squabbles or to obtain national aggrandizement, and were never waged in order to free the slave or protect the oppressed without any ulterior selfish ends.

Wallace then turns to domestic affairs, with the damning charge that our capitalist system has taken the wealth accrued from technological progress, and distributed the bounty to a few owners of the means of production, while actually increasing both the absolute and relative poverty of ordinary working people. In short, the rich get richer and the poor get poorer:

One of the most prominent features of our century has been the enormous and continuous growth of wealth, without any corresponding increase in the well-being of the whole people; while there is ample evidence to show that the number of the very poor— of those existing with a minimum of the bare necessities of life— has enormously increased, and many indications that they constitute a larger proportion of the whole population than in the first half of the century, or in any earlier period of our history.

At his best, Wallace writes with passion and indignation, as in this passage on preventable industrial poisoning of workers:

> Let every death that is clearly traceable to a dangerous trade be made manslaughter, for which the owners . . . are to be punished by imprisonment. . . . and ways will soon be found to carry away or utilize the noxious gases, and provide the automatic machinery to carry and pack the deadly white lead and bleaching powder; as would certainly be done if the owners' families, or persons of their own rank of life, were the only available workers. Even more horrible than the white-lead poisoning is that by phosphorus, in the match-factories. Phosphorus is not necessary to make matches, but it is a trifle cheaper and a little easier to light (and so more dangerous), and is therefore still largely used; and its effect on the workers is terrible, rotting away the jaws with the agonizing pain of cancer followed by death. Will it be believed in future ages that this horrible and unnecessary manufacture, the evils of which were thoroughly known, was yet allowed to be carried on to the very end of this century, which claims so many great and beneficent discoveries, and prides itself on the height of civilization it has attained?

Wallace offers few specific suggestions for a new social order, but he does state a general principle:

> The capitalists as a class have become enormously richer. . . . And so it must remain till the workers learn what alone will save them, and take the matter into their own hands. The capitalists will consent to nothing but a few small ameliorations, which may improve the condition of select classes of workers, but will leave the great mass just where they are.

I doubt that Wallace harbored any muscular or martial fantasies about armed revolt sweeping through the streets of London, with the apostles of a new and better world, himself included, leading a vanguard, rifles held high. Wallace was far too gentle a man even to contemplate such a style of renewal. At most, he looked to electoral reform and unionization as means for workers to take "the matter into their own hands." His final chapter, entitled "The Remedy for Want," goes little beyond a naive proposal for free bread on demand, financed by a voluntary (albeit strongly suggested) governmental tax upon people with the highest incomes.

Wallace's summary of the nineteenth century—a steady inexorability of technological progress derailed by failure of our moral and social sensibilities to keep pace—underscores the second evolutionary theme of this essay, while undermining the entire genre of fin-de-siècle (or millennium) summations: the unpredictability of human futures, and the futility of thinking that past trends will forecast coming patterns. The trajectory of technology might offer some opportunity for prediction—as science moves through networks of implication, and each discovery suggests a suite of following steps. (But even the "pure" history of science features unanticipated findings, and must also contend with nature's stubborn tendency to frustrate our expectations—factors that will cloud anyone's crystal ball.) Moreover, any forecast about the future must also weigh the incendiary instability generated by interaction between technological change and the weird ways of human conduct, both individual and social. How, then, can the accidents that shaped our past give any meaningful insight into the next millennium?

I think that the past provides even dimmer prospects for prediction than Wallace's model of history implies—for another destabilizing factor must be added to Wallace's claim for discordance between technological and moral change. Wallace missed the generality of an important pattern in nature because he remained so committed to Lyellian (and Darwinian) gradualism as the designated way of life on earth. His book devotes an entire chapter (in the first section on scientific progress) to arguing that the replacement of catastrophism by uniformitarian geology—the notion that major features of the earth's history and topography "are found to be almost wholly due to the slow action of the most familiar everyday causes" and should not be "almost always explained as being due to convulsions of nature"—"constitutes one of the great philosophical landmarks of the 19th century."

Wallace knew that the discordance of technological and moral change could produce catastrophic disruption in human history, but he viewed such a result as exceptional among the ways of nature, and not subject to generalization. Now that our modern sensibilities have restored catastrophism as an important option (though not an exclusive pattern) for nature as well, this theme gains ground as a powerful argument against predictability. Not only as an anomaly of human history, but also as a signature of nature, pasts can't imply futures because a pattern inherent in the structure of nature's materials and laws—"the great asymmetry" in my terminology—too often disrupts an otherwise predictable unfolding of historical sequences.

Any complex system must be constructed slowly and sequentially, adding steps one (or a few) at a time, and constantly coordinating along the way. But the same complex systems, once established, can be destroyed in a tiny fraction of the necessary building time—often in truly catastrophic moments—thus engendering the great asymmetry between building up and tearing down. A day of fire destroyed a millennium of knowledge in the library of Alexandria, and centuries of building in the city of London. The last blaauwbock of southern Africa, the last moa of New Zealand, perished in a momentary blow or shot from human hands, but took millions of years to evolve.

The discordance between technological and moral advance acts as a destabilizing factor to feed the great asymmetry, and prevents us from extrapolating past trends into future predictions—for we never know when and how the ax of the great asymmetry will fall, sometimes purging the old to create a better world by revolution, but more often (I fear) simply cutting a swath of destruction, and requiring a true rebirth from the ashes of old systems (as life has frequently done—in a wondrously unpredictable way—following episodes of mass extinction).

Thus, I am even less sanguine than Wallace about possibilities for predicting the future—even though I think that he overstated his case in an important way. I don't fully agree with Wallace's major premise that technology has progressed while morality stagnated. I rather suspect that general levels of morality have improved markedly as well, at least during the last millennium of Western history—though I don't see how we could quantify such a claim. In most of the world, we no longer keep slaves, virtually imprison women, mock the insane, burn witches, or slaughter rivals with such gleeful abandon or such unquestioned feelings of righteousness. Rather, our particular modern tragedy—and our resulting inability to predict the future—resides largely in the great asymmetry, and the consequential, if unintended, power of science to enhance the effect. I suspect that twenty Hitlers ruled over small groups of Europeans a thousand years ago. But what could such petty monsters accomplish with bows and arrows, battering rams, and a small cadre of executioners? Today, one evil man can engineer the murder of millions in months.

Finally, a fascinating effect of scale defeats all remaining hope for meaningful predictability. Yes, if one stands way, way back and surveys the history of

human technology, I suppose that one might identify a broad form of sensible order offering some insight into future possibilities. The invention of agriculture does imply growth in population and construction of villages; gunpowder does move warfare away from the besieging of walled cities; and computers must exert some effect upon printed media. Unless the great asymmetry wipes the slate clean (or even frees the earth from our presence entirely), some broad patterns of technological advance should be discernible amidst all the unpredictable wriggles of any particular moment.

Yes, but almost all our agonized questions about the future focus upon the wriggles, not the broader patterns of much longer scales. We want to know if our children will be able to live in peace and prosperity, or if the Statue of Liberty will still exist to intrigue (or bore) our grandchildren on their school trips, or to greet yet another wave of immigrants. At most, we ask vague and general questions about futures not really very distant, and not truly very different, from what we already know or suspect.

Just consider the most widely discussed pattern of human history since the invention of writing: the rise, spread, and domination of the European world, thanks largely to the auxiliary technologies of gunpowder and navigation. Traditions of Western explanation, largely self-serving of course, have focused upon two successive causes—strikingly different claims to be sure, but strangely united in viewing European domination as predictable, if not foreordained.

The first, as old as our lamentable self-aggrandizement, simply trumpets the inherent superiority of European people, a claim made even uglier in the last few centuries by grafting the phony doctrine of scientific racism upon old-fashioned xenophobia. The second—arising largely from a desire to reject the falsity and moral evil of racism, while still viewing history as predictably sensible—holds that people are much of a muchness throughout the world, but that certain climates, soils, and environments must inspire technological advance, and European people just happened to live in the right place.

This second argument holds much merit, and almost has to be valid at a scale of explanation that steps way back and looks at broadest patterns. Indeed, no other explanation in the determinist mode makes any sense, once we recognize the multitude of recent genetic studies that reveal only trivial differences among human groups, based on an enormous weight of shared attributes and the great variability existing *within* each of our groups.

But again, I ask most readers of this essay (originally published in a Western land and language, and initially read mostly by people of European descent) to look into their guts and examine the basis of their question: are you really asking about an admittedly broad inevitability based on soils and latitudes, or are you wondering about a wriggle lying within the realm of unpredictability? I suspect that most of us are really asking about wriggles, but looking at the wrong scale and thinking about predictability.

Yes, complex technology probably had to emerge from mid-latitude people living in lands that could support agriculture—not from Eskimos or Laplanders in frozen terrains with limited resources, and not from the hottest tropics, with vegetation too dense to clear, and a burden of disease too great to

bear. But which mid-latitude people? Or to be more honest (and for the majority of Anglophones who read this essay in its original form), why among people of my group and not of yours?

In honest and private moments, I suspect that most readers of European
descent regard the spread of European domination as a sensible and predictable
event, destined to happen again if we could rewind time's tape, say to the birth
of Jesus, and let human history unroll on a second and independent run. But I
wouldn't bet a hoplite's shield or a Frenchman's musket on a rerun with
European domination. The little wriggles of a million "might have beens"
make history, not the predictabilities of a few abstract themes lying far from our
concerns in a broad and nebulous background.

Can we really argue that Columbus's caravels began an inevitable expansion
of one kind of people? Surely not when the great Chinese admiral Zheng He
(rendered as Cheng Ho in a previously favored system of transliteration), using
a mariner's compass invented by his people, led seven naval expeditions as far
as the shores of eastern Africa between 1405 and 1433. Some of Zheng He's
ships were five times as long as a European caravel, and one expedition may
have included as many as sixty-two ships carrying nearly 28,000 men.

To be sure, Zheng He sailed for the Yung-lo emperor, the only ruler who
ever favored such expansionist activities during the Ming dynasty. His successors suppressed oceanic navigation and instituted a rigidly isolationist policy. (I
also understand, though I can claim no expertise in Chinese history, that Zheng
He's voyages must be viewed more as tributary expeditions for glorifying the

emperor, than as harbingers of imperialistic expansion on Western models. Incidentally, as further evidence for our fascination with differences, I have never read a document about Zheng He that proceeds past the first paragraph before identifying the great admiral as both a Muslim and a eunuch. I could never quite fathom the relevance, for captains don't navigate with their balls, while we know that court eunuchs played a major role throughout Chinese imperial history.)

In any case, suppose that Chinese history had unfolded a bit differently? Suppose that the successors of the Yung-lo emperor had furthered, rather than suppressed, his expansionist policies? Suppose that subsequent admirals had joined another great Chinese invention—gunpowder as weaponry—with their unmatched naval and navigational skills to subdue and occupy foreign lands? May we not suppose that Caucasian Europe would then have become a conquered backwater?

We must also consider dramatic (and entirely believable) alternatives within Caucasian history. Has any force in human affairs ever matched the spreading power of Islam after a local origin in the sixth century A.D.? The preeminent traveler Ibn Battuta surveyed the entire Muslim world during three decades of voyaging in the mid-fourteenth century. Would any companion have bet on Christianity over Islam at that moment in history? (And how would one vote today, despite the intervening success of European doctrines?) The *Encyclopaedia Britannica* comments: "Thomas Aquinas (c. 1224–1274) might have been read from Spain to Hungary and from Sicily to Norway; but Ibn al-'Arabi (1165–1240) was read from Spain to Sumatra and from the Swahili coast to Kazan on the Volga River."

Islam came close to subduing Europe on several occasions that might easily have experienced an opposite outcome. Perhaps that Moors of Iberia never did have designs on all Europe, despite the cardboard tale we once learned in conventional Western history classes—that Islam peaked and began an inevitable decline when Charles Martel defeated the Moors at Poitiers in 732. *Britannica* remarks that "the Andalusian Muslims never had serious goals across the Pyrenees. In 732, Charles Martel encountered not a Muslim army, but a summer raiding party."

But genuine threats persisted for nearly a thousand years. If the great Timur (also known as Tamerlane), the Turkic conqueror of Samarkand, had not turned his sights toward China, and died in 1405 before his eastern move, Europe might also have fallen to his form of Islam. And the Ottoman sultans, with their trained and efficient armies, took Constantinople (now Istanbul) in 1453, and laid powerful siege to the walls of Vienna as late as 1683—a final failure that gave us the croissant as a living legacy, the breakfast roll based on the Turkish symbol of a crescent moon, and first made by Viennese bakers to celebrate their victory. (As a little footnote, remember that I have not even mentioned Attila, Genghis Khan, and several other serious threats to European domination.)

Our history could have been fashioned in a million different credible ways, and we have no adequate sense of where we are heading. But a good moral compass, combined with an intelligent use of scientific achievements, might keep us going—even prospering—for a long time by our standards (however paltry in geological perspective). We do have the resources, but can we muster both the will and judgment to hold first place in a game that can offer only possibilities, never guarantees—a game that spells oblivion for those who win the opportunity but fail to seize the moment, plunging instead into the great asymmetry of history's usual outcome?

IV

Six Little Pieces

on the

Meaning

and

Location

of Excellence

Substrate

and

Accomplishment

11

Drink Deep,
or Taste Not the
Pierian Spring

MOST FAMOUS QUOTATIONS ARE FABRICATED; AFTER all, who can concoct a high witticism at a moment of maximal stress in battle or just before death? A military commander will surely mutter a mundane "Oh shit, here they come" rather than the inspirational "Don't fire until you see the whites of their eyes." Similarly, we know many great literary lines by a standard misquotation rather than an accurate citation. Bogart never said "Play it again, Sam," and Jesus did not proclaim that "he who lives by the sword shall die by the sword." Ironically, the most famous of all quotations about the cardinal subject of our mental lives—learning—bungles the line and substitutes "knowledge" for the original. So let us restore the proper word to Alexander Pope's "Essay on Criticism":

A little learning is a dangerous thing;
Drink deep, or taste not the Pierian Spring;
There shallow draughts intoxicate the brain,
And drinking largely sobers us again.

I have a theory about the persistence of the standard misquotation, "a little knowledge is a dangerous thing," a conjecture that I can support through the embarrassment of personal testimony. I think that writers resist a full and accurate citation because they do not know the meaning of the crucial second line. What the dickens is a "Pierian Spring," and how can you explain the quotation if you don't know? So you extract the first line alone from false memory, and "learning" disappears.

To begin this short essay about learning in science, I vowed to explicate the Pierian spring so I could dare to quote this couplet that I have never cited for fear that someone would ask. And the answer turned out to be joyfully accessible—a two-minute exercise involving one false lead in an encyclopedia (reading two irrelevant articles about artists named Piero), followed by a good turn to *The Oxford English Dictionary*. Pieria, this venerable source tells us, is "a district in northern Thessaly, the reputed home of the muses." Pierian therefore becomes "an epithet of the muses; hence allusively in reference to poetry and learning."

So I started musing about learning. Doesn't my little story illustrate a general case? We fear that something we want to learn will be difficult and that we will never even figure out how to find out. And then, when we actually try, the answer comes easily—with joy in discovery, for no delight can exceed the definitive solution to a little puzzle. Easy, that is, so long as we can master the tools (not everyone enjoys immediate access to *The Oxford English Dictionary;* more sadly, most people never learned how to use this great compendium or even know that it exists). Learning can be easy because the human mind works as an intellectual sponge of astonishing porosity and voracious appetite, that is, if proper education and encouragement keep the spaces open.

A commonplace of our culture, and the complaint of teachers, holds that, of all subjects, science ranks as the most difficult to learn and therefore the scariest and least accessible of all disciplines. Science may occupy the center of our practical lives, but its content remains mysterious to nearly all Americans, who must therefore take its benefits on faith (turn on your car or computer and pray that the damned thing will work) or fear its alien powers and intrusions (will my clone steal my individuality?). We suspect that public knowledge of science may be extraordinarily shallow, both because few people show any interest or familiarity for the subject (largely through fear or from assumptions of utter incompetence) and because those who profess concern have too superficial an understanding. Therefore, to invoke Pope's topsy-turvy metaphor again, Americans shun the deep drink that sobriety requires and maintain dangerously little learning about science.

But I strongly suspect that this common, almost mantralike belief among educators represents a deep and (one might almost say, given the vital importance and fragility of education) dangerous fallacy, arising as the product of a common error in the sciences of natural history, including human sociology in this case—a false taxonomy. I believe that science is wonderfully accessible, that most people show a strong interest, and that levels of general learning stand quite high (within an admittedly anti-intellectual culture overall), but that we have mistakenly failed to include the domains of maximal public learning within the scope of science. (And like Pope, I do distinguish learning, or visceral understanding by long effort and experience, from mere knowledge, which can be mechanically copied from a book.)

I do not, of course, hold that most people have developed the highly technical skills that lead to professional competence in science. But this situation prevails for any subject or craft, even in the least arcane and mathematical of the humanities. Few Americans can play the violin in a symphony orchestra, but nearly all of us can learn to appreciate the music in a seriously intellectual way. Few can read ancient Greek or medieval Italian, but all can revel in a new translation of Homer or Dante. Similarly, few can do the mathematics of particle physics, but all can understand the basic issues behind deep questions about the ultimate nature of things and even learn the difference between a charmed quark and a top quark.

For the false taxonomy, we don't restrict adequate knowledge of music to professional players; so why do we limit understanding of science to those who live in laboratories, twirl dials, and publish papers? Taxonomies are theories of knowledge, not objective pigeonholes, hatracks, or stamp albums with places preassigned. A false taxonomy based on a bogus theory of knowledge can lead us badly astray. When Guillaume Rondelet, in his classic monograph of 1555 on the taxonomy of fishes, began his list of categories with "flat and compressed fishes," "those that dwell among the rocks," "little fishes" *(pisciculi)*, "genera of lizards," and "fishes that are almost round," he pretty much precluded any deep insight into the truly genealogical basis of historical order.

Millions of Americans love science and have learned the feel of true expertise in a chosen expression. But we do not honor these expressions by categorization within the realm of science, although we certainly should, for they encompass the chief criteria of detailed knowledge about nature and critical thinking based on logic and experience. Consider just a small list, spanning all ages and classes and including a substantial fraction of our population. If all these folks understood their engagement in *doing* science actively, democracy would shake hands with the academy, and we might learn to harvest a deep and wide-spread fascination in the service of more general education. (I thank Philip Morrison, one of America's wisest scientists and humanists, for making this argument to me many years ago, thus putting my thinking on the right track.)

1. Sophisticated knowledge about underwater ecology among tropical fish enthusiasts, mainly blue-collar males and therefore mostly invisible to professional intellectuals who tend to emerge from other social classes.

2. The horticultural experience of millions of members in thousands of garden clubs, mostly tenanted by older middle-class women.

3. The upper-class penchant for birding, safaris, and ecotourism.

4. The intimate knowledge of local natural history among millions of hunters and fishermen.

5. The astronomical learning (and experience in fields from practical lens grinding to theoretical optics) of telescope enthusiasts, with their clubs and journals.

6. The technological intuitions of amateur car mechanics, model builders, and weekend sailors.

7. Even the statistical knowledge of good poker players and racetrack touts. (The human brain works especially poorly in reasoning about probability, and I can cite no greater impediment to truly scientific thinking. But many Americans have learned to understand probability through the ultimate challenge of the pocketbook.)

8. In my favorite and clinching example, the dinosaur lore so lovingly learned by America's children. How I wish that we could quantify the mental might included in all the correct spellings of hideously complex dinosaur names among all the five-year-old children in America. Then we could truly move mountains.

Common belief could not be more ass-backward. We think that science is intrinsically hard, scary, and arcane, and that teachers can only beat the necessary knowledge, by threat and exhortation, into a small minority blessed with innate propensity. No. Most of us begin our education with an inborn love of science (which is, after all, only a method of learning the facts and principles of the natural world surrounding us, and how can anyone fail to be stirred by such an intimate subject?). This love has to be beaten *out* of us if we later fall by the wayside, perversely led to say that we hate or fear the subject. But the same love burns brightly throughout the lives of millions, who remain amateurs in the precious, literal sense of the word ("those who love") and who pursue "hobbies" in scientific fields that we falsely refuse to place within the taxonomic compass of the discipline.

And so, finally, the task of nurture and rescue falls to those people who represent what I have often called the most noble word in our language, the teachers. (*Parent* holds second place on my list; but teachers come first because parents, after an initial decision, have no further choice.) Rage (and scheme) against the dying of the light of childhood's fascination. And then emulate English literature's first instructor, the clerk of Oxenford in Chaucer's *Canterbury Tales,* the man who opened *both* ends of his mind and heart, for "gladly wolde he lerne, and gladly teche."

1 2

Requiem Eternal*

In 1764, the English savant Daines Barrington tested a visiting musical prodigy for his skills in memory, performance, composition, and improvisation. The amazed listener expressed great skepticism about his subject's stated age of eight, and wondered if father Leopold had been passing off a well-trained adult midget as his young son. Barrington therefore delayed his written account for six years until he could obtain proof, in the form of a birth certificate for Johannes Chrysostomus Wolfgangus Theophilus Mozart (later shortened to Wolfgang Amadeus by the composer himself), from an unimpeachable source: "his excellence," in Barrington's description, "Count Haslang, envoy extraordinary and minister plenipotentiary of the electors of Bavaria and Palatine."

Barrington noted that many precocious geniuses die young, and ended his article with a prayer that Mozart might live as long as

---

*Originally written for the pamphlet accompanying a Penguin CD of Mozart's *Requiem*.

Britain's most celebrated German import, G. F. Handel, who had died five years previously at age seventy-four. Barrington wrote: "It may be hoped that little Mozart may possibly attain to the same advanced years as Handel, contrary to the common observation that such *ingenia praecocia* are generally short lived."

Well, Mozart lived long enough to become Mozart, while failing to attain even half Handel's age. He died in 1791, at age thirty-five, with his greatest and final work, his *Requiem,* unfinished. He wrote the very last note of his life to the painfully appropriate text: *Lachrymosa dies illa*—this day full of tears. No musical composition has ever moved more people to tears, or inspired more mythological nonsense, including tales of a masked man commissioning the piece in secret, and a mysterious poisoner using the opportunity to end an opponent's life. (Peter Shaffer, of course, wove all these fictions together into a sublime play full of psychological truth—*Amadeus.*)

I have been a choral singer all my active life, and I love the *Requiem* with all my heart. I have sung the work at least a dozen times, spanning more years than Mozart lived (from a first undergraduate performance at age nineteen to a latest effort at age fifty-five). I do not even care to imagine how much poorer life would be without such music. As with any truly great work of human genius (I have, for example, read Darwin's *Origin of Species* once a decade, each time as an entirely fresh and different book), the *Requiem* never fails to instruct and inspire. As Shakespeare said of Cleopatra: "Age cannot wither her, nor custom stale her infinite variety."

Unpredictable contingency, not lawlike order, rules the pathways of history. A little branch called *Homo sapiens* inhabits this earth by good luck built upon incalculably small probability. Do we not all yearn for the power to tweak those probabilities just a teeny little bit—to replay the tape of history with an apparently inconsequential change that cascades to colossal effect in subsequent times? Suppose, following this greatest of undoable thought experiments, that we alter nothing until 1791, but then let Mozart live to 1830, thus matching Handel's age. Can we even contemplate the added joy, measured in quanta of pleasure for billions of people, provided by another forty symphonies and a dozen operas, perhaps on such sublime texts as *Hamlet, Faust,* or *Lear*? Can we even imagine how differently the history of music, and of human creativity in general, might have run in this ever-so-slightly altered circumstance?

We should, I think, count our blessings instead. Let us not lament an early death at less than half Handelian age. Let us rejoice that smallpox, typhoid, or rheumatic fever (all of which he suffered as a child) did not extinguish Barrington's prodigy before he could grow up to become Mozart. If he had died after *Mitridate* (a teenaged opera of indifferent status), Mozart might only have become a footnote for lamentation. Instead, we received the most sublime swan song ever written—this *Requiem,* fitted with a closing text that might well be read as a prayer of thanks for the sublime gift that Mozart gave to all humanity, and for all time, with his music: *lux aeterna,* eternal light.

13

More Power
to Him

IN 1927, WHEN MY FATHER TURNED TWELVE, AL JOLSON inaugurated the era of sound movies with *The Jazz Singer,* Jerome Kern and Oscar Hammerstein's *Show Boat* opened on Broadway, Charles Lindbergh flew the *Spirit of Saint Louis* across the Atlantic nonstop to Paris, the state of Massachusetts executed Sacco and Vanzetti, and Babe Ruth hit 60 home runs in a single season.

Roger Maris bested the Babe with 61 in 1961, the summer of my nineteenth birthday—with teammate Mickey Mantle batting just afterward, and reaching 54 in one of the two greatest home run derbies in baseball history. This summer, Mark McGwire will surely break 61, and may even reach 70* (with Sammy Sosa just behind, or perhaps in front, in the other greatest derby ever). My two sons, both fans in their different ways, will turn twenty-nine and twenty-five.

---

*I wrote this piece for *The Wall Street Journal* to honor McGwire's sixtieth homer and the certainty of his fracturing Maris's old record of 61. Since nearly every forecast I have ever made has been ludicrously wrong, I do take some pride in the only example I can cite of a personal prediction that, for reasons of pure dumb luck, happened to come up golden. McGwire ended his season with exactly 70 dingers, Sosa with 66.

This magic number, this greatest record in American sports, obsesses us for at least three good reasons. First, baseball has changed no major rule in a century, and we can therefore look and compare, in genuine continuity, across the generations. The seasons of our lives move inexorably forward. As my father saw Ruth, I followed Maris, and my sons watch McGwire. But the game also cycles in glorious sameness, as each winter of our discontent yields to another spring of opening day.

Second, baseball records have clear meaning as personal accomplishments, while marks in most team sports can be judged only as peculiar amalgams. Wilt Chamberlain once scored one hundred points in a single basketball game, but only because his teammates, for that day, elected the odd strategy of feeding him on essentially every play. Home runs are *mano a mano,* batter against pitcher.

Third, and how else can I say this, baseball is just one helluva terrific game, long paramount in American sporting myths and athletic traditions—with the power and definitiveness of a home run as the greatest icon of all. You might argue that Babe Ruth failed to catch the ecumenical spirit when he said, in his famous and moving speech at Yankee Stadium in 1947, as Major League Baseball gathered to honor its dying hero: "The only real game in the world, I think, is baseball. . . . You've got to start from way down . . . when you're six or seven . . . you've got to let it grow up with you." But who would deny the Babe's heartfelt sentiment?

As a veteran and close student of the 1961 Mantle-Maris derby, I thrill to the detailed similarity of McGwire versus Sosa. The two Yankees of 1961 embodied different primal myths about great accomplishments: Mantle, the deserving hero working all his life toward his year of destiny; Maris, the talented journeyman enjoying that one sweet interval in each man's life when everything comes together in some oddly miraculous way. (Maris never hit more than 39 in any other season.) That year, the miracle man won—and more power to him (and shame on his detractors). Fluke or destiny doesn't matter; Roger Maris did the deed.

Mark McGwire is this year's Mantle. No one since Ruth has been so destined, and no one has ever worked harder, and more single-mindedly, to harness and fulfill his gifts of brawn. He is the real item, and this is his year. No one, even Ruth, ever hit more than 50 homers in three successive seasons—as McGwire has now done. (But will anyone ever break Ruth's feat of hitting more than 40 in every year from 1920 to 1932, except for two seasons when injuries caused him to miss more than forty games? Hank Aaron, on the other hand, played as a marvel of consistency over twenty-three seasons. But he never hit more than 47 in a single year, and only once did he hit 40 or more in two successive seasons.) Sammy Sosa is this year's Maris, rising from who-knows-where to challenge the man of destiny. More power to both men.

But we rightly focus on McGwire for the eerie and awesome quality of his particular excellence. Most great records descend in small and even increments from the leader, and no single figure stands leagues ahead of all other mere mortals. The home run record used to follow this conventional pattern: Maris with 61, Ruth with 60, Ruth again with 59, Foxx, Greenberg, and McGwire (last season) with 58, and Wilson and Griffey (also last season) with 56.

But a few champions stand so far above the second-place finisher that we almost have to wonder whether such a leader really belongs within *Homo sapiens*. Consider DiMaggio's fifty-six-game hitting streak in 1941 (regarded by most sports statisticians, myself included, as the most improbable figure in the history of American athletics),* compared with second-place Keeler and Rose, both far away at 44; or Jim Thorpe's lopsided victories in both the pentathlon and decathlon of the 1912 Olympics; or, marking a single man's invention of the art of home run hitting, Babe Ruth's first high figure of 54 in 1920—for with this number he exceeded, all by his lonesome, the sum total for any other entire team in the American League!

McGwire belongs to this most select company of superhuman achievers. He may well hit 70, thus creating the same sweep of empty space that separates DiMaggio and Thorpe from their closest competitors. Moreover, the character of his blasts almost defies belief. A 400-foot home run, while not rare, deserves notice and inspires pride. The vast majority of Major League dingers fall between 300 and 400. Well, only 18 of McGwire's first 60 failed to reach 400 feet, and several have exceeded 500—a figure previously achieved only once every few years among all players combined.

When faced with such an exceptional accomplishment, we long to discover particular reasons. But I think that such a search only denotes a deep fallacy of human thought. No special reason need be sought beyond the good fortune of many effectively random moments grafted upon the guaranteed achievements of the greatest home run hitter in the history of baseball. I don't care if the thin air of Colorado encourages home runs. I don't care if expansion has diluted pitching. I don't care if the ball is livelier or the strike zone smaller. And I deeply don't care if McGwire helps himself to train by taking an over-the-counter

---

*For the details and documentation of this claim, see chapter 15.

substance regarded as legal by Major League Baseball.* (What nonsense to hold McGwire in any way accountable—simply because we fear that kids may ape him as a role model—for an issue entirely outside his call, and fully in the province of baseball's rule-makers. Let no such hypocrisy dim this greatest moment in our sporting life!)

Mark McGwire has prevailed by creating, in his own person, an ultimate fusion between the two great natural forces of luck and dedicated effort: the gift of an extraordinary body, with the skill of a steadfast dedication to training and study that can only merit the literal meaning of a wonderful word—*enthusiasm,* or "the intake of God."

---

*Don't get me started on the illogic and hypocrisy of public attitudes to drugs—a real and tragic problem fueled, rather than helped, by our false taxonomies and hyped moralisms that suppress and paralyze effective thought. McGwire (and many other ballplayers) takes androstenedione, now sold at nutrition stores, entirely legally and over the counter (and overtly advertised, not hidden in drawers and available only by request—as druggists sold condoms in my youth). If baseball eventually decides to ban the substance because it may raise testosterone levels, shall we retrospectively denounce McGwire for obeying the law of his time? Do we annul the records of all artists, intellectuals, politicians, and actors who thought that smoking enhanced their performance by calming their nerves?

De Mortuis
When
Truly Bonum

14

Bright Star
Among Billions*

As Saul despised David for receiving ten thou-
sand cheers to his own mere thousand, scientists often stigmatize, for
the same reason of simple jealousy, the good work done by colleagues for our common benefit. We live in a philistine nation filled
with Goliaths, and we know that science feeds at a public trough.
We therefore all give lip service to the need for clear and support-
ive popular presentation of our work. Why then do we downgrade
the professional reputation of colleagues who can convey the power
and beauty of science to the hearts and minds of a fascinated, if generally uninformed public?

This narrow-minded error—our own philistinism—arises in
part from our general ignorance of the long and honorable literary
tradition of popular presentation for science, and our consequent
mistake in equating popularization with trivialization, cheapening,

---

*Originally written as an editorial for *Science*, the leading professional journal of the trade—hence
the mode of address to professional researchers, rather than to the general public.

or inaccuracy. Great scientists have always produced the greatest populariza-
tions, without compromising the integrity of subject or author. In the seventeenth century, Galileo wrote both his major books as dialogues in Italian for
generally literate readers, not as formal Latin treatises designed only for scholars. In the eighteenth century, the Swiss savant J. J. Scheuchzer produced the
beautifully elaborate eight-volume *Physica sacra,* with 750 full-page copperplate
engravings illustrating the natural history behind all biblical events. In the nineteenth century, Charles Darwin wrote *The Origin of Species,* the most important
and revolutionary of all scientific works, as a book for general readers. (My students often ask me where they can find the technical monograph that served
as the basis of Darwin's popular work; I tell them that *The Origin of Species* ful-
fills both allied, not opposing, functions.)

    With the death of Carl Sagan, we have lost both a fine scientist and the
greatest popularizer of the twentieth century, if not of all time. In his many
books, and especially in his monumental television series *Cosmos*—our century's equivalent of Scheuchzer's *Physica sacra* and the most widely accessed presentation of our subject in all human history—Carl explained the method and
content of our discipline to the general public. He also conveyed the excitement of discovery with an uncanny mix of personal enthusiasm and clear presentation unequaled by any predecessor. I mourn his passing primarily because
I have lost a dear friend, but I am also sad that many scientists never appreciated his excellence or his importance to all of us, while a few of the best of us
(in a shameful incident at the National Academy of Sciences) actively rejected
him. (Carl was a remarkably sanguine man, but I know that this incident hurt
him deeply.) Too many of us never grasped his legendary service to science.

    I would epitomize Carl Sagan's excellence and integrity in three points.
First, in an age characterized by the fusion of high and pop culture, Carl moved
comfortably across the entire spectrum while never compromising scientific
content. He could joke with Johnny Carson, compose a column for *Parade,* and
write a science fiction novel while maintaining an active laboratory and publishing technical papers. He had foibles aplenty; don't we all? We joked about
his emphatic pronunciation of "billions," and my young son (much to Carl's
amusement) called *Cosmos* the "stick-head-up show" because Carl always
looked up dreamily into the heavens. But the public watched, loved, and
learned. Second, for all his pizzazz and charisma, Carl always spoke for true science against the plethora of irrationalisms that surround us. He conveyed one
consistent message: real science is so damned exciting, transforming, and prov-
able; why would anyone prefer the undocumentable nonsense of astrology, alien

abductions, and so forth? Third, he bridged the gaps between our various cultures by showing the personal, humanistic, and artistic side of scientific activity. I will never, for example, forget his excellent treatment of Hypatia, a great woman, philosopher, and mathematician, martyred in Alexandria in A.D. 415.

You had a wonderful life, Carl, but far too short. You will, however, always be with us, especially if we as a profession can learn from you how the common touch enriches science while extending an ancient tradition that lies at the heart of Western humanism, and does not represent (when properly done) a journalistic perversion of the "sound bite" age. In the words that John Dryden wrote about another great artist, the composer Henry Purcell, who died even younger in 1695: "He long ere this had tuned the jarring spheres and left no hell below."

15

The Glory
of His Time
and Ours

IN OUR SAGAS, MOURNING MAY INCLUDE CELEBRATION when the hero dies, not young and unfulfilled on the battlefield, but rich in years and replete with honor. And yet for me, the passing of Joe DiMaggio has evoked a primary feeling of sadness for something precious that cannot be restored—a loss not only of the man, but also of the splendid image that he represented.

I first saw DiMaggio play near the end of his career in 1950, when I was eight and Joe had his last great season, batting .301 with 32 homers and 122 RBIs. He became my hero, my model, and my mentor, all rolled up into one remarkable man. (I longed to be his replacement in center field, but a guy named Mantle came along and beat me out for the job.) DiMaggio remained my primary hero to the day of his death, and through all the vicissitudes of Ms. Monroe, Mr. Coffee, and Mrs. Robinson.

Even with my untutored child's eyes, I could sense something supremely special about DiMaggio's play. I didn't even know the words or their meanings, but I grasped his gracefulness in some visceral way, and I knew that an aura of majesty surrounded all his actions. He played every aspect of baseball with a fluid beauty in minimal motion, a spare elegance that made even his swinging strikeouts look beautiful (an infrequent occurrence in his career; no other leading home run hitter has ever posted more than twice as many lifetime walks as strikeouts or, even more amazingly, nearly as many homers as whiffs—361 dingers versus 369 Ks. Compare this with his two great Yankee long-ball compatriots: 714 homers and 1330 Ks for Ruth, 536 homers and 1710 Ks for Mantle).

His stance, his home run trot, those long flyouts to the cavernous left-center space in Yankee Stadium, his apparently effortless loping run—no hot dog he—to arrive under every catchable fly ball at exactly the right place and time for an "easy" out. If the sports cliché of "poetry in motion" ever held real meaning, DiMaggio must have been the intended prototype.

One cannot extract the essence of DiMaggio's special excellence from the heartless figures of his statistical accomplishments. He did not play long enough to amass leading numbers in any category—only thirteen full seasons from 1936 to 1951, with prime years lost to war, and a fierce pride that led him to retire the moment his skills began to erode.

DiMaggio sacrificed other records to the customs of his time. He hit a career high .381 in 1939, but would probably have finished well over .400 if manager Joe McCarthy hadn't insisted that he play every day in the season's meaningless last few weeks, long after the Yanks had clinched the pennant, while DiMaggio (batting .408 on September 8) then developed such serious sinus problems that he lost sight in one eye, could not visualize in three dimensions, and consequently slipped nearly 30 points in batting average. In those different days, if you could walk, you played.

DiMaggio's one transcendent numerical record—his fifty-six-game hitting streak in 1941—deserves the usual accolade of most remarkable sporting episode of the century, Mark McGwire notwithstanding. Several years ago, I performed a fancy statistical analysis on the data of slumps and streaks, and found that only DiMaggio's shouldn't have happened. All other streaks fall within the expectations for great events that should occur once as a consequence of probabilities, just as an honest coin will come up heads ten times in a row once in a very rare while. But no one should ever have hit in fifty-six straight games. Second place stands at a distant forty-four, a figure reached by Pete Rose and Wee Willie Keeler.

DiMaggio's greatest record therefore embodies pure heart, not the rare expectation of luck. We must also remember that third baseman Ken Keltner robbed DiMaggio of two hits in the fifty-seventh game, and that he then went on to hit safely in sixteen straight games thereafter. DiMaggio also compiled a sixty-one-game hit streak when he played for the San Francisco Seals in the minor Pacific Coast League.

One afternoon in 1950, I sat next to my father near the third base line in Yankee Stadium. DiMaggio fouled a ball in our direction, and my father caught it. We mailed the precious relic to the great man, and sure enough, he sent it back with his signature. That ball remains my proudest possession to this day. Forty years later, during my successful treatment for a supposedly incurable cancer, I received a small square box in the mail from a friend and book publisher in San Francisco, and a golfing partner of DiMaggio. I opened the box and found another ball, signed to me by DiMaggio (at my friend's instigation) and wishing me well in my recovery. What a thrill and privilege—to tie my beginning and middle life together through the good wishes of this great man.

Ted Williams is, appropriately, neither a modest nor a succinct man. When asked recently to compare himself with his rival and contemporary DiMaggio, the greatest batter in history simply replied: "I was a better hitter, he was a better player."

Simon and Garfunkel captured the essence of this great man in their famous lyric about the meaning and loss of true stature: "Where have you gone, Joe DiMaggio? A nation turns its lonely eyes to you."*

He was the glory of a time that we will not see again.

---

*DiMaggio, so wholly possessed of integrity and refinement both on and off the field, was also a very concrete man of few words. In his op-ed obituary for *The New York Times,* Paul Simon tells a wonderful story of his only meeting with DiMaggio and their contretemps over Mrs. Robinson:

A few years after "Mrs. Robinson" rose to No. 1 on the pop charts, I found myself dining at an Italian restaurant where DiMaggio was seated with a party of friends. I'd heard a rumor that he was upset with the song and had considered a lawsuit, so it was with some trepidation that I walked over and introduced myself as its composer. I needn't have worried: he was perfectly cordial and invited me to sit down, whereupon we immediately fell into conversation about the only subject we had in common.

"What I don't understand," he said, "is why you ask where I've gone. I just did a Mr. Coffee commercial, I'm a spokesman for the Bowery Savings Bank and I haven't gone anywhere."

1 6

This Was
a Man

WHEN MEL ALLEN, "VOICE OF THE YANKEES," DIED LAST
week,* I lost the man who ranked second only to my father for sheer
volume of attention during my childhood. (My dad, by the way, was
a Dodger fan and a Red Barber devotee.) As I considered the surprising depth of my sadness, I realized that I was mourning the
extinction of a philosophy as much as the loss of a dear man—and I
felt that most of the warm press commentary had missed the essence
of Mel Allen's strength. The eulogies focused on his three signature
phrases: his invariable opening line, "Hello there, everybody"; his
perennial exclamation of astonishment, "How about that"; and his
inevitable home run mantra, "It's going . . . going . . . gone."

But I would characterize his immense appeal by two singular
statements, one-off comments that I heard in passing moments during a distant childhood. These comments have stayed with me all
my life, for integrity in one case, and for antic humor in the other.

*This piece originally appeared in the *New York Times* on June 26, 1996.

One exemplifies the high road, the other an abyss, however charming. The comments could not be more different, but they embody, when taken together, something precious, something fragile, and something sadly lost when institutions become so large that the generic blandness of commercial immensity chokes off both spontaneity and originality. This phenomenon of modern life, by the way, is entirely general and not confined to broadcasting. In my own academic world, textbooks have become longer, duller, and entirely interchangeable for the same reason. Idiosyncratic works cannot sell sufficiently, for curricula have been standardized (partly by the sameness of conventional textbooks)—and originality guarantees oblivion. Authors have become cogs in an expensive venture that includes, among others, the photo researcher, the slide maker, the teacher's guide preparer, and the publicist. The great texts of the past defined fields for generations because they promulgated the distinctive views of brilliant authors—Lyell's geology, or Marshall's economics—but modern writers are faceless servants of a commercial machine that shuns anything unique.

One day in 1952, as Mickey Mantle struggled in center field the year after Joe DiMaggio's retirement, many fans began to boo after Mickey struck out for the second time in a row. In the midst of his play-by-play broadcast, an infuriated Mel Allen leaned out of the press box and shouted at a particularly raucous fan: "Why are you booing him?" The fan shot back: "Because he's not as good as DiMaggio." And Mel Allen busted a gut, delivering a ferocious dressing-down to the fan for his indecency in razzing an enormously talented but unformed twenty-year-old kid just because he could not yet replace the greatest player of the age.

Ballantine beer and White Owl cigars sponsored the Yankees in those years—and Mel never lost an opportunity for additional endorsement. Home runs, for example, became "Ballantine Blasts" or "White Owl Wallops," depending on the sponsor of the inning. When a potential home run passed just to the wrong side of the foul pole, Allen would exclaim, "Why that ball was just foul by a bottle of Ballantine beer." One day Mickey Mantle hit one that seemed destined for success, and Allen began his mantra: "It's going . . . going . . ." And then he stopped short as the ball went foul by no more than an inch or two. An astonished Allen exclaimed: "Why, I've never seen one miss by so little. That ball was foul by no more than a bottle of Bal—" And then he paused in mid phrase, thought for a fraction of a moment, and exclaimed: "No, that ball was foul by the ash on a White Owl cigar!"

A man of grace and integrity; a shameless huckster of charming originality. But above all, a man who could only be his wonderful cornball self—Mel Allen, the singular, inimitable, human Voice of the Yankees. So take my two stories, titrate them to the optimal distinctness of lost individuality, and let us celebrate Shakespeare's judgment in *Julius Caesar:* "The elements so mix'd in him that Nature might stand up and say to the world, 'This was a man!'"

V

Science
in
Society

17

A Tale of Two
Work Sites

CHRISTOPHER WREN, THE LEADING ARCHITECT OF LONDON'S
reconstruction after the great fire of 1666, lies buried beneath the
floor of his most famous building, St. Paul's cathedral. No elaborate
sarcophagus adorns the site. Instead, we find only the famous epitaph written by his son and now inscribed into the floor: *"si monu-
mentum requiris, circumspice"*—if you are searching for his monument,
look around. A tad grandiose perhaps, but I have never read a finer
testimony to the central importance—one might even say sacred-
ness—of actual places, rather than replicas, symbols, or other forms
of vicarious resemblance.

An odd coincidence of professional life turned my thoughts to
this most celebrated epitaph when, for the second time, I received
an office in a spot laden with history, a place still redolent of ghosts
of past events both central to our common culture and especially
meaningful for my own life and choices.

In 1971, I spent an academic term as a visiting researcher at

Oxford University. I received a cranny of office space on the upper floor of the University Museum. As I set up my books, fossil snails, and microscope, I noticed a metal plaque affixed to the wall, informing me that this reconfigured space of shelves and cubicles had been, originally, the site of the most famous public confrontation in the early history of Darwinism. On this very spot, in 1860, just a few months after Darwin published *The Origin of Species,* T. H. Huxley had drawn his rhetorical sword, and soundly skewered the slick but superficial champion of creationism, Bishop "Soapy Sam" Wilberforce.

(As with most legends, the official version ranks as mere cardboard before a much more complicated and multifaceted truth. Wilberforce and Huxley did put on a splendid, and largely spontaneous, show—but no clear victor emerged from the scuffle, and Joseph Hooker, Darwin's other champion, made a much more effective reply to the bishop, however forgotten by history. See my essay on this debate, entitled "Knight Takes Bishop?" and published in an earlier volume of this series, *Bully for Brontosaurus.*)

I can't claim that the lingering presence of these Victorian giants increased my resolve or improved my work, but I loved the sense of continuity vouchsafed to me by this happy circumstance. I even treasured the etymology—for *circumstance* means "standing around" (as Wren's *circumspice* means "looking around"), and here I stood, perhaps in the very spot where Huxley had said, at least according to legend, that he preferred an honest ape for an ancestor to a bishop who would distort a known truth for rhetorical advantage.

Not so long ago, I received a part-time appointment as visiting research professor of biology at New York University. I was given an office on the tenth floor of the Brown building on Washington Place, a nondescript early-twentieth-century structure now filled with laboratories and other academic spaces. As the dean took me on a casual tour of my new digs, he made a passing remark, intended as little more than "tour-guide patter," but producing an electric effect upon his new tenant. Did I know, he asked, that this building had been the site of the infamous Triangle Shirtwaist fire of 1911, and that my office occupied a corner location on one of the affected floors—in fact, as I later discovered, right near the escape route used by many workers to safety on the roof above. The dean also told me that, each year on the March 25 anniversary of the fire, the International Ladies' Garment Workers Union still holds a ceremony at the site and lays wreaths to memorialize the 146 workers killed in the blaze.

If the debate between Huxley and Wilberforce defines a primary legend of my chosen profession, the Triangle Shirtwaist fire occupies an even more central place in my larger view of life. I grew up in a family of Jewish immigrant garment workers, and this holocaust (in the literal meaning of a thorough sacrifice by burning) had set their views and helped to define their futures.

The shirtwaist—a collared blouse designed on the model of a man's shirt and worn above a separate skirt—had become the fashionable symbol of more independent women. The Triangle Shirtwaist Company, New York City's largest manufacturer of shirtwaists, occupied three floors (eighth through tenth) of the Asch Building (later bought by New York University and rechristened Brown, partly to blot out the infamy of association with the fire). The company employed some five hundred workers, nearly all young women who had recently arrived either as Jewish immigrants from eastern Europe or as Catholics from Italy. Exits from the building, in addition to elevators, included only two small stairways and one absurdly inadequate fire escape. But the owners had violated no codes, both because general standards of regulation were then so weak, and because the structure was supposedly fireproof—as the framework proved to be (for the building, with my office, still stands), though inflammable walls and ceilings could not prevent an internal blaze on floors crammed full of garments and cuttings. The Triangle company was, in fact, a deathtrap—for fire hoses of the day could not pump above the sixth floor, while nets and blankets could not sustain the force of a human body jumping from greater heights.

The fire broke out at quitting time. Most workers managed to escape by the elevators, down one staircase (we shall come to the other staircase later), or by running up to the roof. But the flames trapped 146 employees, nearly all young women. About fifty workers met a hideous, if dramatic, end by jumping in terror from the ninth-floor windows, as a wall of fire advanced from behind. Firemen and bystanders begged them not to jump, and then tried to hold improvised nets of sheets and blankets. But these professionals and good Samaritans could not hold the nets against the force of fall, and many bodies plunged right through the flimsy fabrics onto the pavement below, or even through the "hollow sidewalks" made of opaque glass circles designed to transmit daylight to basements below, and still a major (and attractive) feature of my SoHo neighborhood. (These sidewalks carry prominent signs warning delivery trucks not to back in.) Not a single jumper survived, and the memory of these forced leaps to death remains the most searing image of America's prototypical sweatshop tragedy.

All defining events of history develop simplified legends as official versions—primarily, I suppose, because we commandeer such events for shorthand moral instruction, and the complex messiness of actual truth always blurs the clarity of a pithy epigram. Thus, Huxley, representing the righteousness of scientific objectivity, must slay the dragon of ancient and unthinking dogma. The equally oversimplified legend of the Triangle fire holds that workers became trapped because management had locked all the exit doors to prevent pilfering, unscheduled breaks, or access to union organizers—leaving only the fire escape as a mode of exit. All five of my guidebooks to New York architecture tell this "official" version. My favorite book, for example, states: "Although the building was equipped with fire exits, the terrified workers discovered to their horror that the ninth-floor doors had been locked by supervisors. A single fire-escape was wholly inadequate for the crush of panic-stricken employees."

These traditional (indeed, virtually "official") legends may exaggerate for moral punch, but such interpretations emerge from a factual basis of greater ambiguity—and this reality, as we shall see in the Triangle case, often embodies a deeper and more important lesson. Huxley did argue with Wilberforce, after all, even if he secured no decisive victory, and Huxley did represent the side of the angels—the true angels of light and justice. And although many Triangle workers escaped by elevators and one staircase, another staircase (that might have saved nearly everyone else) was almost surely locked.

If Wilberforce and his minions had won, I might be a laborer, a linguist, or a lawyer today. But the Triangle fire might have blotted me out entirely. My grandmother arrived in America in 1910. On that fatal March day in 1911, she was working as a sixteen-year-old seamstress in a sweatshop—but, thank God, not for the Triangle Shirtwaist Company. My grandfather, at the same moment, was cutting cloth in yet another nearby factory.

These two utterly disparate stories—half a century and an ocean apart, and with maximal contrast between an industrial tragedy and an academic debate— might seem to embody the most unrelatable of items: the apples and oranges, or chalk and cheese (the British version), of our mottoes. Yet I feel that an intimate bond ties these two stories together in illustrating opposite poles of a central issue in the history of evolutionary theory: the application of Darwinian thought to the life and times of our own troubled species. I claim nothing beyond personal meaning—and certainly no rationale for boring anyone else— in the accidental location of my two offices in such sacred spots of history. But the emotion of a personal prod often dislodges a general theme well worth sharing.

The application of evolutionary theory to *Homo sapiens* has always troubled Western culture deeply—not for any reason that might be called scientific (for humans are biological objects, and must therefore take their place with all other living creatures on the genealogical tree of life), but only as a consequence of ancient prejudices about human distinctiveness and unbridgeable superiority. Even Darwin tiptoed lightly across this subject when he wrote *The Origin of Species* in 1859 (though he plunged in later, in 1871, by publishing *The Descent of Man*). The first edition of the *Origin* says little about *Homo sapiens* beyond a cryptic promise that "light will be thrown on the origin of man and his history." (Darwin became a bit bolder in later editions and ventured the following emendation: "Much light will be thrown . . .")

Troubling issues of this sort often find their unsurprising resolution in a bit of wisdom that has permeated our traditions from such sublime sources as Aristotle's *aurea mediocritas* (or golden mean) to the vernacular sensibility of Goldilocks's decisions to split the difference between two extremes, and find a solution "just right" in the middle. Similarly, one can ask either too little or too much of Darwinism in trying to understand "the origin of man and his history." As usual, a proper solution lies in the intermediary position of "a great deal, but not everything." Soapy Sam Wilberforce and the Triangle Shirtwaist fire gain their odd but sensible conjunction as illustrations of the two extremes that must be avoided—for Wilberforce denied evolution altogether and absolutely, while the major social theory that hindered industrial reform (and permitted conditions that led to such disasters as the Triangle Shirtwaist fire) followed the most overextended application of biological evolution to patterns of human history—the theory of "Social Darwinism." By understanding the fallacies of Wilberforce's denial and social Darwinism's uncritical and total embrace, we may find the proper balance between.

They didn't call him Soapy Sam for nothing. The orotund bishop of Oxford saved his finest invective for Darwin's attempt to apply his heresies to human origins. In his review of *The Origin of Species* (published in the *Quarterly Review,* England's leading literary journal, in 1860), Wilberforce complained above all: "First, then, he not obscurely declares that he applies his scheme of the action of the principle of natural selection to Man himself, as well as to the animals around him." Wilberforce then uncorked a passionate argument for a human uniqueness that could only have been divinely ordained:

> Man's derived supremacy over the earth; man's power of articulate speech; man's gift of reason; man's free-will and responsibility; man's fall and man's redemption; the incarnation of the Eternal Son; the indwelling of the Eternal Spirit,—all are equally and

utterly irreconcilable with the degrading notion of the brute origin of him who was created in the image of God, and redeemed by the Eternal Son.

But the tide of history quickly engulfed the good bishop. When Wilberforce died in 1873, from a head injury after a fall from his horse, Huxley acerbically remarked that, for once, the bishop's brains had come into contact with reality—and the result had been fatal. Darwinism became the reigning intellectual novelty of the late nineteenth century. The potential domain of natural selection, Darwin's chief explanatory principle, seemed nearly endless to his devotees (though not, interestingly, to the master himself, as Darwin remained cautious about extensions beyond the realm of biological evolution). If a "struggle for existence" regulated the evolution of organisms, wouldn't a similar principle also explain the history of just about anything—from the cosmology of the universe, to the languages, economics, technologies, and cultural histories of human groups?

Even the greatest of truths can be overextended by zealous and uncritical acolytes. Natural selection may be one of the most powerful ideas ever developed in science, but only certain kinds of systems can be regulated by such a process, and Darwin's principle cannot explain all natural sequences that develop historically. For example, we may talk about the "evolution" of a star through a predictable series of phases over many billion years from birth to explosion, but natural selection—a process driven by the differential survival and reproductive success of some individuals in a variable population—cannot be the cause of stellar development. We must look, instead, to the inherent physics and chemistry of light elements in such large masses.

Similarly, although Darwinism surely explains many universal features of human form and behavior, we cannot invoke natural selection as the controlling cause of our cultural changes since the dawn of agriculture—if only because such a limited time of some ten thousand years provides so little scope for any general biological evolution at all. Moreover, and most importantly, human cultural change operates in a manner that precludes a controlling role for natural selection. To mention the two most obvious differences: first, biological evolution proceeds by continuous division of species into independent lineages that must remain forever separated on the branching tree of life. Human cultural change works by the opposite process of borrowing and amalgamation. One good look at another culture's wheel or alphabet may alter the course of a civilization forever. If we wish to identify a biological analog for cultural change, I suspect that infection will work much better than evolution.

Second, human cultural change runs by the powerful mechanism of Lamarckian inheritance of acquired characters. Anything useful (or alas, destructive) that our generation invents can be passed directly to our offspring by direct education. Change in this rapid Lamarckian mode easily overwhelms the much slower process of Darwinian natural selection, which requires a Mendelian form of inheritance based on small-scale and undirected variation that can then be sifted and sorted through a struggle for existence. Genetic variation is Mendelian, so Darwinism rules biological evolution. But cultural variation is largely Lamarckian, and natural selection cannot determine the recent history of our technological societies.

Nonetheless, the first blush of high Victorian enthusiasm for Darwinism inspired a rush of attempted extensions to other fields, at least by analogy. Some efforts proved fruitful, including the decision of James Murray, editor of *The Oxford English Dictionary* (first volume published in 1884, but under way for twenty years before then), to work strictly by historical principles and to treat the changing definitions of words not by current preferences in use (as in a truly normative dictionary), but by the chronology and branching evolution of recorded meanings (making the text more an encyclopedia about the history of words than a true dictionary).

But other extensions proved both invalid in theory, and also (or so most of us would judge by modern moral sensibilities) harmful, if not tragic, in application. As the chief offender in this category, we must cite a highly influential theory that acquired the inappropriate name of "Social Darwinism." (As many historians have noted, this theory should really be called "social Spencerism," since Herbert Spencer, chief Victorian pundit of nearly everything, laid out all the basic postulates in his *Social Statics* of 1850, nearly a decade before Darwin published *The Origin of Species.* Darwinism did add the mechanism of natural selection as a harsher version of the struggle for existence, long recognized by Spencer. Moreover, Darwin himself maintained a highly ambivalent relationship to this movement that borrowed his name. He felt the pride of any creator toward useful extensions of his theory—and he did hope for an evolutionary account of human origins and historical patterns. But he also understood only too well why the mechanism of natural selection applied poorly to the causes of social change in humans.)

Social Darwinism often serves as a blanket term for any genetic or biological claim made about the inevitability (or at least the "naturalness") of social inequalities among classes and sexes, or military conquests of one group by another. But such a broad definition distorts the history of this important subject—although pseudo-Darwinian arguments have long been advanced, prominently and forcefully, to cover all these sins. Classical Social Darwinism operated as a more specific theory about the nature and origin of social classes in the modern industrial world. The *Encyclopaedia Britannica* article on this subject correctly emphasizes this restriction by first citing the broadest range of potential meaning, and then properly narrowing the scope of actual usage:

> *Social Darwinism:* the theory that persons, groups, and races are sub-ject to the same laws of natural selection as Charles Darwin had per-ceived in plants and animals in nature. . . . The theory was used to support laissez-faire capitalism and political conservatism. Class stratification was justified on the basis of "natural" inequalities among individuals, for the control of property was said to be a cor-relate of superior and inherent moral attributes such as industrious-ness, temperance, and frugality. Attempts to reform society through state intervention or other means would, therefore, interfere with natural processes; unrestricted competition and defense of the status quo were in accord with biological selection. The poor were the "unfit" and should not be aided; in the struggle for existence, wealth was a sign of success.

Spencer believed that we must permit and welcome such harshness to unleash the progressive development that all "evolutionary" systems undergo if allowed to follow their natural course in an unimpeded manner. As a central principle of his system, Spencer believed that progress—defined by him as movement from a simple undifferentiated homogeneity, as in a bacterium or a "primitive" human society without social classes, to complex and structured heterogeneity, as in "advanced" organisms or industrial societies—did not arise as an inevitable property of matter in motion, but only through interaction between evolving systems and their environments. These interactions must therefore not be obstructed.

The relationship of Spencer's general vision to Darwin's particular theory has often been misconstrued or overemphasized. As stated above, Spencer had published the outline (and most of the details) of his system nearly ten years before Darwin presented his evolutionary theory. Spencer certainly did wel-come the principle of natural selection as an even more ruthless and efficient mechanism for driving evolution forward. (Ironically, the word *evolution,* as a description for the genealogical history of life, entered our language through Spencer's urgings, not from Darwin. Spencer favored the term for its vernacu-lar English meaning of "progress," in the original Latin sense of *evolutio,* or "unfolding." At first, Darwin resisted the term—he originally called his process "descent with modification"—because his theory included no mechanism or rationale for general progress in the history of life. But Spencer prevailed, largely because no society has ever been more committed to progress as a central notion or goal than Victorian Britain at the height of its colonial and industrial expansion.)

Spencer certainly used Darwin's mechanism of natural selection to buttress his system. Few people recognize the following historical irony: Spencer, not Darwin, coined the term "survival of the fittest," now our conventional catchphrase for Darwin's mechanism. Darwin himself paid proper tribute in a statement added to later editions of *The Origin of Species:* "I have called this principle, by which each slight variation, if useful, is preserved, by the term Natural Selection. . . . But the expression often used by Mr. Herbert Spencer of the Survival of the Fittest is more accurate, and is sometimes equally convenient."

As a mechanism for driving his universal "evolution" (of stars, species, languages, economics, technologies, and nearly anything else) toward progress, Spencer preferred the direct and mechanistic "root, hog, or die" of natural selection (as William Graham Sumner, the leading American social Darwinian, epitomized the process), to the vaguer and largely Lamarckian drive toward organic self-improvement that Spencer had originally favored as a primary cause. (In this colorful image, Sumner cited a quintessential American metaphor of self-sufficiency that my dictionary of catchphrases traces to a speech by Davy Crockett in 1834.) In a post-Darwinian edition of his *Social Statics,* Spencer wrote:

> The lapse of a third of a century since these passages were published, has brought me no reason for retreating from the position taken up in them. Contrariwise, it has brought a vast amount of evidence strengthening that position. The beneficial results of the survival of the fittest, prove to be immeasurably greater than [I formerly recognized]. The process of "natural selection," as Mr. Darwin called it . . . has shown to be a chief cause . . . of that evolution through which all living things, beginning with the lower, and diverging and re-diverging as they evolved, have reached their present degrees of organization and adaptation to their modes of life.

But putting aside the question of Darwin's particular influence, the more important, underlying point remains firm: the theory of Social Darwinism (or social Spencerism) rests upon a set of analogies between the causes of change and stability in biological and social systems—and on the supposedly direct applicability of these biological principles to the social realm. In his founding document, the *Social Statics* of 1850, Spencer rests his case upon two elaborate analogies to biological systems.

1. The struggle for existence as purification in biology and society. Darwin recognized the "struggle for existence" as metaphorical shorthand for any strategy that promotes increased reproductive success, whether by outright battle, cooperation, or just simple prowess in copulation under the old principle of "early and often." But many contemporaries, including Spencer, read "survival of the fittest" only as overt struggle to the death—what T. H. Huxley later dismissed as the "gladiatorial" school, or the incarnation of Hobbes's *bellum omnium contra omnes* (the war of all against all). Spencer presented this stark and limited view of nature in his *Social Statics:*

> Pervading all Nature we may see at work a stern discipline which is a little cruel that it may be very kind. That state of universal warfare maintained throughout the lower creation, to the great perplexity of many worthy people, is at bottom the most merciful provision which the circumstances admit of. . . . Note that carnivorous enemies, not only remove from herbivorous herds individuals past their prime, but also weed out the sickly, the malformed, and the least fleet or powerful. By the aid of which purifying process . . . all vitiation of the race through the multiplication of its inferior samples is prevented; and the maintenance of a constitution completely adapted to surrounding conditions, and therefore most productive of happiness, is ensured.

Spencer then compounds this error by applying the same argument to human social history, without ever questioning the validity of such analogical transfer. Railing against all governmental programs for social amelioration—Spencer opposed state-supported education, postal services, regulation of housing conditions, and even public construction of sanitary systems—he castigated such efforts as born of good intentions but doomed to dire consequences by enhancing the survival of social dregs who should be allowed to perish for the good of all. (Spencer insisted, however, that he did not oppose private charity, primarily for the salutary effect of such giving upon the moral development of

donors. Does this discourse remind you of arguments now advanced as refor-
matory and spanking-new by our "modern" ultraconservatives? Shall we not
profit from Santayana's famous dictum that those ignorant of history must be
condemned to repeat it?) In his chapter on poor laws (which he, of course,
opposed), Spencer wrote in the *Social Statics:*

> We must call those spurious philanthropists who, to prevent present
> misery, would entail greater misery on future generations. That rig-
> orous necessity which, when allowed to operate, becomes so sharp
> a spur to the lazy and so strong a bridle to the random, these pau-
> pers' friends would repeal, because of the wailings it here and there
> produces. Blind to the fact that under the natural order of things
> society is constantly excreting its unhealthy, imbecile, slow, vacillat-
> ing, faithless members, these unthinking, though well-meaning, men
> advocate an interference which not only stops the purifying process,
> but even increases the vitiation—absolutely encouraging the mul-
> tiplication of the reckless and incompetent by offering them an
> unfailing provision. . . . Thus, in their eagerness to prevent the salu-
> tary sufferings that surround us, these sigh-wise and groan-foolish
> people bequeath to posterity a continually increasing curse.

2. The stable body and the stable society. In the universal and progressive
"evolution" of all systems, organization becomes increasingly more complex
by division of labor among the growing number of differentiating parts. All
parts must "know their place" and play their appointed role, lest the entire system collapse. A primitive hydra, constructed of simple "all purpose" modules,
can regrow any lost part, but nature gives a man only one head, and one
chance. Spencer recognized the basic inconsistency in validating social stability by analogy to the integrated needs of a single organic body—for he recognized the contrary rationales of the two systems: the parts of a body serve
the totality, but the social totality (the state) supposedly exists only to serve the
parts (individual people). But Spencer never allowed himself to be fazed by
logical or empirical difficulties when pursuing such a lovely generality.
(Huxley was speaking about Spencer's penchant for building grandiose systems when he made his famous remark about "a beautiful theory, killed by a
nasty, ugly little fact.") So Spencer barged right through the numerous absur-
dities of such a comparison, and even claimed that he had found a virtue in
the differences. In his famous 1860 article, "The Social Organism," Spencer

described the comparison between a human body and a human society: "Such, then, are the points of analogy and the points of difference. May we not say that the points of difference serve but to bring into clearer light the points of analogy."

Spencer's article then lists the supposed points of valid comparison, including such far-fetched analogies as the historical origin of a middle class to the development, in complex animals, of the mesoderm, or third body layer between the original ectoderm and endoderm; the likening of the ectoderm itself to the upper classes, for sensory organs that direct an animal arise in ectoderm, while organs of production, for such activities as digesting food, emerge from the lower layer, or endoderm; the comparison of blood and money; the parallel courses of nerve and blood vessels in higher animals with the side-by-side construction of railways and telegraph wires; and finally, in a comparison that even Spencer regarded as forced, the likening of a primitive all-powerful monarchy with a simple brain, and an advanced parliamentary system with a complex brain composed of several lobes. Spencer wrote: "Strange as this assertion will be thought, our Houses of Parliament discharge in the social economy, functions that are in sundry respects comparable to those discharged by the cerebral masses in a vertebrate animal."

Spencer surely forced his analogies, but his social intent could not have been more clear: a stable society requires that all roles be filled and well executed—and government must not interfere with a natural process of sorting out and allocation of appropriate rewards. A humble worker must toil, and may remain indigent forever, but the industrious poor, as an organ of the social body, must always be with us:

> Let the factory hands be put on short time, and immediately the colonial produce markets of London and Liverpool are depressed. The shopkeeper is busy or otherwise, according to the amount of the wheat crop. And a potato-blight may ruin dealers in consols. . . . This union of many men into one community—this increasing mutual dependence of units which were originally independent—this gradual segregation of citizens into separate bodies with reciprocally-subservient functions—this formation of a whole consisting of unlike parts—this growth of an organism, of which one portion cannot be injured without the rest feeling it—may all be generalized under the law of individuation.

Social Darwinism grew into a major movement, with political, academic, and journalistic advocates for a wide array of particular causes. But as historian Richard Hofstadter stated in the most famous book ever written on this subject—*Social Darwinism in American Thought,* first published in 1944, in press ever since, and still full of insight despite some inevitable archaisms—the primary impact of this doctrine lay in its buttressing of conservative political philosophies, particularly through the central (and highly effective) argument against state support of social services and governmental regulation of industry and housing:

> One might, like William Graham Sumner, take a pessimistic view of the import of Darwinism, and conclude that Darwinism could serve only to cause men to face up to the inherent hardship of the battle of life; or one might, like Herbert Spencer, promise that, whatever the immediate hardships for a large portion of mankind, evolution meant progress and thus assured that the whole process of life was tending toward some very remote but altogether glorious consummation. But in either case the conclusions to which Darwinism was at first put were conservative conclusions. They suggested that all attempts to reform social processes were efforts to remedy the irremediable, that they interfered with the wisdom of nature, that they could lead only to degeneration.

The industrial magnates of America's gilded age ("robber barons," in a terminology favored by many people) adored and promoted this argument against regulation, evidently for self-serving reasons, and however frequently they mixed their lines about nature's cruel inevitability with standard Christian piety. John D. Rockefeller stated in a Sunday school address:

> The growth of a large business is merely a survival of the fittest. . . . The American Beauty rose can be produced in the splendor and fragrance which bring cheer to its beholder only by sacrificing the early buds which grow up around it. This is not an evil tendency in business. It is merely the working-out of a law of nature and a law of God.

And Andrew Carnegie, who had been sorely distressed by the apparent failure of Christian values, found his solution in Spencer's writings, and then sought out the English philosopher for friendship and substantial favors. Carnegie wrote about his discovery of Spencer's work: "I remember that light came as in a flood and all was clear. Not only had I got rid of theology and the supernatural, but I had found the truth of evolution. 'All is well since all grows better' became my motto, and true source of comfort." Carnegie's philanthropy, primarily to libraries and universities, ranks as one of the great charitable acts of American history, but we should not forget his ruthlessness and resistance to reforms for his own workers (particularly his violent breakup of the Homestead strike of 1892) in building his empire of steel—a harshness that he defended with the usual Spencerian line that any state regulation must derail an inexorable natural process eventually leading to progress for all. In his most famous essay (entitled "Wealth," and published in *North American Review* for 1889), Carnegie stated:

> While the law may be sometimes hard for the individual, it is best for the race, because it insures the survival of the fittest in every department. We accept and welcome, therefore, as conditions to which we must accommodate ourselves, great inequality of environment, the concentration of wealth, business, industrial and commercial, in the hands of a few, and the law of competition between these, as being not only beneficial, but essential for the future progress of the race.

I don't want to advocate a foolishly grandiose view about the social and political influence of academic arguments—and I also wish to avoid the common fallacy of inferring a causal connection from a correlation. Of course I do not believe that the claims of Social Darwinism directly caused the ills of unrestrained industrial capitalism and the suppression of workers' rights. I know that most of these Spencerian lines functioned as mere window dressing for social forces well in place, and largely unmovable by any academic argument.

On the other hand, academic arguments should not be regarded as entirely impotent either—for why else would people in power invoke such claims so forcefully? The general thrust of social change unfolded in its own complex manner without much impact from purely intellectual rationales, but many particular issues—especially the actual rates and styles of changes that would have eventually occurred in any case—could be substantially affected by academic discourse. Millions of people suffered when a given reform experienced years of legislative delay, and then became vitiated in legal battles and compromises. The Social Darwinian argument of the superrich and the highly conservative did stem, weaken, and slow the tides of amelioration, particularly for workers' rights.

Most historians would agree that the single most effective doctrine of Social Darwinism lay in Spencer's own centerpiece—the argument against state-enforced standards for industry, education, medicine, housing, public sanitation, and so on. Few Americans, even the robber barons, would go so far, but Spencerian dogma did become a powerful bludgeon against the regulation of industry to ensure better working conditions for laborers. On this particular point—the central recommendation of Spencer's system from the beginning—we may argue for a substantial effect of academic writing upon the actual path of history.

Armed with this perspective, we may return to the Triangle Shirtwaist fire, the deaths of 146 young workers, and the palpable influence of a doctrine that applied too much of the wrong version of Darwinism to human history. The battle for increased safety of workplaces, and healthier environments for workers, had been waged with intensity for several decades. The trade union movement put substantial priority upon these issues, and management had often reacted with intransigence, or even violence, citing their Spencerian rationale for the perpetuation of apparent cruelty. Government regulation of industry had become a major struggle of American political life—and the cause of benevolent state oversight had advanced from the Sherman Anti-Trust Act of 1890 to the numerous and crusading reforms of Theodore Roosevelt's presidency (1901–9). When the Triangle fire broke out in 1911, regulations for the health and safety of workers were so weak, and so unenforceable by tiny and underpaid staffs, that the company's managers—cynically and technically "up to code" in their firetrap building—could pretty much impose whatever the weak and nascent labor union movement couldn't prevent.

If the standard legend were true—and the Triangle workers died because all the doors had been locked by cruel owners—then this heart-wrenching story might convey no moral beyond the personal guilt of management. But the loss of 146 lives occurred for much more complicated reasons, all united by the pathetic weakness of legal regulations for the health and safety of workers. And I do not doubt that the central thrust of Social Darwinism—the argument that governmental regulation can only forestall a necessary and natural process—exerted a major impact in slowing the passage of laws that almost everyone today, even our archconservatives, regard as beneficial and humane. I accept that these regulations would eventually have been instituted even if Spencer had never been born—but life or death for the Triangle workers rode upon the "detail" that forces of pure laissez-faire, buttressed by their Spencerian centerpiece, managed to delay some implementations to the 1920s, rather than acceding to the just demands of unions and social reformers in 1910.

One of the two Triangle stairways almost surely had been locked on that fateful day—although lawyers for company owners won acquittal of their clients on this issue, largely by using legal legerdemain to confuse, intimidate, and draw inconsistencies from young witnesses with poor command of English. Two years earlier, an important strike had begun at the Triangle company, and had spread to shirtwaist manufacturers throughout the city. The union won in most factories but not, ironically, at Triangle—where management held out, and compelled the return of workers without anything gained. Tensions remained high at Triangle in 1911, and management had become particularly suspicious, even paranoid, about thefts. Therefore, at quitting time (when the fire erupted, and against weakly enforced laws for maintaining multiple active exits), managers had locked one of the doors to force all the women to exit by the Greene Street stairwell, where a supervisor could inspect every handbag to guard against thefts of shirtwaists.

But if the bosses broke a weak and unenforceable law in this instance, all other causes of death can be traced to managerial compliance with absurdly inadequate standards, largely kept so weak by active political resistance to legal regulation of work sites, buttressed by the argument of Social Darwinism. Fire hoses could not pump above the sixth floor, but no law prevented the massing of workers into crowded floors above. No statute required fire drills or other forms of training for safety. In other cases, weak regulations were risibly inadequate, easy to flaunt, and basically unenforced in any case. For example, by law, each worker required 250 cubic feet of air space—a good rule to prevent crowding. But companies had managed to circumvent the intent of this law, and maintain their traditional (and dangerous) density of workers, by moving into large loft buildings with high ceilings and substantial irrelevant space that could be included in calculating the 250-cubic-foot minimum.

When the Asch Building opened in 1900, an inspector for the Buildings Department informed the architect that a third staircase should be provided. But the architect sought and received a variance, arguing that the single fire escape could count as the missing staircase required by law for structures with more than ten thousand square feet per floor. Moreover, the single fire escape— which buckled and fell during the fire, as a result of poor maintenance and the weight of too many workers trying to escape—led only to a glass skylight in a closed courtyard. The building inspector had also complained about this arrangement, and the architect had promised to make the necessary alterations.

But no changes had been made, and the falling fire escape plunged right through the skylight, greatly increasing the death toll.

Two final quotations highlight the case for inadequate legal protection as a primary cause for the unconscionable death toll in the Triangle Shirtwaist fire (Leon Stein's excellent book, *The Triangle Fire,* J. B. Lippincott Company, 1962, served as my chief source for information about this event). Rose Safran, a survivor of the fire and supporter of the 1909 strike, said: "If the union had won we would have been safe. Two of our demands were for adequate fire escapes and for open doors from the factories to the street. But the bosses defeated us and we didn't get the open doors or the better fire escapes. So our friends are dead." A building inspector who had actually written to the Triangle management just a few months before, asking for an appointment to discuss the initiation of fire drills, commented after the blaze: "There are only two or three factories in the city where fire drills are in use. In some of them where I have installed the system myself, the owners have discontinued it. The neglect of factory owners in the matter of safety of their employees is absolutely criminal. One man whom I advised to install a fire drill replied to me: 'Let 'em burn. They're a lot of cattle, anyway.'"

The Triangle fire galvanized the workers' reform movement as never before. An empowered force, now irresistible, of labor organizers, social reformers, and liberal legislators pressed for stronger regulation under the theme of "never again." Hundreds of laws passed as a direct result of this belated agitation. But nothing could wash the blood of 146 workers from the sidewalks of New York.

This tale of two work sites—of a desk situated where Huxley debated Wilberforce, and an office built on a floor that burned during the Triangle Shirtwaist fire—has no end, for the story illustrates a theme of human intellectual life that must always be with us, however imbued with an obvious and uncontroversial solution. Extremes must usually be regarded as untenable, even dangerous places on complex and subtle continua. For the application of Darwinian theory to human history, Wilberforce's "none" marks an error of equal magnitude with the "all" of an extreme Social Darwinism. In a larger sense, the evolution of a species like *Homo sapiens* should fill us with notions of glory for our odd mental uniqueness, and of deep humility for our status as a tiny and accidental twig on such a sturdy and luxuriantly branching tree of life. Glory *and* humility! Since we can't abandon either feeling for a unitary stance in the middle, we had best make sure that both attitudes *always* walk together, hand in hand, and secure in the wisdom of Ruth's promise to Naomi: "Whither thou goest, I will go; and where thou lodgest, I will lodge."

18

The
Internal Brand of
the Scarlet W

As a setting for an initial welcome to a new home, the international arrivals hall at Kennedy airport pales before the spaciousness, the open air, and the symbol of fellowship in New York's harbor. But the plaque that greets airborne immigrants of our time shares one feature with the great lady who graced the arrival of so many seaborne ancestors, including all my grandparents in their childhood. The plaque on Kennedy's wall and the pedestal of the Statue of Liberty bear the same inscription: Emma Lazarus's poem "The New Colossus"—but with one crucial difference. The Kennedy version reads:

> *Give me your tired, your poor,*
> *Your huddled masses yearning to breathe free . . .*
> *Send these, the homeless, tempest-tossed to me:*
> *I lift my lamp beside the golden door.*

One might be excused for supposing that the elision represents a large and necessary omission to fit the essence of a longer poem onto a smallish plaque. But only one line, easily accommodated, has been cut—and for a reason that can only reflect thoughtless (as opposed to merely ugly) censorship, therefore inviting a double indictment on independent charges of stupidity *and* cowardice. (As a member of the last public school generation trained by forced memorization of a holy historical canon, including the Gettysburg Address, the preamble to the Constitution, Mr. Emerson on the rude bridge that arched the flood, and Ms. Lazarus on the big lady with the lamp, I caught the deletion right away, and got sufficiently annoyed to write a *New York Times* op-ed piece a couple of years ago. Obviously, I am still seething, but at least I now have the perverse pleasure of using the story for my own benefit to introduce this essay.) I therefore restore the missing line (along with Emma Lazarus's rhyming scheme and syntax):

> *The wretched refuse of your teeming shore*

Evidently, the transient wind of political correctness precludes such a phrase as "wretched refuse," lest any visitor read the line too literally or personally. Did the authorities at our Port Authority ever learn about metaphor, and its prominence in poetry? Did they ever consider that Ms. Lazarus might be describing the disdain of a foreign elite toward immigrants whom we would welcome, nurture, and value?

This story embodies a double irony that prompted my retelling. We hide Emma Lazarus's line today because we misread her true intention, and because contemporary culture has so confused (and often even equated) inappropriate words with ugly deeds. But the authorities of an earlier generation invoked the false and literal meaning—the identification of most immigrants as wretched refuse—to accomplish a deletion of persons rather than words. The supposed genetic inferiority of most refugees (an innate wretchedness that American opportunity could never overcome) became an effective rallying cry for a movement that did succeed in imposing strong restrictions upon immigration, beginning in the 1920s. These laws, strictly enforced despite pleas for timely exception, immured thousands of Europeans who sought asylum because Hitler's racial laws had marked them for death, while our national quotas on immigration precluded any addition of their kind. These two stories of past exclusion and truncated present welcome surely illustrate the familiar historical dictum that significant events tend to repeat themselves with an ironic difference—the first time as tragedy, the second as farce.

In 1925, Charles B. Davenport, one of America's foremost geneticists, wrote to his friend Madison Grant, the author of a best-selling book, *The Passing of the Great Race,* on the dilution of America's old (read northern European, not Indian) blood by recent immigration: "Our ancestors drove Baptists from Massachusetts Bay into Rhode Island, but we have no place to drive the Jews to." Davenport faced a dilemma. He sought a genetic argument for innate Jewish undesirability, but conventional stereotypes precluded the usual claim for inherent stupidity. So Davenport opted for weakness in moral character rather than intellect. He wrote in his 1911 book, *Heredity in Relation to Eugenics*—not, by the way, a political tract, but his generation's leading textbook in the developing science of genetics:

> In earning capacity both male and female Hebrew immigrants rank high and the literacy is above the mean of all immigrants. . . . On the other hand, they show the greatest proportion of offenses against chastity and in connection with prostitution . . . The hordes of Jews that are now coming to us from Russia and the extreme southeast of Europe, with their intense individualism and ideals of gain at the cost of any interest, represent the opposite extreme from the early English and the more recent Scandinavian immigration, with their ideals of community life in the open country, advancement by the sweat of the brow, and the uprearing of families in the fear of God and love of country.

The rediscovery and publication of Mendel's laws in 1900 initiated the modern study of genetics. Earlier theories of heredity had envisaged a "blending" or smooth mixture and dilution of traits by interbreeding with partners of different constitution, whereas Mendelism featured a "particulate" theory of inheritance, with traits coded by discrete and unchanging genes that need not be expressed in all offspring (especially if "recessive" to a "dominant" form of the gene carried on the other chromosome of a given pair), but that remain in the hereditary constitution, independent and undiluted, awaiting expression in some future generation.

In an understandable initial enthusiasm for this great discovery, early geneticists committed their most common and consistent error in trying to identify single genes as causes for nearly every human trait, from discrete bits of anatomy to complex facets of personality. The search for single genetic determinants seemed reasonable (and testable by analysis of pedigrees) for simple, discrete, and discontinuous characters and contrasts (like blue versus brown eyes). But the notion that complex behaviors and temperaments might also emerge from a similar root in simple heredity of single genes never made much sense, for two major reasons: (1) a continuity in expression that precludes any easy definition of traits supposedly under analysis (I may know blue eyes when I see them, but where does a sanguine personality end and melancholia take over?); and (2) a virtual certainty that environments can substantially mold such characters, whatever their underlying genetic influence (my eyes may become blue whatever I eat, but my inherently good brain may end up residing in a stupid adult if poor nutrition starved my early growth, and crushing poverty denied me an education).

Nonetheless, most early human geneticists searched for "unit characters"— supposed traits that could be interpreted as the product of a single Mendelian factor—with abandon, even in complex, continuous, environmentally labile, and virtually undefinable features of personality or accomplishment in life. (These early analyses proceeded primarily by the tracing of pedigrees. I can envisage accurate data, and reliable results, for a family chart of eye color, but how could anyone trace the alleged gene for "optimism," "feeble inhibition," or "wanderlust"—not to mention such largely situational phenomena as "pauperism" or "communality"? Was Great-uncle George a jovial backslapper or a reclusive cuss?)

Whatever the dubious validity of such overextended attempts to reduce complex human behaviors to effects of single genes, this strategy certainly served the aims and purposes of the early twentieth century's most influential social crusade with an allegedly scientific foundation: the eugenics movement, with its stated aim of "improving" America's hereditary stock by preventing procreation among the supposedly unfit (called "negative eugenics") and encouraging more breeding among those deemed superior in bloodline ("positive eugenics"). The abuses of this movement have been extensively documented in many excellent books covering such subjects as the hereditarian theory of mental testing, and the passage of legislation for involuntary sterilization and restriction of immigration from nations deemed inferior in hereditary stock.

Many early geneticists played an active role in the eugenics movement, but none more zealously than the aforementioned Charles Benedict Davenport (1866–1944), who received a Ph.D. in zoology at Harvard in 1892, taught at the University of Chicago, and then became head of the Carnegie Institution's Station for Experimental Evolution at Cold Spring Harbor, New York, where he also established and directed the Eugenics Record Office, beginning in

1910. This office, with mixed aims of supposedly scientific documentation and overt political advocacy, existed primarily to establish and compile detailed pedigrees in attempts to identify the hereditary basis of human traits. The hyperenthusiastic Davenport secured funding from several of America's leading (and in their own judgment, therefore eugenically blessed) families, particularly from Mrs. E. H. Harriman, the guardian angel and chief moneybags for the entire movement.

In his 1911 textbook, dedicated to Mrs. Harriman "in recognition of the generous assistance she has given to research in eugenics," Davenport stressed the dependence of effective eugenics upon the new Mendelian "knowledge" that complex behavioral traits may be caused by single genes. Writing of the five thousand immigrants who passed through Ellis Island every day, Davenport stated:

> Every one of these peasants, each item of that "riff-raff" of Europe, as it is sometimes carelessly called, will, if fecund, play a role for better or worse in the future history of this nation. Formerly, when we believed that factors blend, a characteristic in the germ plasm of a single individual among thousands seemed not worth considering: it would soon be lost in the melting pot. But now we know that unit characters do not blend; that after a score of generations the given characteristic may still appear, unaffected by repeated unions. . . . So the individual, as the bearer of a potentially immortal germ plasm with innumerable traits, becomes of the greatest interest.

—that is, of *our* "greatest interest" to exclude by vetting and restricting immigration, lest American heredity be overwhelmed with a deluge of permanent bad genes from the wretched refuse of foreign lands.

To illustrate Davenport's characteristic style of argument, and to exemplify his easy slippage between supposed scientific documentation and overt political advocacy, we may turn to his influential 1915 monograph entitled *The Feebly Inherited* (publication number 236 of his benefactors, the Carnegie Institute of Washington), especially to part 1 on "Nomadism, or The Wandering Impulse, With Special Reference to Heredity." The preface makes no bones about either sponsorship or intent. With three of America's wealthiest and most conservative families on board, one could hardly expect disinterested neutrality toward the full range of possible results. The Carnegies had endowed the general show, while Davenport paid homage to specific patrons: "The cost of training the field-workers was met by Mrs. E. H. Harriman, founder and principal patron of the Eugenics Record Office, and Mr. John D. Rockefeller, who paid also the salaries of many of the field-workers."

Davenport's preface also boldly admits his political position and purposes. He wishes to establish "feeble inhibition" as a category of temperament leading to inferior morality. Such a formulation will provide a one-two punch for identification of the eugenically unfit—bad intellect *and* bad morals. According to Davenport, the genetic basis of intelligence had already been documented in numerous studies of the feebleminded. But eugenics now needed to codify the second major reason for excluding immigrants and discouraging or denying reproductive rights to the native unfit—bad moral character (as in Davenport's fallback position, documented earlier in this essay, for restricting Jewish immigration when he could not invoke the usual charge of intellectual inferiority). Davenport writes:

> A word may be said as to the term "feebly inhibited" used in these studies. It was selected as a fit term to stand as co-ordinate with "feeble-minded" and as the result of a conviction that the phenomena with which it deals should properly be considered apart from those of feeble-mindedness.

To allay any doubt about his motivations, Davenport then makes his political point up front. Feeble inhibition, leading to immorality, may be more dangerous than feeblemindedness, leading to stupidity:

> I think it helps to consider separately the hereditary basis of the intellect and the emotions. It is in this conviction that these studies are submitted for thoughtful consideration. For, after all, the chief problem in administering society is that of disordered conduct, conduct is controlled by emotions, and the quality of the emotions is strongly tinged by the hereditary constitution.

Davenport then selects "nomadism" as his primary example of a putatively simple Mendelian trait—the product of a single gene—based on "feeble inhibition" and leading almost inevitably to immoral behavior. He encounters a problem of definition at the very outset of his work, as expressed in an opening sentence that must be ranked as one of the least profound in the entire history of science! "A tendency to wander in some degree is a normal characteristic of man, as indeed of most animals, in sharp contrast to most plants."

How then shall the "bad" form of wanderlust, defined as a compulsion to flee from responsibility, be distinguished from the meritorious sense of bravery and adventure—leading to "good" wanderlust—that motivated our early (and largely northern European) immigrants to colonize and subdue the frontier? Davenport had warmly praised the "good" form in his 1911 book as "the enterprising restlessness of the early settlers . . . the ambitious search for better conditions. The abandoned farms of New England point to the trait in our blood that entices us to move on to reap a possible advantage elsewhere."

In a feeble attempt to put false labels on segments of complex continua, Davenport identified the "bad" form as "nomadism," defined as an inability to inhibit the urge we all feel (from time to time) to flee from our duties, but that folks of normal and decent morality suppress. Nomads are society's tramps, bums, hoboes, and gypsies—"those who, while capable of steady and effective work, at more or less regular periods run away from the place where their duties lie and travel considerable distances."

Having defined his quarry (albeit in a fatally subjective way), Davenport then required two further arguments to make his favored link of a "bad" trait (rooted in feeble inhibition and leading to immoral behavior) to a single gene that eugenics might labor to breed down and out: he needed to prove the hereditary basis, and then to find the "gene," for nomadism.

His arguments for a genetic basis must be judged as astonishingly weak, even by the standards of his own generation (and despite the renown of his work, attributable, we must assume in retrospect, to its consonance with what most readers wanted to believe rather than to the quality of Davenport's logic or data). He simply argued, based on four dubious analogies, that features akin to nomadism emerge whenever situations veer toward "raw" nature (where genetics must rule), and away from environmental refinements of modern human society. Nomadism must be genetic because analogous features appear as "the wandering instinct in great apes," "among primitive peoples," in children (then regarded as akin to primitives under the false view that ontogeny recapitulates phylogeny), and in adolescents (where raw instinct temporarily overwhelms social inhibition in the *Sturm and Drang* of growing up). The argument about "primitive" people seems particularly weak, since a propensity for wandering might be regarded as well suited to a lifestyle based on hunting mobile game, rather than identified as a mark of inadequate genetic constitution (or any kind of genetic constitution at all). But Davenport, reversing the probable route of cause and effect, pushed through any difficulty to his desired conclusion:

> If we regard the Fuegians, Australians, Bushmen and Hottentots as the most primitive men, then we may say that primitive man is nomadic. . . . It is frequently assumed that they are nomadic because they hunt, but it is more probable that their nomadic instincts force them to hunting rather than agriculture for a livelihood.

Davenport then pursues his second claim—nomadism as the product of a single gene—by tracing pedigrees stored in his Eugenics Record Office. On the subjective criterion of impressions recorded by fieldworkers, or written descriptions of amateur informants (mostly people who had submitted their family trees in response to a general appeal for data), Davenport marked all nomads in his table with a scarlet *W* (for *Wanderlust,* the common German term for "urge to roam"). He then examined the distribution of *W*'s through families and generations to reach one of the most peculiar and improbable conclusions ever advanced in a famous study: nomadism, he argued, is caused by *a single gene,* a sex-linked recessive located on what would later be identified as the female chromosome.

Davenport reached this conclusion by arguing that nomadism occurred in families with the same distribution as hemophilia, colorblindness, and other truly sex-linked recessive traits. Such a status can be legitimately inferred from several definite patterns of heredity. For example, fathers with the trait do not pass it to their sons (since the relevant gene resides on the X-chromosome and males only pass a Y-chromosome to their sons). Mothers with the trait pass it to all their sons, but none of their daughters when the father lacks the trait. (Since the feature is recessive, an afflicted mother must carry the gene on both X-chromosomes. She passes a single X to her son, who must then express the trait, for he has no other X-chromosome. But a daughter will receive one afflicted X-chromosome from her mother and one normal X-chromosome from her father; she will therefore not express the trait because her father's normal copy of the gene is dominant.) Davenport knew these rules, so his study didn't fail on this score. Rather, his criteria for identifying "nomadism" as a discrete and scorable "thing" remained so subjective, and so biased by his genetic assumptions, that his pedigree data can only be judged as worthless.

Davenport's summary reached (and preached) a eugenic crescendo: "The wandering instinct," he stated, "is a fundamental human instinct, which is, however, typically inhibited in intelligent adults of civilized peoples." Unfortunately, however, people who express the bad gene *W* (the scarlet letter of wanderlust) cannot achieve this healthy inhibition, and become feckless nomads who run from responsibility by literal flight. The trait is genetic, racial, and undesirable. Immigrants marked by *W* should be excluded (and many immigrants must be shiftless wanderers rather than brave adventurers), while nomadic natives should be strongly encouraged, if not compelled, to desist from breeding. Davenport concludes:

> The new light brought by our studies is this: The nomadic impulse is, in all the cases, one and the same unit character. Nomads, of all kinds, have a special racial trait—are, in a proper sense, members of the nomadic race. This trait is the absence of the germinal determiner that makes for sedentariness, stability, domesticity.

Of course, no one would now defend Davenport's extreme view that single genes determine nearly every complex human behavior. Most colleagues eventually rejected Davenport's theory during his own career, especially since he lived into the 1940s, long past the early flush of Mendelian enthusiasm, and well past our modern recognition that complex traits usually record the operation of many genes, each with a small and cumulative effect (not to mention a strong, and often predominant influence from nongenetic environmental contexts of growth and expression). A single gene for anger, conviviality, contemplation, or wanderlust now seems as absurd as a claim that one assassin's bullet, and nothing else, caused World War I, or that Darwin discovered evolution all by himself, and we would still be creationists if he had never been born.

Nonetheless, in our modern age of renewed propensity for genetic explanations (a valid and genuine enthusiasm when properly pursued), Davenport's general style of error resurfaces on an almost daily basis, albeit in much more subtle form, but with all the vigor of his putative old gene—yes, he did propose one—for stubbornly persistent behavior.

No sensible critic of biological determinism denies that genes influence behavior; of course they do. Moreover, no honorable skeptic would argue that genetic explanations should be resisted because they entail negative political, social, or ethical connotations—a charge that must be rejected for two primary reasons. First, nature's facts stand neutral before our ethical usages. We have, to be sure, often made dubious, even tragic decisions based on false genetic claims. But in other contexts, valid arguments about the innate and hereditary basis of human attributes can be profoundly liberating. Consider only the burden lifted from loving parents who raise beautiful and promising children for twenty years, and then "lose" them to the growing ravages of schizophrenia—almost surely a genetically based disease of the mind, just as many congenital diseases of other bodily organs also appear in the third decade of life or even later. Generations of psychologists had subtly blamed parents for unintentionally inducing such a condition, then viewed as entirely "environmental" in origin. What could be more cruel than a false weight of blame added to such an ultimate tragedy? Second, we will never get very far, either in our moral deliberations or in our scientific inquiries, if we disregard genuine facts because we dislike their implications. In the most obvious case, I cannot think of a more unpleasant fact than the inevitable physical death of each human body, but no sane person would bet on extended stability for a society built on the premise that King Prospero will reign in his personal flesh forever.

However, if we often follow erroneous but deeply rooted habits of thinking to generate false conclusions about the role of heredity in human behavior, then these habits should be exposed and corrected—all the more vigorously if such arguments usually lead to recommendations for action that most people would also regard as ethically wrong (involuntary sterilization of the mentally retarded, for example). I believe that we face such a situation today, and that the genetic fallacies underlying our misusages bear a striking similarity in style and logic to Davenport's errors, however much we have gained in subtlety of argument and factual accuracy.

Throughout the history of genetics, political misuse has most frequently originated from claims for "biological determinism"—the argument that a given behavior or social situation can't be altered because people have been "built that way" by their genes. Once we attribute something we don't like to "genes," we tend either to make excuses, or to make less effort for change. In the most obvious, egregious, and persisting example, many people still argue that we should deny educational benefits and social services to groups (usually races or social classes) falsely judged as genetically inferior on average, because their poverty and misfortune lie in their own heredity and cannot be significantly ameliorated by social intervention. Thus, history shows a consistent linkage between genetic claims cast in this mold and conservative political arguments for maintenance of an unjust status quo of great benefit to people currently in power.

Of course, no serious student of either genetics or politics would now advance this argument in Davenport's style of "one gene, one complex behavior." That is, no one talks today about *the* gene for stupidity, promiscuity, or lack of ambition. But a series of three subtle—and extremely common—errors lead all too often to the same eugenical style of conclusion. Somehow we remain fascinated with the idea that complex social behaviors might be explained, at least in large part, by inherited "atoms" of behavioral propensity lying deeply within individuals. We seem so much more satisfied, so much more intrigued, by the claim that a definite gene, rather than a complex and inextricable mix of heredity and social circumstances, causes a particular phenomenon. We feel that we have come much nearer to a real or essential cause when we implicate a particle within an individual, rather than a social circumstance built of multiple components, as the reason behind a puzzling behavior. We will avidly read a front-page headline entitled "gay gene found," but newspapers will not even bother to report an equally well documented story on other components of homosexual preference with a primary social root and no correlated genetic difference.

The common source of these errors lies much deeper than any crude correlation to a political utility that most of us do not even recognize and would disavow if we did. The source lies, I believe, in a general view about causality that has either been beaten into us by a false philosophy about science and the natural world, or may even record an unfortunate foible in our brain's evolved mode of operation. We favor simple kinds of explanations that flow in one direction from small, independent, constituent atoms of being, to complex and messy interactions among large bodies or organizations. In other words, and to use the technical term, we prefer to be "reductionists" in our causal schemes— to explain the physical behavior of large objects as consequences of atoms in motion, or to explain the social behavior of large animals by biological atoms called genes.

But the world rarely matches our simplistic hopes, and the admittedly powerful methods of reductionism don't always apply. Wholes *can* be bigger than the sums of their parts, and interactions among objects cannot always be disaggregated into rules of action for each object considered separately. The rules and randomnesses of particular situations must often be inferred from direct and overt study of large objects and their interactions, not by reduction to constituent "atoms" and their fundamental properties. The three common errors of genetic explanation all share the same basic fallacy of reductionist assumptions.

1. We regard ourselves as sophisticated when we acknowledge that *both* genes and environment produce a given outcome, but we err in assuming that we can best express such a correct principle by assigning percentages and

stating, for example, that "behavior A is 40 percent genetic and 60 percent environmental." Such reductionist expressions pass beyond the status of simple error into the even more negative domain of entirely meaningless statements. Genetics and environment do interact to build a totality, but we need to understand why resulting wholes are unbreakable and irreducible to separate components. Water cannot be explained as two-thirds of the separate properties of hydrogen gas mixed with one-third of oxygen's independent traits—just as wanderlust cannot be analyzed as 30 percent of a gene for feeble inhibition mixed with 70 percent of social circumstances that abet an urge to hit the road.

2. We think that we have reached some form of subtle accuracy in saying that many genes, not just a Davenportian unity, set the hereditary basis of complex behaviors. But we then take this correct statement and reintroduce the central error of reductionism by asserting that if 10 genes influence behavior A, and if the causes of A may be regarded as 50 percent genetic (the first error), then each gene must contribute roughly 5 percent to the totality of behavior A. But complex interactions cannot be calculated as the sum of independent parts considered separately. I cannot be understood as one-eighth of each of my great-grandparents (though my genetic composition may be roughly so determined); I am a unique product of my own interactive circumstances of social setting, heredity composition, and all the slings and arrows of individual and outrageous natural fortune.

3. We suppose that we have introduced sufficient caution in qualifying statements about "genes for" traits by admitting their only partial, and often quite small, contribution to an interactive totality. Thus, we imagine that we may legitimately talk of a "gay gene" so long as we add the proviso that only 15 percent of sexual preference records this cause. But we need to understand why such statements have no meaning and therefore become (as for the first argument above) worse than merely false. Many genes interact with several other factors to influence sexual preference, but no unitary and separable "gay gene" exists. When we talk about a "gene for" 10 percent of behavior A, we simply commit the old Davenportian fallacy on the "little bit pregnant" analogy.

As a concrete example of how a good and important study can be saddled with all these errors in public reporting (and also by less than optimally careful statements of some participating researchers), *The New York Times* greeted 1996 with a headline on the front page of its issue for January 2: "Variant Gene Tied to a Love of New Thrills." The article discussed two studies published in the

January 1996 issue of *Nature Genetics*. Two independent groups of researchers, one working with 124 Ashkenazi and Sephardic Jews from Israel, the other with a largely male sample of 315 ethnically diverse Americans, both found a clearly significant, if weak, association between propensity for "novelty-seeking" behavior (as ascertained from standard survey questionnaires) and possession of a variant of a gene called the D4 dopamine receptor, located on the eleventh chromosome, and acting as one of at least five receptors known to influence the brain's response to dopamine.

This gene exists in several forms, defined by differing lengths recording the number (anywhere from two to ten) of repeated copies of a particular DNA subunit within the gene. Individuals with a high number of repeated copies (that is, with a longer gene) tended to manifest a greater tendency for "novelty-seeking" behavior—perhaps because the longer form of the gene somehow acts to enhance the brain's response to dopamine.

So far, so good—and very interesting. We can scarcely doubt that heredity influences broad and basic aspects of temperament—a bit of folk wisdom that surely falls into the category of "what every parent with more than one child knows." No one should feel at all offended or threatened by the obvious fact that we are not all born entirely blank, or entirely the same, in our mixture of the broad behavioral propensities defining what we call "temperament." Certain genes evidently influence particular aspects of brain chemistry; and brain chemistry surely affects our moods and behaviors. We know that basic and powerful neurotransmitters like dopamine strongly impact our moods and feelings (particularly, for dopamine, our sensations of pleasure). Differing forms of genes that affect the brain's response to dopamine may influence our behaviors—and a form that enhances the response may well incline a person toward "novelty-seeking" activities.

But the long form of the D4 receptor does not therefore become *the* (or even a) "novelty-seeking" gene, and these studies do not show that novelty seeking can be quantified and explained as a specified percent "genetic" in origin—although statements in this form dominated popular reports of these discoveries. Even the primary sources—the two original reports in *Nature Genetics* and the accompanying editorial feature entitled "Mapping Genes for Human Personality"—and the excellent *Times* story (representing the best of our serious press) managed, amidst their generally careful and accurate accounts, to propagate all three errors detailed above.

The *Times* reporter committed the first error of assigning separable percentages by writing "that about half of novelty-seeking behavior is attributable to genes, the other half to as yet ill defined environmental circumstances." Dr. R. P. Ebstein, principal author of one of the reports, then stated the second error of adding up effects without considering interactions when he argued that the long form of the D4 gene accounts for only about 10 percent of novelty-seeking behavior. If, by the first error, the totality of novelty seeking can be viewed as 50 percent genetic, and if D4 accounts for 10 percent of the totality, then we can infer that about four other genes must be involved (each contributing its 10 percent for the grand total of 50 percent genetic influence). Ebstein told the *Times* reporter: "If we assume that there are other genes out there that we haven't looked at yet, and that each gene exerts more or less the same influence as the D4 receptor, then we would expect maybe four or five genes are involved in the trait."

But the most significant errors, as always, fall into the third category of mis-proclaiming "genes for" specific behaviors—as in the title of the technical report from *Nature Genetics,* previously cited: "Mapping Genes for Human Personality." (If our professional journals so indulge and err, imagine what the popular press makes of "gay genes," "thrill genes," "stupidity genes," and so on.) First of all, the D4 gene by itself exerts only a weak potential influence upon novelty-seeking behavior. How can a gene accounting for only 10 percent of the variance in a trait be proclaimed as a "gene for" the trait? If I decide that 10 percent of my weight gain originated from the calories in tofu (because I love the stuff and eat it by the ton), this item, generally regarded as nutritionally benign, does not become a "fatness food."

More importantly, genes make enzymes, and enzymes control the rates of chemical processes. Genes do not make "novelty-seeking" or any other complex and overt behavior. Predisposition via a long chain of complex chemical reactions, mediated through a more complex series of life's circumstances, does not equal identification or even causation. At most, the long form of D4 induces a chemical reaction that can, among other possible effects, generate a mood leading some people to greater openness toward behaviors defined by some questionnaires as "novelty seeking."

In fact, a further study, published in 1997, illustrated this error in a dramatic way by linking the same long form of D4 to greater propensity for heroin addiction. The original *Times* article of 1996 had celebrated the "first known report of a link between a specific gene and a specific normal personality trait." But now the same gene—perhaps via the same route of enhanced dopamine response—also correlates with a severe pathology in other personalities. So what shall we call D4—a "novelty-seeking" gene in normal folk, or an "addiction" gene in troubled people? We need instead to reform both our terminology and our concepts. The long form of D4 induces a chemical response. This response may correlate with many different overt behaviors in people with widely varying histories and genetic constitutions.

The deepest error of this third category lies in the reductionist, and really rather silly, notion that we can even define discrete, separable, specific traits within the complex continua of human behaviors. We encounter enough difficulty in trying to identify characters with clear links to particular genes in the much clearer and simpler features of human anatomy. I may be able to specify genes "for" eye color, but not for leg length or fatness. How then shall I parse the continuous and necessarily subjective categories of labile personalities? Is "novelty seeking" really a "thing" at all? Can I even talk in a meaningful way about "genes for" such nebulous categories? Have I not fallen right back into the errors of Davenport's search for the internal scarlet letter *W* of wanderlust?

I finally realized what had been troubling me so much about the literature on "genes for" behavior when I read the *Times*'s account of C. R. Cloninger's theory of personality (Cloninger served as principal author of the *Nature Genetics* editorial commentary):

> Novelty seeking is one of four aspects that Dr. Cloninger and many other psychologists propose as the basic bricks of normal temperament, the other three being avoidance of harm, reward dependence and persistence. All four humors are thought to be attributable in good part to one's genetic makeup.

The last line crystallized my distress—"all four humors"—for I grasped, with the emotional jolt that occurs when all the previously unconnected pieces of an argument fall suddenly into place, why the canny reporter (or the scientist himself) had used this old word. Consider the theory in outline: four independent components of temperament, properly in balance in "normal" folks, with each individual displaying subtly different proportions, thus determining our individual temperaments and building our distinct personalities. But if our body secretes too much, or too little, of any particular humor, then a pathology may develop.

But why four, and why these four? Why not five, or six, or six hundred? Why any specific number? Why try to parse such continua into definite independent "things" at all? I do understand the mathematical theories and procedures that lead to such identifications (see my book *The Mismeasure of Man*), but

I regard the entire enterprise as a major philosophical error of our time (while I view the mathematical techniques, which I use extensively in my own research, as highly valuable when properly applied). Numerical clumps do not identify physical realities. A four-component model of temperament may act as a useful heuristic device, but I don't believe for a moment that four homunculi labeled *novelty seeking, avoidance of harm, reward dependence,* and *persistence* reside in my brain, either vying for dominance or cooperating in balance.

The logic of such a theory runs in uncanny parallel—hence the clever choice of "humor" as a descriptive term for the proposed modules of temperament—with the oldest and most venerable of gloriously wrong theories in the history of medicine. For more than a thousand years, from Galen to the dawn of modern medicine, prevailing wisdom regarded the human personality as a balance among four humors—blood, phlegm, choler, and melancholy. *Humor,* from the Latin word for "liquid" (a meaning still preserved in designating the fluids of the human eye as aqueous and vitreous humor), referred to the four liquids that supposedly formed the chyle, or digested food in the intestine just before it entered the body for nourishment. Since the chyle formed, on one hand, from a range of choices in the food we eat and, on the other hand, from constitutional differences in how various bodies digest this food, the totality recorded both innate and external factors—an exact equivalent to the modern claim that both genes and environment influence our behavior.

The four humors of the chyle correspond to the four possible categories of a double dichotomy—that is, to two axes of distinction based on warm–cold and wet–dry. The warm and wet humor forms blood; cold and wet generates phlegm; warm and dry makes choler; while cold and dry builds melancholy. I regard such a logically abstract scheme as a heuristic organizing device, much like Cloninger's quadripartite theory of personality. But we make a major error if we elevate such a scheme to a claim for real and distinct physical entities inside the body.

In the medical theory of humors, good health results from a proper balance among the four, while distinctive personalities emerge from different proportions within the normal range. But too much of any one humor may lead to oddness or pathology. As a fascinating linguistic remnant, we still use the names of all four humors as adjectives for types of personality—sanguine (dominance of the hot–wet blood humor) for cheerful people, phlegmatic for stolid folks dominated by the cold–wet humor of phlegm, choleric for angry individuals saddled with too much hot–dry choler, and melancholic for sad people overdosed with black bile, the cold–dry humor of melancholia. Does the modern quadripartite theory of personality really differ in any substantial way from this older view in basic concepts of number, balance, and the causes of both normal personality and pathology?

In conclusion, we might imagine two possible reasons for such uncanny similarity between a modern conception of four components to temperament, and the old medical theory of humors. Perhaps the similarity exists because the ancients had made a great and truthful discovery, while the modern version represents a major refinement of a central fact that our ancestors could only glimpse through a glass darkly. But alternatively—and ever so much more likely in my judgment—the stunning similarities exist because the human mind has remained constant throughout historical time, despite all our growth of learning and all the tumultuous changes in Western culture. We therefore remain sorely tempted by the same easy fallacies of reasoning.

I suspect that we once chose four humors, and now designate four end members of temperament, because something deep in the human psyche leads us to impose simple taxonomic schemes of distinct categories upon the world's truly complex continua. After all, our forebears didn't invoke the number four only for humors. We parsed many other phenomena into schemes with four end members—four compass points, four ages of man, and four Greek elements of air, earth, fire, and water. Could these similarities of human ordering be coincidental, or does the operation of the human brain favor such artificial divisions? Carl G. Jung, for reasons that I do not fully accept, strongly felt that division by four represented something deep and archetypal in human proclivities. He argued that we inherently view divisions by three as incomplete and leading onward (for one triad presupposes another for contrast), whereas divisions by four stand in optimal harmony and internal balance. He wrote: "Between the three and the four there exists the primary opposition of male and female, but whereas fourness is a symbol of wholeness, threeness is not."

I think that Jung correctly discerned an inherent mental attraction to divisions by four, but I suspect that the true basis for this propensity lies in our clear (and probably universal) preference for dichotomous divisions. Division by four may denote an ultimate and completed dichotomization—a dichotomy of dichotomies: two axes (each with two end members) at right angles to each other. We may experience four as an ultimate balance because such schemes fill our mental space with two favored dichotomies in perfect and opposite coordination.

In any case, if this second reason explains why we invented such eerily similar theories as four bodily humors and four end members of temperament, then such quadripartite divisions reflect biases of the mind's organization, not "real things" out there in the physical world. We can hardly talk about "genes for" the components of such artificial and prejudicial parsings of a much more complex reality. Interestingly, the greatest literary work ever written on the theory of humors, the early-seventeenth-century *Anatomy of Melancholy* by the English divine and scholar Robert Burton, properly recognized the four humors as just one manifestation of a larger propensity to divide by four. This great man who used the balm of literature to assuage his own lifelong depression, wrote of his condition: "Melancholy, cold and drie, thicke, blacke, and sowre . . . is a bridle to the other two hot humors, bloode and choler, preserving them in the blood, and nourishing the bones: These foure humors have some analogie with the foure elements, and to the foure ages in man."

I would therefore end—and where could an essayist possibly find a more appropriate culmination—with some wise words from Montaigne, the sixteenth-century founder of the essay as a literary genre. Perhaps we should abandon our falsely conceived and chimerical search for a propensity to wander, or to seek novelty (perhaps a spur to wandering), in a specific innate sequence of genetic coding. Perhaps, instead, we should pay more attention to the wondrous wanderings of our mind. For until we grasp the biases and propensities of our own thinking, we will never see through the humors of our vision into the workings of nature beyond. Montaigne wrote:

> It is a thorny undertaking, and more so than it seems, to follow a movement so wandering as that of our mind, to penetrate the opaque depths of its innermost folds, to pick out and immobilize the innumerable flutterings that agitate it.

19

Dolly's Fashion
and Louis's Passion

NOTHING CAN BE MORE FLEETING OR CAPRICIOUS THAN fashion. What, then, can a scientist, committed to objective description and analysis, do with such a haphazardly moving target? In a classic approach, analogous to standard advice for preventing the spread of an evil agent ("kill it before it multiplies"), a scientist might say, "quantify it before it disappears."

Francis Galton, Charles Darwin's charmingly eccentric and brilliant cousin, and a founder of the science of statistics, surely took this prescription to heart. He once decided to measure the geographic patterning of female beauty. He therefore attached a piece of paper to a small wooden cross that he could carry, unobserved, in his pocket. He held the cross at one end in the palm of his hand and, with a needle secured between thumb and forefinger, made pinpricks on the three remaining projections (the two ends of the cross bar and the top). He would rank every young woman he passed on the street into one of three categories, as beautiful, average, or substandard (by his admittedly subjective preferences)—and he would then place a pinprick for each woman into the designated domain of this cross. After a hard day's work, he tabulated the relative percentages by counting pinpricks. He concluded, to the dismay of Scotland, that beauty followed a simple trend from north to south, with the highest proportion of uglies in Aberdeen, and the greatest frequency of lovelies in London.

Some fashions (body piercings, perhaps?) flower once and then disappear, hopefully forever. Others swing in and out of style, as if fastened to the end of a pendulum. Two foibles of human life strongly promote this oscillatory mode. First, our need to create order in a complex world, begets our worst mental habit: dichotomy (see chapter 3), or our tendency to reduce a truly intricate set of subtle shadings to a choice between two diametrically opposed alternatives (each with moral weight and therefore ripe for bombast and pontification, if not outright warfare): religion versus science, liberal versus conservative, plain versus fancy, "Roll Over Beethoven" versus the "Moonlight" Sonata. Second, many deep questions about our loves and livelihood, and the fates of nations, truly have no answers—so we cycle the presumed alternatives of our dichotomies, one after the other, always hoping that, this time, we will find the nonexistent key to an elusive solution.

Among oscillating fashions governed primarily by the swing of our social pendulum, no issue can claim greater prominence for an evolutionary biologist, or hold more relevance to a broad range of political questions, than genetic versus environmental sources of human abilities and behaviors. This issue has been falsely dichotomized for so many centuries that English even features a mellifluous linguistic contrast for the supposed alternatives: nature versus nurture.

As any thoughtful person understands, the framing of this question as an either-or dichotomy verges on the nonsensical. Both inheritance and upbringing matter in crucial ways. Moreover, an adult human being, built by interaction of these (and other) factors, cannot be disaggregated into separate components with attached percentages (see chapter 18 for detailed arguments on this vital issue). Nonetheless, a preference for either nature or nurture swings back and forth into fashion as political winds blow, and as scientific breakthroughs grant transient prominence to one or another feature in a spectrum of vital influences. For example, a combination of political and scientific factors favored an emphasis upon environment in the years just following World War II: an understanding that Hitlerian horrors had been rationalized by claptrap genetic theories about inferior races; the heyday of behaviorism in psychology.

Today genetic explanations have returned to great vogue, fostered by a similar mixture of social and scientific influences: a rightward shift of the political pendulum (and the cynical availability of "you can't change them, they're made that way" as a bogus argument for reducing government expenditures on social programs); and an overextension to all behavioral variation of genuinely exciting results in identifying the genetic basis of specific diseases, both physical and mental.

Unfortunately, in the heat of immediate enthusiasm, we often mistake transient fashion for permanent enlightenment. Thus, many people assume that the current popularity of genetic determinism represents a permanent truth, finally wrested from the clutches of benighted environmentalists of previous generations. But the lessons of history suggest that the worm will soon turn again. Since both nature and nurture can teach us so much—and since the fullness of our behavior and mentality represents such a complex and unbreakable combination of these and other factors—a current emphasis on nature will no doubt yield to a future fascination with nurture as we move toward better understanding by lurching upward from one side to another in our quest to fulfill the Socratic injunction: know thyself.

In my Galtonian desire to measure the extent of current fascination with genetic explanations (before the pendulum swings once again and my opportunity evaporates), I hasten to invoke two highly newsworthy items of recent times. The subjects may seem quite unrelated—Dolly the cloned sheep, and Frank Sulloway's book on the effects of birth order upon human behavior*— but both stories share a curious common feature offering striking insight into the current extent of genetic preferences. In short, both stories have been reported almost entirely in genetic terms, but both cry out (at least to me) for

---

*This essay, obviously, represents my reaction to the worldwide storm of news and ethical introspection launched by the public report of Dolly, the first mammal cloned from an adult cell, in early 1997. In collecting several years of essays together to make each of these books, I usually deepsix the rare articles keyed to immediate news items of "current events"—for the obvious reason of their transiency under the newsman's adage that "yesterday's paper wraps today's garbage." But in rereading this essay, I decided that it merited reprinting on two counts: first, I don't think that its relevance has at all faded (while Dolly herself also persists firmly in public memory); second, I fancy that I found something general and original to say by linking Dolly to Sulloway's book, and by relating both disparate events to a common theme that had puzzled me enormously by being so blessedly obvious, yet so totally unreported in both the serious and popular press. As King Lear dis covered to his sorrow, the absence of an expected statement can often be far more meaningful than an anticipated and active pronouncement. Since these essays experience a three-month "lead time" between composition and original publication, I must always treat current events in a more general context potentially meriting republication down the line—for ordinary fast-breaking news can only become rock-hard stale in those interminable ninety days.

a radically different reading as proof of strong environmental influences. Yet no one seems to be drawing (or even mentioning) this glaringly obvious inference. I cannot imagine that anything beyond current fashion for genetic arguments can explain this puzzling silence. I am convinced that exactly the same information, if presented twenty years ago in a climate favoring explanations based on nurture, would have elicited a strikingly different interpretation. Our world, beset by ignorance and human nastiness, contains quite enough background darkness. Should we not let both beacons shine all the time?

CREATING SHEEP

Dolly must be the most famous sheep since John the Baptist designated Jesus in metaphor as "lamb of God, which taketh away the sin of the world" (John 1:29). She has certainly edged past the pope, the president, Madonna, and Michael Jordan as the best-known mammal of the moment. And all this brouhaha for a carbon copy, a Xerox! I don't mean to drip cold water on this little lamb, cloned from a mammary cell of her adult mother, but I remain unsure that she's worth all the fuss and fear generated by her unconventional birth.

When one reads the technical article describing Dolly's manufacture (I. Wilmut, A. E. Schnieke, J. McWhir, A. J. Kind, and K. H. S. Campbell, "Viable offspring derived from fetal and adult mammalian cells," *Nature,* February 27, 1997, pages 810–13), rather than the fumings and hyperbole of so much public commentary, one can't help feeling a bit underwhelmed, and left wondering whether Dolly's story tells less than meets the eye.

I don't mean to discount or underplay the ethical issues raised by Dolly's birth (and I shall return to this subject in a moment), but we are not about to face an army of Hitlers or even a Kentucky Derby run entirely by genetically identical contestants (a true test for skills of jockeys and trainers!). First, Dolly breaks no theoretical ground in biology, for we have known how to clone in principle for at least two decades, but had developed no techniques for reviving the full genetic potential of differentiated adult cells. (Still, I admit that a technological solution can pack as much practical and ethical punch as a theoretical breakthrough. I suppose one could argue that the first atomic bomb only realized a known possibility.)

Second, my colleagues have been able to clone animals from embryonic cell lines for several years, so Dolly does not rank as the first mammalian clone, but rather as the first clone from an adult cell. Ian Wilmut and his coworkers also cloned sheep from cells of a nine-day embryo and a twenty-six-day fetus—with much greater success. They achieved fifteen pregnancies (though not all proceeded to term) in thirty-two "recipients" (that is, surrogate mothers for transplanted cells) for the embryonic cell line, and five pregnancies in sixteen recipients for the fetal cell line, but only Dolly (one pregnancy in thirteen tries) for the adult cell line. This experiment cries out for confirming repetition. (Still, I allow that current difficulties will surely be overcome, and cloning from adult cells, if doable at all, will no doubt be achieved more routinely as techniques and familiarity improve.)*

Third, and more seriously, I remain unconvinced that we should regard Dolly's starting cell as adult in the usual sense of the term. Dolly grew from a cell taken from the "mammary gland of a 6-year-old ewe in the last trimester of pregnancy" (to quote the technical article of Wilmut et al.). Since the breasts of pregnant mammals enlarge substantially in late stages of pregnancy, some mammary cells, though technically adult, may remain unusually labile or even "embryolike," and thus able to proliferate rapidly to produce new breast tissue at an appropriate stage of pregnancy. Consequently, we may only be able to clone from unusual adult cells with effectively embryonic potential, and not from any stray cheek cell, hair follicle, or drop of blood that happens to fall into the clutches of a mad Xeroxer. Wilmut and colleagues admit this possibility in a sentence written with all the obtuseness of conventional scientific prose, and therefore almost universally missed by journalists: "We cannot exclude the possibility that there is a small proportion of relatively undifferentiated stem cells able to support regeneration of the mammary gland during pregnancy."

But if I remain relatively unimpressed by achievements thus far, I do not discount the monumental ethical issues raised by the possibility of cloning from adult cells. Yes, we have cloned fruit trees for decades by the ordinary process of grafting—and without raising any moral alarms. Yes, we may not face the evolutionary dangers of genetic uniformity in crop plants and livestock, for I trust that plant and animal breeders will not be stupid enough to eliminate all but one genotype from a species, and will always maintain (as plant breeders now do) an active pool of genetic diversity in reserve. (But then, I suppose we should never underestimate the potential extent of human stupidity—and

---

*Science moves fast, especially when spurred by immense public interest and pecuniary possibilities. These difficulties have been much mitigated, if not entirely overcome, in the three years between my original writing and this republication. Cloning from adult cells has not become, by any means, routine, but undoubted clones have been produced from adult cells in several mammalian species. Moreover, initial doubts about Dolly herself (mentioned in this essay) have largely been allayed, and her status as a clone from an adult cell now seems secure.

localized reserves could be destroyed by a catastrophe, while genetic diversity spread throughout a species guarantees maximal evolutionary robustness.)

Nonetheless, while I regard many widely expressed fears as exaggerated, I do worry deeply about potential abuses of human cloning, and I do urge an open and thorough debate on these issues. Each of us can devise a personal worst-case scenario. Somehow I do not focus upon the specter of a future Hitler making an army of 10 million identical robotic killers—for if our society ever reaches a state where someone in power could actually realize such an outcome, we are probably already lost. My thoughts run to localized moral quagmires that we might actually have to face in the next few years—for example, the biotech equivalent of ambulance-chasing slimeballs among lawyers: a hustling firm that scans the obits for reports of children who died young, and then goes to grieving parents with the following offer: "So sorry for your loss; but did you save a hair sample? We can make you another for a mere fifty thou."

However, and still on the subject of ethical conundrums, but now moving to my main point about current underplaying of environmental sources for human behaviors, I do think that the most potent scenarios of fear, and the most fretful ethical discussions of radio talk shows, have focused on a nonexistent problem that all human societies solved millennia ago. We ask: is a clone an individual? Would a clone have a soul? Would a clone made from my cell negate my unique personhood?

May I suggest that these endless questions—all variations on the theme that clones threaten our traditional concept of individuality—have already been answered empirically, even though public discussion of Dolly seems blithely oblivious to this evident fact.* We have known human clones from the dawn of our consciousness. We call them identical twins—and they constitute far better clones than Dolly and her mother. Dolly shares only nuclear DNA with her genetic mother—for only the nucleus of her mother's mammary cell was

---

*My deep puzzlement over public surprise at this obvious point, and at the failure of media to grasp and highlight the argument immediately, has only grown since I wrote this essay. (I believe that I was the first to stress or even to mention—in commentary to journalists before I published this article—the clonal nature of identical twins as an ancient and conclusive disproof of the major ethical fears that Dolly had so copiously inspired. No argument of such a basic and noncryptic nature should ever be first presented by a magazine essayist with a lead time of several months, rather than the next day by a journalist, or the next minute in cyberspace.) I can only conclude that public misunderstanding of environmental impact upon human personalities, emotions, and distinctivenesses runs much deeper than even I had realized, and that barriers to recognition of this self-evident truth stand even higher than I had suspected in the light of current fashions for genetic explanations.

inserted into an embryonic stem cell (whose own nucleus had been removed) of a surrogate female. Dolly then grew in the womb of this surrogate.

Identical twins share at least four additional (and important) attributes that differ between Dolly and her mother. First, identical twins also carry the same mitochondrial genes. (Mitochondria, the "energy factories" of cells, contain a small number of genes. We obtain our mitochondria from the cytoplasm of the egg cell that made us, not from the nucleus formed by the union of sperm and egg. Dolly received her nucleus from her mother but her egg cytoplasm, and hence her mitochondria, from her surrogate.) Second, identical twins share the same set of maternal gene products in the egg. Genes don't grow embryos all by themselves. Egg cells contain protein products of maternal genes that play a major role in directing the early development of the embryo. Dolly's embryology proceeded with her mother's nuclear genes but her surrogate's gene products in the cytoplasm of her founding cell.

Third—and now we come to explicitly environmental factors—identical twins share the same womb. Dolly and her mother gestated in different places. Fourth, identical twins share the same time and culture (even if they fall into the rare category, so cherished by researchers, of siblings separated at birth and raised, unbeknownst to each other, in distant families of different social classes). The clone of an adult cell matures in a different world. Does anyone seriously believe that a clone of Beethoven, grown today, would sit down one fine day to write a tenth symphony in the style of his early-nineteenth-century forebear?

So identical twins are truly eerie clones—much more alike on all counts than Dolly and her mother. We do know that identical twins share massive similarities, not only of appearance, but also in broad propensities and detailed quirks of personality. Nonetheless, have we ever doubted the personhood of each member in a pair of identical twins? Of course not. We know that identical twins are distinct individuals, albeit with peculiar and extensive similarities. We give them different names. They encounter divergent experiences and fates. Their lives wander along disparate paths of the world's complex vagaries. They grow up as distinctive and undoubted individuals, yet they stand forth as far better clones than Dolly and her mother.

Why have we overlooked this central principle in our fears about Dolly? Identical twins provide sturdy proof that inevitable differences of nurture guarantee the individuality and personhood of each human clone. And since any future human Dolly must depart far more from her progenitor (in both the nature of mitochondria and maternal gene products, and the nurture of different wombs and surrounding cultures) than any identical twin differs from her sibling clone, why ask if Dolly has a soul or an independent life when we have never doubted the personhood or individuality of far more similar identical twins?

Literature has always recognized this principle. The Nazi loyalists who cloned Hitler in *The Boys from Brazil* understood that they had to maximize similarities of nurture as well. So they fostered their little Hitler babies in families maximally like Adolf's own dysfunctional clan—and not one of them grew up anything like history's quintessential monster. Life has always verified this principle as well. Eng and Chang, the original Siamese twins and the (literally) closest clones of all, developed distinct and divergent personalities. One became a morose alcoholic, the other remained a benign and cheerful man. We may not attribute much individuality to sheep in general—they do, after all, establish our icon of blind following and identical form as they jump over fences in the mental schemes of insomniacs—but Dolly will grow up to be as unique and as ornery as any sheep can be.

KILLING KINGS

My friend Frank Sulloway recently published a book that he had fretted over, nurtured, massaged, and lovingly shepherded toward publication for more than two decades. Frank and I have been discussing his thesis ever since he began his studies. I thought (and suggested) that he should have published his results twenty years ago. I still hold this opinion—for while I greatly admire his book, and do recognize that such a long gestation allowed Frank to strengthen his case by gathering and refining his data, I also believe that he became too committed to his central thesis, and tried to extend his explanatory umbrella over too wide a range, with arguments that sometimes smack of special pleading and tortured logic.

Born to Rebel documents a crucial effect of birth order in shaping human personalities and styles of thinking. Firstborns, as sole recipients of parental attention until the arrival of later children, and as more powerful than their subsequent siblings (by virtue of age and size), generally cast their lot with parental authority and with the advantages of incumbent strength. They tend to grow up competent and confident, but also conservative and unlikely to favor quirkiness or innovation. Why threaten an existing structure that has always offered you clear advantages over siblings? Later children, however, are (as Sulloway's title proclaims) born to rebel. They must compete against odds for parental attention long focused primarily elsewhere. They must scrap and struggle, and learn to make do for themselves. Laterborns therefore tend to be flexible, innovative, and open to change. The business and political leaders of stable nations are usually firstborns, but the revolutionaries who have discombobulated our cultures and restructured our scientific knowledge tend to be laterborns.

Sulloway defends his thesis with statistical data on the relationship of birth order and professional achievements in modern societies, and by interpreting historical patterns as strongly influenced by characteristic differences in behaviors between firstborns and laterborns. I found some of his historical arguments fascinating and persuasive when applied to large samples (but often uncomfortably overinterpreted in trying to explain the intricate details of individual lives, for example the effect of birth order on the differential success of Henry VIII's various wives in overcoming his capricious cruelties).

In a fascinating case, Sulloway chronicles a consistent shift in relative percentages of firstborns among successive groups in power during the French Revolution. The moderates initially in charge tended to be firstborns. As the revolution became more radical, but still idealistic and open to innovation and free discussion, laterborns strongly predominated. But when control then passed to the uncompromising hard-liners who promulgated the Reign of Terror, firstborns again ruled the roost. In a brilliant stroke, Sulloway tabulates the birth orders for several hundred delegates who decided the fate of Louis XVI in the National Convention. Among hard-liners who voted for the guillotine, 73 percent were firstborns; but 62 percent of laterborns opted for the compromise of conviction with pardon. Since Louis lost his head by a margin of one vote, an ever-so-slightly different mix of birth orders among delegates might have altered the course of history.

Since Frank is a good friend (though I don't accept all details of his thesis), and since I have been at least a minor midwife to this project for two decades, I took an unusually strong interest in the delayed birth of *Born to Rebel*. I read the text and all the prominent reviews that appeared in many newspapers and journals. And I have been puzzled—*stunned* would not be too strong a word—by the total absence from all commentary of the simplest and most evident inference from Frank's data, the one glaringly obvious point that everyone should have stressed, given the long history of issues raised by such information.

Sulloway focuses nearly all his interpretation on an extended analogy (broadly valid in my judgment, but overextended as an exclusive device) between birth order in families and ecological status in a world of Darwinian competition. Children vie for limited parental resources, just as individuals struggle for existence (and ultimately for reproductive success) in nature. Birth order places children in different "niches," requiring disparate modes of competition for maximal success. While firstborns shore up incumbent advantages, laterborns must grope and grub by all clever means at their disposal—leading to the divergent personalities of stalwart and rebel. Alan Wolfe, in my favorite negative review in *The New Republic*, writes (December 23, 1996; Jared Diamond stresses the same themes in my favorite positive review in *The New York Review of Books*, November 14, 1996): "Since firstborns already occupy their own niches, laterborns, if they are to be noticed, have to find unoccupied niches. If they do so successfully, they will be rewarded with parental investment."

As I said, I am willing to follow this line of argument up to a point. But I must also note that restriction of commentary to this Darwinian metaphor has diverted attention from the foremost conclusion revealed by a large effect of birth order upon human behavior. The Darwinian metaphor smacks of biology; we also (albeit erroneously) tend to regard biological explanations as intrinsically genetic. I suppose that this common but fallacious chain of argument leads us to stress whatever we think that Sulloway's thesis might be teaching us about "nature" (our preference, in any case, during this age of transient fashion for genetic causes) under our erroneous tendency to treat the explanation of human behavior as a debate between nature and nurture.

But consider the meaning of birth-order effects for environmental influences, however unfashionable at the moment. Siblings differ genetically of course, but no aspect of this genetic variation correlates in any systematic way with birth order. Firstborns and laterborns receive the same genetic shake within a family. *Systematic* differences in behavior between firstborns and laterborns therefore cannot be ascribed to genetics. (Other biological effects may correlate with birth order—if, for example, the environment of the womb changes systematically with numbers of pregnancies—but such putative influences bear no relationship whatever to genetic differences among siblings.) Sulloway's substantial birth-order effect therefore provides our best and ultimate documentation of nurture's power. If birth order looms so large in setting the paths of history and the allocation of people to professions, then nurture cannot be denied a powerfully formative role in our intellectual and behavioral variation. To be sure, we often fail to see what stares us in the face; but how can the winds of fashion blow away such an obvious point, one so relevant to our deepest and most persistent questions about ourselves?

In this case, I am especially struck by the irony of fashion's veil. As noted above, I urged Sulloway to publish his data twenty years ago—when (in my judgment) he could have presented an even better case because he had already documented the strong and general influence of birth order upon personality, but had not yet ventured upon the slippery path of trying to explain too many details with forced arguments that sometimes lapse into self-parody. If Sulloway had published in the mid-1970s, when nurture rode the pendulum of fashion in a politically more liberal age (probably dominated by laterborns!), I am confident that this obvious argument about effects of birth order as proof of nurture's power would have won primary attention, rather than consignment to a limbo of invisibility.

Hardly anything in intellectual life can be more salutatory than the separation of fashion from fact. Always suspect fashion (especially when the moment's custom matches your personal predilection); always cherish fact (while remembering that an apparent jewel of pure and objective information may only record the biased vision of transient fashion). I have discussed two subjects that couldn't be "hotter," but cannot be adequately understood because a veil of genetic fashion now conceals the richness of full explanation by relegating a preeminent environmental theme to invisibility. Thus, we worry whether the first cloned sheep represents a genuine individual at all, while we forget that we have never doubted the distinct personhood guaranteed by differences in nurture to clones far more similar by nature than Dolly and her mother—identical twins. And we try to explain the strong effects of birth order solely by invoking a Darwinian analogy between family status and ecological niche, while forgetting that these systematic effects cannot arise from genetic differences, and can therefore only demonstrate the predictable power of nurture. Sorry, Louis. You lost your head to the power of family environments upon head children. And hello, Dolly. May we forever regulate your mode of manufacture, at least for humans. But may genetic custom never stale the infinite variety guaranteed by a lifetime of nurture in the intricate complexity of our natural world—this vale of tears, joy, and endless wonder.

20

Above All,
Do No Harm

LONG, STAGNANT, AND COSTLY WARS TEND TO BEGIN
in idealistic fervor and end in cynical misery. Our own Civil War
inflicted a horrendous toll of death and seared our national consciousness with a brand that has only become deeper with time. In
1862, the Union Army rejoiced in singing the year's most popular
ditty:

> *Yes we'll rally round the flag, boys, we'll rally once again*
> *Shouting the battle cry of Freedom,*
> *We will rally from the hillside, we'll gather from the plain,*
> *Shouting the battle cry of Freedom . . .*
> *So we're springing to the call from the East and from the West*
> *And we'll hurl the rebel crew from the land we love the best.*

By 1864, Walter Kittredge's "Tenting on the Old Camp
Ground" had become the favorite song of both sides. The chorus, with its haunting (if naive) melody, summarizes the common
trajectory:

Many are the hearts that are weary tonight,
Wishing for the war to cease;
Many are the hearts looking for the right
To see the dawn of peace.

But nothing can quite match the horrors of World War I, the conflict that the French still call *la grande guerre* (the Great War) and that we labeled "the war to end all wars." America entered late and suffered relatively few casualties as a consequence—so we rarely appreciate the extent of carnage among soldiers or the near certainty of death or serious maiming along lines of stagnant trenches, where men fought back and forth month after month to take, and then lose again, a few shifting feet of territory. I feel chills up and down my spine whenever I look at the "honor roll" posted on the village green or main square of any small town in England or France. Above all else, I note the much longer lists for 1914–18 (often marking the near extermination of a generation of males) than for 1941–45. Rupert Brooke could write his famous poems of resignation and patriotism because he died in 1915, during the initial blush of enthusiasm:

If I should die, think only this of me:
That there's some corner of a foreign field
That is for ever England. There shall be
In that rich earth a richer dust concealed.

An actual gas attack in World War I.

His fellow poet Siegfried Sassoon, who survived and became a pacifist (a condition first attributed to shell shock and leading to his temporary confinement in a sanatorium), caught the drift of later realism:

> *And when the war is done and youth stone dead*
> *I'll toddle safely home and die—in bed.*

Sassoon met Wilfred Owen, the third member of this famous trio of British war poets, in the sanatorium. But Owen went back to the front, where he fell exactly one week before Armistice Day. Sassoon published his friend's single, slim volume of posthumous poetry, containing the most famous and bitter lines of all:

> *What passing-bells for these who died as cattle?*
> *Only the monstrous anger of the guns.*
> *Only the stuttering rifles' rapid rattle*
> *Can patter out their hasty orisons.*

Among the horrors of World War I, we remember not only the carnage caused by conventional tactics of trench warfare with bombs and bullets but also the first effective and large-scale use of newfangled chemical and biological weapons—beginning with a German chlorine gas attack along four miles of the French line at Ypres on April 22, 1915, and ending with 100,000 tons of various chemical agents used by both sides. The Geneva Protocol, signed in 1925 by most major nations (but not by the United States until much later), banned both chemical and biological weapons—a prohibition followed by all sides in World War II, even amid some of the grimmest deeds in all human history, including, and let us never forget, the most evil acts ever committed with poisonous gases in executing the "final solution" of the Holocaust in Nazi concentration camps. (A few violations have occurred in local wars: by the Italian army in Ethiopia in 1935–36, for example, and in recent fighting in Iran and Iraq.) The Geneva Protocol prohibited "the use in war of asphyxiating, poisonous or other gases, and of all analogous liquids, materials or devices."

A recent contribution to *Nature* (June 25, 1998), the leading British professional journal of science, recalled this episode of twentieth-century history in a remarkable letter entitled "Deadly relic of the Great War." The opening paragraph reads:

The curator of a police museum in Trondheim, Norway, recently discovered in his archive collection a glass bottle containing two irregularly shaped sugar lumps. A small hole had been bored into each of these lumps and a glass capillary tube, sealed at its tip, was embedded into one of the lumps. A note attached to the exhibit translated as follows: "A piece of sugar containing anthrax bacilli, found in the luggage of Baron Otto Karl von Rosen, when he was apprehended in Karasjok in January 1917, suspected of espionage and sabotage."

Modern science to the rescue, even in pursuit of a mad scheme that came to naught in a marginal and forgotten outpost of a great war—the very definition of historical trivia, however intriguing, in the midst of great pith and moment. The authors of the letter removed the capillary tube and dumped the contents ("a brown fluid") onto a petri dish. Two columns of conventional scientific prose then detailed the procedures followed, with all the usual rigor of long chemical names and precise amounts: "After incubation, 200μ1 of these cultures were spread on 7% of horse-blood agar and L-agar medium (identical to L-broth but solidified by the addition of 2% Difco Bacto agar)." The clear results may be stated more succinctly, as the authors both grew some anthrax bacilli in their cultures and then confirmed the presence of DNA from the same organism by PCR (polymerase chain reaction for amplifying small amounts of DNA to levels that can be analyzed). They write: "We therefore confirmed the presence of *B. anthracis* [scientific name of the anthrax bacillus] in the specimen by both culture and PCR. It proved possible to revive a few surviving organisms from the brink of extinction after they had been stored, without any special precautions, for 80 years."

But what was the good baron, an aristocrat of German, Swedish, and Finnish extraction, doing in this forsaken area of northeastern Norway in the middle of winter? Clearly up to no good, but to what form of no good? The authors continue:

When the Sheriff of Kautokeino, who was present at the group's arrest, derisively suggested that he should prepare soup from the contents of the tin cans labeled "Svea kott" (Swedish meat), the baron felt obliged to admit that each can actually contained between 2 and 4 kilograms of dynamite.

The baron's luggage also yielded some bottles of curare, various microbial cultures, and nineteen sugar cubes, each containing anthrax. The two cubes in Trondheim are, apparently, the only survivors of this old incident. The baron claimed that he was only an honorable activist for Finnish independence, out to destroy supply lines to Russian-controlled areas. (Finland had been under loose control of the Russian czar and did win independence after the Bolshevik revolution.) Most historians suspect that he had traveled to Norway at the behest, and in the employ, of the Germans, who had authorized a program for infecting horses and reindeer with anthrax to disrupt the transport of British arms (on sleds pulled by these animals) through northern Norway.

The baron, expelled after a few weeks in custody, never carried out his harebrained scheme. The authors of the *Nature* letter, Caroline Redmond, Martin J. Pearce, Richard J. Manchee, and Bjorn P. Berdal, have inferred his intent:

> The grinding of the sugar and its glass insert between the molar teeth of horses would probably result in a lethal infection as the anthrax spores entered the body, eventually facilitated through the small lesions produced in the wall of the alimentary tract by the broken glass. It is not known whether reindeer eat sugar lumps but presumably the baron never had the chance to carry out this piece of research.

As anthrax cannot be transmitted directly from animal to animal, the scheme probably would not have worked without a large supply of sugar cubes and very sweet teeth in the intended victims. But the authors do cite a potential danger to other participants: "However, if the meat from a dying animal had been consumed without adequate cooking, it is likely that human fatalities from gastrointestinal anthrax would have followed." The authors end their letter with a frank admission:

> This small but relatively important episode in the history of biological warfare is one of the few instances where there is confirmation of the intent to use a lethal microorganism as a weapon, albeit 80 years after the event. It did not, however, make any significant difference to the course of the Great War.

We may treat this botched experiment in biological warfare as light relief in a dark time, but the greatest evils often begin as farcical and apparently harmless escapades, while an old motto cites eternal vigilance as the price of liberty. If Hitler had been quietly terminated after his ragtag band failed to seize local power in their Beer Hall Putsch of 1923 in Munich—even the name of this incident marks the derision then heaped on the protagonists— the history of our century might have unfolded in a much different, and almost surely happier, manner. Instead, Hitler spent a mere nine months in jail, where he wrote *Mein Kampf* and worked out his grisly plans.

We humans may be the smartest objects that ever came down the pike of life's history on earth, but we remain outstandingly inept in certain issues, particularly when our emotional arrogance joins forces with our intellectual ignorance. Our inability to forecast the future lies foremost among these ineptitudes—not, in this case, as a limitation of our brains, but more as a principled consequence of the world's genuine complexity and indeterminism (see chapter 10 for a general discussion of our inability to predict coming events and patterns). We could go with this flow, but our arrogance intercedes, leading us to promote our ignorant intuitions into surefire forecasts about things to come.

I know only one antidote to the major danger arising from this incendiary mixture of arrogance and ignorance. Given our inability to predict the future, particularly our frequent failures to forecast the later and dire consequences of phenomena that seem impotent, or even risible, at their first faltering steps (a few reindeer with anthrax today, an entire human population with plague tomorrow), moral restraint may represent our only sure salvation. The wisdom of the Geneva Protocol lies in understanding that some relatively ineffective novelties of 1925 might become the principal horrors of a not-so-distant future. If such novelties can be nipped in the bud of their early ineffectiveness, we may be spared. We need only remember the legend of Pandora to recognize that some boxes, once opened, cannot be closed again.

The good sense in this vital form of moral restraint has been most seriously and effectively challenged by scientists who stand at the cutting edge of a developing technology and therefore imagine that they can control, or at least accurately forecast, any future developments. I dwell in the camp of scientists, but I want to illustrate the value of moral restraint as a counterweight to dangerous pathways forged either by complacency or active pursuit and fueled by false confidence about forecasting the future.

I told a story about aristocratic bumbling with ineffective biological weapons in World War I—but we might be in quite a fix today if we had assumed that this technology could never transcend such early ineptitude, and if we had not worked hard for international restriction. But a much deeper lesson may be drawn from the second innovative, and much more effective, technology later banned by the Geneva Protocol: chemical weaponry in World War I. The primary figure for this lesson became one of the founders of my own field of modern evolutionary biology—J.B.S. Haldane (1892–1964), called "the cleverest man I ever knew" by Sir Peter Medawar, who was certainly the cleverest man I have ever known.

Haldane mixed so many apparently contradictory traits into his persona that one word stands out in every description of him I have ever read: *enigmatic*. He could be shy and kind or blustering and arrogant, elitist (and viciously dismissive of underlings who performed a task poorly) or egalitarian. (Haldane became a prominent member of the British Communist Party and wrote volumes of popular essays on scientific subjects for their *Daily Worker*. Friends, attributing his political views to a deep personal need for iconoclastic and contrarian behavior, said that he would surely have become a monarchist if he had lived in the Soviet Union.) Haldane held no formal degree in science but excelled in several fields, largely as a consequence of superior mathematical ability. He remains most famous, along with R. A. Fisher and Sewall Wright, as one of the three founders of the modern theory of population genetics,

J.B.S. Haldane in his World War I military outfit.

especially for integrating the previously warring concepts of Mendelian rules for heredity with Darwinian natural selection.

But a different contradiction motivates Haldane's appearance as the focus of this essay. Haldane, a man of peace and compassion, adored war—or at least his role on the front lines in World War I, where he was twice wounded (both times seriously) and mighty lucky to come home in one piece. Some people regarded him as utterly fearless and courageous beyond any possible call of duty; others, a bit more cynically perhaps (but also, I suspect, more realistically), viewed him as a latter-day Parsifal—a perfect fool who survived in situations of momentary danger (usually created as a result of his own bravado and appalling recklessness) by a combination of superior intelligence joined with more dumb luck than any man has a right to expect. In any case, J. B. S. Haldane had a good war—every last moment of it.

He particularly enjoyed a spell of trench warfare against Turkish troops near the Tigris River, where, away from the main European front and unencumbered by foolish orders from senior officers without local experience, men could fight *mano a mano* (or at least gun against gun). Haldane wrote: "Here men were pitted against individual enemies with similar weapons, trench mortars or rifles with telescopic sights, each with a small team helping him. This was war as the great poets have sung it. I am lucky to have experienced it." Haldane then offered a more general toast to such a manly occupation: "I enjoyed the comradeship of war. Men like war because it is the only socialized activity in which they have ever taken part. The soldier is working with comrades for a great cause (or so at least he believes). In peacetime he is working for his own profit or someone else's."

Haldane's contact with chemical warfare began in great disappointment. After the first German gas attack at Ypres, the British War Office, by Lord Kitchener's direct command, dispatched Haldane's father, the eminent respiratory physiologist John Scott Haldane, to France in a desperate effort to overcome this new danger. The elder Haldane, who had worked with his son on physiological experiments for many years, greatly valued both J.B.S.'s mathematical skills and his willingness to act as a human guinea pig in medical experiments (an ancient tradition among biologists and a favorite strategy of the elder Haldane, who never asked his son to do anything he wouldn't try on himself). So J.B.S., much to his initial disgust, left the front lines he loved so well and moved into a laboratory with his father.

J.B.S. already knew a great deal about toxic gases, primarily through his role as father's helper in self-experimentation. He recalled some early work with his father on firedamp (methane) in mines:

To demonstrate the effects of breathing firedamp, my father told me
to stand up and recite Mark Antony's speech from Shakespeare's
Julius Caesar, beginning "Friends, Romans, countrymen." I soon
began to pant, and somewhere about "the noble Brutus" my legs
gave way and I collapsed on to the floor, where, of course, the air
was all right. In this way I learnt that firedamp is lighter than air and
not dangerous to breathe.

(Have you ever read a testimony more congruent with the stereotype of British
upper-class intellectual dottiness?)

The Haldanes, *père et fils,* led a team of volunteer researchers in vitally
important work (no doubt saving many thousands of lives) on the effects of
noxious substances and the technology of gas masks. As always, they performed
the most unpleasant and dangerous experiments on themselves. J.B.S. recalled:

We had to compare the effects on ourselves of various quantities,
with and without respirators. It stung the eyes and produced a tendency to gasp and cough when breathed. . . . As each of us got sufficiently affected by gas to render his lungs duly irritable, another
would take his place. None of us was much the worse for the gas,
or in any real danger, as we knew where to stop, but some had to

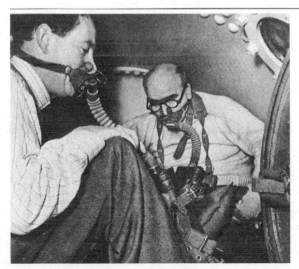

*J.B.S. Haldane
engaged in
physiological self-
experimentation on
the effects of various
gases.*

go to bed for a few days, and I was very short of breath and incapable of running for a month or so.

Thus, we cannot deny Haldane's superior knowledge or his maximal experience in the subject of chemical warfare. He therefore becomes an interesting test for the proposition that such expertise should confer special powers of forecasting—and that the technical knowledge of such people should therefore be trusted if they advocate a path of further development against the caution, the pessimism, even the defeatism of others who prefer moral restraint upon future technological progress because they fear the power of unforeseen directions and unanticipated consequences.

In 1925, as nations throughout the world signed the Geneva Protocol to ban chemical and biological warfare, J. B. S. Haldane published the most controversial of all his iconoclastic books: a slim volume of eighty-four pages entitled *Callinicus: A Defense of Chemical Warfare,* based on a lecture he had given in 1924. (Callinicus, a seventh-century Jewish refugee in Constantinople, invented Greek fire, an incendiary liquid that could be shot from siphons toward enemy ships or troops. The subsequent flames, almost impossible to extinguish, helped save the Byzantine empire from Islamic conquest for several centuries. The formula, known only to the emperor and to Callinicus' family, who held an exclusive right of manufacture, remained a state secret and still elicits controversy among scholars of warfare.)

Haldane's argument can be easily outlined. He summarized the data, including death tolls and casualty rates, from gas attacks in World War I and proclaimed the results more humane than the consequences of conventional weaponry.

A case can be made out for gas as a weapon on humanitarian grounds, based on the very small proportion of killed to casualties from gas in the War, and especially during its last year [when better gas masks had been made and widely distributed].

Haldane based this conclusion on two arguments. He first listed the chemical agents used in the war and branded most of them as not dangerous for having only transient effects (making the assumption that temporarily insensate soldiers would be passed by or humanely captured rather than slaughtered). He regarded the few chemicals that could induce more permanent harm—mustard gas, in particular—as both hard to control and relatively easy to avoid, with proper equipment. Second, he called upon his own frequent experience with poison gases and stated a strong preference for these agents over his equally personal contact with bullets:

> Besides being wounded, I have been buried alive, and on several occasions in peacetime I have been asphyxiated to the point of unconsciousness. The pain and discomfort arising from the other experiences were utterly negligent compared with those produced by a good septic shell wound.

Haldane therefore concluded that gas, for reasons of effectiveness as a weapon and relative humaneness in causing few deaths compared with the number of temporary incapacitations, should be validated and further developed as a primary military tactic:

> I certainly share their [pacifists'] objection to war, but I doubt whether by objecting to it we are likely to avoid it in future, however lofty our motives or disinterested our conduct. . . . If we are to have more wars, I prefer that my country should be on the winning side. . . . If it is right for me to fight my enemy with a sword, it is right for me to fight him with mustard gas; if the one is wrong, so is the other.

I do not flinch before this last statement from the realm of ultimate realpolitik. The primary and obvious objection to Haldane's thesis in *Callinicus*—not only as raised now by me in the abstract, but also as advanced by Haldane's

World War I paraphernalia for protection from poison gas attacks.

numerous critics in 1925—holds that, whatever the impact of poison gas in its infancy in World War I (and I do not challenge Haldane's assessment), unrestrained use of this technology may lead to levels of effectiveness and numbers of deaths undreamed of in earlier warfare. Better the devil we know best than a devil seen only as an ineffective baby just introduced into our midst. If we can squelch this baby now, by moral restraint and international agreement, let's do so before he grows into a large and unstoppable adult potentially far more potent than any devil we know already.

(I should offer the proviso that, in making this general argument for moral restraint, I am speaking only of evident devils, or destructive technologies with no primary role in realms usually designated as human betterment: healing the sick, increasing agricultural yields, and so on. I am *not* talking about the more difficult, and common, problem of new technologies—cloning comes to mind as the current topic of greatest interest [see chapter 19]—with powerfully benevolent intended purposes but also some pretty scary potential misuses in the wrong hands, or in the decent hands of people who have not pondered the unintended consequences of good deeds. Such technologies may be regulated, but surely should not be banned.)

Haldane's response to this obvious objection reflects all the arrogance described in the first part of this essay: I have superior scientific knowledge of this subject and can therefore be trusted to forecast future potentials and dangers; from what I know of chemistry, and from what I have learned from the data of World War I, chemical weapons will remain both effective and relatively humane and should therefore be further developed. In other words, and in epitome: trust me.

> One of the grounds given for objection to science is that science is responsible for such horrors as those of the late War. "You scientific men (we are told) never think of the possible applications of your discoveries. You do not mind whether they are used to kill or to cure. Your method of thinking, doubtless satisfactory when dealing with molecules and atoms, renders you insensible to the difference between right and wrong." . . . The objection to scientific weapons such as the gases of the late War, and such new devices as may be employed in the next, is essentially an objection to the unknown. Fighting with lances or guns, one can calculate, or thinks one can calculate, one's chances. But with gas or rays or microbes one has an altogether different state of affairs.

. . . What I have said about mustard gas might be applied, *mutatis mutandis*, to most other applications of science to human life. They can all, I think, be abused, but none perhaps is always evil; and many, like mustard gas, when we have got over our first not very rational objections to them, turn out to be, on the whole, good.

In fact, Haldane didn't even grant moral arguments—or the imposition of moral restraints—any role at all in the prevention of war. He adopted the same parochial and arrogant position, still all too common among scientists, that war can be ended only by rational and scientific research: "War will be prevented only by a scientific study of its causes, such as has prevented most epidemic diseases."

I am no philosopher, and I do not wish to combat Haldane's argument on theoretical grounds here. Let us look instead at the basic empirical evidence, unwittingly presented by Haldane himself in *Callinicus*. I therefore propose the following test: if Haldane's argument should prevail, and scientific recommendations should be trusted because scientists can forecast the future in areas of their expertise, then the success of Haldane's own predictions will validate his approach.

I propose that two great impediments generally stand in the way of successful prediction: first, our inability, in principle, to know much about complex futures along the contingent and nondeterministic pathways of history; and second, the personal hubris that leads us to think we act in a purely and abstractly rational manner, when our views really arise from unrecognized social and personal prejudices.

Callinicus contains an outstanding example of each error, and I rest my case for moral restraint here. Haldane does consider the argument that further development of chemical and biological weapons might prompt an investigation into even more powerful technologies of destruction—in particular, to unleashing the forces of the atom. But he dismisses this argument on scientific grounds of impossible achievement:

> Of course, if we could utilize the forces which we now know to exist inside the atom, we should have such capacities for destruction that I do not know of any agency other than divine intervention which would save humanity from complete and peremptory annihilation. . . . [But] we cannot utilize subatomic phenomena. . . . We cannot make apparatus small enough to disintegrate or fuse atomic nuclei. . . . We can only bombard them with particles of which

perhaps one in a million hit, which is like firing keys at a safe-door from a machine gun a mile away in an attempt to open it. . . . We know very little about the structure of the atom and almost nothing about how to modify it. And the prospect of constructing such an apparatus seems to me to be so remote that, when some successor of mine is lecturing to a party spending a holiday on the moon, it will still be an unsolved (though not, I think, an ultimately unsolvable) problem.

To which, we need only reply: Hiroshima, 1945; Mr. Armstrong on the Moon, 1969. And we are still here—in an admittedly precarious atomic world—thanks to moral and political restraint.

But the even greater danger of arrogant and "rational" predictions unwittingly based on unrecognized prejudice led Haldane to the silliest statement he ever made—one that might be deemed socially vicious if our laughter did not induce a more generous mood. Haldane tries to forecast the revised style of warfare that mustard gas must impose upon future conflicts. He claims that some people have a natural immunity, differently distributed among our racial groups. He holds that 20 percent of whites, but 80 percent of blacks, are unaffected by the gas. Haldane then constructs a truly dotty scenario for future gas warfare: vanguards of black troops will lead the attack; German forces, with less access to this aspect of human diversity, might suffer some disadvantage, but their superior chemical knowledge should see them through, and balances should therefore be maintained:

> It seems, then, that mustard gas would enable an army to gain ground with far less killed on either side than the methods used in the late War, and would tend to establish a war of movement leading to a fairly rapid decision, as in the campaigns of the past. It would not upset the present balance of power, Germany's chemical industry being counterposed by French negro troops. Indians [that is, East Indians available to British forces] may be expected to be nearly as immune as negroes.

But now Haldane sees a hole in his argument. He steps back, breathes deeply, and finds a solution. Thank God for that 20 percent immunity among whites!

The American Army authorities made a systematic examination of the susceptibility of large numbers of recruits. They found that there was a very resistant class, comprising 20% of the white men tried, but no less than 80% of the negroes. This is intelligible, as the symptoms of mustard gas blistering and sunburn are very similar, and negroes are pretty well immune to sunburn. It looks, therefore, as if, after a slight preliminary test, it should be possible to obtain colored troops who would all be resistant to mustard gas blistering in concentrations harmful to most white men. Enough resistant whites are available to officer them.

I am simply astonished (and also bemused) that this brilliant man, who preached the equality of humankind in numerous writings spanning more than fifty years, could have been so mired in conventional racial prejudices, and so wedded to the consequential and standard military practices of European and American armies, that he couldn't expand his horizons far enough even to imagine the possibility of competent black officers—and therefore had to sigh in relief at the availability of a few good men among the rarely resistant whites. If Haldane couldn't anticipate even this minor development in human relationships and potentialities, why should we trust his judgments about the far more problematical nature of future wars?

(This incident should carry the same message for current discussions about underrepresentation of minorities as managers of baseball teams or as quarterbacks in football. I also recall a famous and similar episode of ridiculously poor prediction in the history of biological determinism—the estimate by a major European car manufacturer, early in the century, that his business would be profitable but rather limited. European markets, he confidently predicted, would never require more than a million automobiles—for only so many men in the lower classes possessed sufficient innate intellectual ability to work as chauffeurs! Don't you love the triply unacknowledged bias of this statement—that poor folks rarely rank high in fixed genetic intelligence and that neither women nor rich folks could ever be expected to drive a car?)

The logic of my general argument must lead to a truly modest proposal. Wouldn't we all love to fix the world in one fell swoop of proactive genius? We must, of course, never stop dreaming and trying. But we must also temper our projects with a modesty born of understanding that we cannot predict the future and that the best-laid plans of mice and men often founder into a deep pit dug by unanticipated consequences. In this context, we should honor what might be called the "negative morality" of restraint and consideration, a principle that wise people have always understood (as embodied in the golden rule) and dreamers have generally rejected, sometimes for human good but more often for the evil that arises when demagogues and zealots try to impose their "true belief" upon all humanity, whatever the consequences.

The Hippocratic oath, often misunderstood as a great document about general moral principles in medicine, should be read as a manifesto for protecting the secret knowledge of a guild and for passing skills only to designated initiates. But the oath also includes a preeminent statement, later recast as a Latin motto for physicians, and ranking (in my judgment) with the Socratic dictum "know thyself" as one of the two greatest tidbits of advice from antiquity. I can imagine no nobler rule of morality than this single phrase, which every human being should engrave into heart and mind: *primum non nocere*—above all, do no harm.

VI

Evolution

at

All Scales

21

Of Embryos
and Ancestors

"EVERY DAY, IN EVERY WAY, I'M GETTING BETTER
and better." I had always regarded this famous phrase as a primary
example of the intellectual vacuity that often passes for profundity in
our current era of laid-back, New Age bliss—a verbal counterpart to
the vapidity of the "have a nice day" smiley face. But when I saw this
phrase chiseled in stone on the pediment of a French hospital built in
the early years of our century, I knew that I must have missed a longer
and more interesting pedigree. This formula for well-being, I then discovered, had been devised in 1920 by Emile Coué (1857–1926), a
French pharmacist who made quite a stir in the pop-psych circles of
his day with a theory of self-improvement through autosuggestion
based on frequent repetition of this mantra—a treatment that received
the name of Couéism. (In a rare example of improvement in translation, this phrase gains both a rhyme and better flow, at least to my ears,
when converted to English from Coué's French original—*"tous les
jours, à tous les points de vue, je vais de mieux en mieux."*)

I don't doubt the efficacy of Coué's mantra, for the "placebo effect" (its only possible mode of action) should not be dismissed as a delusion, but cherished as a useful strategy for certain forms of healing—a primary example of the influence that mental attitudes can wield upon our physical sense of well-being. However, as a general description for the usual style and pacing of human improvement, the constant and steady incrementalism of Coué's motto—a twentieth-century version of an ancient claim embodied in the victory cry of Aesop's tortoise, "slow and steady wins the race"—strikes me as only rarely applicable, and surely secondary to the usual mode of human enlightenment, either attitudinal or intellectual: that is, *not* by global creep forward, inch by subsequent inch, but rather in rushes or whooshes, usually following the removal of some impediment, or the discovery of some facilitating device, either ideological or technological.

The glory of science lies in such innovatory bursts. Centuries of vain speculation dissolved in months before the resolving power of Galileo's telescope, trained upon the full range of cosmic distances, from the moon to the Milky Way (see chapter 2). About 350 years later, centuries of conjecture and indirect data about the composition of lunar rocks melted before a few pounds of actual samples, brought back by Apollo after Mr. Armstrong's small step onto a new world.

In the physical sciences, such explosions of discovery usually follow the invention of a device that can, for the first time, penetrate a previously invisible realm—the "too far" by the telescope, the "too small" by the microscope, the imperceptible by X-rays, or the unreachable by spaceships. In the humbler world of natural history, episodes of equal pith and moment often follow a eureka triggered by continually available mental, rather than expensively novel physical, equipment. In other words, great discovery often requires a map to a hidden mine filled with gems then easily gathered by conventional tools, not a shiny new space-age machine for penetrating previously inaccessible worlds.

The uncovering of life's early history has featured several such cascades of discovery following a key insight about proper places to look—and I introduce this year's wonderful story by citing a previous incident of remarkably similar character from the last generation of our science (literally so, for this year's discoverer wrote his Ph.D. dissertation under the guidance of the first innovator).

When, as a boy in the early 1950s, I first became fascinated with paleontology and evolution, the standard dogma about the origin of life proclaimed such an event inherently improbable, but achieved on this planet only because the immensity of geological time must convert the nearly impossible into the virtually certain. (With no limit on the number of tries, you will eventually flip fifty heads in a row with an honest coin.) As evidence for asserting the exquisite specialness of life in the face of overwhelmingly contrary odds, these conventional sources cited the absence of any fossils representing the first half of the earth's existence—a span of more than two billion years, often formally designated on older geological charts as the Azoic (literally "lifeless") era. Although scientists do recognize the limitations of such "negative evidence" (the first example of a previously absent phenomenon may, after all, turn up tomorrow), this failure to find any fossils for geology's first two billion years did seem fairly persuasive. Paleontologists had been searching assiduously for more than a century and had found nothing but ambiguous scraps and blobs. Negative results based on such sustained effort over so many years do begin to inspire belief.

But the impasse broke in the 1950s, when Elso Barghoorn and Stanley Tyler reported fossils of unicellular life in rocks more than two billion years old. Paleontologists, to summarize a long and complex story with many exciting turns and notable heroes, had been looking in the wrong place—in conventional sediments that rarely preserve the remains of single-celled bacterial organisms without hard parts. We had not realized that life had remained so simple for so long, or that the ordinary sites for good fossil records could not preserve such organisms.

Barghoorn and colleagues dispelled a century of frustration by looking in a different place, where cellular remains of bacteria might be preserved—in chert beds. Chert has the same chemical formula (with a different molecular arrangement) as quartz: silicon dioxide. Paleontologists rarely think of looking for fossils in silicate rocks—for the perfectly valid and utterly obvious reason that silicates form by the cooling of volcanic magmas and therefore cannot contain organic remains. (Life, after all, doesn't flourish in bubbling lavas, and anything falling in gets burnt to a crisp.) But cherts can form at lower temperatures and be deposited amid layers of ordinary sediments in oceanic waters. Bacterial cells, when trapped in this equivalent of surrounding glass, can be preserved as fossils.

This cardinal insight—that we had been searching in the wrong venue of ordinary (and barren) sediments rather than in fruitful cherts—created an entire field of study: collecting data for the first two-thirds or so of life's full history! Forty years later, we may look back with wonder at the flood of achievement and the complete overturn of established wisdom. We now possess a rich fossil record of early life, extending right back to the earliest potential source for cellular evidence. (The oldest rocks on earth that could preserve such data do contain abundant fossils of bacterial organisms. These 3.5-to-3.6-billion-year-old rocks from Australia and South Africa are the most ancient strata on earth that have not been sufficiently altered by subsequent heat and pressure to destroy all anatomical evidence of life.)

Such ubiquity and abundance has forced a reversal of the old view. The origin of life at simplest bacterial grade now seems inevitable rather than improbable. As a mantra for memory, may I suggest: "Life on earth is as old as it could be." I realize, of course, that an earliest possible appearance builds no proof for inevitability. After all, even a highly improbable event might occur, by good fortune, early in a series of trials. (You might flip those fifty successive heads on your tenth attempt—but don't count, or bet, on it!) Nonetheless, faced with the data we now possess—that life appeared as soon as environmental conditions permitted such an event and then remained pervasive forever after—our thoughts must move to ideas about almost predictive inevitability. Given a planet of earthly size, distance, and composition, life of simplest grade probably originates with virtual certainty as a consequence of principles of organic chemistry and the physics of self-organizing systems.

But whatever the predictability of life's origin, subsequent pathways of evolution have run in mighty peculiar directions, at least with respect to our conventional hopes and biases. The broadest pattern might seem to confirm our usual view of generally increasing complexity, leading sensibly to human consciousness: after all, the early earth sported only bacteria, while our planet now boasts people, ant colonies, and oak trees. Fair enough, but closer scrutiny of general timings or particular details can promote little faith in any steady pattern. If greater size and complexity provide such Darwinian blessings, why did life take so long to proceed "onward," and why do most of the supposed steps occur so quirkily and so quickly? Consider the following epitome of major events:

Fossils, as stated above, appear as soon as they possibly could in the geological record. But life then remains exclusively at this simplest so-called prokaryotic grade (unicells without any internal organelles, that is, no nuclei, chromosomes, mitochondria, and so on) for about half its subsequent history. The first unicells of the more complex eukaryotic grade (with the conventional organelles of our high school textbook pictures of an amoeba or paramecium) do not appear in the fossil record until about two billion years ago. The three great multicellular kingdoms of plants, fungi, and animals arise subsequently (and, at least for algae within the plant kingdom, more than once and independently) from eukaryotic unicells. Fossils of simple multicellular algae extend back fairly reliably to about one billion years, and far more conjecturally to as many as 1.8 billion years.

But the real enigma—at least with respect to our parochial concerns about the progressive inevitability of our own lineage—surrounds the origin and early history of animals. If life had always been hankering to reach a pinnacle of expression as the animal kingdom, then organic history seemed in no hurry to initiate this ultimate phase. About five-sixths of life's time had elapsed before animals made their first appearance in the fossil record some 600 million years ago. Moreover, the earth's first community of animals, which held nearly exclusive sway from an initial appearance some 600 million years ago right to the dawn of the Cambrian period 543 million years ago, consisted of enigmatic species with no clear relation to modern forms.

These so-called Ediacaran animals (named for the locality of first discovery in Australia, but now known from all continents) could grow quite large (up to a few feet in length), but apparently contained neither complex internal organs, nor even any recognizable body openings for mouth or anus. Many Ediacaran creatures were flattened, pancakelike forms (in a variety of shapes and sizes), built of numerous tubelike sections, complexly quilted together into a single structure. Theories about the affinities of Ediacaran organisms span the full gamut—from viewing them, most conventionally, as simple ancestors for several modern phyla; to interpreting them, most radically, as an entirely separate (and ultimately failed) experiment in multicellular animal life. An intermediate position, now gaining favor—a situation that should lead to no predictions about the ultimate outcome of this complex debate—treats Ediacaran animals as a bountiful expression of the range of possibilities for diploblastic animals (built of two body layers), a group now so reduced in diversity (and subsisting only as corals, jellyfishes, and their allies) that living representatives provide little understanding of full potentials.

Modern animals—except for sponges, corals, and a few other minor groups—are all triploblastic, or composed of three body layers: an ectoderm, forming nervous tissue and other organs; mesoderm, forming reproductive structures and other parts; and endoderm, building the gut and other internal organs. (If you learned a conventional list of phyla back in high school biology, all groups from the flatworms on "up," including all the "big" phyla of annelids, arthropods, mollusks, echinoderms, and vertebrates, are triploblasts.) This three layered organization seems to act as a prerequisite for the formation of conventional, complex, mobile, bilaterally symmetrical organisms with body cavities, appendages, sensory organs, and all the other accoutrements that set our standard picture of a "proper" animal. Thus, in our aimlessly parochial manner (and ignoring such truly important groups as corals and sponges), we tend to equate the problem of the beginning of modern animals with the origin of triploblasts. If the Ediacaran animals are all (or mostly) diploblasts, or even more genealogically divergent from triploblast animals, then this first fauna does not resolve the problem of the origin of animals (in our conventionally limited sense of modern triploblasts).

The story of modern animals then becomes even more curious. The inception of the Cambrian period, 543 million years ago, marks the extinction, perhaps quite rapidly, of the Ediacara fauna, and the beginning of a rich record for animals with calcareous skeletons easily preserved as fossils. But the first phase of the Cambrian, called Manakayan and lasting from 543 to 530 million years ago, primarily features a confusing set of spines, plates, and other bits and pieces called (even in our technical literature) the SSF, or "small shelly fossils" (presumably the disarticulated fragments of skeletons that had not yet evolved to large, discrete units covering the entire organism).

The next two phases of the Cambrian (called Tommotian and Atdabanian, and ranging from 530 to about 520 million years ago) mark the strangest, most important, and most intriguing of all episodes in the fossil record of animals— the short interval known as the "Cambrian explosion," and featuring the first fossils of all animal phyla with skeletons subject to easy preservation in the fossil record. (A single exception, a group of colonial marine organisms called the

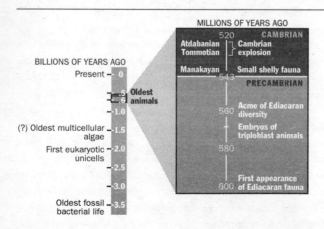

Time charts for major events in the history of life (left) and for details of the Cambrian Explosion and other events in the origin of multicellular animals (right).

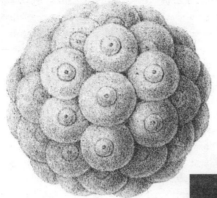

Haeckel's theoretical drawings of ancestral animals (left) compared with fossils of a Precambrian embryo (below).

Bryozoa, make their appearance at the beginning of the next or Ordovician period. Many intriguing inventions, including human consciousness and the dance language of bees, have arisen since then, but no new phyla, or animals of starkly divergent anatomical design.)

The Cambrian explosion ranks as such a definitive episode in the history of animals that we cannot possibly grasp the basic tale of our own kingdom until we achieve better resolution for both the antecedents and the unfolding of this cardinal geological moment. A major discovery, announced in February 1998, and also based on learning to look in a previously unsuspected place, has thrilled the entire paleontological community for its promise in unraveling the previously unknown history of triploblast animals before the Cambrian explosion.

If the Cambrian explosion inspires frustration for its plethora of data—too much, too confusing, and too fast—the Precambrian history of triploblast animals engenders even more chagrin for its dearth. The complex animals of the explosion, so clearly assignable to modern phyla, obviously didn't arise *ex nihilo* at their first moment of fossilization, but who (and where) are their antecedents in Precambrian times? What were the forebears of modern animals doing for 50 million prior years, when Ediacaran diploblasts (or stranger creatures) ruled the animal world?

Up to now, we have engaged in much speculation, while possessing only a whiff or two of data. Ediacaran strata also contain trails and feeding traces presumably made by triploblast organisms of modern design (for the flattened and mostly immobile Ediacaran animals could not crawl, burrow, or feed in a manner so suggestive of activities now confined to triploblast organisms). Thus, we do have evidence for the existence, and even the activities, of precursors for modern animals before the Cambrian explosion, but no data at all about their anatomy and appearance—a situation akin to the frustration we might feel if we could hear birdsong but had never seen a bird.

A potential solution—or at the very least, a firm and first source of anatomical data—has just been discovered by applying the venerable motto (so beloved by people, including yours truly, of shorter-than-average stature): good things often come in small packages, or to choose a more literary and inspirational expression, Micah's statement (5:2) taken by the later evangelists as a prophecy of things to come: "But thou, Bethlehem . . . though thou be little among the thousands of Judah, yet out of thee shall he come forth unto me that is to be ruler in Israel."

In short, paleontologists had been looking for conventional fossils in the usual (and visible) size ranges of adult organisms: fractions to few inches. But a solution had been lurking at the smaller size of creatures just barely visible (in principle) but undetectable in conventional practice—in the domain of embryos. Who would ever have thought that delicate embryos might be preserved as fossils, when presumably hardier adults left no fragments of their existence? The story, a fascinating lesson in the ways of science, has been building for more than a decade, but has only just been extended to the problem of Precambrian animals.

Fossils form in many modes and styles—as original hard parts preserved within entombing sediments, or as secondary structures formed by impressions of bones or shells (molds) that may then become filled with later sediments (casts). But original organic materials may also be replaced by percolating minerals—a process called petrifaction, or literally "making into stone," a phenomenon perhaps best represented in popular knowledge by gorgeous specimens from the Petrified Forest of Arizona, where multicolored agate (another form of silicon dioxide) has replaced original carbon so precisely that the wood's cellular structure can still be discerned. (Petrifaction enjoys sufficient public renown that many people mistakenly regard such replacement as the primary definition of a fossil. But any bit of an ancient organism qualifies as a fossil, whatever the style of preservation. In almost any circumstance, a professional would much prefer to work with unaltered hard parts than with petrified replacements.)

In any case, one poorly understood style of petrifaction leads to replacement of soft tissues by calcium phosphate—a process called phosphatization. This style of replacement can occur within days of death, thus leading to the rare and precious phenomenon of petrifaction before decay of soft anatomy. Phosphatization might provide a paleontologist's holy grail if all soft tissues could thus be preserved at any size in any kind of sediment. Alas, the process seems to work in detail only for tiny objects up to about two millimeters in length (26.4 millimeters make an inch, so we are talking about barely visible dots, not even about bugs large enough to be designated as "yucky" when found in our dinner plates or beds).

Still, on the good old principle of not looking gift horses (or unexpected bounties) in the mouth (by complaining about an unavailable better deal), let us rejoice in the utterly unanticipated prospect that tiny creatures—which are, after all, ever so abundant in nature, however much they may generally pass beneath our exalted notice—might become petrified in sufficient detail to preserve their bristles, hairs, or even their cellular structure. The recognition that phosphatization may open up an entire world of tiny creatures, previously never considered as candidates for fossilization at all, may spark the greatest burst of paleontological exploration since the discovery that two billion years of Precambrian life lay hidden in chert.

The first hints that exquisite phosphatization of tiny creatures might resolve key issues in the early evolution of animals date to a discovery made in the mid-1970s and then researched and reported in one of the most elegant, but rather sadly underappreciated, series of papers ever published in the history of paleontology: the work of two German scientists, Klaus J. Müller and Dieter Walossek, on the fauna of distinctive upper Cambrian rocks in Sweden, known as Orsten beds. In these layers of limestone concretions, tiny arthropods (mostly larvae of crustaceans) have been preserved by phosphatization in exquisite, three-dimensional detail. The photography and drawings of Walossek and Müller have rarely been equaled in clarity and aesthetic brilliance, and their papers are a delight both to read and to see. (For a good early summary, consult K. J. Müller and D. Walossek, "A remarkable arthropod fauna from the Upper Cambrian 'Orsten' of Sweden," *Transactions of the Royal Society of Edinburgh,* 1985, volume 76, pages 161–72; for a recent review, see Walossek and Müller, "Cambrian 'Orsten'-type arthropods and the phylogeny of Crustacea," in R. A. Fortey and R. H. Thomas, eds., *Arthropod Relationships,* London: Chapman and Hall, 1997.)

By dissolving the limestone in acetic acid, Walossek and Müller can recover the tiny phosphatized arthropods intact. They have collected more than one hundred thousand specimens following this procedure and have summarized their findings in a recent paper of 1997:

> The cuticular surface of these arthropods is still present in full detail, revealing eyes and limbs, hairs and minute bristles, . . . gland openings, and even cellular patterns and grooves of muscle attachments underneath. . . . The maximum size of specimens recovered in this type of preservation does not exceed 2 mm.

From this beginning, other paleontologists have proceeded backward in time, and downward in growth from larvae to early embryonic stages containing just a few cells. In 1994, Xi-guang Zhang and Brian R. Pratt found balls of presumably embryonic cells measuring 0.30 to 0.35 millimeter in length and representing, perhaps, the earliest stages of adult trilobites also found in the same Middle Cambrian strata (Zhang and Pratt, "Middle Cambrian arthropod embryos with blastomeres," *Science,* 1994, volume 266, pages 637–38). In 1997, Stefan Bengston and Yue Zhao then reported even earlier phosphatized embryos from basal Cambrian strata in China and Siberia. In an exciting addition to this growing literature, these authors traced a probable growth series, from embryos to tiny near adults, for two entirely different animals: a species from an enigmatic extinct group, the conulariids; and a probable segmented worm (Bengston and Zhao, "Fossilized metazoan embryos from the earliest Cambrian," *Science,* 1997, volume 277, pages 1645–48).

When such novel technologies first encounter materials from a truly unknown or unsuspected world, genuinely revolutionary conclusions often emerge. In what may well be regarded by subsequent historians as the greatest paleontological discovery of the late twentieth century, Shuhai Xiao, a postdoctoral student in our paleontological program, Yun Zhang of Beijing University, and my colleague, and Shuhai Xiao's mentor, Andrew H. Knoll, have just reported their discovery of the oldest triploblastic animals, preserved as phosphatized embryos in rocks from southern China estimated at 570 million years in age—and thus even older than the best-preserved Ediacaran faunas, found in strata about 10 million years younger (see Xiao, Zhang, and Knoll, "Three-dimensional preservation of algae and animal embryos in a Neoproterozoic phosphorite," *Nature,* 1998, volume 391, pages 553–58). These phosphatized fossils include a rich variety of multicellular algae, showing, according to the authors, that "by the time large animals enter the fossil record, the three principal groups of multicellular algae had not only diverged from other protistan [unicellular] stocks but had evolved a surprising degree of the morphological complexity exhibited by living algae."

Still, given our understandably greater interest in our own animal kingdom, most attention will be riveted upon some smaller and rarer globular fossils, averaging half a millimeter in length, and found phosphatized in the same strata: an exquisite series of earliest embryonic stages, beginning with a single fertilized egg and proceeding through two-cell, four-cell, eight-cell, and sixteen-cell stages to small balls of cells representing slightly later phases of early development. These embryos cannot be assigned to any particular group (as more distinctive later stages have not yet been found), but their identification as earliest stages of triploblastic animals seems secure, both from characteristic features (especially the unchanging overall size of the embryo during these earliest stages, as average cell size decreases to pack more cells into a constant space), and uncanny resemblance to particular traits of living groups. (Several embryologists have told Knoll and colleagues that they would have identified these specimens as embryos of living crustaceans had they not been informed of their truly ancient age!)

Elso Barghoorn, Knoll's thesis adviser, opened up the world of earliest life by discovering that bacteria could be preserved in chert. Now, a full generation later, Knoll and colleagues have penetrated the realm of earliest known animals of modern design by accessing a new domain where phosphatization preserves minute embryonic stages, but no known process of fossilization can reliably render potentially larger phases of growth. When I consider the cascade of knowledge that proceeded from Barghoorn's first report of Precambrian bacteria to our current record spanning three billion Precambrian years and hundreds of recorded forms, I can only conclude that the discovery of Xiao, Zhang, and Knoll places us at a gateway of equal promise for reconstructing the earliest history of modern animals, before their overt evolutionary burst to large size and greatly increased anatomical variety in the subsequent Cambrian explosion. If we can thereby gain any insight into the greatest of all mysteries surrounding the early evolution of animals—the causes of both the anatomical explosion itself and the "turning off" of evolutionary fecundity for generating new phyla thereafter—then paleontology will shake hands with evolutionary theory in the finest merger of talents ever applied to the resolution of a historical enigma.

A closing and more general commentary may help to set a context of both humility and excitement at the threshold of this new quest. First, we might be able to coordinate the direct evidence of fossils with a potentially powerful indirect method for judging the times of origin and branching for major animal groups: the measurement of relative degrees of detailed genetic similarity among living representatives of diverse animal phyla. Such measurements can be made with great precision upon large masses of data, but firm conclusions do not always follow because various genes evolve at different rates that also maintain no constancy over time—and most methods applied so far have made simplifying (and probably unjustified) assumptions about relatively even ticking of supposed molecular clocks.

For example, in a paper that received much attention upon publication in 1996, G. A. Wray, J. S. Levinton, and L. H. Shapiro used differences in the molecular sequences of seven genes in living representatives of major phyla to derive an estimate of roughly 1.2 billion years for the divergence time between chordates (our phylum) and the three great groups on the other major genealogical branch of animals (arthropods, annelids, and mollusks), and 1.0 billion years for the later divergence of chordates from the more closely related phylum of echinoderms (Wray, Levinton, and Shapiro, "Molecular evidence for deep Precambrian divergences among metazoan phyla," *Science,* 1996, volume 274, pages 568–73).

This paper sowed a great deal of unnecessary confusion when several uncomprehending journalistic reports, and a few careless statements by the authors, raised the old and false canard that such an early branching time for animal phyla disproves the reality of the Cambrian explosion by rendering this apparent burst of diversity as the artifact of an imperfect fossil record (signifying, perhaps, only the invention of hard parts, rather than any acceleration of anatomical innovation). For example, Wray et al. write: "Our results cast doubt on the prevailing notion that the animal phyla diverged explosively during the Cambrian or late Vendian [Ediacaran times], and instead suggest that there was an extended period of divergence . . . commencing about a billion years ago."

But such statements confuse the vital distinction, in both evolutionary theory and actual results, between times of initial branching and subsequent rates of anatomical innovation or evolutionary change in general. Even the most vociferous advocates of a genuine Cambrian explosion have never argued that this period of rapid anatomical diversification marks the moment of origin for animal phyla—if only because we all acknowledged the evidence for Precambrian tracks and trails of triploblasts even before the recent discovery of embryos. Nor do these same vociferous advocates imagine that only one worm-like species crawled across the great Cambrian divide to serve as an immediate common ancestor for all modern phyla. In fact, I can't imagine why anyone would care (for adjudicating the reality of the explosion, though one would care a great deal for discussions of some other evolutionary issues) whether one wormlike species carrying the ancestry of all later animals, or ten similar worm-like species already representing the lineages of ten subsequent phyla, crossed this great divide from an earlier Precambrian history. The Cambrian explosion represents a claim for a rapid spurt of *anatomical innovation* within the animal kingdom, not an argument about times of *genealogical divergence*.

The following example should clarify the fundamental distinction between times of genealogical splitting and rates of change. Both rhinoceroses and horses may have evolved from the genus *Hyracotherium* (formerly called *Eohippus*). A visitor to the Eocene earth about 50 million years ago might determine that the basic split had already occurred. He might be able to identify one species of *Hyracotherium* as the ancestor of all later horses, and another species of the same genus as the progenitor of all subsequent rhinos. But this visitor would be laughed to justified scorn if he then argued that later divergences between horses and rhinos must be illusory because the two lineages had already split. After all, the two Eocene species looked like kissing cousins (as evidenced by their placement in the same genus), and only gained their later status as pro-genitors of highly distinct lineages by virtue of a subsequent history, utterly unknowable at the time of splitting. Similarly, if ten nearly identical wormlike forms (the analogs of the two *Hyracotherium* species) crossed the Cambrian boundary, but only evolved the anatomical distinctions of great phyla during the subsequent explosion, then the explosion remains as real, and as vitally important for life's history, as any advocate has ever averred.

This crucial distinction has been recognized by most commentators on the work of Wray et al. Geerat J. Vermeij, in his direct evaluation (*Science*, 1996, page 526), wrote that "this new work in no way diminishes the significance of the Vendian-Cambrian revolution." Fortey, Briggs, and Wills added (*BioEssays*, 1997, page 433) that "there is, of course, no necessary correspondence between morphology and genomic change." In any case, a later publication by Ayala, Rzhetsky, and Ayala (*Proceedings of the National Academy of Sciences*, 1998,

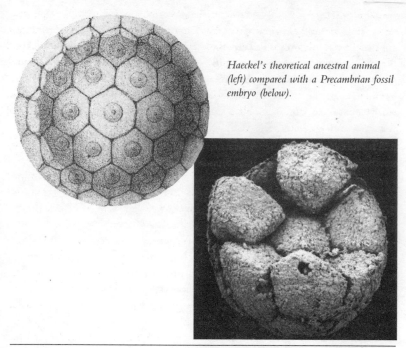

Haeckel's theoretical ancestral animal (left) compared with a Precambrian fossil embryo (below).

volume 95, pages 606–11) presents a powerful rebuttal to Wray et al.'s specific conclusions. By correcting statistical errors and unwarranted assumptions, and by adding data for twelve additional genes, these authors provide a very different estimate for initial diversification in late Precambrian times: about 670 million years ago for the split of chordates from the line of arthropods, annelids, and mollusks; and 600 million years for the later divergence of chordates from echinoderms.

We are left, of course, with a key mystery (among many others): where are Precambrian adult triploblasts "hiding," now that we have discovered their embryos? An old suggestion, first advanced in the 1870s in the prolific and often highly speculative work of the German biologist Ernst Haeckel (who was nonetheless outstandingly right far more often than random guesswork would allow), held that Precambrian animals had evolved as tiny forms not much larger than, or very different from, modern embryos—and would therefore be very hard to find as fossils. (The similarity between Haeckel's hypothetical ancestors and Xiao, Zhang, and Knoll's actual embryos is almost eerie—see figures on pages 323 and 330.) Moreover, E. H. Davidson, K. J. Peterson, and R. A. Cameron (*Science*, 1995, volume 270, pages 1319–25) have made a powerful case, based on genetic and developmental arguments, that Precambrian animals did originate at tiny sizes, and that the subsequent Cambrian explosion featured the evolution of novel embryological mechanisms for substantially increasing cell number and body size, accompanied by consequent potential for greatly enhanced anatomical innovation. If Haeckel's old argument, buttressed by Davidson's new concepts and data, has validity, we then gain genuine hope, even realistic expectation, that Precambrian adult triploblasts may soon be discovered, for such animals will be small enough to be preserved by phosphatization.

As a final point, this developing scenario for the early history of animals might foster humility and generate respect for the complexity of evolutionary pathways. To make the obvious analogy: we used to regard the triumph of "superior" mammals over "antediluvian" dinosaurs as an inevitable consequence of progressive evolution. We now realize that mammals originated at the same time as dinosaurs and then lived for more than 100 million years as marginal, small-bodied creatures in the nooks and crannies of a dinosaur's world. Moreover, mammals would never have expanded to dominate terrestrial ecosystems (and humans would surely never have evolved) without the supreme good fortune (for us) of a catastrophic extraterrestrial impact that, for some set of unknown reasons, eliminated dinosaurs and gave mammals an unanticipated opportunity.

Does this essay's tale from an earlier time—Ediacaran "primitives" versus contemporary Precambrian ancestors of modern animals—differ in any substantial way? We now know (from the evidence of Xiao, Zhang, and Knoll's embryos) that animals of modern design had already originated before the Ediacara fauna evolved into full bloom. Yet "primitive" Ediacara dominated the world of animal life, perhaps for 100 million years, while modern triploblasts waited in the proverbial wings, perhaps as tiny animals of embryonic size, living in nooks and crannies left over by much larger Ediacaran dominants. Only a mass extinction of unknown cause, a great dying that wiped out Ediacara and initiated the Cambrian transition 543 million years ago, gave modern triploblasts an opportunity to shine—and so we have.

In evolution, as well as in politics, incumbency offers such powerful advantages that even a putatively more competent group may be forced into a long period of watchful waiting, always hoping that an external stroke of good luck will finally grant an opportunity for picking up the reins of power. If fortune continues to smile, the new regime may eventually gain enough confidence to invent a comforting and commanding mythology about the inevitability of its necessary rise to dominance by gradually growing better and better—every day and in every way.

22

The
Paradox of the
Visibly Irrelevant

AN ODD PRINCIPLE OF HUMAN PSYCHOLOGY, WELL
known and exploited by the full panoply of prevaricators, from
charming barkers like Barnum to evil demagogues like Goebbels,
holds that even the silliest of lies can win credibility by constant repetition. In current American parlance, these proclamations of
"truth" by Xeroxing fall into the fascinating domain of "urban legends."

My favorite bit of nonsense in this category intrudes upon me
daily, and in very large type, thanks to a current billboard ad campaign by a company that will remain nameless. The latest version
proclaims: "Scientists say we use 10 percent of our brains. That's way
too much." Just about everyone regards the "truth" of this proclamation as obvious and incontrovertible—though you might still
start a barroom fight over whether the correct figure should be 10,

15, or 20 percent. (I have heard all three asserted with utter confidence.) But this particular legend can only be judged as even worse than false: for the statement is truly meaningless and nonsensical. What do we mean by "90 percent unused"? What is all this superfluous tissue doing? The claim, in any case, can have no meaning until we develop an adequate theory about how the brain works. For now, we don't even have a satisfactory account for the neurological basis of memory and its storage—surely the sine qua non for formulating any sensible notion about unused percentages of brain matter! (I think that the legend developed because we rightly sense that we ought to be behaving with far more intelligence than we seem willing to muster—and the pseudoquantification of the urban legend acts as a falsely rigorous version of this legitimate but vague feeling.)

In my field of evolutionary biology, the most prominent urban legend—another "truth" known by "everyone"—holds that evolution may well be the way of the world, but one has to accept the idea with a dose of faith because the process occurs far too slowly to yield any observable result in a human lifetime. Thus, we can document evolution from the fossil record and infer the process from the taxonomic relationships of living species, but we cannot see evolution on human timescales "in the wild."

In fairness, we professionals must shoulder some blame for this utterly false impression about evolution's invisibility in the here and now of everyday human life. Darwin himself—though he knew and emphasized many cases of substantial change in human time (including the development of breeds in his beloved pigeons)—tended to wax eloquent about the inexorable and stately slowness of natural evolution. In a famous passage from *The Origin of Species,* he even devised a striking metaphor about clocks to underscore the usual invisibility:

> It may be said that natural selection is daily and hourly scrutinizing, throughout the world, every variation, even the slightest; rejecting that which is bad, preserving and adding up all that is good; silently and invisibly working. . . . We see nothing of these slow changes in progress until the hand of time has marked the long lapse of ages.

Nonetheless, the claim that evolution must be too slow to see can only rank as an urban legend—though not a completely harmless tale in this case, for our creationist incubi can then use the fallacy as an argument against evolution at any scale, and many folks take them seriously because they just "know" that evolution can never be seen in the immediate here and now. In fact, a precisely opposite situation actually prevails: biologists have documented a veritable glut of cases for rapid and eminently measurable evolution on timescales of years and decades.

However, this plethora of documents—while important for itself, and surely valid as a general confirmation for the proposition that organisms evolve—teaches us rather little about rates and patterns of evolution at the geological scales that build the history and taxonomic structure of life. The situation is wonderfully ironic—a point that I have tried to capture in the title of this article. The urban legend holds that evolution is too slow to document in palpable human lifetimes. The opposite truth has affirmed innumerable cases of measurable evolution at this minimal scale—but to be visible at all over so short a span, evolution must be far too rapid (and transient) to serve as the basis for major transformations in geological time. Hence the "paradox of the visibly irrelevant"—or, "if you can see it at all, it's too fast to matter in the long run!"

Our best and most numerous cases have been documented for the dominant and most evolutionarily active organisms on our planet—bacteria. In the most impressive of recent examples, Richard E. Lenski and Michael Travisano (*Proceedings of the National Academy of Sciences,* 1994, volume 91, pages 6808–14) monitored evolutionary change for ten thousand generations in twelve laboratory populations of the common human gut bacterium *Escherichia coli.* By placing all twelve populations in identical environments, they could study evolution under ideal experimental conditions of replication—a rarity for the complex and unique events of evolutionary transformation in nature. In a fascinating set of results, they found that each population reacted and changed differently, even within an environment made as identical as human observers know how to do. Yet Lenski and Travisano did observe some important and repeated patterns within the diversity. For example, each population increased rapidly in average cell size for the first two thousand generations or so, but then remained nearly stable for the last five thousand generations.

A cynic might still reply: fine, I'll grant you substantial observable evolution in the frenzied little world of bacteria, where enormous populations and new generations every hour allow you to monitor ten thousand episodes of natural selection in a manageable time. But a similar "experiment" would consume thousands of years for multicellular organisms that measure generations in years or decades rather than minutes or hours. So we may still maintain that evolution cannot be observed in the big, fat, furry, sexually reproducing organisms that serve as the prototype for "life" in ordinary human consciousness. (A reverse cynic would then rereply that bacteria truly dominate life, and that vertebrates only represent a latecoming side issue in the full story of evolution, however falsely promoted to centrality by our own parochial focus. But we must leave this deep issue to another time.)

I dedicate this essay to illustrating our cynic's error. Bacteria may provide our best and most consistent cases for obvious reasons, but measurable (and substantial) evolution has also, and often, been documented in vertebrates and other complex multicellular organisms. The classic cases have not exactly been hiding their light under a bushel, so I do wonder why the urban legend of evolution's invisibility persists with such strength. Perhaps the firmest and most elegant examples involve a group of organisms named to commemorate our standard-bearer himself—Darwin's finches of the Galápagos Islands, where my colleagues Peter and Rosemary Grant have spent many years documenting fine-scale evolution in such adaptively important features as size and strength of the bill (a key to the mechanics of feeding), as rapid climatic changes force an alteration of food preferences. This work formed the basis for Jonathan Weiner's excellent book, *The Beak of the Finch*—so the story has certainly been well and prominently reported in both the technical and popular press.

Nonetheless, new cases of such short-term evolution still maintain enormous and surprising power to attract public attention—for interesting and instructive but utterly invalid reasons, as I shall show. I devote this essay to the three most prominent examples of recent publications that received widespread attention in the popular press as well. (One derives from my own research, so at least I can't be accused of sour grapes in the debunking that will follow—though I trust that readers will also grasp the highly positive twist that I will ultimately impose upon my criticisms.) I shall briefly describe each case, then present my two general critiques of their prominent reporting by the popular press, and finally explain why such cases teach us so little about evolution in the large, yet remain so important for themselves, and at their own equally legitimate scale.

1. Guppies from Trinidad. In many drainage systems on the island of Trinidad, populations of guppies live in downstream pools, where several species of fish can feed upon them. "Some of these species prey preferentially on large, mature-size classes of guppies." (I take all quotes from the primary technical article that inspired later press accounts—"Evaluation of the rate of evolution in natural populations of guppies *(Poecilia reticulata),*" by D. N. Reznick, F. H. Shaw, F. H. Rodd, and R. G. Shaw, published in *Science,* 1977, volume 275, pages

1934–37). Other populations of the same species live in "upstream portions of each drainage" where most "predators are excluded . . . by rapids or waterfalls, yielding low-predation communities."

In studying both kinds of populations, Reznick and colleagues found that "guppies from high-predation sites experience significantly higher mortality rates than those from low-predation sites." They then reared both kinds of guppies under uniform conditions in the laboratory, and found that fishes from high-predation sites in lower drainages matured earlier and at a smaller size. "They also devote more resources to each litter, produce more, smaller offspring per litter, and produce litters more frequently than guppies from low-predation localities."

This combination of observations from nature and the laboratory yields two important inferences. First, the differences make adaptive sense, for guppies subjected to greater predation would fare better if they could grow up fast and reproduce both copiously and quickly before the potential boom falls—a piscine equivalent of the old motto for electoral politics in Boston: vote early and vote often. On the other hand, guppies in little danger of being eaten might do better to bide their time and grow big and strong before engaging their fellows in any reproductive competition. Second, since these differences persist when both kinds of guppies are reared in identical laboratory environments, the distinction must record genetically based and inherited results of divergent evolution between the populations.

In 1981, Reznick had transferred some guppies from high-predation downstream pools into low-predation upstream waters then devoid of guppies. These transplanted populations evolved rapidly to adopt the reproductive strategy favored by indigenous populations in neighboring upstream environments: delayed sexual maturity at larger size, and longer life. Moreover, Reznick and colleagues made the interesting observation that males evolved considerably more rapidly in this favored direction. In one experiment, males reached their full extent of change within four years, while females continued to alter after eleven years. Since the laboratory populations had shown higher heritability for these traits in males than in females, these results make good sense. (Heritability may be roughly defined as the correlation between traits in parents and offspring due to genetic differences. The greater the heritable basis of a trait, the faster the feature can evolve by natural selection.)

This favorable set of circumstances—rapid evolution in a predictable and presumably adaptive direction based on traits known to be highly heritable—provides a "tight" case for well-documented (and sensible) evolution at scales well within the purview of human observation, a mere decade in this case. The headline for the news report on this paper in *Science* magazine (March 28, 1997) read: "Predator-free guppies take an evolutionary leap forward."

2. Lizards from the Exuma Cays, Bahama Islands. During most of my career, my fieldwork has centered on biology and paleontology of the land snail *Cerion* in the Bahama Islands. During these trips, I have often encountered fellow biologists devoted to other creatures. In one major program of research, Tom Schoener (a biology professor at the University of California, Davis) has, with numerous students and colleagues, been studying the biogeography and evolution of the ubiquitous little lizard *Anolis*—for me just a fleeting shadow running across a snail-studded ground, but for them a focus of utmost fascination (while my beloved snails, I assume, just blend into their immobile background).

In 1977 and 1981, Schoener and colleagues transplanted groups of five or ten lizards from Staniel Cay in the Exuma chain to fourteen small and neighboring islands that housed no lizards. In 1991, they found that the lizards had thrived (or at least survived and bred) on most of these islands, and they collected samples of adult males from each experimental island with an adequate population. In addition, they gathered a larger sample of males from areas on Staniel Cay that had served as the source for original transplantation in 1977 and 1981.

This study then benefits from general principles learned by extensive research on numerous *Anolis* species throughout the Bahama Islands. In particular, relatively longer limbs permit greater speed, a substantial advantage provided that preferred perching places can accommodate long-legged lizards. Trees and other "thick" perching places therefore favor the evolution of long legs. Staniel Cay itself includes a predominant forest, and the local *Anolis* tend to be long-legged. But when lizards must live on thin twigs in bushy vegetation, the agility provided by shorter legs (on such precarious perches) may outweigh the advantages in speed that longer legs would provide. Thus, lizards living on narrow twigs tend to be shorter-legged. The small cays that received the fourteen transported populations have little or no forest growth and tend instead to be covered with bushy vegetation (and narrow twigs).

J. B. Losos, the principal author of the new study, therefore based an obvious prediction on these generalities. The populations had been transferred from forests with wide perches to bushy islands covered with narrow twigs. "From the kind of vegetation on the new islands," Losos stated, "we predicted that the lizards would develop shorter hindlimbs." Their published study validates this expected result: a clearly measurable change, in the predicted and adaptive direction, in less than twenty years. (See details in J. B. Losos, K. I. Warheit, and T. W. Schoener, "Adaptive differentiation following experimental island colonization in *Anolis* lizards," *Nature,* 1997, volume 387, pages 70–73). A news report appeared in *Science* magazine (May 2, 1997) under the title "Catching lizards in the act of adapting."

This study lacks a crucial piece of documentation that the Trinidadian guppies provided—an absence immediately noted by friendly critics and fully acknowledged by the authors. Losos and colleagues have not studied the heritability of leg length in *Anolis sagrei* and therefore cannot be certain that their results record a genetic process of evolutionary change. The growth of these lizards may feature extensive flexibility in leg length, so that the same genes yield longer legs if lizards grow up on trees, and shorter legs if they always cavort in the bushes (just as the same genes can lead to a thin or fat human being depending upon a personal history of nutrition and exercise). In any case, however, a sensible and apparently adaptive change in average leg length has occurred within twenty years on several islands, whatever the cause of modification.

3. Snails from Great Inagua, Bahama Islands. Most of Great Inagua, the second-largest Bahamian Island (Andros wins first prize), houses a large and ribby *Cerion* species named *C. rubicundum.* But fossil deposits of no great age lack this species entirely and feature instead an extinct form named *Cerion excelsior,* the largest of all *Cerion* species. Several years ago, on a mudflat in the southeastern corner of Great Inagua, David Woodruff (of the University of California, San Diego) and I collected a remarkable series of shells that seemed to span (and quite smoothly) the entire range of form from extinct *C. excelsior* to modern *C. rubicundum.* Moreover, and in general, the more eroded and "older looking" the shell, the closer it seemed to lie to the anatomy of extinct *C. excelsior.*

This situation suggested a local evolutionary transition by hybridization, as *C. rubicundum,* arriving on the island from an outside source, interbred with indigenous *C. excelsior.* Then, as *C. excelsior* declined toward extinction, while *C. rubicundum* thrived and increased, the average anatomy of the population transformed slowly and steadily in the direction of the modern form. This hypothesis sounded good and sensible, but we could devise no way to test our idea—for all the shells had been collected from a single mudflat (analogous to a single bedding plane of a geological stratum), and we could not determine their relative ages. The pure *C. excelsior* shells "looked" older, but such personal impressions count for less than nothing (subject as they are to a researcher's bias) in science. So we got stymied and put the specimens in a drawer.

Several years later, I teamed up with paleontologist and geochemist Glenn A. Goodfriend from the Carnegie Institution of Washington. He had refined a dating technique based on changes in the composition of amino acids in the shell over time. By keying these amino acid changes to radiocarbon dates for some of the shells, we could estimate the age of each shell. A plot of shell age versus position on an anatomical spectrum from extinct *C. excelsior* to modern *C. rubicundum* produced a beautiful correlation between age and anatomy: the younger the specimen, the closer to the modern anatomy.

This ten- to twenty-thousand-year transition by hybridization exceeds the time period of the Trinidad and Exuma studies by three orders of magnitude (that is, by a factor of 1,000), but even ten thousand years represents a geological eye-blink in the fullness of evolutionary time—while this transformation in our snails marks a full change from one species to another, not just a small decrement of leg length, or a change in the timing of breeding, within a single species. (For details, see G. A. Goodfriend and S. J. Gould, "Paleontology and chronology of two evolutionary transitions by hybridization in the Bahamian land snail *Cerion*," *Science,* 1996, volume 274, pages 1894–97). Harvard University's press release (with no input from me) carried the headline "Snails caught in act of evolving."

A scanning of any year's technical literature in evolutionary biology would yield numerous and well-documented cases of such measurable, small-scale evolutionary change—thus disproving the urban legend that evolution must always be too slow to observe in the geological microsecond of a human lifetime. These three studies, all unusually complete in their documentation and in their resolution of details, do not really rank as "news" in the journalist's prime sense of novelty or deep surprise. Nonetheless, each of these three studies became subjects for front-page stories in either *The New York Times* or *The Boston Globe.*

Now please don't get me wrong. I do not belong to the cadre of rarefied academics who cringe at every journalistic story about science for fear that the work reported might become tainted with popularity thereby. And in a purely "political" sense, I certainly won't object if major newspapers choose to feature any result of my profession as a lead story—especially, if I may be self-serving for a moment, when one of the tales reports my own work! Nonetheless, this degree of public attention for workaday results in my field (however elegantly done) does fill me with wry amusement—if only for the general reason that most of us feel a tickle in the funny bone when we note a gross imbalance between public notoriety and the true novelty or importance of an event, as when Hollywood spinmeisters manage to depict their client's ninth marriage as the earth's first example of true love triumphant and permanent.

Of course I'm delighted that some ordinary, albeit particularly well done studies of small-scale evolution struck journalists as front-page news. But I still feel impelled to ask why these studies, rather than a hundred others of equal care and merit that appear in our literature every month, caught this journalistic fancy and inspired such prime attention. When I muse over this issue, I can only devise two reasons—both based on deep and interesting fallacies well worth identifying and discussing. In this sense, the miselevation of everyday good work to surprising novelty may teach us something important about public attitudes toward evolution, and toward science in general. We may, I think, resolve each of the two fallacies by contrasting the supposed meaning of these studies, as reported in public accounts, with the significance of such work as viewed by professionals in the field.

1. The fallacy of the crucial experiment

In high school physics classes, we all learned a heroically simplified version of scientific progress based upon a model that does work sometimes but by no means always—the *experimentum crucis,* or crucial experiment. Newton or Einstein? Ptolemy or Copernicus? Special Creation or Darwin? To find out, perform a single decisive experiment with a clearly measurable result replete with decisive power to decree yea or nay.

The decision to treat a limited and particular case as front-page news must be rooted in this fallacy. Reporters must imagine that evolution can be proved by a single crucial case, so that any of these stories may provide decisive confirmation of Darwin's truth—a matter of some importance given the urban legend that evolution, even if valid, must be invisible on human timescales.

But two counterarguments vitiate this premise. First, as a scientific or intellectual issue, we hardly need to "prove" evolution by discovering new and elegant cases. We do not, after all, expect to encounter a page-one story with the headline "New experiment proves earth goes around sun, not vice versa. Galileo vindicated." The fact of evolution has been equally well documented for more than a century.

Second, and more generally, single "crucial" experiments rarely decide major issues in science—especially in natural history, where nearly all theories require data about "relative frequencies" (or percentage of occurrences), not pristine single cases. Of course, for a person who believes that evolution never occurs at all, one good case can pack enormous punch—but science resolved this basic issue more than one hundred years ago. Nearly every interesting question in evolutionary theory asks "how often?" or "how dominant in setting the pattern of life?"—not "does this phenomenon occur at all?" For example, on the most important issue of all—the role of Darwin's own favored mechanism of natural selection—single examples of selection's efficacy advance the argument very little. We already know, by abundant documentation and rigorous theorizing, that natural selection can and does operate in nature. We need to determine the *relative strength* of Darwin's mechanism among a set of alternative modes for evolutionary change—and single cases, however elegant, cannot establish a relative frequency.

Professionals also commit this common error of confusing well-documented single instances with statements about relative strength among plausible alternatives. For example, we would like to know how often small and isolated populations evolve differences as adaptive responses to local environments (presumably by Darwin's mechanism of natural selection), and how often such changes occur by the random process known as "genetic drift"—a potentially important phenomenon in small populations (just as a small number of coin flips can depart radically from fifty-fifty for heads and tails, while a million flips with an honest coin cannot stray too far from this ideal). Losos's study on lizard legs provides one vote for selection (if the change turns out to have a genetic basis)—because leg length altered in a predicted direction toward better adaptation to local environments on new islands. But even such an elegant case cannot prove the domination of natural selection in general. Losos has only shown the power of Darwin's process in a particular example. Yet the reporter for *Science* magazine made this distressingly common error in concluding: "If it [change in leg length] is rooted in the genes, then the study is strong evidence that isolated populations diverge by natural selection, not genetic drift as some theorists have argued." Yes, strong evidence for these lizards on that island during those years—but not proof for the general domination of selection over drift. Single cases don't establish generalities, so long as alternative mechanisms retain their theoretical plausibility.

2. *The paradox of the visibly irrelevant*

As a second reason for overstating the centrality of such cases in our general understanding of evolution, many commentators (and research scientists as well) ally themselves too strongly with one of the oldest (and often fallacious) traditions of Western thought: reductionism, or the assumption that laws and mechanics of the smallest constituents must explain objects and events at all scales and times. Thus, if we can render the behavior of a large body (an organism or a plant, for example) as a consequence of atoms and molecules in motion, we feel that we have developed a "deeper" or "more basic" understanding than if our explanatory principles engage only large objects themselves, and not their constituent parts.

Reductionists assume that documenting evolution at the smallest scale of a few years and generations should provide a general model of explanation for events at all scales and times—so these cases should become a gold standard for the entire field, hence their status as front-page news. The authors of our two studies on decadal evolution certainly nurture such a hope. Reznick and colleagues end their publication on Trinidadian guppies by writing: "It is part of a growing body of evidence that the rate and patterns of change attainable through natural selection are sufficient to account for the patterns observed in the fossil record." Losos and colleagues say much the same for their lizards: "Macroevolution may just be microevolution writ large—and, consequently, insight into the former may result from study of the latter."

We tend to become beguiled by such warm and integrative feelings (for science rightly seeks unity and generality of explanation). But does integration by reduction of all scales to the rates and mechanisms of the smallest really work for evolution—and do we crave this style of unification as the goal of all science? I think not, and I also regard our best general reason for skepticism as conclusive for this particular subject—however rarely appreciated, though staring us in the face.

These shortest-term studies are elegant and important, but they cannot represent the general mode for building patterns in the history of life. The reason for their large-scale impotence strikes most people as deeply paradoxical, even quite funny—but the argument truly cannot be gainsaid. Evolutionary rates as measured for guppies and lizards are *vastly too rapid* to represent the general modes of change that build life's history through geological ages.

But how can I say such a thing? Isn't this statement ridiculous a priori? How could these tiny, minuscule changes—a little less leg, a minimally larger size—represent too much of anything? Doesn't the very beauty of these studies lie in their minimalism? We have always been taught that evolution is wondrously slow and cumulative—a grain-by-grain process, a penny a day toward the domain of Bill Gates. Doesn't each of these studies document a grain? Haven't my colleagues and I found the "atom" of evolutionary incrementation?

I believe that these studies have discerned something important, but they have discovered no general atom. These measured changes over years and decades are too fast by several orders of magnitude to build the history of life by simple cumulation. Reznick's guppy rates range from 3,700 to 45,000 darwins (a standard metric for evolution, expressed as change in units of standard deviation—a measure of variation around the mean value of a trait in a population—per million years). By contrast, rates for major trends in the fossil record generally range from 0.1 to 1.0 darwin. Reznick himself states that "the estimated rates [for guppies] are . . . four to seven orders of magnitude greater than those observed in the fossil record" (that is, ten thousand to ten million times faster).

Moreover and with complete generality—thus constituting the "paradox of the visibly irrelevant" in my title—we may say that any change measurable *at all* over the few years of an ordinary scientific study must be occurring far too rapidly to represent ordinary rates of evolution in the fossil record. The culprit of this paradox, as so often, can be identified as the vastness of time (a concept that we can appreciate "in our heads" but seem quite unable to place into the guts of our intuition). The key principle, however ironic, requires such a visceral understanding of earthly time: if a case of evolution proceeds with sufficient speed to be discerned by our instruments in just a few years—that is, if the change becomes substantial enough to stand out as a genuine and directional effect above the random fluctuations of nature's stable variation and our inevitable errors of measurement—then we have witnessed something far too substantial to serve as an atom of steady incrementation in a paleontological trend. Thus, to restate the paradox: if we can measure it at all (in a few years), it is too powerful to be the stuff of life's history.

If large-scale evolution proceeded by stacking Trinidad guppy rates end to end, then any evolutionary trend would be completed in a geological moment, not over the many million years actually observed. "Our face from fish to man," to cite the title of a famous old account of evolution for popular audiences, would run its course within a single geological formation, not over more than 400 million years, as our fossil record demonstrates.

Evolutionary theory must figure out how to slow down these measured rates of the moment, not how to stack them up! In fact, most lineages are stable (*non*changing) nearly all the time in the fossil record. When lineages do change, their alteration usually occurs "momentarily" in a geological sense (that is, confined to a single bedding plane) and usually leads to the origin of a new species by branching. Evolutionary rates during these moments may match the observed speed of Trinidadian guppies and Bahamian lizards—for most bedding planes represent several thousand years. But during most of a typical species's lifetime, no change accumulates, and we need to understand why. The sources of stasis have become as important for evolutionary theory as the causes of change.

(To illustrate how poorly we grasp this central point about time's immensity, the reporter for *Science* magazine called me when my *Cerion* article, coauthored with Glenn Goodfriend, appeared. He wanted to write an accompanying news story about the exception I had found to my own theory of punctuated equilibrium—an insensibly gradual change over ten to twenty thousand years. I told him that, although exceptions abound, this case does not lie among them, but actually represents a strong confirmation of punctuated equilibrium! We found all twenty thousand years' worth of snails on a single mudflat—that is, on what would become a single bedding plane in the geological record. Our *entire* transition occurred in a geological moment and represented a punctuation, not a gradual sequence of fossils. We were able to "dissect" the punctuation in this unusual case—hence the value of our publication—because we could determine ages for the individual shells. The reporter, to his credit, completely revised his originally intended theme and published an excellent account.)

In conclusion, I suspect that most cases like the Trinidadian guppies and Bahamian lizards represent transient and momentary blips and fillips that "flesh out" the rich history of lineages in stasis, not the atoms of substantial and steadily accumulated evolutionary trends. Stasis is a dynamic phenomenon. Small local populations and parts of lineages make short and temporary forays of transient adaptation, but these tiny units almost always die out or get reintegrated into the general pool of the species. (Losos himself regards the new island populations of lizards as evolutionarily transient in exactly this sense—for such tiny and temporary colonies are almost always extirpated by hurricanes in the long run. How, then, can such populations represent atoms of a major evolutionary trend? The news report in *Science* magazine ends by stating: "But whether the lizards continue to evolve depends largely on the winds of fate, says Losos. These islets are periodically swept by hurricanes that could whisk away every trace of anolian evolution.")

But transient blips and fillips are no less important than major trends in the total "scheme of things." Both represent evolution operating at a standard and appropriate measure for a particular scale and time—Trinidadian blips for the smallest and most local moment, faces from fish to human for the largest and most global frame. One scale doesn't translate into another. No single scale can be deemed more important than any other; and none operates as a basic model for all the others. Each scale embodies something precious and unique to teach us; none can be labeled superior or primary. (Guppies and lizards, in their exposition of momentary detail, give us insight, unobtainable at broader scales, into the actual mechanics of adaptation, natural selection, and genetic change.)

The common metaphor of the science of fractals—Mandelbrot's familiar argument that the coast of Maine has no absolute length, but depends upon the scale of measurement—epitomizes this principle well (see chapter 23). When we study guppies in a pond in Trinidad, we are operating at a scale equivalent to measuring the coastline by wrapping our string around every boulder on every headland of Acadia National Park. When we trace the increase in size of the human brain from Lucy (about four million years ago) to Lincoln, we are measuring the coastline as depicted on my page of Maine in *Hammond's Atlas*. Both scales are exactly right for their appropriate problems. You would be a fool to spend all summer measuring the details in one cove in Acadia, if you just wanted to know the distance from Portland to Machiasport for your weekend auto trip.

I find a particular intellectual beauty in such fractal models—for they invoke hierarchies of inclusion (the single cove embedded within Acadia, embedded within Maine) to deny hierarchies of worth, importance, merit, or meaning. You may ignore Maine while studying the sand grain, and be properly oblivious of the grain while perusing the map of Maine on the single page of your atlas. But you can love and learn from both scales at the same time. Evolution does not lie patent in a clear pond on Trinidad any more than the universe (*pace* Mr. Blake) lies revealed in a grain of sand. But how poor would be our understanding—how bland and restricted our sight—if we could not learn to appreciate the rococo details that fill our immediate field of vision, while forming, at another scale, only some irrelevant and invisible jigglings in the majesty of geological time.

23

Room of One's Own

GOLGOTHA, THE SITE OF CHRIST'S CRUCIFIXION, appears in most paintings as a substantial hill in the countryside, far from the city walls of Jerusalem depicted in a distant background. In fact, if the traditional spot has been correctly identified, Golgotha is a tiny protuberance located just next to the old city limits but now inside the walls built by Suleiman the Magnificent in the early sixteenth century. These walls extended the boundaries of Jerusalem, and the old town now sits as a small "jewel" at the center of a much bigger, modern city. Golgotha is small and low enough to fit *within* the Church of the Holy Sepulchre, located *within* Suleiman's city walls. Visitors just have to climb an internal staircase to reach the top of Golgotha, located on the church's second story. (Several theories compete to explain the derivation of the name, for *golgotha* means "skull" in Aramaic, while the alternative label of *calvary* has the same definition in Latin. Most scholars think that the name designates the shape of the small hill, not the mortal remains of executions.)

As one of the most sacred sites on earth, the Church of the Holy Sepulchre might be expected to exude dignity, serenity, and a spirit of transcendence above merely earthly cares. Yet in maximal, almost perverse contrast, the church is a site of constant bickering and division. The etymology of *religion* may refer to "tying together," but the actual experience, given the propensities of *Homo sapiens,* the earth's most various and curmudgeonly species, tends more often to separation and anathematization. The precious space is "shared" (in this case, a euphemism for "wrangled over") by six old Christian groups—Greek Orthodox, Roman Catholic, Armenian, Syrian, Coptic, and Abyssinian. (The various Protestant denominations came upon the scene a few centuries too late and didn't even get a pew.)

Before visiting the church several years ago, I had encountered the Latin phrase *status quo* only as a general description for leaving well enough alone. But I learned that the phrase can also be used as a proper noun—capital *S,* capital *Q.* In 1852, after centuries of more serious bickering, the six groups signed an agreement, called *the* Status Quo, to regulate every move and every square inch in the building. At this point, I will yield to Baedeker's *Guide to Jerusalem* (1982 edition), a publication generally known for authoritative and stodgy prose but uncharacteristically pungent in this case:

> No lamp, no picture, nothing whatsoever may be moved without its giving rise to a complaint. The rules governing when and where each community may celebrate Mass are minutely prescribed as are the times when the lamps may be lit and the windows may be opened. Everything must be done in accordance with the originally agreed rules, i.e. the "status quo." . . . Modifications to this are persistently being sought and just as persistently rejected—they even cropped up in the negotiations for the Treaty of Versailles and in the League of Nations. . . . Anyone hoping to find harmony and quiet contemplation . . . is due for a disappointment—the sects are on a Cold War footing. Even the background noise can be put down to psychological warfare—the sound of the blows of hammers and chisels constantly engaged on improvement work mingles with the chanting of Greek plainsong, blasts from the Franciscan organ and the continual tinkling of Armenian bells.

And lest anyone hope that equality might reign among the six groups, I hasten to point out that the Status Quo assigned 65 percent of the church to the

Greek Orthodox, while granting the Abyssinians—the only black African group by ethnicity—just the tomb of Joseph of Arimathea ("a tiny cavity that can only be reached by passing through Coptic territory," to quote Baedeker's one more time). Adding insult to this injury, the poor Abyssinians can't even reside within the church but must live instead in tiny cells built on the roof! (And let me tell you, it was really hot up there the day I visited.)

To move from a ridiculous story about a sublime place to the fully ridiculous all around, I got the idea for this essay from an English newspaper story of July 9, 1997: "Punch-up Between Brewery Rivals Over Future of Historic Hostelry." One of London's most interesting pubs, the Punch Tavern on Fleet Street, bears a name that reflects a former role as the favorite watering hole for staff members of the famous humor magazine. These ghosts of the past could have filed quite a story on the current situation. Bass, a large national brewery, owns two-thirds of the property, including the only toilets. But Samuel Smith, a smaller, regional operation, bought the other third, including the passageway for delivery of beer to the Bass side. The two businesses have coexisted in constant tension and bickering but have now opted for something closer to the Holy Sepulchre solution of strict division. A new wall now rises within the pub, and the Bass people are building "a new cellar drop so workers can move beer supplies without using Samuel Smith's passageway." We must assume that the Smith folks will construct some new toilets, for we all know that such items rank second only to what comes in the other end as a necessary fixture in these establishments.

One last item, ridiculous but personal this time, will serve to establish this theme as a generality. My brother and I shared a small room throughout our childhood. We usually coexisted reasonably well, but we did have our battles from time to time. One day, following our worst blowup, Peter decided that we would just have to divide the room right down the middle, and each of us promise never to set so much as a toe into the other's territory. He proceeded to gather all his possessions and move them to his side. But I just lay on my bed laughing—as he got progressively angrier at my lighthearted approach to such a serious situation. When he finished all the moving and shoving, he confronted me in a fury: "What are you laughing about?" I didn't say a word, but only lifted my finger and pointed at the room's single door—located on my side. Fortunately, Peter started to laugh too; so we made up and amalgamated all our stuff again.

If people, representing a mere few billion souls within a single species spread throughout the planet, can generate so much strife about divvying a space, what can nature possibly do with millions of species, gazillions of individuals, and nothing with the ability or power to negotiate, or even to understand, a status quo? Much of ecological theory has been devoted to debating concepts that may usually be framed differently, but really represent variants of this fundamental question.

Consider just two examples that generally make the rounds in any basic college course on evolution. In discussing the crucial question of how many species can coexist in a single habitat (obviously an ever more important issue as natural spaces shrink before human onslaught, and many species face imminent extinction), students invariably hear about something called the "competitive exclusion" principle, or the notion that two species cannot occupy the same "niche." This conclusion follows more as a logical consequence of natural selection than an observation from nature. If two species lived in truly identical environments, sharing all the same spaces and resources, then one of the two would surely hold some advantage, however slight, over the other, and the relentless force of natural selection, acting on even the tiniest differential over countless generations, should secure total victory for the species with a small edge in the competitive struggle for existence.

But this principle probably says less than its weighty words seem to imply, for niches do not exist independently of the species that inhabit them. Niches are not comparable to houses in a suburban development, built "on spec" and fully decked out with all furnishings and utilities before people come to buy under a strict rule of "one lot for one family." Niches are constructed by organisms as they interact with complex environments—and how could two different species read an environment in exactly the same way for all particulars?

A related principle (and second example) called "limiting similarity" attempts to put this theme into a more reasonable and testable light. If two separate species cannot be identical in appearance and behavior, and cannot read the surrounding environment in exactly the same way, then how close can they be? What are the limits to their similarity? How many species of beetles can live in a tropical tree? How many species of fishes in a temperate pond?

We can at least pose such a question without logical contradiction, and we can test certain ideas about minimal discrepancies in body size, feeding preferences, and so on. Much useful research has been done on this subject, but no general answers have emerged. And none may be possible (at least in such simplistic form as "no more similar than a 10 percent difference in body weight on average"), given the irreducible uniqueness of each species and each group of organisms. Beetle rules will almost surely not work as fish rules, not to mention the vastly more different world of rules for bacteria.

But if we cannot generate quantitative laws of nature about numbers of species in a single place, we can at least state some general principles. And the rule behind Jerusalem's Status Quo, whatever its moral dubiety in the ethical systems of *Homo sapiens*, provides a good beginning: large numbers of species can be crammed into a common territory only if each can commandeer some room of its own and not always stand in relentless competition with a maximally similar form.

Two general strategies may be cited, the second far more interesting than the first, for acquiring the requisite "breathing room"—a little bit of unique space that no other species contests in exactly the same way. In the first strategy—the "Holy Sepulchre solution" if you will—two species perceive the surrounding environment in basically the same manner and therefore must divide the territory to keep out of each other's way. Division may be strictly spatial, as in my fraternal dispute about our single common room. But organisms may also use nature's other prime dimension and construct temporal separations as well. The Status Quo divides the space within the Church of the Holy Sepulchre, but the agreement also decrees when the unitary domain of sound belongs to the masses, instruments, and voices of various competing groups.

To make an ugly analogy, based on cruel social practices now thankfully abandoned, but in force not long ago, I encountered both spatial and temporal modes of segregation when I began my college studies in southwestern Ohio during the late 1950s. The town movie theater placed whites in the orchestra and blacks in the balcony, while the local skating rinks and bowling alleys maintained different "white" and "Negro" nights. (Student and community activism, spurred by the nascent civil rights movement, fought and vanquished these cruelties during my watch. I remember my own wholehearted and, in retrospect, pretty inconsequential participation with pride.)

An instructive evolutionary example of this first strategy arises from a classical argument about modes of speciation, or the origin of a new species by branching from an ancestral population. Such branching may occur if a group of organisms can become isolated from the parental population and begin to breed only among themselves in a different environment that might favor the evolution of new features by natural selection. (If members of the separating group continue to interact and breed with individuals of the parental population, then favorable new features will be lost by dilution and diffusion, and the two groups will probably reamalgamate, thus precluding the origin of a new species by branching.)

The conventional theory for speciation —called allopatric, and meaning "living in another place"— holds that a population can gain the potential to form a new species only by becoming geographically isolated from the ancestral group, for only strict spatial separation can guarantee the necessary cutoff from contact with members of the parental population. Much research into the process of speciation has focused on the modes of attaining such geographic isolation—new islands rising in the sea, continents splitting, rivers changing their courses, and so on.

A contrasting idea—called sympatric speciation, or "living in the same place"—holds that new groups may speciate while continuing to inhabit the same geographic domain as the parental population. The defense of sympatric speciation faces a classic conundrum, and most research on the subject has been dedicated to finding solutions: if isolation from members of the parental population is so crucial to the formation of a new species, how can a new species arise within the geographic range of the parents?

This old issue in evolutionary theory remains far from resolution, but we should note, in the context of this essay, that proposed mechanisms usually follow the Holy Sepulchre principle of granting the new group a room of its own within the spatial boundaries of the parental realm—and that such "internal isolation" may be achieved by either the spatial or the temporal route. The best-documented cases of the spatial strategy invoke a process with the technical name of *host specificity,* or the restriction of a population to a highly specific site within a general area. For example, to cite an actual (although still controversial) case, flies of the genus *Rhagoletis* tend to inhabit only one species of tree as an exclusive site for breeding and feeding. Suppose that some individuals within a species that lives on apple trees experience a mutation that leads them to favor hawthorns. A new population, tied exclusively to hawthorns, may soon arise and may evolve into a separate species. The hawthorn flies live within the same geographic region as the apple flies, but members of the two groups never interbreed because each recognizes only a portion of the total area as a permissible home—just as the six sects of the Holy Sepulchre never transgress into one another's territory.

The same principle may work temporally as well. Suppose that two closely related species of frogs live and reproduce in and around the same pond, but one species uses the day-lengthening cues of spring to initiate breeding, while the other waits for the day-shortening signals of fall. The two populations share the same space and may even (metaphorically) wave and wink at each other throughout the year, but they can never interbreed and can therefore remain separate as species.

In the second, and philosophically far more interesting, strategy for securing a requisite room of one's own, species may share the same region but avoid the need for a natural equivalent of the Status Quo, because they do not perceive each other at all and therefore cannot interfere or compete—blessedly benign ignorance rather than artfully negotiated separation. This fascinating form of imperception, which can also be achieved by either spatial or temporal routes, raises one of the most illuminating issues of intellectual life and nature's construction: the theme of scaling, or strikingly different ways of viewing the world from disparate vantage points of an observer's size or life span, with no single way either universally "normal" or "better" than any other.

To begin with a personal story, I share my Harvard office with about a hundred thousand trilobites, all fossils at least 250 million years old, and now housed in cabinets lining the perimeter of my space. For the most part, we coexist in perfect harmony. They care little for my eye-blink of a forty-year career, and I view them with love and respect to be sure, but also as impassive, immobile pieces of rock. They cause me no trouble because I just move the appropriate drawers to an adjacent room when a visiting paleontologist needs to study a genus or two. But one week, about ten years ago, two British visitors wanted to look at *all* Ordovician trilobites, an endeavor that required exploratory access to all drawers. I had no choice but to abandon my office for several days—a situation made worse by the stereotypical politeness of my visitors, as they apologized almost hourly: "Oh, I do hope we're not disturbing you too much." I wanted to reply: "You bloody well are, but there's nothing I can do about it," but I just shut up instead. I relaxed when I finally figured out the larger context. My visitors, of course, had been purposely sent by the trilobites to teach me the following lesson of scaling: we will let you borrow this office for a milli-moment of our existence; this situation troubles us not at all, but we do need to remind you about the room's true ownership once every decade or so, just to keep you honest.

Species can also share an environment without conflict when each experiences life on such a different temporal scale that no competitive interaction ever occurs. A bacterial life cycle of half an hour will pass beneath my notice and understanding, unless the population grows big enough to poison or crowd out something of importance to me. And how can a fruit fly ever experience me as a growing, changing organism if I manifest such stability throughout the fly's full life cycle of two weeks or so? The pre-Darwinian Scottish evolutionist Robert Chambers devoted a striking metaphor to this point when he wondered if the adult mayfly, during its single day of earthly life, might mistake the active metamorphosis of a tadpole into a frog for proof of the immutability of species, since no visible change would occur during the mayfly's entire lifetime. (And so, Chambers argued by extension, we might miss the truth of evolution if the process unrolled so slowly that we could never notice any changes during the entire history of potential human observation.) Chambers wrote in 1844:

> Suppose that an ephemeron [a mayfly], hovering over a pool for its one April day of life, were capable of observing the fry of the frog in the waters below. In its aged afternoon, having seen no change upon them for such a long time, it would be little qualified to conceive that the external branchiae [gills] of these creatures were to decay, and be replaced by internal lungs, that feet were to be developed, the tail erased, and the animal then to become a denizen of the land.

Since organisms span such a wide range of size, from the invisible bacterium to the giant blue whale (or to the fungus that underlies a good part of Michigan), the second, or spatial, strategy of coexistence by imperception achieves special prominence in nature. This concept can best be illustrated by an example that has become something of a cliché (by repetition for appropriateness) in intellectual life during the past decade.

To illustrate his concept of "fractals," mathematical curves that repeat an identical configuration at successively larger or smaller scales ad infinitum, mathematician Benoit Mandelbrot asked a disarmingly simple question with a wonderfully subtle nonanswer: how long is the coastline of Maine? The inquiry sounds simple but cannot be resolved without ambiguity, for solutions depend upon the scale of inquiry, and no scale can claim a preferred status. (In this respect, the question recalls the classic anecdote, also told about folks "down East" in Maine, of a woman who asks her neighbor, "How's your husband?"— and receives the answer, "Compared to what?")

If I'm holding an atlas with a page devoted to the entire state of Maine, then I may measure a coastline at the level of resolution permitted by my source. But if I use a map showing every headland in Acadia National Park, then the equally correct coastline becomes much longer. And if I try to measure the distance around every boulder in every cove of Acadia, then the length becomes ever greater (and increasingly less meaningful as tides roll and boulders move). Maine has no single correct coastline; any proper answer depends upon the scale of inquiry.

Similarly for organisms. Humans rank among the largest animals on earth, and we view our space as one might see all of Maine on a single page. A tiny organism, living in a world entirely circumscribed by a single boulder in a cove, will therefore be completely invisible at our scale. But neither of us sees "the world" any better or any more clearly. The atlas defines my appropriate world, while the boulder defines the space of the diatom or rotifer (while the rotifer then builds the complete universe of any bacterium dwelling within).

We need no Status Quo to share space with a bacterium, for we dwell in different worlds of a common territory—that is, unless we interfere or devise a way to intrude: the bacterium by generating a population large enough to incite our notice or cause us harm; *Homo sapiens* by inventing a microscope to penetrate the world of the invisible headland on a one-page map of the earth.

Frankly, given our aesthetic propensities, we would not always wish to perceive these smaller worlds within our domain. About 40 percent of humans house eyebrow mites, living beneath our notice at the base of hair follicles above our eyes. By ordinary human standards, and magnified to human size, these mites are outstandingly ugly and fearsome. I would just as soon let them go their way in peace, so long as they continue the favor of utter imperceptibility. And do we really want to know the details of ferocious battles between our antibodies and bacterial invaders—a process already distasteful enough to us in the macroscopic consequence of pus? (Don't get me wrong. As a dedicated scientist, I do assert the cardinal principle that we always want to know intellectually, both to understand the world better and to protect ourselves. I am just not sure that we should always crave visceral perception of phenomena that don't operate at our scale in any case.)

Finally, this theme of mutually invisible life at widely differing scales bears an important implication for the "culture wars" that supposedly now envelop our universities and our intellectual discourse in general (but that have, in my opinion, been grossly oversimplified and exaggerated for their perceived newsworthiness). One side of this false dichotomy features the postmodern relativists who argue that all culturally bound modes of perception must be equally valid, and that no factual truth therefore exists. The other side includes the benighted, old-fashioned realists who insist that flies truly have two wings, and that Shakespeare really did mean what he thought he was saying. The principle of scaling provides a resolution for the false parts of this silly dichotomy. Facts are facts and cannot be denied by any rational being. (Often, facts are also not at all easy to determine or specify—but this question raises different issues for another time.) Facts, however, may also be highly scale dependent—and the perceptions of one world may have no validity or expression in the domain of another. The one-page map of Maine cannot recognize the separate boulders of Acadia, but both provide equally valid representations of a factual coastline.

Why should we privilege one scale over another, especially when a fractal world can express the same form at every scale? Is my hair follicle, to an eyebrow mite, any less of a universe than our entire earth to the Lord of Hosts (who might be a local god as tiny as a mite to the great god of the whole universe—who then means absolutely nothing in return to the mite on my eyebrow)? And yet each denizen of each scale may perceive an appropriate universe with impeccable, but local, factual accuracy.

We don't have to love or even to know about all creatures of other scales (although we have ever so much to learn by stretching our minds to encompass, however dimly and through our own dark glasses, their equally valid universes). But it is good and pleasant for brethren to dwell together in unity—each with some room of one's own.

ILLUSTRATION CREDITS

Grateful acknowledgment is made to the following for permission to reproduce the images herein:

pages v, 14	American Museum of Natural History, photograph by Jackie Beckett
page 37	American Museum of Natural History, photograph by Stephanie Bishop
page 38	The Granger Collection, New York
page 96	Rare Book Collection, Skillman Library, Lafayette College
page 127	Courtesy of Jonathan A. Hill
page 128	Christie's Images
page 193	American Museum of Natural History, photographs by Jackie Beckett
pages 204, 208, 212, 213, 214	All images from the Edward Arnold Collection, courtesy of the Eiffel Tower Millennial Exhibition
page 300	Corbis-Bettmann
page 309	Corbis-Bettmann
page 322	Courtesy of Joyce Pendola
pages 323, 330 (bottom images)	Courtesy of Andrew Knoll

All other images appearing throughout are from the author's collection.

INDEX

Time Out Digital Ltd
4th Floor
125 Shaftesbury Avenue
London WC2H 8AD
United Kingdom
Tel: +44 (0)20 7813 3000
Fax: +44 (0)20 7813 6001
Email: guides@timeout.com
www.timeout.com

Published by Time Out Digital Ltd, a wholly owned subsidiary
of Time Out Group Ltd. Time Out and the Time Out logo are
trademarks of Time Out Group Ltd.

© Time Out Group Ltd 2015
Previous editions 1998, 2000, 2002, 2004, 2006, 2010.

10 9 8 7 6 5 4 3 2 1

This edition first published in Great Britain in 2015 by Ebury Publishing
20 Vauxhall Bridge Road, London SW1V 2SA

Ebury Publishing is part of the Penguin Random House group of companies
whose addresses can be found at global.penguinrandomhouse.com

Distributed in the US and Latin America by Publishers Group West
(1-510-809-3700)

For further distribution details, see www.timeout.com

ISBN: 978-1-90504-299-9

A CIP catalogue record for this book is available from the British Library.

Printed and bound in China by Leo Paper Products Ltd.

MIX
Paper from
responsible sources
FSC
www.fsc.org   FSC® C018179

Time Out

Edinburgh

www.timeout.com/edinburgh

Contents

118

32

TimeOut Edinburgh

Editorial
Editor Keith Davidson
Copy Editors Cath Phillips
Proofreader Tamsin Shelton

Editorial Director Sarah Guy
Group Finance Manager Margaret Wright

Design
Senior Designer Kei Ishimaru
Designer Darryl Bell
Group Commercial Senior Designer Jason Tansley

Picture Desk
Picture Editor Jael Marschner
Deputy Picture Editor Ben Rowe
Picture Researcher Lizzy Owen

Advertising
Managing Director St John Betteridge
Advertising Sales Deborah Maclaren, Helen Debenham
@ The Media Sales House

Marketing
Senior Publishing Brand Manager Luthfa Begum
Head of Circulation Dan Collins

Production
Production Controller Katie Mulhern-Bhudia

Time Out Group
Founder Tony Elliott
Chief Executive Officer Tim Arthur
Managing Director Europe Noel Penzer
Publisher Alex Batho

Contributors
The Editor would like to thank all contributors to previous editions of *Time Out Edinburgh*, whose work forms the basis for parts of this book.

Maps JS Graphics Ltd (john@jsgraphics.co.uk)

Cover and pull-out map photography Guido Cozzi/4Corners

Back Cover Photography Clockwise from top left: Bildagentur Zoonar GmbH/Shutterstock.com; Heartland Arts/Shutterstock.com; Skully/Shutterstock.com; Alessandro Storniolo/Shutterstock.com; © G&V Royal Mile Hotel, Edinburgh

Photography Pages 2/3 Albert Pego/Shutterstock.com; 4 (top), 118 SurangaSL/Shutterstock.com; 4 (bottom), 32 (top) Krizek Vaclav/Shutterstock.com; 5 (top), 71 Marketing Edinburgh/Greater Grassmarket; 5 (bottom right), 261 (top and bottom) James Crawford Photography; 7, 16 (top left) trotalo/Shutterstock.com; 10 Wonge Bergmann 10/11, 38/39 (top), 47, 78/79, 105, 142, 192/193, 201, 204, 216 Brendan Howard/Shutterstock.com; 11 WELBURNSTUART/Shutterstock.com; 13 (top), 139 Sally Jubb Photography; 13 (bottom) Chris G. Walker/Shutterstock.com; 14 (top) National Galleries of Scotland/Chris Watt; 14/15 (middle) Bildagentur Zoonar GmbH/Shutterstock.com; 14/15 (bottom) Andrew Cowan/Scottish Parliament/© Scottish Parliamentary Corporate Body – 2015; 15 (top) Antonia Reeve; 15 (bottom) © National Museums Scotland; 16 (bottom) Keith Hunter Photography; 17 (top left), 23 Olivia Rutherford; 17 (right) Steffen Zahn/Wikimedia Commons; 18 Jonathan Smith/Getty Images/Lonely Planet Images; 20 (top) Calum Hutchinson/Wikimedia Commons; 20 (bottom) REX; 21 © Findlay/Alamy; 22, 22/23 (main), 25 (bottom), 26, 58 (right), 86, 123, 126, 157, 219, 226/227, 244 (bottom), 249 Ivica Drusany/Shutterstock.com; 22/23 (top) lowsun/Shutterstock.com; 24 (top), 87, 102, 113 National Galleries of Scotland/Eoin Carey; 24 (bottom) Doug Stacey/Shutterstock.com; 25 (top), 90 (top), 94 Paul Johnston - Copper Mango Ltd; 28/29 Jasper Schwartz; 31 Fraser Iain Cameron; 32 (middle) © Edinburgh Festival Fringe Society; 32 (bottom) Eoin Carey; 36 Skully/Shutterstock.com; 38/39 (middle) Madrugada Verde/Shutterstock.com; 42/43, 58 (left), 67, 74, 114, 116, 202, 215, 239 Heartland Arts/Shutterstock.com; 44 Ioannis Lachanis; 44/45 MyImages - Micha/Shutterstock.com; 50 Alessandro Storniolo/Shutterstock.com; 51 Ad Meskens/Wikimedia Commons; 53 (left) TEEMEE/Shutterstock.com; 53 (middle) Christy Nicholas/Shutterstock.com; 53 (right) Stuart Caie/Wikimedia Commons; 54, 146 StockCube/Shutterstock.com; 57 Stefan Schaefer/Wikimedia Commons; 64 Bill McKelvie/Shutterstock.com; 65 TanArt/Shutterstock.com; 68 Ben Roberts; 72 (top) Shannon Tofts; 72 (bottom) Michael Wolchover; 78, 89 (right) © National Galleries of Scotland; 80 Nathan Danks/Shutterstock.com; 81 roy henderson/Shutterstock.com; 84 Delta-NC/Wikimedia Commons; 88 Bill Spiers/Shutterstock.com; 89 (left) National Galleries of Scotland/Andrew Lee; 90 (middle) Jane Barlow; 91, 244 (top), 256 Paul Zanre Photography; 93 Matt Davis; 101 FCG/Shutterstock.com; 109 (bottom) Gilles Moulin/Marketing Edinburgh/Greater Grassmarket; 111 John Lord/Wikimedia Commons; 112 jean morrison/Shutterstock.com; 114/115 Saffron Blaze/Wikimedia Commons; 119 Rafa Irusta/Shutterstock.com; 120 © Crown Copyright reproduced courtesy of Historic Scotland; 126/127 Richie Chan/Shutterstock.com; 128, 141 Magnus Hagdorn/Wikimedia Commons; 130 Abi Radford; 130/131 SergeBertasiusPhotography/Shutterstock.com; 132 Ham/Wikimedia Commons; 133, 257 Gareth Easton; 144/145 Allan Pollok-Morris; 148 (left) Craig Stephen; 151 Chris Brown/Wikimedia Commons; 152/153 Christophe Jossic/Shutterstock.com; 154 Andrea Obzerova/Shutterstock.com; 158 (top) Chris Jenner/Shutterstock.com; 158 (bottom), 159 Marc Millar; 162 Patrik Dietrich/Shutterstock.com; 164 Victor Denovan/Shutterstock.com; 165 Aga & Tomek Adameczek/Shutterstock.com; 166/167, 186 Marco Borggreve; 169 www.tonymarshphotography.com; 171 Deborah Mullen Photography; 175 Courtesy Everett Collection/REX; 178 The Mainz Company; 184 Gavin Brown; 185 Andy Laing; 187 KK Dundas; 189 (top) DECARLO; 191 Eamonn McGoldrick; 194 AAR Studio/Shutterstock.com; 195 rubiphoto/Shutterstock.com; 198 Cornfield/Shutterstock.com; 199 (top) Frank Cornfield; 203 donsimon/Shutterstock.com; 206 Stuart Jenner/Shutterstock.com; 209 AdamEdwards/Shutterstock.com; 210/211 PlusONE/Shutterstock.com; 212/213 Georgios Kollidas/Shutterstock.com; 214 Wikimedia Commons; 217 Patten's The Expedicione into Scotlande of Edward Duke of Somerset/Wikimedia Commons; 220 Scott Liddell/Shutterstock.com; 223 Thomas Keith/Wikimedia Commons; 228 ArTono/Shutterstock.com; 229, 230, 232 Kim Traynor/Wikimedia Commons; 234 (bottom) Everett Historical/Shutterstock.com; 236 (top) ITV/REX; 236 (bottom) Little, Brown Book Group; 237 (left) Szymon Sokół; 237 (middle) The Herald Evening Times Picture Archive; 237 (right) Simon Bradshaw/Wikimedia Commons; 240 Bikeworldtravel/Shutterstock.com; 241 Scottish National Gallery/Wikimedia Commons; 260 DrimaFilm/Shutterstock.com; 261 (middle) Mr Droogle; 276/277 Philip Cormack/Shutterstock.com.

The following images were supplied by the featured establishments: 5 (bottom left), 16 (top right), 17 (bottom left), 27, 30, 38/39 (bottom), 60, 61, 62, 63, 77, 85, 90 (bottom), 92, 97, 98, 102/103, 109 (top), 110, 121, 124, 136, 143, 144, 148 (right), 149, 152, 160, 161, 168, 170, 172, 173, 176, 177, 179, 180, 182, 183, 188, 189 (bottom), 199 (bottom), 234 (top), 238, 242/243, 245, 247, 248, 250, 251, 252, 253, 254, 255, 258/259

About the Guide

GETTING AROUND

Each sightseeing chapter contains a street map of the area marked with the locations of sights and museums (●), restaurants (●), cafés and bars (●) and shops (●). There are also street maps of Edinburgh at the back of the book, along with an overview map of the city. In addition, there is a detachable fold-out street map.

THE ESSENTIALS

For practical information, including visas, disabled access, emergency numbers, lost property, websites and local transport, see the Essential Information section (page 242).

THE LISTINGS

Addresses, phone numbers, websites, transport information, hours and prices are all included in our listings, as are selected other facilities. All were checked and correct at press time. However, business owners can alter their arrangements day to day, and fluctuating economic conditions can cause prices to change rapidly. The very best venues in the city, the must-sees and must-dos in every category, have been marked with a red star (★). In the sightseeing

chapters, we've also marked venues with free admission with a FREE symbol.

PHONE NUMBERS

The area code for Edinburgh is 0131. You don't need to use this code when calling from Edinburgh on a landline. From outside the city, dial your country's access code (00 from the UK, 011 from the US) or a plus symbol, followed by the UK country code (44), then the area code for Edinburgh (dropping the initial zero) and the number. So, to reach Edinburgh Castle, dial + 44 131 225 9846. For more on phones, see p269.

FEEDBACK

We welcome feedback on this guide, both on the venues we've included and on any other locations that you'd like to see featured in future editions. Please email us at guides@timeout.com.

To keep up to date with the latest goings-on, check out **www.timeout.com/ edinburgh**. To get in touch, email helloedinburgh@timeout.com.

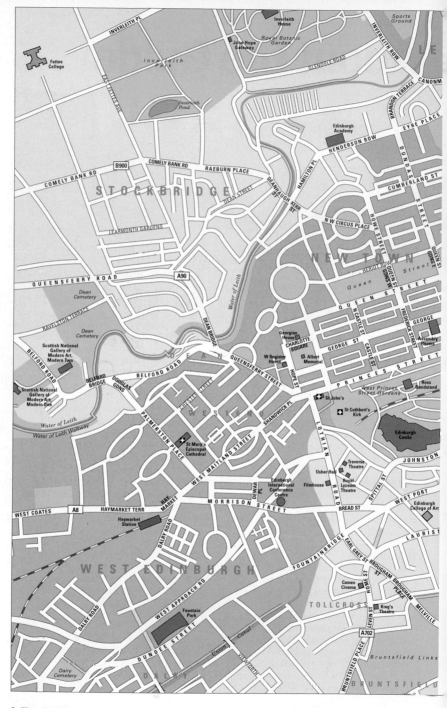

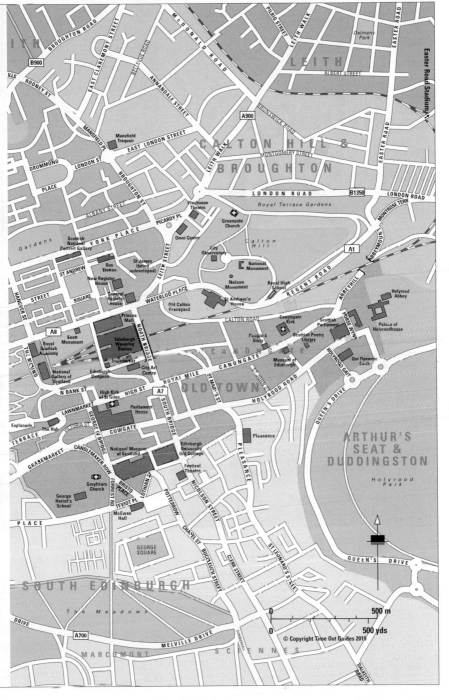

Edinburgh's Top **20**

*Counting down the
very best in Scotland's
number one city.*

1 The festivals
(pages 32-37)

August sees not just one but several
significant, international arts festivals
running concurrently. The city
becomes the planet's cultural epicentre,
the population skyrockets and the
entertainments are legion. For those
few weeks each year, nowhere else in
the world can match Edinburgh for
vibrancy and excitement.

2 National Museum of Scotland
(page 74)

Everything you ever wanted to know about Scotland, from its geological origins to its post-industrial identity, under one roof. It's a national treasure, admission is free and there's a unique view of the city and surrounding area from the roof terrace as a bonus.

3 Edinburgh Castle
(page 55)

How many capitals have an authentic medieval castle perched on a tough, basalt rock defining the city centre? Containing St Margaret's Chapel, the Scottish crown jewels, the National War Museum, the deeply affecting Scottish National War Memorial and more, this is a Scottish icon.

artmag

EAST OF SCOTLAND GALLERY GUIDE

Gallery guide

**Complete directories & maps
for Edinburgh, East Lothian,
Fife & the Borders**

Free in galleries, museums
& art spaces

art**news**
art**world**

Previews
50+ exhibitions
in every issue

www.artmag.co.uk

4 High Kirk of St Giles (page 57)

The Mother Church of Presbyterianism and a hardy survivor that has endured and adapted since the medieval period. On a prominent Royal Mile site with a distinctive crown spire, its stonework is gravid with authenticity, drama and history. Don't miss the Thistle Chapel.

5 Scottish National Gallery (page 88)

The National Gallery building, its Royal Scottish Academy counterpart and the Gardens Entrance linking them at basement level form the Scottish National Gallery complex. Together they offer everything from late medieval religious painting and Renaissance masters to Post-Impressionism, contemporary artworks and even a decent lunch.

6 Food

For a city of just under half a million inhabitants, Edinburgh has five Michelin-starred restaurants, several more deserving of that accolade, as well as many other excellent eateries and a resuscitated street market scene with fabulous food stalls. Outside London, it's the best place to eat in the British Isles.

7 Edinburgh Zoo (page 170)

There's plenty to please children here: cuddly entertainments, celebrity pandas and waddling penguins. But look at the giant anteater, the Malayan tapir or the white-faced saki monkey and you'll realise just how alien and unpredictable natural selection can be.

8 Holyrood Park (page 120)

In the middle of Edinburgh, this park offers 650 acres of wild land with crags, a hill rising to 823 feet, three lochs and assorted wildlife. There are cyclists, joggers, raptors and rare plants, but the city is never far away.

9 Royal Botanic Garden Edinburgh (page 105)

This is an important centre for botany and horticulture as well as an educational resource. It's also a beautiful green haven in the city, providing respite from Edinburgh's bustle, its sandstone tenements and traffic-busy streets – somewhere to lose yourself for half an hour contemplating the fearsome Brazilian rhubarb.

10 Scottish National Portrait Gallery (page 88)

From the frieze of famous Scots around the Great Hall to the galleries telling tales of entertainers, Jacobites, princesses and more, this excellent venue deserves more visitors than it gets. Since reopening in 2011 after a major refurbishment, it's looking better and more thematically coherent than ever.

11 Forth Bridge (page 205)

Not the new bridge, due for completion in 2016, nor the other bridge that was the height of modernity when built in 1964, but the original late Victorian wonder from 1890. It's worth the trip to Queensferry just to stand and marvel at its ambition, scale and red oxide magnificence.

12 Scottish Parliament (page 67)

Born amid a project management controversy, completed late, over-budget and with a radical design, the Scottish Parliament faced up to its toughest challenges at the outset. A decade later, Scotland now has an accessible parliament fit for the 21st century, with an interior that's even more fascinating than the exterior.

13 Scottish National Gallery of Modern Art (page 113)

Housed in two impressive buildings, called Modern One and Modern Two, located opposite each other on Belford Road, the gallery has big names from these islands including Bacon, Freud and Hockney, home-grown Scottish talent such as Gordon, Fergusson and Paolozzi, as well as your Picassos and Pollocks.

14 National War Museum (page 55)

Although it's within the complex of Edinburgh Castle (see no.3), the National War Museum is administratively separate, run by National Museums of Scotland. It's not a simple list of battles, but more about how Scotland and its people were shaped by the experience of war over centuries. Aside from the permanent displays, covering everything from recruitment to weapons, military art to memorabilia, there is also space for travelling exhibitions.

15 Palace of Holyroodhouse
(page 66)

The palace, with its early 16th-century core, is the British monarch's official residence in Scotland (the Queen is usually in residence for a week in summer, around late June). There's a ruined medieval abbey in the grounds plus the adjacent Queen's Gallery showcasing art from the Royal Collection.

16 Royal Yacht Britannia
(page 158)

From the optimistic lines of the exterior to the suburban quaintness of the state rooms and apartments, this is a fascinating relic of post-war Britain, commissioned in 1954, decommissioned in 1997. Whether you visit for the regal fairy dust, or to revile it, you can still have tea and scones on board.

17 Whisky

Single-malt Scotch may be a mystery when you arrive in Edinburgh, but thanks to the **Scotch Whisky Experience** (p58), shops **Cadenhead's** (p70) and **Royal Mile Whiskies** (p63), plus venues such as the **Bow Bar** (p75), expert status is only a matter of time.

18 Calton Hill (page 119)

There are monuments to an architect, a moral philosopher, the Napoleonic War dead and a naval hero, an old observatory partly occupied by an art gallery, an annual spring festival of bacchanalian exuberance, and all-round views of Edinburgh and the Forth. If you're not tempted, check your pulse.

19 Jupiter Artland (page 151)

This art park features site-specific sculpture and other works that virtually defy categorisation. With names like Goldsworthy, Gormley, Kapoor and Kovats, it's an important venue not just for Edinburgh or Scotland, but in international terms.

20 Scott Monument (page 89)

It's on the main street, it's 200 feet high and it looks like a steampunk space rocket. It doesn't commemorate a monarch, a politician or a warrior, however, but novelist and poet Sir Walter Scott, 1771-1832. Edinburgh: a city with its own priorities.

Edinburgh Today

In a place of change.

TEXT: KEITH DAVIDSON

Go back a decade and Edinburgh was different. There were no trams on Princes Street, the *Scotsman* newspaper was established at its modern headquarters on Holyrood Road, the appearance of Polish and other east European nationals in the city was still a novelty and the locally based Royal Bank of Scotland (RBS) was one of the biggest companies in the world.

But now? After a controversial and expensive gestation period, Edinburgh has a tram service. The continuing slump in the sales of the *Scotsman*, once one of Scotland's newspapers of record, saw it vacate its offices and move to more modest premises outside the city centre. The Polish, Baltic and other workers who moved here after EU enlargement in 2004 are now well settled; their children are being born in Scotland. RBS, meanwhile, was at the eye of the credit crunch and had to be nationalised. Although the plan is to sell the bank back into the private sector, when the time is right, currently it is owned by the UK state.

Clockwise from top left: **Scottish Parliament**; **tram on Princes Street**; **voting in the 2014 referendum**.

DISASTER STRIKES

The credit crunch had implications for Edinburgh that went far beyond the reputation of a single bank. It may be a long way behind London, but the Scottish capital is still an important financial centre; home to banks, insurance companies and investment houses. Before 2008 there was also an unprecedented frenzy of development in the city, with plans for thousands of new homes and other buildings along the Forth from Granton to Leith. The financial crisis, and its aftermath, was therefore devastating on various counts. All kinds of jobs were lost, developers went bust and the majority of homes on the coast were never built.

The recession – and the time it took for some sort of recovery to kick in – would have been enough on its own to preoccupy the city, never mind the trams farrago, the decline of a once iconic newspaper, demographic changes or the specifics of the RBS situation. Except, of course, the most important event in Edinburgh and Scotland's recent history hasn't even been mentioned yet: the independence referendum of September 2014.

YES, YES, NO

Outside Scotland, this seemed to appear on people's radar as a surprise, but meaningful moves towards self-determination had been going on here since the 1970s. There was a marginal 'yes' vote in a devolution referendum

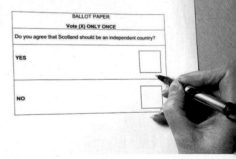

in 1979, and an overwhelming 'yes' vote in a second devolution referendum in 1997; the Scottish Parliament was convened in 1999 and took up residence at Holyrood in 2004.

The Scottish people then elected a Scottish National Party (SNP) minority government in 2007, and an SNP majority government in 2011. Given that the raison d'être of that party is independence – stated in its manifesto – a third referendum followed in 2014. This time the ballot paper said, simply, 'Should Scotland be an independent country?' The people's answer was, 'No.'

The numbers were 55 per cent against and 45 per cent in favour; separation from the UK was clearly rejected, but the figures hardly added up to a fulsome endorsement of the constitutional status quo. In the following weeks and months, pro-independence parties received a boost in membership, particularly

now there are two, in Edinburgh and London, and two significant assemblies, in Belfast and Cardiff. Although serious political power still resides in London – over the economy, foreign policy, the military, nuclear weapons and tax – the overall historical trajectory doesn't look promising for a centralised British state, whether or not that means full Scottish independence.

TIMES THEY ARE A-CHANGIN'

It's beyond glib to shrug one's shoulders at this point and say, 'Hey, things change.' But in 1981, when Charles and Di went on honeymoon in the Mediterranean, on the Royal Yacht *Britannia*, no one imagined the vessel would end up in Leith as a tourist attraction with a tacked-on tearoom. Similarly, once upon a time you pretty much had to go to London if you wanted to fly anywhere outside Britain. Now you can go direct to Abu Dhabi, Doha, New York and all over Europe from Edinburgh, thanks to structural shifts in the airline industry.

When the *Scotsman* moved to Barclay House, its new Holyrood headquarters, in 1999, it assumed a bright future because no one understood the implications of broadband internet for a Scottish broadsheet, let alone interactivity, smartphones and tablets. These days, Barclay House is occupied by Rockstar North: responsible for the *Grand Theft Auto* series of computer games.

The list goes on. Before EU expansion a decade ago, kids born in Edinburgh were not routinely called Mateusz and Maja, but they are now. In 2005, who, at the lavish opening of the new RBS headquarters, in the west of the city, would have dreamed of the bank being crisis-ridden and state-owned within a few years?

Changes come radically, unpredictably and according to a much deeper grammar than the one informing spats between political parties. Edinburgh today is certainly in a state of constitutional flux, as is Scotland. Predicting what's going to come next is a mug's game – but one bet does look reasonable. The future will no more look like the stolid British state of the latter 20th century than *Grand Theft Auto V* resembles an old broadsheet newspaper.

With its parliament and its referendum-prodded consciousness, Edinburgh has a ringside seat for the coming circus.

the SNP but also the Scottish Green Party. Conversely, the pro-union parties went back to squabbling over who should run the UK government in London. Counter-intuitively, the uplift effect of the referendum went to the losers. The vote was statistically decisive, but it was far from politically decisive.

As it stands, there is a definite strand of opinion in Scotland that is vehemently anti-nationalist, viewing that ideology as pernicious and regressive, come what may. Others see nationalism as merely the vehicle that might deliver a way out of what they believe to be an increasingly sclerotic and dysfunctional British polity. Alongside these views is the altogether geekier conjecture that changes are afoot far beyond the control of self-interested party politicians.

Up to 1922, the whole of Ireland was part of the British union, for example. Up to 1945, Britain still had an empire, but it evaporated in a matter of decades. Until the late 1990s, there was only one parliament in the UK but

Itineraries

You have only two days in the city? Let's make the best of it then...

8AM

9.30AM

Day 1

8AM Before exploring the historic and attraction-packed Old Town, go back in time even further by taking a brisk hike up **Arthur's Seat** (p128), formed some 350 million years ago as the igneous plug to a once active volcano. It's an invigorating and geologically dramatic start to the day.

9.30AM Back at the bottom of the Royal Mile, the **Palace of Holyroodhouse** (p66) gives an excellent grounding in Edinburgh's royal history

going back 500 years, covering both Scottish and British monarchs. Art-lovers should divert to the **Queen's Gallery**, which shows paintings from the Royal Collection. If the concept of an unelected head of state troubles you, tour the modern **Scottish Parliament** (p67) at Holyrood instead, a functioning legislature since 1999, based at this ambitious site since 2004.

11.30AM Low-key attractions dot the Canongate, among them the **Museum of Edinburgh** (p65), the

3PM

6PM

Clockwise from top left: **Arthur's Seat**; **Queen's Gallery**; **Bow Bar**; **Edinburgh Castle**.

People's Story (p66) and the **Museum of Childhood** (p65). Dip into them as you wander the Royal Mile before stopping for lunch: regional French at **La Garrigue** (p69), vegetarian at **David Bann** (p68) or Modern European at **Wedgwood** (p69).

2PM After lunch, continue east to the **High Kirk of St Giles** (p57), its role in Scottish history assured, its Thistle Chapel an ornate must-see. Then head for the **Writers' Museum** (p59) dedicated to Burns, Scott and Stevenson, or explore the atmospheric labyrinth of closes and wynds (p52) that run off the Royal Mile, giving a flavour of the medieval city.

3PM Edinburgh Castle (p55) dominates the landscape from the top of Castlehill. Its precincts have many attractions from the simplicity of the 12th-century St

Margaret's Chapel to the haunting Scottish National War Memorial; it's an extraordinary place.

6PM After the castle, head for the **Bow Bar** (p75) on West Bow, to sample some excellent single malt Scotch while you reflect on your sightseeing and consider the options for the evening.

8PM For some stellar cooking, head immediately south of the castle to where **Dominic Jack** runs his **Castle Terrace** restaurant (p132) while **Ondine** (p61) on George IV Bridge is your seafood option. For more informal eats wander to Teviot Place and **Ting Thai Caravan** (p75).

10PM & LATER For clubbing options, or late-night drinks, the **Nightlife** chapter (pp179-185) has suggestions that will see you through until the wee, small hours.

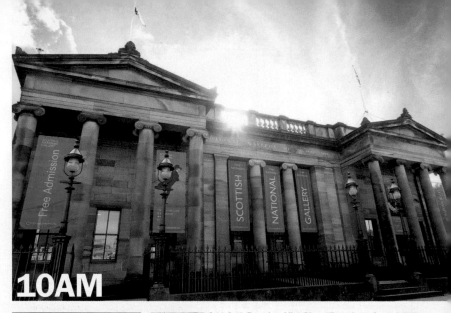

10AM

Day 2

10AM At the foot of the Mound, the **Scottish National Gallery** (p88) is a gentle way to start a day's exploration with everything from late medieval religious art to French Impressionism. There's yet more art at the achingly modern **Fruitmarket Gallery** (p56) while the **City Art Centre** (p55) often has excellent exhibitions; both of these are found along Market Street. Alternatively, for some exercise with a view, head north into Princes Street Gardens and climb the **Scott Monument** (p89). It has 287 steps, but from the top you can see everything from the Pentlands to the castle, Princes Street to Fife.

NOON Head further north into the **New Town** (pp78-101) and explore its handsome Georgian architecture. Dundas Street has private art galleries; Heriot Row maintains a patrician air;

2.30PM

Northumberland Street is as understated as Great King Street or Moray Place are grand. When you've seen enough, head into the genteel neighbourhood of Stockbridge adjacent.

1PM Looking for lunch in Stockbridge, you'll be spoiled for choice; two very different pub options would be the more traditional **Baillie** (p108) or the popular **Scran & Scallie**

(p110), the latter run by the same people who have Michelin-star restaurants elsewhere in the city. The area is also full of independent shops worth a browse. The Raeburn Estate up Leslie Place is another impressive neo-classical neighbourhood.

2.30PM Dean Terrace, parallel to Leslie Place, runs along the Water of Leith, an unexpected presence in

6.30PM

these urban surrounds. The small river winds its way all the way north-east to Leith itself. After a little bankside stroll, divert north into the **Royal Botanic Garden Edinburgh** (p105), which pleases both serious horticulturalists and casual wanderers. As well as the greenery and the various glasshouses, you'll find fine art exhibitions in **Inverleith House** (p104). Check the opening hours before setting out: the RBGE's closing time varies by season.

6.30PM After this gentle, cultured day, it's time for a cocktail. Stop for some informal pre-dinner mixology at the **Lucky Liquor Co** (p95), then try **Contini Ristorante** (p90) or **Restaurant Mark Greenaway** (p93) for dinner; The **Dogs** (p90) is a more relaxed and economical option for eats.

10PM & LATER One last something before bedtime? Visit the pub where Detective Inspector Rebus spends his time, the **Oxford Bar** (p96). Some nights you might even see author Ian Rankin in there himself.

Clockwise from left: **Scottish National Gallery; Restaurant Mark Greenaway; Royal Botanical Garden Edinburgh**.

FREE CAPITAL
Getting in gratis.

Not everything in Edinburgh costs money. It's worth noting that major attractions such as the **National Museum of Scotland**, the **High Kirk of St Giles**, the **Scottish National Gallery**, the **Royal Botanic Garden Edinburgh** and tours of the **Scottish Parliament** are all free – although some venues appreciate donations.

The museums and galleries run by City of Edinburgh Council are largely free too. They include the municipal art collection at the **City Art Centre**, the **Museum of Childhood**, **Museum of Edinburgh**, **People's Story** and **Writers' Museum**.

Some prominent city chuches have regular free concerts, including the **High Kirk of St Giles** and **St Mary's Episcopal Cathedral**; the University of Edinburgh's **Reid Concert Hall** also has free classical recitals at lunchtime, twice a week.

Finally, there are the city parks. **Holyrood Park**, with **Arthur's Seat** in the middle, offers a stretch of wilderness in the heart of the capital; **Princes Street Gardens** couldn't be more central; the **Meadows** is a great place to stretch out with a book on a sunny day, or to walk along the cherry blossom avenues in spring. The **Water of Leith** and the **Union Canal** are waterways rather than parks, but free to walk beside, and Edinburgh has a great deal of coastline from the foreshore at **Cramond** to the beach at **Portobello**.

Princes Street Gardens.

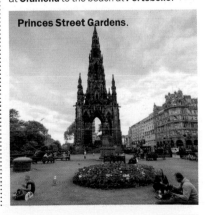

HISTORY
Pass the pass.

If you plan to visit a number of venues run by **Historic Scotland**, it may be worth getting an Explorer Pass. There are two versions: one covering three days of sightseeing (valid over a five-day period; £23.20, £13.60-£19.20 reductions, £46.40 family) and one covering seven days (valid over a 14-day period; £30.40, £17.60-£24.80 reductions, £60.80 family).

You can buy the pass at any Historic Scotland attraction or online from www.historic-scotland.gov.uk. Edinburgh attractions covered include Edinburgh Castle (where cardholders get fast-track access) and Craigmillar Castle. Many of the sites featured in the **Escapes & Excursions** chapters (pp194-209) also qualify.

Above: **bus tour**. Right: **Cadies & Witchery Tours**.

Guided Tours

BUS TOURS
Edinburgh Bus Tours
220 0770, www.edinburgh tour.com. Rates Single bus tour £14. Four bus tours £18. Bus & Boat tour £20.
Lothian Buses (p263) operates a number of double-decker bus tours under its Edinburgh Bus Tours brand: City Sightseeing, a sights primer in nine languages; the Edinburgh Tour, the same but with live commentary in English; and the Majestic Tour, which runs to Holyroodhouse and Ocean Terminal. In summer, there's also a Bus & Boat tour, which includes a boat trip on the Firth of Forth; and MacTours, exploring the city on vintage buses. Tours start at 9-10am and finish for the day at 3-4pm in winter, 7-8pm in high summer. All tours depart regularly from Waverley Bridge, and booking isn't required. Riders can jump on or off the bus at any point over a 24-hour period.

WALKING TOURS
Auld Reekie Tours
557 4700, www.auldreekie tours.com. Tours Original Underground Tour noon, 2pm, 4pm, 5pm daily. Ghost & Torture Tour 6pm, 8pm daily. Terror Tour 10pm daily. Haunted Vaults & Graveyard Tour 7pm, 9pm Fri, Sat. Meeting point Tron Kirk, High Street, Old Town. Tickets Original Underground Tour £9. Ghost & Torture Tour £10. Terror Tour, Haunted Vaults & Graveyard Tour £12. Map p284 H7.
Sacrificing some factual accuracy for engaging pantomime instead, guides lead you through the streets and cemeteries of Edinburgh,

then head underground. Some tours are not suitable for children.

Cadies & Witchery Tours

225 6745, www.witcherytours. com. Tours Murder & Mystery Jan-Apr, Sept-Dec 7pm daily. May-Aug 9pm daily. Ghosts & Gore May-Aug 7pm daily. Meeting point outside Witchery restaurant, 352 Castlehill, Old Town. Tickets £7.50. Map p284 G7.

Moving from the castle through the Cowgate, a character guide entertains in an agreeably light-hearted fashion, with other characters popping up to startle the unwary tourist. The summer-only Ghosts & Gore tour is a daylight reworking of the same event.

City of the Dead Tours

225 9044, www.cityofthedead tours.com. Tours Underground City of the Dead Tour Easter-Oct 1.30pm, 3.30pm, 8.30pm daily. Nov-Easter 3.30pm, 8pm daily. City of the Dead Haunted Graveyard Tour Easter-Oct 9pm daily. Nov-Easter 8.30pm daily.

The Double Dead Walk Tour Easter-Oct 8pm, 10pm daily. Nov-Easter 7.30pm, 9.30pm daily. Meeting point Mercat Cross, High Kirk of St Giles, High Street, Old Town. Tickets Underground City of the Dead Tour, City of the Dead Haunted Graveyard Tour £10. The Double Dead Walk Tour £13. Map p284 H7.

The Underground tour may introduce you to the 'South Bridge Entity', while the Haunted Graveyard jaunt takes in Greyfriars, home to an active poltergeist called MacKenzie. Feel the fear. Age limits apply on the later and more dramatic tours.

Edinburgh Literary Pub Tour

0800 169 7410, www. edinburghliterarypubtour. co.uk. Tours May-Sept 7.30pm daily. Apr, Oct 7.30pm Thur Sun. Jan-Mar 7.30pm Fri, Sun. Nov, Dec 7.30pm Fri. Meeting point Beehive Inn, Grassmarket, Old Town. Tickets £14. Map p284 G8.

A lively guide to the city's literary history, from Burns to Welsh via points in between.

Mercat Walking Tours

225 5445, www.mercattours. com. Ghost tours: Doomed Dead & Buried Tour Apr-Oct 8.30pm daily. Nov-Mar 8.30pm Thur-Sat. Ghostly Underground Tour Apr-Oct 3pm, 5pm daily. Nov-Mar 3pm Mon-Fri, Sun; 5pm Sat. Ghosts & Ghouls Tour 7pm, 8pm daily. Hidden & Haunted Tour Apr-Oct 9pm, 9.30pm, 10pm daily. Nov-Mar 9pm, 9.30pm Mon-Thur, Sun; 10pm Fri, Sat. History tours: Descend & Discover Tour 10.30am, 11.30am daily. Edinburgh Outlander Experience Tour Apr-Oct 11am Sat. Historic Vaults Tour Apr-Oct 2pm, 4pm daily. Nov-Mar 2pm Mon-Fri; 4pm Sat, Sun. Secrets

of the Royal Mile & Edinburgh Castle Tour 1.30pm daily. Meeting point Mercat Cross, High Kirk of St Giles, High Street, Old Town. Tickets Doomed Dead & Buried Tour £14. Ghostly Underground Tour £11. Ghosts & Ghouls Tour £11-£14. Hidden & Haunted Tour £11. Descend & Discover Tour £8. Edinburgh Outlander Experience Tour £14. Historic Vaults Tour £11. Secrets of the Royal Mile & Edinburgh Castle Tour £11-£27. Map p284 H7.

The longest established of all the tour companies, Mercat prides itself on historical accuracy. It also conducts tours in German and Spanish. Ghostly goings-on may not be suitable for children.

Real Mary King's Close

0845 070 6244, www.real marykingsclose.com. Tours every 15mins Easter-Oct 10am-9pm daily. Nov-Easter 10am-5pm Mon-Thur, Sun; 10am-9pm Fri, Sat. Meeting point Mary King's Close, via Warriston Close, Old Town. Admission £13.25. Map p284 H7.

The remains of a street pre-dating the City Chambers, which were built on top, have been turned into an historical attraction, overseen by a costumed guide. It's a fascinating look at life in the Old Town centuries ago. Not suitable for under-5s.

Rebus Tours

553 7473, www.rebustours. com. Tours noon Sat. Meeting point Royal Oak, Infirmary Street, Old Town. Tickets £10. No credit cards. Map p285 J8.

Ian Rankin's Edinburgh is uncovered on a walking tour run by knowledgeable enthusiast Colin Brown. Booking is required (call, or see website); payment is on the day and in cash only.

Diary

*What to do when it's
not August.*

The concentration of cultural events that takes place in Edinburgh during August captures the headlines (and gets its own chapter in this guide; *see pp32-37*), but plenty of other festivals brighten up the rest of the year. The revival of Beltane has turned into a late-night spectacular with an audience of thousands; summer sees Scotland's best breweries take over the Corn Exchange for a few days for the Scottish Real Ale Festival; Doors Open Day is a fascinating glimpse behind the scenes of otherwise inaccessible buildings in the autumn; and winter brings Edinburgh's Christmas and Edinburgh's Hogmanay. All are worth taking into account when planning a trip. Dates can vary to some extent each year; always check the respective website before making plans.

Beltane Fire Festival.

Spring

TradFest Edinburgh
Various venues (556 9579, www.tracscotland.org).
Date late Apr/early May.
A showcase for Celtic arts and traditions, updated for the 21st century, featuring dance, music, song, storytelling and more. Launched in 2013, it features more than 80 shows and events in 12 days spread across a couple of dozen Edinburgh venues.

Edinburgh International Harp Festival
Merchiston Castle School, Colinton Road, South Edinburgh (554 0212, www.harpfestival.co.uk).
Date mid Apr.
Six days of classes, concerts and workshops with the harp – or clarsach – taking centre stage.

★ Edinburgh International Science Festival
0844 557 2686, www.sciencefestival.co.uk.
Date Apr.
This hugely enjoyable festival held over 19 days gives an accessible slant on difficult subjects without dumbing down. Running for more than 25 years, it's the UK's largest science jamboree, attracting tens of thousands of people to numerous talks, events and workshops. The outdoor photography exhibition in St Andrew Square is usually excellent.

Beltane Fire Festival
Calton Hill (www.beltanefiresociety.wordpress.com).
Date 30 Apr.
An ancient tradition marking the transition from winter to spring, Beltane was revived in 1988 and has grown into quite a drama: body-paint, costume, dancing, drumming, fire and ritual, late at night on Calton Hill. If the weather holds, it can attract more people than the average Hibernian FC home game. Tickets cost £7.50 in advance, £10 after 26 Apr, £10.50 on the night, and are available from the Hub (*see p58*), Ripping Records (*see p77*) and Tickets Scotland (127 Rose Street, New Town, 220 3234, www.tickets-scotland.com).

Summer

★ Royal Highland Show
Royal Highland Centre, Ingliston, West Edinburgh (335 6200, www.royalhighlandshow.org.uk). **Date** mid June.
The charm of the farm for four days in June (*see p148* **A Festival on the Hoof**).

Scottish Real Ale Festival
Edinburgh Corn Exchange, 10 New Market Road, West Edinburgh (www.scottishbeerfestival.org.uk).
Date June or July.
Beer enthusiasts take over the Corn Exchange for four days in high summer, getting to grips with cask

Clockwise: **Edinburgh International Science Festival** (see p29); **Edinburgh's Christmas**; **TradFest Edinburgh** (see p29).

ales from dozens of Scottish breweries based as far afield as Orkney and the Borders. For the latest on timings and prices, check the website.

Autumn

Doors Open Day

557 8686, www.cockburnassociation.org.uk.
Date last weekend Sept.
Each year, heritage body the Cockburn Association collaborates with other organisations and individuals to allow public access to buildings that few people usually get to see – everything from private homes to lighthouses. All for free.

Edinburgh Independent Radical Book Fair

Out of the Blue Drill Hall, 36 Dalmeny Street, Leith (662 9112, www.word-power.co.uk).
Date late Oct/early Nov.
Five days of readings and discussions, focusing on independent and radical publishing houses. For more information, contact Word Power (*see p139*).

Samhuinn (Hallowe'en)

High Street, Old Town (www.beltanefiresociety. wordpress.com). **Date** 31 Oct.
Samhuinn marks the end of summer, six months after Beltane (*see p29*), so the Beltane Fire Society takes to the streets once again, parading from the High Street, down Cockburn Street to the Mound. En route, the interior shadow stalking the Green Man of summer manifests as the Horned God of Winter, all overseen by the Divine Hag. With its costume and fire, set against the impeccable Old Town backdrop at night, it's a seductive spectacle.

Bonfire Night

Date 5 Nov.
Guy Fawkes was a Yorkshire-born terrorist, part of a Catholic group who were opposed to the Protestant British monarchy; they tried to blow up Parliament

while James I & VI was inside. The plot was foiled on 5 Nov 1605; ever since, bonfires have been lit on that date in commemoration. The big municipal fireworks display is traditionally held at Meadowbank Stadium (661 5351, www.edinburgh leisure.co.uk); it focuses on pyrotechnics, not Jacobean politics. Private bonfire parties take place all across the city, mostly for the benefit of children.

St Andrew's Day

Date 30 Nov.
St Andrew's Day (Andrew is the the patron saint of Scotland) is nowhere near as debauched for the Scots as St Patrick's Day is for the Irish, but it's still a decent reason to raise a glass of whisky. In 2007, the Scottish Parliament designated it as a holiday of sorts and efforts are now being made to make the day a little more festive than it's been in the past.

Winter

Edinburgh's Christmas

www.edinburghschristmas.com. **Date** late Nov-early Jan.
What started out as a few casual events has grown into a large and popular festival, largely based in St

PUBLIC HOLIDAYS

New Year's Day
1 Jan

2 January Holiday
2 Jan

Good Friday
varies with Easter

Easter Monday
varies with Easter

Spring Public Holiday
3rd Mon in Apr

May Day Holiday
1st Mon in May

Victoria Day
Mon before 24 May

Autumn Public Holiday
3rd Mon in Sept

Christmas Day
25 Dec

Boxing Day
26 Dec

Andrew Square and East Princes Street Gardens. There are fairground rides, food stalls, a Christmas market, a big wheel, an ice rink and, in recent years, a quite terrifying Star Flyer fairground ride. In 2014/15, the event claimed an audience of over 3.6 million with more than half a million tickets sold for attractions and shows in its six-week run. Yes, Christmas is now officially that long these days.

★ Edinburgh's Hogmanay

www.edinburghshogmanay.org. **Date** 30 Dec-1 Jan.
Edinburgh's modern Hogmanay celebration is a three-day festival featuring a torchlight procession, bands, fireworks and even, for the hardy, a New Year's Day charity swim in the Forth. The main event is the big street party on the evening of 31 December, when the city centre is cordoned off for safety reasons and a maximum of 100,000 people admitted. Some locals love it; others view it as an open-air binge-drinking session for young singles and couples. The crowning gig is in Princes Street Gardens as the old year turns into the new – although this is weather-dependent; winter storms have seen it cancelled. Headline acts have included Primal Scream, the Pet Shop Boys and Lily Allen. Admission to the city centre and to the main gig require tickets; for details and prices, see website.

Burns Night

www.visitscotland.com. **Date** 25 Jan.
In his short life, Robert Burns (1759-96) built a reputation as a fine poet, but it was only after his death that he came to be regarded as Scotland's national bard. Burns Suppers – dinners in his memory – have been run for more than 200 years by Burns societies, churches, masonic lodges and other community groups. Held around the anniversary of his birth (25 January, Burns Night), they follow a traditional format, including recitation of the poem 'Address to a Haggis', and a meal including haggis, 'chappit neeps' (mashed swede), mashed potatoes and whisky. There's no single, set-piece Edinburgh celebration, but watch out for themed evenings in hotels and restaurants, charity ceilidhs and other such events.

IT'S COMPLICATED...
Edinburgh public holidays explained.

In Scotland, both **1** and **2 Jan** are public holidays, so normal working life in the New Year doesn't start until 3 Jan. In years when 1 or 2 Jan (or both) fall on a weekend, normal life doesn't get going again until later.

The two **Easter** days vary with Easter, of course, never a fixed date (because its timing is linked to the first full moon after 21 March). The **Spring** and **Autumn Public Holiday** is happily rational, while **May Day** is the first Mon of the month in order to pre-empt the question, 'What if 1 May falls on a weekend?' **Victoria Day** is not a universal holiday, but is taken in a number of Scottish towns and cities to celebrate the birthday of the long-departed Queen Victoria (24 May 1819), fixed to the Mon before that date.

Finally, if **Christmas Day** or **Boxing Day** (or both) fall on a weekend, then 27 or 28 Dec (or both) are taken as holidays instead.

Festival Edinburgh

A city transformed.

TEXT: KEITH DAVIDSON

Above: **Edinburgh Festival Fringe**. Below: **Edinburgh International Festival**.

For 11 months of the year Edinburgh is a typical northern European city of around half a million people, content with the capital status and critical mass that brings great galleries, museums, restaurants, gigs, clubs and more. But come August, it's all change. The population doubles, the atmosphere becomes almost continental, and the stone façades burst into colour. It's the largest arts celebration in the world, drawing performers from Abuja to Zagreb and all points in between.

Not everyone is enamoured of this dramatic shift in the city's character. Many locals jump ship and rent their properties to visitors through agencies or Airbnb, while others stay and grumble. However, they're in the minority; for most, the array of August festivals comprises the highlight of Edinburgh's cultural calendar. Unique is an overused word, but it absolutely applies here. August in Edinburgh is like nowhere else in the world.

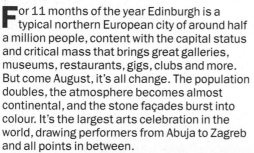

A FESTIVALS PRIMER

The thousands of performances and events that take place in Edinburgh in August are not part of a single cultural festival. The primacy of the **Edinburgh International Festival** (EIF) has meant that August's events are often referred to as, simply, 'the festival'. However, there are a number of administratively separate jamborees going on at the same time. This chapter attempts to make sense of them all – even those that don't happen in August.

It all began in 1947 with the launch of the aforementioned Edinburgh International Festival, established to provide Europe with the best in performing arts in the aftermath of World War II. Thanks to the programme, and the charms of the city, organisers achieved their ambitions many times over.

The **Edinburgh Festival Fringe** (or simply the Fringe) began in modest circumstances in 1947, when eight companies not invited to the

EIF decided to hold their own unofficial shows. Today, it dwarfs its grander rival: it routinely stages in excess of 3,000 shows across 300 or so venues during its 25-day run. Anyone can appear, provided they can pay for a listing in the programme (approximately £100-£400 depending on the number of performances), hire a venue and find accommodation. This attracts an array of aspiring talent, crowd-pleasers, amateurs and wily old pros. Comedy and theatre are the mainstays, but dance, music and, to a lesser degree, visual art all feature. Venues vary from professional multi-theatre operations to people's living rooms.

The **Edinburgh International Film Festival** (EIFF) also arrived in 1947, after the Edinburgh Film Guild were piqued that the EIF had omitted film from its remit. Held in June now, it's an important date in the industry calendar, popular with film fans who get to see a selection of flicks before their official UK release dates.

The **Royal Edinburgh Military Tattoo** made its entrance in 1950, a parade of music, dance and athleticism with added showbiz elements, on the Castle Esplanade.

The **Edinburgh International Television Festival**, founded in 1976, runs for three days in late August, but is more of an industry bun fight than public entertainment. Those who might attend out of some vague interest in the subject are deterred by its high cost. The **Edinburgh Jazz & Blues Festival**, which joined the fray in 1979, is very much aimed at the public, but is held in July these days to give it room to breathe.

The **Edinburgh International Book Festival** was launched in 1983, first as a biennial affair

but annual since 1997. It's now the largest and most successful event of its type in the world. The **Edinburgh Mela**, a two-day outdoor shindig towards the end of August, celebrates Asian culture with performances, food stalls and more. It was launched in 1995.

Finally, in 2004, another couple of festivals threw their hats into the ring. The **Edinburgh Art Festival** brings together venues big and small to exhibit an enormous range of artworks. The **Free Festival** stages comedy, cabaret, theatre, music and children's entertainments over 24 days. All shows have free admission; audience members can make a voluntary donation on the way out.

For other festivals and events during the Edinburgh year, see **Diary** (*pp28-31*) or check this book's complementary website (www.timeout.com/edinburgh).

CULTURE DURING THE FESTIVALS

Art

Edinburgh Art Festival

The **Edinburgh Art Festival** has grown in profile, with more than 40 exhibitions in galleries, museums and other spaces each August, as well as special events and other projects. It's city-wide and runs for one calendar month.

Comedy

Edinburgh Festival Fringe, Free Festival

The **Fringe** is the largest comedy gathering of its kind in the world, which is great for audiences, but gruelling for performers; ambitious comedians commonly take to the stage six days a week for a three-week stretch, trying to make a name for themselves. Aided by their existing reputations, the lucky land peak evening slots at prestigious venues. Others resign themselves to the fact that, if the sun comes out, their daily 3pm pub basement gig may not sell out. Or at all.

Although the **Free Festival** hardly offers an antidote to this joyless career grind for stand-ups, it has been a refreshing change for audiences. Instead of paying a tenner to see some unknown, you simply go to a venue, buy a beer and see what happens. If it's funny, you pay what's fair.

Film & television

Edinburgh International Film Festival, Edinburgh International Television Festival

The **Edinburgh International Film Festival** presents a mix of new features and shorts, animated flicks and factual programming, retrospectives and revivals. Film director John Huston described it in 1972 as 'the only film festival worth a damn'. Since moving from August to June in 2008, the event has found it easier not only to distinguish itself from the city's other big festivals, but also to strike a clearer identity on the international circuit. Its audience-friendliness stands in contrast to the short **Edinburgh International Television Festival**, aimed almost exclusively at industry insiders.

Literature

Edinburgh International Book Festival

The **Edinburgh International Book Festival** is a cultured haven that carefully sets itself apart from the chaos in the Old Town, pitching its marquees in Charlotte Square Gardens. The programme comprises a range of talks, readings and discussions; organisers are keen to encourage participation. It does attract the big international names, though: George RR Martin, Haruki Murakami and Salman Rushdie have appeared in recent years. Many of its events and readings are notably child-friendly.

> ### IN THE KNOW
> ### WHAT HAPPENED TO THE
> ### PERRIER?
>
> The Perrier Awards for comedy and cabaret at the Fringe were launched in 1981 and subsequently gave a boost to the careers of Frank Skinner, Steve Coogan, Dylan Moran, Al Murray and many others. A best newcomer prize was added in 1992, and and a judging panel award in 2006. The sponsor has also changed over the years; since 2010, it's been Foster's. The awards are now generally known as the Eddies (Edinburgh Comedy Awards).

Edinburgh Jazz & Blues Festival.

Music

Edinburgh International Festival, Edinburgh Festival Fringe, Free Festival, Edinburgh Jazz & Blues Festival

Classical music is at the heart of the **Edinburgh International Festival**. The Usher Hall and Queen's Hall stage concerts almost daily, and various venues have opera throughout August. The enormously popular Fireworks Concert in Princes Street Gardens marks the official end of the EIF each year. There's also a surprising amount of classical music on the **Fringe**. Young musicians dominate, with one-off concerts or short-run recital series, but there's a plethora of other offerings, from Scottish folk to bump'n'grind burlesque. The **Free Festival**, although comedy-oriented, also has some music.

The **Edinburgh Jazz & Blues Festival** takes place over ten days in July, with well-respected names from the UK circuit, local musicians and international headliners.

Theatre & dance

Edinburgh International Festival, Edinburgh Festival Fringe

Major international theatre and world-class dance are part of the reason the **EIF** was created in the first place – with shows that are often grand and expensive to stage. On the **Fringe**, meanwhile, you can find everything from cutting-edge theatre and amateur productions to *Glee*-like song and dance material by youth theatre groups.

VISITING THE FESTIVAL

The spontaneity of Edinburgh in August is part of the experience, but some forward planning is crucial. (Particularly when it comes to accommodation; *see pp244-262*.) Look at the programmes before you reach the city, and keep your ears open. Arriving without any tickets, and just seeing what happens, won't get you into the hottest shows.

As well as a printed programme, each festival has its own website; the Fringe and the Film Festival have apps too. The **Edinburgh Festivals** website (www. edinburghfestivalcity.com) has sections on the city, trip planning and useful links.

Press coverage really counts in August, so read the reviews, especially from the Edinburgh-based press (*Scotsman*, *Evening News*), other Scottish titles (*Herald*) and the UK-national *Guardian*. Edinburgh magazine the *List* usually offers comprehensive coverage: it's published every two months, but makes an exception for August when it appears weekly. For more on media, *see p267*.

Booking tickets

Buying tickets is straightforward, whether online, by phone or in person. The easiest way is obviously online. If you like the personal touch, however, box offices are everywhere.

Advance booking will be necessary for the biggest shows. Many events staged as part of the EIF, the International Book Festival and the International Film Festival are one-offs and sell out well ahead of time.

DIRECTORY

Although information below is correct at time of going to press, always check the relevant websites for the latest details. Venues are various unless specified.

Edinburgh Art Festival

City Art Centre, 2 Market Street, Old Town, EH1 1DE (226 6558, www.edinburghartfestival. com). **Tickets** usually follow gallery and museum admission fees, but most exhibitions are free. **Date** one calendar month, end July-end Aug.

Edinburgh Festival Fringe

Edinburgh Festival Fringe Society, 180 High Street, Old Town, EH1 1QS (information 226 0026, box office 226 0000, www.edfringe. com). **Tickets** typically £10-£20; some are more expensive, others are free. Advance booking online for some shows starts in late Jan; full programme available for booking online, by phone and at box office from mid June. **Date** 25 days, Aug.

Edinburgh International Book Festival

5 Charlotte Square, New Town, EH2 4DR (administration 718 5666, box office 0845 373 5888, www.edbookfest.co.uk). **Venue** Charlotte Square Gardens, New Town. **Tickets** mostly £8-£15 (£5 for children's events), plus some free events. Box office sales online and by phone mid June-late Aug. Walk-up sales from the Hub (Castlehill, Old Town) mid June-start of festival. Walk-up sales during festival from Charlotte Square site. **Date** 17 days, ending last Mon in Aug.

Edinburgh International Festival

The Hub, Castlehill, Old Town, EH1 2NE (administration 473 2099, box office 473 2000, www.eif.co.uk). **Tickets** £8-£76; some free events. Advance booking online, by phone and in person from late Apr. **Date** 25 days, Aug.

Edinburgh International Film Festival

Filmhouse, 88 Lothian Road, Old Town, EH3 9BZ (administration 228 4051, box office 623 8030, www.edfilmfest.org.uk). **Tickets** mostly £5-£10; opening and closing galas £10-£15.

Advance booking online, by phone and in person from early June. **Date** 12 days, mid-late June.

Edinburgh International Television Festival

Guardian Edinburgh International Television Festival, 117 Farringdon Road, London, EC1R 3BX (registration enquiries 020 7843 0146, www.geitf.co.uk). **Venue** Edinburgh International Conference Centre. **Tickets** £269-£825. Advance booking online from early in the year. **Date** 3 days (Wed-Fri), late Aug.

Edinburgh Jazz & Blues Festival

The Hub, Castlehill, Old Town, EH1 2NE (information 467 5200, box office 473 2000, www.edinburghjazzfestival.co.uk). **Tickets** mostly £10-£20. Advance booking online and by phone from May. Individual venues also sell tickets. **Date** ten days, mid-late July.

Edinburgh Mela

Unit 14, Abbeymount Techbase, 2 Easter Road, Calton Hill, EH7 5AN (661 7100, www. edinburgh-mela.co.uk). **Venue** Leith Links, Leith. **Tickets** £4 day pass covers most events; headline shows £12-£15. Advance booking online. **Date** two days (Sat, Sun), late Aug.

Free Festival

020 3287 5533, 07768 584 881, www.free festival.co.uk. Information, Aug: c/o Espionage, India Buildings, Victoria Street, Old Town. **Tickets** free, donations welcome. **Date** 25 days, Aug.

Leith Festival

17 Academy Street, Leith, EH6 7EE (629 1214, www.leithfestival.com). **Tickets** free £10. **Date** nine days, mid June. Comedy, dance, film, music, theatre and more. It starts with a gala day and ends with the Leith Tattoo.

Royal Edinburgh Military Tattoo

32-34 Market Street, Old Town, EH1 1QB (225 1188, www.edintattoo.co.uk). **Venue** Castle Esplanade, Old Town. **Tickets** £27-£65; royal gallery seats £300. Advance booking online, by phone, by post and in person from Apr. **Date** 23 days, Aug.

Edinburgh's Best

A selection of the city's highlights.

Sightseeing

VIEWS
Arthur's Seat p128
The highest point in the city centre at 823 feet.
Blackford Hill p140
Where to go if you want to look at Arthur's Seat.
Calton Hill p119
Access steps three minutes' walk from Princes Street.
Scott Monument p89
A soaring Gothic spaceship of a memorial on Princes Street.

BY THE COAST
Cramond p150
Beach, beautiful foreshore, tidal island and bonus riverside walk.
Forth Bridge p205
Victorian engineering wonder, born 1890.

Portobello Beach p164
A proper beach with bars, cafés and promenade.
Royal Yacht Britannia p158
The monarch's former floating holiday runabout, 1954-97.

CHILDREN
Camera Obscura & The World of Illusion p55
A child-friendly hoot.
Edinburgh Zoo p170
Penguins of Corstorphine. We say, 'Better than *Penguins of Madagascar*.'
Gorgie City Farm p170
Farm animals and a pet lodge.
Our Dynamic Earth p66
Evolution, geology, space, volcanoes and other woo stuff.
Scottish Storytelling Centre p67
Once upon a time, an expert got the kids to listen quietly…

From top: **Cramond**; **High Kirk of St Giles**; **Paul Kitching 21212**.

CHURCHES & GRAVEYARDS

Greyfriars Kirkyard p73
Atmospheric, haunted, once an open-air prison.
High Kirk of St Giles p57
A medieval survivor with a key role in Scottish history.
Old Calton Burial Ground p116
Philosopher David Hume is here, and many others.
St John's Episcopal Church p87
A fine interior, shops in the basement.
St Mary's Episcopal Cathedral p100
Triple-spired landmark with a stellar musical reputation.

DAYS OUT

East Lothian beaches p202
Bonny sands au bord de la Forth.
Glasgow p194
It's bigger, but is it better? Go and decide for yourself.
Linlithgow p206
Canal, eateries, loch, pubs, ruined palace.
North Berwick p203
East Lothian seaside resort par excellence; attractions, beach and harbour.
Pentland Hills p204
Rising to 1,900 feet – proper topography.

PARKS & WALKS

Holyrood Park p126
The big green one, with lochs and three dimensions.
The Meadows p140
The flat one, with cherry blossom avenues.
Royal Botanic Garden Edinburgh p105
The world-class scientific and horticultural one; wander and wonder.
Union Canal p146
Edinburgh backstreets and bucolic bursts of loveliness.

UNIQUELY SCOTTISH

Edinburgh Castle p55
Nowhere else has got one like this. Except Stirling.
National Museum of Scotland p74
Everything you ever wanted to know about Scotland.
Palace of Holyroodhouse p66
A site with a royal pedigree dating from the 13th century.
Scotch Whisky Experience p58
Does exactly what it says on the label.
Scottish Parliament p67
A functioning alternative to more ossified legislatures.
Writers' Museum p59
The holy triumvirate of ScotLit: Burns, Scott, Stevenson.

Eating & drinking

CLASSIC FINE DINING

Castle Terrace p132
Kitchin p160
Number One p93
Plumed Horse p160
Restaurant Martin Wishart p161

COCKTAILS

Bond No.9 p162
Bramble p95
Dragonfly p136
Last Word Saloon p110
Lucky Liquor Co p95

CREATIVE CUISINE

Aizle p138
Neo-bistro with no menu, just ingredients.
Gardener's Cottage p121
Local, seasonal, actually in a gardener's cottage.
Paul Kitching 21212 p121
Imagination, execution and flavour.

Restaurant Mark Greenaway p93
Some dishes smoke here, others have lozenges. You have fun.
Timberyard p133
Food of the future.

FROM THE FAR EAST
Chop Chop p148
Regional Chinese dumplings.
Dusit p92
Thai of distinction, since 2002.
Passorn p133
Angelic Thai cooking.
Sushiya p150
Beautifully formed sushi.
Ting Thai Caravan p75
Thai familiars, street food style but indoors.

SCOTTISH
Dubh Prais p68
Scran & Scallie p110
Stac Polly p00
Wildest Drams p61

SEAFOOD
C-Shack p164
Fishers in Leith p159
Ondine p61
Ship on the Shore p161

WHISKY IN THE BAR
Bennet's p136
Bow Bar p75
Canny Man's p142
Scotch Malt Whisky Society p163
Whiski p69

Shopping

BOOKS & COMICS
Armchair Books p137
Deadhead Comics p76
Old Town Bookshop p77
Transreal Fiction p77
Word Power p139

FOOD & DRINK
Archipelago Bakery p97
Superior bread and snacks.

Cadenhead's p70
Where to buy rare whisky.
Coco p143
Designer chocolate.
IJ Mellis p76
The epicentre of cheese.
Royal Mile Whiskies p63
The other place to buy rare whisky.
Valvona & Crolla p122
Italian grotto of yum, since 1934.

MADE IN SCOTLAND
21st Century Kilts p96
Contemporary kilts for today's city gent.
Anta p96
Sort of like Habitat but better.
Arran Aromatics p97
Bath and body smellies for gents and ladies.
Geoffrey the Tailor p70
Genuine kilted tailoring in Touristville.
Hawico p76
Cashmere classics.
Ness p63
Quietly elegant womenswear and accessories with a Scottish aesthetic.

Nightlife

BARS WITH MUSIC
Bannermans p180
Banshee Labyrinth p180
Henry's Cellar Bar p180
Jam House p180
Whistlebinkies p69

CLUBS & MUSIC VENUES
La Belle Angèle p182
Bongo Club p182
Cabaret Voltaire p182
Mash House p183
Sneaky Pete's p183
Studio 24 p183

FOLK, ROOTS & JAZZ
Edinburgh Folk Club p184
Weekly, Wednesdays, in a Fringe-style cabaret bar.

Jazz Bar p185
Daily jazz in a basement; opens late.
Leith Folk Club p184
Weekly, Tuesdays, in a Leith hotel.
Royal Oak p75
Traditional pub with regular folk sessions.
Sandy Bell's p76
Another traditional pub with regular folk sessions.

Arts & culture

FESTIVALS
Edinburgh Festival Fringe p37
Edinburgh International Book Festival p37
Edinburgh International Festival p37
Royal Edinburgh Military Tattoo p37

FILM
Cameo Picturehouse p174
Dominion p174
Filmhouse p174

THEATRE
Bedlam Theatre p189
Home of the University of Edinburgh Theatre Company.
Brunton p190
Art and populism in Musselburgh.
King's Theatre p190
Comedy, drama, music and panto in a handsome venue, from 1906.
Royal Lyceum p190
Victorian venue that's home to a significant producing company.
Scottish Storytelling Centre p191
Arts space with family-friendly theatre shows.
Traverse Theatre p191
New writing and important productions for more than 50 years.

Explore

The Old Town

Historians sometimes talk of the 'herringbone' pattern that defines the structure of old Scottish towns and villages, and Edinburgh's Old Town is essentially no different. With the castle up on its rock and Holyrood Abbey located a mile to the east, a medieval backbone was created that endures today: the Royal Mile. The closes, streets and wynds falling away north and south complete a picture that now takes in more historic attractions and key sites than anywhere else in Scotland. Here, you find tales of politics, religion and death made manifest, ranging across nearly 1,000 years of history. The actors have moved on, but the stage endures; at locations such as Edinburgh Castle, Greyfriars Kirkyard or the Palace of Holyroodhouse, visitors can follow in the actual footsteps of St Margaret, Mary Queen of Scots, Bonnie Prince Charlie and other significant figures from the past.

EXPLORE

Scotch Whisky Experience.

Don't Miss

1 National Museum of Scotland Setting the scene for your visit (p74).

2 Edinburgh Castle Built of authenticity (p55).

3 High Kirk of St Giles Scottish history incarnate (p57).

4 Scottish Parliament A 21st-century democracy at work (p67).

5 Scotch Whisky Experience Educational, but with eats, shopping and *uisge beatha* (p58).

INTRODUCING THE OLD TOWN

Edinburgh loosened her stays and expanded in the late 18th century, when the New Town was built. Since then, further urban development has seen it transformed into a sprawling city of nearly 500,000 people. As a result of this expansion, the boundaries of the Old Town can seem compromised or mysterious, but the area still retains its own identity beneath these hidden borders.

In 1513, the Scots suffered a catastrophic defeat to an English army at Flodden in Northumberland. Edinburgh's citizens feared a subsequent invasion, and threw up the Flodden Wall to protect themselves. The course of the wall, or what remains of it, gives a good idea of the city boundaries at the time. More than 500 years later, the most apparent parts of the wall still standing are beside George Heriot's School and at the east end of Cowgate (*see p70*).

To the north, where Waverley Station and Princes Street Gardens now sit, the town was traditionally protected by the Nor' Loch. However, the wall provided extra insurance, running from the eastern bounds of the loch, up and over the Royal Mile, south up the Pleasance and west along what's now Drummond Street, then on to Teviot Place and north past George Heriot's and the West Port to the Castle Rock. The enclosed area, plus the environs of Holyrood Abbey, essentially was Edinburgh from the medieval period onwards.

When finding your way around the streets and narrow thoroughfares of the Old Town, bear in mind that a 'close' is a narrow alleyway that usually opens up into a courtyard of some sort (check out Trunk's Close or Lady Stair's Close, and read the Scottish poetry extracts carved into the flagstones in the latter); a 'wynd' is a narrow winding lane leading off the main thoroughfare; a 'pend' is a narrow, covered entryway to the backcourt of a block of houses; and a 'vennel' is simply a narrow alley. Indeed, narrow is the key word in the Old Town. The buildings were

traditionally crammed together, and the streets off the Royal Mile are outlandishly skinny.

In the 15th and 16th centuries, the Old Town was home to two main types of resident: those with money and those without. For residents in the former category, life could be comfortable, if not ostentatious. Even the better houses were sparsely furnished, but at least they were prettified by painted rafters and wall tapestries. The poor, though, suffered, especially as the population grew and sanitation worsened. The Nor' Loch and the Burgh Loch (now the Meadows; *see p140*) were dumping grounds for all kinds of waste; householders got rid of slops by heaving them out of the window on to the street below with a warning cry, 'Gardy loo!'

Typhus was rife in the Old Town and, inevitably, plague struck. By the end of the 16th century, it had wiped out a third of the population. One visitor likened Edinburgh to a comb: filthy at the teeth, but with some clean parts in between. Planning for the expansion of the city began in 1752, and construction of the New Town started in the following decade, but it wasn't until well into the 19th century that the cramped environs of the Old Town really became too much for its population, and 'the Great Flitting' north to the New Town began.

Two centuries later, the residential population of the Old Town is now comparatively small, but tourists make up the numbers; parts of the area resemble a huge open-air museum and UNESCO declared the entire area a World Heritage site in 1995 (*see p47* **UNESCO-Approved**). The **Royal Mile**, the central thoroughfare, runs between the Castle Esplanade in the west and the Palace of Holyroodhouse in the east. It's not a single street, however, but a catch-all term for a route that comprises four distinct streets: **Castlehill**, the **Lawnmarket**, the **High Street** (divided by the Bridges) and the **Canongate**. Key events in Scotland's history have been played out here: kings and queens have sallied from the castle to the palace and back, men have been hanged for criminality and treason, women have been burned for witchcraft. It's now more sedate, at least for 11 months of the year.

The Royal Mile is busy with tourist foot-traffic most of the time; unsurprisingly so, since many of Edinburgh's main attractions sit on or near it. It's flanked on both sides by around 60 closes and wynds, which are home to tightly packed clusters of tenements, pubs, restaurants and cafés. These atmospheric streets also provide an eerie setting for many of Edinburgh's popular ghost tours (*see p26*).

Between George IV Bridge and the Tron, the Royal Mile becomes the focal point of August's **Edinburgh Festival Fringe** (*see pp32-37*), as hundreds of street performers and souvenir-hawkers descend upon it to advertise their shows

UNESCO-APPROVED

But what does it mean to be a World Heritage Site?

In 1995, the United Nations Educational, Scientific and Cultural Organisation (UNESCO) awarded World Heritage status to both Edinburgh's medieval Old Town and the Georgian New Town. It wasn't awarded to a special street, or a specific site. UNESCO just looked at the entirety of the city centre and decided that all of it was worthy of the global accolade.

Since then, travel guides, the city authorities, Scotland's official tourist agency and no end of hotels and attractions have made great play of Edinburgh's World Heritage status – you have to admit, it sounds pretty darn important. What generally fails to happen, however, is the follow-up. What exactly is World Heritage status?

In the aftermath of the Great War, and its toll of catastrophic death and destruction, there were discussions about creating some sort of international organisation to protect the world's heritage. Given the circumstances of the century – another world war, a cold war – it took until the late 1950s for UNESCO to launch the kind of international action in the manner envisaged: moving ancient temples in Egypt, threatened by flooding because of the construction of the Aswan High Dam. Some 50 countries contributed half of the $80 million cost of the project and the practical protection of the world's cultural and natural heritage took a major step forward.

Both the United States of America and the International Union for Conservation of Nature (IUCN) put forward practical proposals of their own about world heritage, but it wasn't until 1972, and a UNESCO general conference, that a single text was agreed upon. Since then, virtually every United Nations member state, with a few minor and eccentric exceptions, has ratified this 1972 convention. These days, UNESCO runs a World Heritage Committee, which administers the programme, identifying sites of special cultural or natural significance to the common heritage of humanity as a whole.

Despite the bureaucracy, this is stirring stuff. UNESCO is basically saying, 'It's in everyone's interest that we look after these places properly. In a very important sense, they belong to all of us, wherever we come from.' There are just over 1,000 sites on the World Heritage list at the moment, from archaeological remains in Afghanistan to

the Victoria Falls in Zimbabwe. The UK has 28 sites, spread across England, Northern Ireland, Scotland and Wales. In Scotland, the Neolithic remains on Orkney, the volcanic archipelago of St Kilda out in the North Atlantic, the 18th-century planned industrial village of New Lanark and Edinburgh's Old and New Towns have all been deemed internationally significant.

World Heritage status is therefore a great deal more than a marketing tool. Given the size of this planet, the number of people on it and the depth of its history, World Heritage status is a rare and privileged thing. It's also a salutary reminder to Edinburgh citizens, who see the Royal Mile and elegant Georgian terraces on a daily basis, that not everyone has a castle in the city centre, a medieval cathedral that you can walk into for free, or a half-millennium-old royal palace sitting around the neighbourhood. In cultural terms, it puts Edinburgh on a par with Vatican City, the Acropolis in Athens, Chartres Cathedral and the historic centre of St Petersburg.

As if anyone actually needed to be told, UNESCO provides an external benchmark that confirms the obvious: Edinburgh is a very special place indeed. Be nice to it while you're here.

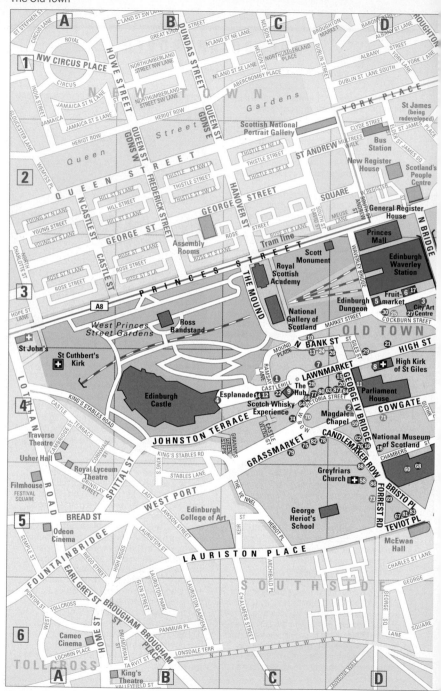

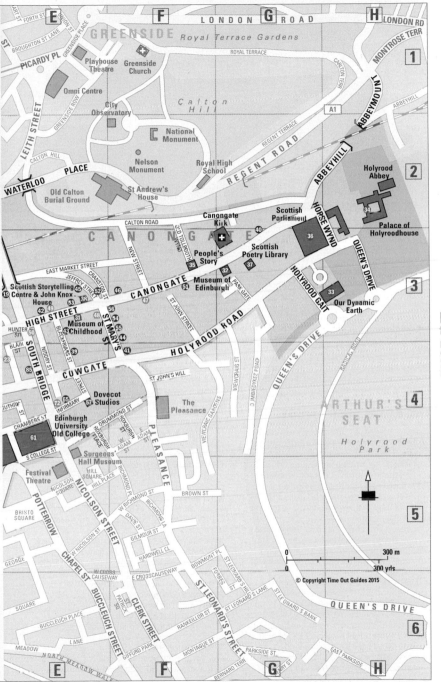

EXPLORE

Edinburgh Castle Esplanade.

and sell their wares. Members of the public not paying total attention may find themselves enlisted as the unwitting star or victim of an amateur juggling act, or the reluctant paperweight for thousands of fliers. The street is also popular year-round with buskers, especially pipers playing reels, airs and even cover versions of pop songs.

In the pages that follow are details of the Old Town's sights and museums, restaurants, pubs and bars, and shops and services. These have been divided into three main areas: the Royal Mile west of the junction with North Bridge and South Bridge; the Royal Mile east of that junction; and everything around Cowgate and the southern stretch of the Old Town.

THE ROYAL MILE – WEST

The western end of the Royal Mile begins at Castlehill, home – naturally – of **Edinburgh Castle**. The fortress is perched on the towering Castle Rock; a surviving basalt plug from a volcano that would have been active around 350 million years ago. It stands 427 feet above sea level, and has long provided Edinburghers with a vantage point over the city. The rock may have been inhabited since the late Bronze Age (evidence is not conclusive); over the centuries, a succession of wooden and stone fortifications were built on the site, although most of the buildings visible today date from the 18th and 19th centuries.

The castle was a royal residence until the Lang Siege of 1571-73, when supporters of Mary, Queen of Scots were bombarded by troops led by James Douglas. At the time, Douglas was the Regent Morton and governed on behalf of Mary's son, the infant King James VI of Scotland. (He later became the King of England too; as England had never before had a king named James, he was styled James I & VI.) The royal

home was then moved to the Palace of Holyroodhouse at the other end of the Mile, and the castle was refurbished and heavily fortified. It last heard shots fired in anger in 1745, when its cannons were used to repel the invading forces of the Young Pretender, Charles Edward Stuart (Bonnie Prince Charlie, the grandson of deposed British monarch King James II & VII).

The approach to the castle runs through the **Esplanade**, where the almost invariably sold-out **Edinburgh Military Tattoo** (*see p37*) has been held every August since 1950. When the Tattoo's temporary seating isn't in place, the castle's imposing Half Moon Battery dominates the Esplanade. Built after the Lang Siege to defend the eastern side of the stronghold (the other sides are sheer), it provided the basis for the castle's massive artillery strength. Behind it on the left are the palace apartments, now one of the castle's museums. A small gate at the eastern end of the north side of the Esplanade leads down winding paths to **Princes Street Gardens** (*see p81*).

The view from the Esplanade's southern parapet looks out over the suburbs of Edinburgh as far as the Pentland Hills, while the northern aspect leads the eye over the New Town and across the Firth of Forth to Fife. Various military memorials on the Esplanade serve as sombre remembrance of the many Scottish soldiers killed in action overseas. Also here is the tomb of Ensign Charles Ewart, who single-handedly captured the standard of the famous French Invincibles at the Battle of Waterloo. His memory is celebrated in the name of a nearby pub (521-523 Lawnmarket).

Look out too for the **Witches' Memorial**, a lasting epitaph to a dreadful historical trend Edinburgh shares with much of Europe. Although it's often associated with the Middle Ages, witch-burning was actually a popular sport of Renaissance and Enlightenment man, and was a paranoid passion of James VI. A bronze, wall-mounted well marks the spot where more than 300 women were burned as witches between 1479 and 1722. It's little comfort to learn that in Scotland, the victims were usually strangled before the fire was lit.

On the extreme left, as you face away from the castle, **Ramsay Gardens** is an irregular complex of romantic baronial buildings bristling with spiral staircases and overhangs. Constructed around the poet Allan Ramsay's octagonal 'goose-pie' house, the buildings were erected (for the most part) in the late 19th century in a bid to lure the upper classes back to the Old Town. Today, they're all in private hands and are some of the city's most desirable and expensive real estate. The low, flat building adjacent is less attractive, but originally performed an important function: built in the mid 19th century, as the Castlehill cistern, it supplied drinking water to Edinburgh. It now houses an extensive tourist emporium selling all kinds of Scottish-themed souvenirs and tartanalia.

The **Camera Obscura**, the city's oldest official tourist attraction, stands next door in a striking black and white tower. Best visited on clear days, it's akin to an 18th-century CCTV system. Nearby **Cannonball House** (now housing the **Contini Cannonball** catery) is named for the cannonball lodged about halfway up the west gable end wall. It's said to have marked the level to which water piped from Comiston Springs in the Pentland Hills would rise, proving that it could be used to feed the Castlehill cistern. Next door to Cannonball House, in a former school building, is the **Scotch Whisky Experience**. Regular tours offer insight into the history of whisky-making;

there's also a well-stocked shop, restaurant and an educational amusement ride.

A few yards away, where Castlehill meets the top of Johnston Terrace, stands a grand building by James Graham and Augustus Pugin (the latter famed for designing London's Houses of Parliament). Completed in 1844 as the Victoria Hall, it later become the Highland Tolbooth St John's Kirk, complete with a towering 240-foot Gothic spire. The church was occupied by the Edinburgh International Festival from 1999 as its base; it's now known as the **Hub**, and has a serviceable café-restaurant with a popular outdoor terrace.

Stretching between the Hub and George IV Bridge, the Lawnmarket draws its name not from a grassy history, but from the fine linen cloth called 'lawn' that was once sold here. The street is now dotted with pubs and souvenir shops, but isn't without its appeal, especially once you get off the main drag and explore some of the surrounding closes.

The Lawnmarket's most handsome building is **Gladstone's Land**, a plush and well-maintained 17th-century townhouse that's open to the public. There's another fine building from the same era down nearby Lady Stair's Close: Lady Stair's House, where Robert Burns stayed on his first visit to Edinburgh. Appropriately, it now holds the **Writers' Museum**, which celebrates the work of a trio of Scottish writers: Burns, Sir Walter Scott and Robert Louis Stevenson. Indeed, thanks in part to the one-time presence of Burns and James Boswell, who used to live on nearby James Court, this area is now known as Makars' Court from 'makar', the Scots word for poet. Quotations in Scots, Gaelic and English from Scottish writers of note are engraved on paving stones approaching the museum.

Bank Street, which winds around to the left before becoming the Mound, is home to the grand former **Bank of Scotland** head office. Originally completed in 1806 in classical style, and substantially revamped by 1878, it now has a baroque look and a domed roof topped with a golden statue. Bank of Scotland merged with

Witches' Memorial.

EXPLORE

Halifax in 2001 to become HBOS, which was then folded into Lloyds Banking Group after the 2008 banking crisis. Regardless, the building remains a key part of the city skyline while, inside, the **Museum on the Mound** still offers its engaging history of banking. From here, the Mound continues down to **Princes Street** (*see p80*).

Opened in 1834 to complement the parallel South Bridge, George IV Bridge heads south from the Lawnmarket and crosses the Cowgate. The latest addition to the area, at the corner with the Lawnmarket, is **G&V Royal Mile Hotel** (*see p247*). Walking along the bridge, libraries dominate: the **National Library of Scotland** and the **Central Library**. Numerous bars and cafés dot the length of the bridge; continuing south will soon lead you to the famous **Greyfriars Kirk** and **Kirkyard** (*see p73*).

Just below the crossroads between George IV Bridge, the Mound and the Royal Mile, the High Street broadens for a short while, with a sense of ordered elegance brought about by Georgian and Victorian planning. A 1997 statue of Edinburgh-born philosopher and historian David Hume (*see p57* **Statuesque**), a key figure in the Scottish Enlightenment (*see p239-241*), watches over proceedings. Although it's modern, the bronze is fashioned on classical lines, lending its subject a pedagogic gravitas. Three innocuous brass bricks laid into the pavement opposite mark the

CLOSE ENCOUNTERS

Out of sight, up a wynd.

Its one-way streets, labyrinthine topography and devious parking restrictions mean that Edinburgh is best traversed on foot. And nowhere is better suited to pedestrian exploration than the Old Town, thanks to the array of narrow closes, courts, lanes, stairs and wynds that cut across the various levels of the High Street, and elsewhere, and lead into the depths of the city.

These cramped, narrow streets originally slipped between Edinburgh's tenements, accommodating taverns, shops and thousands of the city's residents. Their floors were once dirt tracks, with the emphasis on the word 'dirt': all Edinburghers were legally permitted to throw their daily waste out of the window. The closes heaved with humanity; in 1774, it was recorded that they housed more people per square foot than anywhere else in Europe. Living conditions remained dire right up until the early 20th century, and many of the original closes remain condemned these days. However, plenty remain open, and the ability to negotiate them successfully tends to be what separates the seasoned local from the visitor.

Starting from the top of the Lawnmarket at castle level, the **Castle Wynd Steps** will take you right to the bottom of the Grassmarket, while **James's Court**, **Lady Stair's Close** and **Wardrop's Close** all lead in the opposite direction, between the Lawnmarket and the top of the Mound. These latter three closes are a particularly rewarding find, with the **Writers' Museum** (*see p59*) on one side and the Jolly Judge pub (493 Lawnmarket, 225 2669, www.jollyjudge.co.uk) on the other. Directly opposite Lady Stair's Close

is **Fisher's Close**, which heads down on to Victoria Terrace and then, a few steps later, on to Victoria Street.

On the south side of the Lawnmarket, opposite **Gladstone's Land** (*see p56*), run two other closes with interesting histories. Leading into two courtyards, **Riddle's Close** was where philosopher David Hume wrote his *Political Discourses*, while nearby **Brodie's Close** was home to the rather less savoury Deacon Brodie, a respectable member of Edinburgh society who led a double life as a burglar. He was put to death outside St Giles on gallows he had designed himself, an irony remembered in the plaque on the wall of Deacon Brodie's Tavern (435 Lawnmarket, 225 6531, www.nicholsonspubs.co.uk).

With the exception of a few dead-ends, all the closes along the High Street either feed on to the parallel artery that runs at a lower level – Cowgate and the Grassmarket – or run down towards Princes Street. In the centre of the High Street are **Advocate's**, **Roxburgh's** and **Warriston's Closes**, which bring you to the foot of Cockburn Street. In the 15th century, these tributaries would have taken you down to the Nor' Loch, created as part of the city's defences but eventually a breeding ground for the plague. The water was drained in 1817 to make way for the New Town; the closes are now a handy route to the more pleasant Princes Street Gardens.

Perhaps the most infamous of all the Old Town alleys was lost for two centuries. As part of efforts to gentrify the Old Town in 1753, the grand Royal Exchange (now the City Chambers) was built over part of **Mary King's Close**, the lower floors of which acted as the site of the city's last public hanging: on 21 June 1864, murderer George Bruce was put to death in front of 20,000 people. The public hangman was busy with other duties at the time, so another prisoner obliged.

This section of the Royal Mile, east of George IV Bridge but west of the North and South Bridges, is an impressive part of town. It's dominated by the suitably imposing **High Kirk of St Giles**, where Scottish Reformer John Knox once preached and where, in 1639, stallholder Jenny Geddes is said to have hurled her three-legged stool at the Dean of Edinburgh as he read from the Anglican Book of Common Prayer, viewed by many post-Reformation Scots as

being a deal too close to Catholicism. 'Deil colic the wame o' ye, fause thief,' she cried, adding, 'Daur ye say Mass in my lug?'

In the 15th century, the area outside St Giles was crammed with shops and luckenbooths (lockable stalls). Now **Parliament Square**, it's empty save for the occasional ambling tourist. **Parliament House** runs along the back of the square; its plush Parliament Hall is worth a visit. Home to the Scottish Parliament in the 17th century, it's now used by lawyers discussing cases from the adjoining Supreme Courts of Scotland. It's not unusual to find solicitors marching up and down Parliament Hall, usually in pairs, engaged in earnest discussion about cases on which they're

EXPLORE

new building's foundations. The construction was a conscious effort literally to bury the memories of disease, starvation and desperation that lurked in the close, which was quarantined during the plague that struck in 1645. It's said that the whole street was blocked up and its inhabitants left to die; it's now reputed to be one of the most haunted places in Scotland. Forgotten for years, the underground lane was reopened to the public in 2003 and is now open for tours (see p27). Planned renovations will create a new entrance on the Royal Mile, plus a new café and exhibition area.

If you're in a hurry to catch a train, **Fleshmarket Close** cuts through from the High Street right down to Market Street, where a side entrance to Waverley Station sits at the bottom of the steps. Further down the High Street towards Holyrood, you'll find **Carruber's Close**, once home to the theatre of literary great Allan Ramsay. However, magistrates gave it the final curtain only one year after its opening in 1736. Before this short-lived thespian invasion, the close was a refuge for Jacobites in the late 17th century.

Walk past the Old Tolbooth and turn left after the Canongate Kirk, and you'll find an inconspicuous lane that leads to **Dunbar's Close**, perhaps the most rewarding of all the city's closes. At the end is a tranquil garden, with ornamental flower beds and manicured hedges; it's laid out in a 17th-century style, but was created less than 30 years ago by the Mushroom Trust, a local charity.

Nearby, **Lochend**, **Little Lochend** and **Campbell's Closes** all lead on to Calton Road, while picturesque **White Horse Close**, named in honour of Queen Mary's palfrey horse, formed part of the Royal Mews in the 18th century. Further up, on the other side is **Crichton's Close**; although not as lavish as its regal counterpart, it leads on to Holyrood Road and is home to the wonderful **Scottish Poetry Library** (see p67).

Closes even feature in modern design. A fire in 2002 burned down a block of the Old Town where Cowgate passes under South Bridge. The new development there, dubbed SoCo and finally completed in early 2014, allows pedestrian access from Cowgate via stairs and a courtyard space to Chambers Street (but only from 7am to 11pm daily).

working. The area was once St Giles's churchyard, which explains why John Knox's grave can be found in the middle of a car park.

In front of St Giles, in Parliament Square, is the **Heart of Midlothian,** a heart shape set into the cobblestones of the street that marks the spot where Edinburgh's Tolbooth prison stood (not to be confused with the Canongate Tolbooth, which still survives). Built as a town hall, the Tolbooth became a multifunctional civic building and was used by the Scottish Parliament in the 16th and 17th centuries, until Charles I demanded that a new parliamentary building be erected. It continued to house the HQ of the city guard, who often displayed the severed heads of executed criminals outside the prison. After the building was demolished in 1817, its stones went to build the sewerage system of Fettes Row in the New Town. However, the long-held habit of spitting on the Heart of Midlothian, begun by the criminal fraternity when the land still belonged to the Tolbooth, is upheld today by some locals and ambivalent tourists.

To the east of the kirk stands the **Mercat Cross** ('market cross'), identifiable by the white unicorn holding a Saltire flag at the top of its turret. Reconstructed in the 19th century, the cross once stood at the top of Old Fishmarket Close; the site is marked on the pavement with a pattern of bricks. Among the many unfortunates executed here was James Graham, the dashing Marquess of Montrose (*see p219* **Shot by Both Sides**). This part of the High Street is also where you find the **Edinburgh Festival Fringe office** (*see p37*), which sells tickets and souvenirs.

Across the High Street from St Giles are the **City Chambers**, where the city council sits. Completed in 1761 and one of the first truly Georgian buildings in Edinburgh, the premises were originally part of the Royal Exchange. However, it failed to thrive (traders still preferred to do their business in the open air at the Mercat Cross) and the city council moved into the building in 1811. The chambers were built on top of three closes; the most famous, Mary King's Close, is today a spooky attraction accessible only as part of a tour.

Heart of Midlothian

Cockburn Street, winding north off the High Street, was once the city centre's home of alternative culture, ornamented by skiving youths loitering outside record shops. There is still a hint of that today with music shop **Underground Solu'shn,** but the street is now largely populated by eateries, fashion boutiques and gift emporiums. The **Stills Gallery** (23 Cockburn Street, 622 6200, www.stills.org) offers a creative take on photography, and if you go to the bottom of Cockburn Street, turn right on to Market Street, past the **Edinburgh Dungeon,** you'll find more art at the **Fruitmarket Gallery** and the **City Art Centre**.

Back up on the High Street, the **Tron Kirk** sits almost opposite the junction with Cockburn Street. Built in the 17th century, it survived as a church into the 1950s then lay derelict for a time. Excavations in 1974 uncovered a 16th-century thoroughfare named Marlin's Wynd beneath the church floor; it was eventually opened as a kind of unofficial tourist information centre and shop. However, the Tron Kirk is now empty once more, although it does act as an occasional temporary performance venue, particularly in August. The last plan for the Tron envisaged a visitor centre, retail and education hub – the city waits with interest. Just behind the kirk is Hunter Square, a favoured haunt of Edinburgh's homeless fraternity. Some of them can be a little aggressive, so use caution.

At the traffic lights, the High Street is cut in two by North Bridge and South Bridge. Collectively known as the Bridges, they were built to provide access to the south of the city, and hastened expansion into the New Town in the late 18th century. Although South Bridge looks like a

---

IN THE KNOW
WORTH WEIGHTING FOR

The **Tron Kirk** (*see right*) derives its name from a 'salt tron', abbreviated to 'tron': a pillar mounted on steps with a beam and scales that once stood on the High Street. This was the public weighing scale for commodities sold in the city in the 17th century. The kirk was built close by and took its name from the device.

---

continuous street, it's actually supported by 19 massive arches, only one of them visible.

Looking down North Bridge towards the New Town, the last building on the left formerly housed the offices of the *Scotsman* newspaper. Built in 1905, the iconic edifice was vacated by the company 94 years later, since when it has been transformed into a luxury hotel, also called the **Scotsman** (*see p245*). At the front of the building, an enclosed staircase provides a shortcut down to Market Street and Waverley Station. Once grand, later neglected and dank, the Scotsman Steps were refurbished in 2011 by artist Martin Creed, along with blacksmiths, painters, sculptors and stonemasons. The stairway is now a piece of art in its own right, titled *Work 1059*; each of the 104 steps is clad in a different colour of marble.

Sights & Museums

Camera Obscura & World of Illusions
Castlehill (226 3709, www.camera-obscura.co.uk). Bus 23, 27, 41, 42, 67. Open July, Aug 9am-9pm daily. Apr-June 9.30am-7.30pm daily. Sept, Oct 9.30am-7pm daily. Nov-Mar 10am-6pm daily. **Admission** £13.95; £9.95-£11.95 reductions. **Map** p48 C4 ❶

Created by optician Maria Short in the 1850s, the Camera Obscura is a system of mirrors that projects a periscope image of the city on to a white disc in the centre of a darkened room. Major landmarks are pointed out by guides as they pan the lens across Edinburgh. While the camera is no longer as thrilling as it must have been in Victorian times, its innate cleverness is still engaging. (Note that the last camera presentation usually begins an hour before scheduled closing, sometimes earlier in winter depending on the levels of daylight.) Before you reach the camera, you pass through three floors of child-friendly exhibits, including holograms, illusions, a hall of mirrors and other visual and interactive technology.

FREE Central Library
George IV Bridge (242 8000, www.edinburgh. gov.uk). Bus 23, 27, 41, 42, 67. Open 10am-8pm Mon-Wed; 10am-5pm Thur-Sat. Admission free. **Map** p48 D4 ❷

Built in 1870, the headquarters of the city's library service today houses the Edinburgh Room, the Scottish Department, and reference, fiction and lending libraries. The Edinburgh Room contains over 100,000 items pertaining to the city, from newspaper cuttings to historical prints. It's a reference library only, but some of its items are available on loan from the Scottish Department beneath it. In an adjacent building are the Central Children's Library and the Music Library, the latter holding an extensive selection of sheet music, biographies and recordings. You have to be a local resident to borrow items, but anyone is welcome to browse.

FREE City Art Centre
2 Market Street (529 3993, www.edinburgh museums.org.uk). Princes Street buses. **Open** 10am-5pm Mon-Sat; noon-5pm Sun. **Admission** free, except for touring exhibitions. **Map** p48 D3 ❸

Hosting the city's civic collection alongside major visiting shows (for which there is an entrance charge), the City Art Centre treads the line between public entertainment and contemporary art. Over six floors you can find works by major Scottish artists from the 17th century to the present, plus a gift shop and café. A rewarding place to spend a quiet couple of hours.

★ Edinburgh Castle
Castlehill (225 9846, www.edinburghcastle. gov.uk). Bus 23, 27, 41, 42, 67. **Open** *Apr-Sept* 9.30am-6pm daily. *Oct-Mar* 9.30am-5pm daily. Last entry 1hr before closing. **Admission** £16.50; £9.90-£13.20 reductions; free under-5s. **Map** p48 B4 ❹

Military barracks, prison, royal residence, murder scene, birthplace of kings and queens… Edinburgh Castle has served a variety of purposes during the centuries that it's stood high above the city. While its lofty position was employed to military advantage in years gone by, it's now extremely useful as a navigational guide if you get lost in the surrounding warren of streets and closes. For most visitors, however, it's the city's main tourist attraction.

Built on centuries of older structures, the castle now consists of a collection of buildings housed within the protective enclave of the battery walls (the other sides are protected by the sheer drop of the basalt cliffs). Many of the buildings were constructed and altered over several centuries, which can prove confusing for visitors. Although the Great Hall was originally built in 1511 under the instruction of James IV, for instance, almost everything there today dates from an extensive restoration that began in 1886. The main exception is the incredibly ornate hammerbeam roof, one of the foremost architectural treasures within the castle.

The oldest extant building is St Margaret's Chapel. Dating from the 12th-century reign of David I, it fell out of use in the 16th century and was employed as a gunpowder store for years. Its intended use was rediscovered in 1845, and it was restored to a serene simplicity. David's Tower – or, rather, the ruins of it – is another remnant of early royal constructions, although most of what can be seen in the dank vaults dates from rebuilding after the Lang Siege. The Royal Palace in Crown Square (originally the Palace Yard) requires far less imagination to visualise its regal history. The redoubtable Mary of Guise, mother to Mary, Queen of Scots, died here in 1560, while Mary herself gave birth to James VI in the birthing chamber, a small, panelled room. The last sovereign to sleep within this royal residence was Charles I in 1633.

The Honours of Scotland Exhibition is housed in the Royal Palace, in the first floor's Crown Room. Alongside the Crown, commissioned in 1540 by James V from local goldsmith John Mossman, the Sceptre, presented to James IV by Pope Alexander VI around 1494, and the Sword of State, presented to James IV by Pope Julius II in 1507, you can see the Stone of Destiny (also known as the Stone of Scone), on which Scottish kings were crowned for centuries. Or, at least, you can see what staff believe to be the Stone of Destiny. In 1950, four Scots students swiped it from Westminster Abbey, which had been its home since Edward I removed it from Scone Abbey in 1296. Three months later, a similar stone turned up outside Arbroath Abbey, and was taken back to London. It was eventually returned to Scotland in 1996, but opinion is split as to its legitimacy.

The castle is steeped in military history, but it's also still a British Army barracks and houses the regimental headquarters of the Royal Regiment of Scotland. The castle also hosts the National War Museum (0300 123 6789, www.nms.ac.uk), which charts centuries of Scottish involvement in wars in a humbling and largely objective way.

A more sombre military note is sounded by the imposing Scottish National War Memorial on Crown Square. Designed in 1924 by Sir Robert Lorimer and opened in 1927 by the Prince of Wales (later Edward VIII), it's an affecting shrine to Scotland's war dead. Below Crown Square are the castle vaults, where you'll find an effective reconstruction of the conditions endured by prisoners of war from successive skirmishes with France in the 18th and 19th centuries, and even the American War of Independence.

The buildings are the main attractions at the castle, but it's worth keeping your eyes peeled for more ephemeral bits and pieces: the Dog Cemetery on the Upper Ward; the graffiti scrawled by Napoleonic and American POWs (and their banknote forgery equipment); the 'Laird's Lug' spying device in the Great Hall; and Mons Meg, the huge six-ton cannon next to St Margaret's Chapel. Representing the height of technological advancement in her time, she was presented to James II in 1457 and last fired in 1681, when her barrel burst. While you're enjoying the views or scaring yourself with a peep over the drops, spare a thought for Sir Thomas Randolph and his men, who scaled the northern precipice in 1314 in order to wrest the castle from the English.

The most illuminating way of exploring the castle is with one of the audio guides (available in eight languages; £3.50, £1.50-£2.50 reductions). The gift shop covers all bases, from tartan tat to toys, shirts and full-size replica weaponry. Disabled visitors should note that a courtesy mobility vehicle is available and can be pre-booked or requested from Historic Scotland staff at the castle entrance.

Edinburgh Dungeon

31 Market Street (240 1001, www.thedungeons. com). Princes Street buses. **Open** varies, call or check website for details. **Admission** £16.95; £12.95-£15.95 reductions; free under-5s. **Map** p48 D3 ⑤

If you like your history packed with facts, this might not be for you. However, if disease, murder, exaggerated pantomime mayhem and the pornography of violence are more your bag, then a trip to the Edinburgh Dungeon – run by the folks behind similar operations in Blackpool, London and York – is an entertaining way to find out about Scotland's murky past. Exhibits focus on local horrors, from the plague-ridden streets and brutal judicial executions to the murderous trade of Burke and Hare. There are all sorts of diabolical torture instruments on show, and a reconstruction of the cave that was home to notorious (and, quite possibly, wholly mythical) 14th-century cannibal Sawney Bean. Costumed guides do their utmost to scare you through the place. Tickets are often cheaper if you book in advance online.

FREE Fruitmarket Gallery

45 Market Street (225 2383, www.fruitmarket. co.uk). Princes Street buses. **Open** 11am-5.30pm Mon-Sat; noon-4.30pm Sun. **Admission** free. **Map** p48 D3 ⑥

Edinburgh's leading venue for contemporary art was once an actual fruit and veg market. It was first used as a gallery space from 1974 and has been in its present guise since 1984. It has a rotating programme of exhibitions, with names such as Tania Kovats, Louise Bourgeois and Stan Douglas featuring in recent years. The splendid café will buoy you up to meet the intellectual challenge of the work.

Gladstone's Land

477B Lawnmarket (0844 493 2100, www.nts. org.uk). Bus 23, 27, 41, 42, 67. **Open** *July, Aug* 10am-6.30pm daily. *Apr-June, Sept, Oct* 10am-5pm daily. **Admission** £6.50; £5 reductions; £11.50-£16.50 family. **Map** p48 C4 ⑦

Built in 1550, then extensively rebuilt 70 years later by the merchant burgess Thomas Gledstanes, an ancestor of Victorian-era British prime minister William Gladstone, Gladstone's Land is a typical example of the lands (tenements) that once lined the Royal Mile, right down to the high-level entry door up a narrow flight of external stairs. The National Trust for Scotland (NTS) maintains the property in the 17th-century style of its former owner; you can poke around in half a dozen rooms over two floors, including a bedchamber complete with painted wooden ceiling and ornately carved bed. Costumed guides appear on Saturdays in July and August. The gilded bird of prey outside the building links to the original owner's name: 'gled' is an old Scots word for red-tailed kite, a bird that sometimes nests among 'stanes' (stones).

If you're very taken with Gladstone's Land, the NTS rents out the upper floors as self-contained holiday accommodation – see the website for details.

EXPLORE

★ FREE High Kirk of St Giles

High Street (225 9442, www.stgilescathedral.org. uk). Bus 23, 27, 35, 41, 42, 67 or Nicolson Street–North Bridge buses. **Open** *May-Sept* 9am-7pm Mon-Fri; 9am-5pm Sat; 1-5pm Sun. *Oct-Apr* 9am-5pm Mon-Sat; 1-5pm Sun. Call or check website for service times. **Admission** free; donations welcome. **Map** p48 D4 ❽

There's been a church on this site since 854. Nothing remains of the earliest structures, but the four pillars that surround the Holy Table in the centre have stood firm since around 1120, surviving the turmoil of the Reformation. The kirk was considerably refurbished in the 19th century; much of what can be seen today dates from this period. Pedants should note that the fabric of the building is referred to as St Giles, while the church itself is known as the High Kirk of Edinburgh. Either way, it's the mother church of Presbyterianism.

John Knox became minister here in 1560, 12 years before his death. This was a tumultuous time for religion in Scotland, with Edinburgh – and Knox – very much at the heart of the Scottish Reformation.

The kirk has changed status many times through the years, and today is often referred to as a cathedral, even though it's only had two bishops in its history. As a Presbyterian place of worship, it cannot be considered a cathedral at all.

Inside, a great vaulted ceiling shelters a medieval interior dominated by the banners and plaques of Scottish regiments. The main entrance takes visitors past the West Porch screen, originally designed as a royal pew for Queen Victoria. Newer features include the 1911 Thistle Chapel, an intricately decorated chamber built in honour of a chivalric order called the Knights of the Thistle. The intimate panelled room was designed by Robert Lorimer, who also designed the Scottish National War Memorial at Edinburgh Castle. The organ is an even more recent addition, installed in 1992, and features a glass back that reveals its workings.

Memorials and statues pay tribute to the likes of Knox, Robert Louis Stevenson and even Jenny Geddes, but the most notable feature may be the magnificent stained-glass windows. Constructed in the workshops of William Morris, the richly hued

---

STATUESQUE

Five monuments of note.

THE MOST VENERABLE
There are many statues of royalty around the city, but the most interesting is probably that of **Charles II**, behind the High Kirk of St Giles in Parliament Square (see p53). Edinburgh's oldest statue, erected in 1685, it's made of lead with an oak and steel infrastructure, which means it's pretty delicate. It was most recently restored in 2011, so Charlie Stuart – Mary, Queen of Scots' great-grandson – will be around for a while yet.

THE MOST BELATED
Outside the Canongate Kirk (see p64), you'll find a statue of a man hurrying down the road: **Robert Fergusson** *(pictured)*, a talented and prolific poet who died in the city's Bedlam in 1774, aged just 24. Robert Burns had no doubt that Fergusson's fame should have been as great as his own. However, Edinburgh didn't honour him with this monument until 2004.

THE BEST CONCEALED
Visit the Scottish National Gallery of Modern Art, Modern Two (see p113),

formerly known as the Dean Gallery, and you're faced with Sir Eduardo Paolozzi's towering **Vulcan**. The creation is a human-mechanoid that connects past and present, while the subject matter harks back to ancient times (Vulcan was the Roman god of fire).

THE MOST INCONGRUOUS
Officially called **Dreaming Spires**, Helen Denerley's two giraffe sculptures stand outside the Omni Centre (see p123). Made out of scrap metal culled from motorcycles and cars, they're endearing figures, unveiled in 2005. But they also couldn't look much more out of place in their surroundings.

THE OLDEST... OR IS IT?
The statue of **David Hume** on the High Street (see p52), the work of Edinburgh-born sculptor Alexander Stoddard, is a perfect example of neoclassical composition. Although it looks as if it's been there for centuries, it was unveiled as recently as 1997. People touch the big toe for luck.

EXPLORE

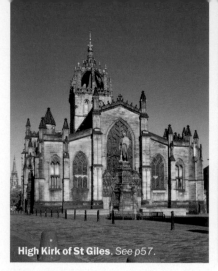

High Kirk of St Giles. *See p57.*

Edward Burne Jones window was designed to be enhanced by the western light it catches. The dazzling West Window, dedicated to Robert Burns by Icelandic artist Leifur Breidfjord in 1984, is also noteworthy, while Douglas Strachan's North Window is a blaze of rich, bold colours and swirling designs.

FREE Hub

348 Castlehill (enquiries 473 2015, tickets 473 2000, café 473 2067, www.thehub-edinburgh.com). Bus 23, 27, 41, 42, 67. **Open** *Café* 9.30am-5.30pm Mon-Sat; 11am-5.30pm Sun. *Box office & festival office* varies, call for details. **Admission** free. **Map** p48 C4 **9**

The Grade A-listed Highland Tolbooth St John's Kirk started life as the Victoria Hall for the Established Church General Assembly in 1844, following the religious 'Disruption' of the previous year. After it was bought by the Edinburgh International Festival in the late 1990s, it was renamed and extensively refurbished; a fantastic mix of Victorian Gothic and contemporary style, it's a prime example of how the architectural approaches of different eras can be sympathetically blended. The bold colour scheme of the Assembly Hall upstairs, said to adhere to Augustus Pugin's original palette, is particularly noteworthy. The rich red stairwell is decorated with 200 plaster statues by Jill Watson, representing people who have performed in the EIF in years gone by. The place is now an information point, a ticket office and a café; if the weather's good, action spills out on to a terrace.

FREE Museum on the Mound

The Mound (243 5464, www.museumonthemound. com). Bus 23, 27, 41, 42, 67. **Open** 10am-5pm Tue-Fri; 1-5pm Sat, Sun (and official Mon holidays). **Admission** free. **Map** p48 C3 **10**

Opened in 2006, the Museum on the Mound is housed within the former Bank of Scotland head office (it's now the Scottish HQ for the Lloyds Banking Group) and is one of only three banking museums in the UK. Showcasing a unique collection of artefacts and memorabilia, it has a mix of static and interactive displays to keep visitors entertained. There's a section on forgers and forging, plus a gallery dedicated to the history of Bank of Scotland at the Mound that's set in context alongside the development of the city. The bank itself was founded by the Scottish Parliament in 1695; it still issues its own banknotes.

FREE National Library of Scotland

George IV Bridge (623 3700, www.nls.uk). Bus 23, 27, 41, 42, 67. **Open** *General & special collections reading rooms* 9.30am-8.30pm Mon-Tue, Thur, Fri; 10am-8.30pm Wed; 9.30am-1pm Sat. *Exhibitions & shop* 10am-8pm Mon-Fri; 10am-5pm Sat; 2-5pm Sun. **Admission** free. **Map** p48 D4 **11**

The NLS is one of the UK's deposit libraries, entitled to request a copy of every printed item published in the UK and Ireland, and also a major European research library. It contains around 14 million printed items, newspaper and magazine titles, manuscripts, and rare books dating from the 16th century (including the first ever printed Scottish history book, *Scotorum historiae a prima gentis origine*). Items can be viewed for research purposes here, or in its Causewayside building in South Edinburgh. Admission is by library card, for which ID is required. Rotating exhibitions cover various Scottish themes.

Scotch Whisky Experience

354 Castlehill (220 0441, www.scotchwhisky experience.co.uk). Bus 23, 27, 41, 42, 67. **Open** *Tours* 10am-5.45pm daily. *Whisky shop* 10am-7pm daily. *Restaurant & whisky bar* noon-7.30pm Mon-Thur, Sun; noon-9pm Fri, Sat. **Admission** *Tours* £14-£60; £7-£22.50 reductions. **Map** p48 C4 **12**

The shop alone makes this whisky centre worth a visit. The array of blended whisky and single malts, some popular and some obscure, often includes nifty limited editions at premium prices: a 1966 Auchentoshan at £4,500 per bottle, for example.

There are various tours at various prices (for £60, no reductions, you get the Taste of Scotland tour with whisky tasting and dinner), but the basic hour-long Silver tour (£14, £7-£12 reductions) remains the main attraction. Visitors are guided through a series of displays, exhibitions and finally a theme park-style ride, which together chart the history of whisky production since the 15th century. Scotland's national drink is shown in all its constituent parts, with the tastes, smells and noises of its production cleverly intertwined in an educational (if light-hearted) sensory journey. If you're over 18, the charge includes a dram, designed in part to entice drinkers down to the well-stocked bar next to Amber, the on-site restaurant. In conclusion, one of the city's top ten attractions.

▶ *There's more whisky nearby at Royal Mile Whiskies (see p63) on High Street and Cadenhead's (see p70) on the Canongate.*

FREE Writers' Museum

Lady Stair's House, Lady Stair's Close, Lawnmarket (529 4901, www.edinburgh museums.org.uk). Bus 23, 27, 41, 42, 67. **Open** *Aug* 10am-5pm Mon-Sat; noon-5pm Sun. *Sept-July* 10am-5pm Mon-Sat. **Admission** free. **Map** p48 C4 ⑬

This museum is based in Lady Stair's House, remarkable for its sharp turnpike staircases and maze-like layout. Built by William Gray in 1622, it was given to the City of Edinburgh in 1907 and now has curios and memorabilia relating to three of Scotland's most famous writers: Robert Burns, Sir Walter Scott and Robert Louis Stevenson. Early editions of their works are supplemented by personal effects including a dining table and rocking horse that once belonged to Scott, a writing desk that belonged to Burns and a ring given to Stevenson by a Samoan chief. The museum shop is well stocked with the appropriate books, while there's also a quaint trip-step staircase that served earlier occupants as a low-tech 17th-century burglar alarm. The courtyard adjacent to the museum, Makars' Court, is worth a look as it celebrates important Scottish literary figures: short quotes from their works are engraved on the paving stones.

Restaurants

Contini Cannonball

Cannonball House, 356 Castlehill (225 1550, www.contini.com). Bus 23, 27, 41, 42, 67. **Open** *Restaurant* noon-2.30pm Mon, Tue; noon-2.30pm, 5.30-10pm Wed-Sat; noon-5pm Sun. *Caffè and ice-cream* 10am-5pm daily. **Main courses** (restaurant) £16-£20. **Map** p48 C4 ⑭ **Italian**

Victor and Carina Contini are leading lights of Edinburgh's restaurant scene. Their latest venture, launched in autumn 2014, may be their most high-profile yet: an Italian food complex in the 17th-century Cannonball House, adjacent to the Castle Esplanade. It's in the epicentre of Edinburgh tourism and has an ice-cream parlour, an upstairs *caffè* and a separate restaurant. *Gelato al cioccolato* or a T-bone steak for two? It's all here.

Cucina

G&V Royal Mile Hotel, 1 George IV Bridge (240 1666, www.quorvuscollection.com). Bus 23, 27, 41, 42, 67. **Open** 12.30-3pm, 6-10pm Mon-Thur, Sun; 12.30-3pm, 6-11pm Fri, Sat. **Main courses** £7-£30. **Map** p48 D4 ⑮ **Italian**

Hotel Missoni arrived in the city in 2009, with a flagship Italian restaurant called Cucina. In 2014, the hotel morphed into the G&V Royal Mile Hotel, but Cucina survived. Its Missoni heritage means it retains the designed-to-the-eyeballs look, while the menu offers the likes of pappardelle with rabbit and cinnamon ragù, or sea bream with fennel, radicchio, endive and bay leaf sauce. People dress up for this one.

▶ *For a review of the hotel, see p247.*

£ Elephant House

21 George IV Bridge (220 5355, www.elephant-house.co.uk). Bus 23, 27, 41, 42, 67. **Open** 8am-10pm Mon-Thur; 8am-11pm Fri; 9am-11pm Sat; 9am-10pm Sun. **Main courses** £6-£8.50. **Map** p48 D4 ⑯ **Café**

This popular café draws everyone from skiving students to grannies out with the grand-kids. The food menu runs to salads, baguettes, casseroles and pizza; some of the teas and coffees are first-rate, and there's also a short and affordable wine list. The main room at the back has fantastic views out over the Old Town. Given the associations with JK Rowling (she wrote some early Harry Potter here), Muggle memento photography outside the front door is an ever-present hazard.

£ Fruitmarket Gallery

45 Market Street (226 1843, www.fruitmarket.co.uk). Princes Street buses. **Open** 11am-5.30pm Mon-Sat; noon-4.30pm Sun. **Main courses** £8-£10. **Map** p48 D3 ⑰ **Café**

EXPLORE

WHISKY GALORE

A brief history of Scotland's most famous export.

A brief history of whisky-making in Scotland runs something like this: centuries of unfettered production, 17th-century taxes, some strife, 18th-century taxes, lots more strife, legislation in 1823, and a commercial industry thereafter. An array of 19th-century technological developments buoyed the industry, which received a major boost after the 1860s when French brandy production started to collapse because of the insect pest phylloxera. By the late 19th century, whisky was the country's spirit of choice.

This was simple blended whisky, mind. The intense and occasionally erratic single-malt varieties produced by small distilleries were deemed too wild, and so various single malts were married to bulk-grain whiskies to create blends. These blended whiskies have dominated the market ever since; today's big-sellers include Bell's, the Famous Grouse, Chivas Regal and Johnny Walker Black Label.

But over the last half century, the whisky industry has rediscovered the wonders of single-malt scotch in its own right. Glenfiddich was the first to dip a toe in the water in the 1960s, since when the floodgates have opened and the business has grown beyond all expectation. There are now around 100 distilleries across the country producing single-malt whisky.

It should follow that these distilleries are producing 100 or so whiskies, but things aren't quite that simple. Take Glenfiddich,

for instance. The label sells huge amounts of its 12-year-old single malt, but its business has expanded greatly through the marketing of countless other expressions, from a 15-year-old to a very expensive 50-year-old. Similarly, Glenmorangie is rightly celebrated for its ten-year-old whisky, now sold as the 'Original', but it also offers more than a dozen others. And then there are the multitudes of specialist expressions and limited-edition single-cask bottlings found at the likes of the **Scotch Malt Whisky Society** (*see p163*) in Leith.

The result of all this divergence has been a through-the-roof increase in the number of single-malt varieties over the last few decades. A recent edition of the annual *Jim Murray's Whisky Bible* (www.whiskybible.com), for instance, had tasting notes for a daunting 2,000-plus varieties. With such an insane level of choice, personal preference comes to the fore.

Sometimes it really is best simply to head to a decent whisky pub, such as the **Bow Bar** (*see p75*) in the Old Town, and try a few, but more hands-on help is also available. The **Scotch Whisky Experience** (*see p58*) on Castlehill at the top of the Royal Mile offers supervised tastings and runs a formal one-day course; the Scotch Malt Whisky Society also runs supervised tastings for members; and you can ask staff at **Cadenhead's** (*see p70*) in the Canongate about the tasting sessions it holds in a local pub. *Sláinte!*

Scotch Whisky Experience.

EXPLORE

This modern art gallery challenges the viewer to ponder what they've seen long after they've left the exhibits. Just as well, then, that this light, airy café is on the premises, providing some space to think along with some excellent salads, sandwiches, daily specials and deli platters. Drinks run to a few wines and beers.

► *For the gallery itself, see p56.*

£ Lucano's Kitchen

37-39 George IV Bridge (225 6690, www.caffelucano.co.uk). Bus 23, 27, 41, 42, 67. **Open** 7am-10pm Mon-Fri; 8am-10pm Sat; 9am-8pm Sun. **Main courses** £7-£8.50. **Map** p48 D4 ⑬ **Italian**

A simple Italian establishment, Lucano's Kitchen offers breakfast (scrambled egg and smoked salmon on toast), filled ciabattas and focaccias, cakes, coffees and even full-on plates of pizza or spaghetti; there are specials every day, priced a cut above average, and the house wines are reasonable. The good-natured staff, sheer lack of whimsy or pretension and groups of retired Italian gents having a natter in the daytime combine to make it a welcome find.

North Bridge Brasserie

Scotsman Hotel, 20 North Bridge (622 2900, www.northbridgebrasserie.com). Nicolson Street–North Bridge buses. **Open** 7am-9.45pm Mon-Sat; 10.30am-9.45pm Sun. **Main courses** £14-£29. **Map** p49 E3 ⑲ **Modern European**

As the street-front brasserie of the Scotsman Hotel, formerly the headquarters of the *Scotsman* newspaper until 1999, the fixtures and fittings here are Edwardian-plush with a modern makeover. The seating is split over two levels, and the cooking is modern-robust (organic beer-battered fish of the day, or wild rabbit cannelloni). If they're not too busy dealing with hotel guests, it's an early and central stop for a decent breakfast.

Ondine

2 George IV Bridge (226 1888, www.ondine restaurant.co.uk). Bus 23, 27, 41, 42, 67. **Open** noon-2.30pm, 5.30-10pm Mon-Sat. **Main courses** £14-£60. **Map** p48 D4 ㉑ **Seafood**

Opened in 2009 and garnering immediate praise, Ondine has been regularly cited as one of Edinburgh's best restaurants, and probably its very best seafood restaurant. It's an attractive first-floor venue with a central bar and impressive contemporary design, and chef Roy Brett picks from across the planet to bring starters as conceptually far-flung as squid tempura with Vietnamese dipping sauce (although it will be Scottish-landed squid) to the down-home simplicity of oysters from Carlingford Lough, Lochailort or Loch Fyne. There is a mighty fruits of the sea option, an equally mighty roasted shellfish platter, and other mains range from fish and chips to grilled whole lemon sole with brown shrimps and capers.

Wildest Drams.

Wildest Drams

209-213 High Street (226 2289, www.wildest drams.co.uk). Nicolson Street–North Bridge buses. **Open** *Food served* noon-4pm, 5-9pm daily. *Bar* noon-1am daily. **Main courses** £15-£19. **Map** p48 D3 ㉒ **Scottish**

With a blink-and-you'll-miss-it entrance on the High Street, this is a recent addition (2014) to the Old Town with a bare-stone-and-benches cellar bar featuring whisky and craft beers – a couple on draught, lots in bottles – and a simply decorated upstairs restaurant specialising in game. There are fish and vegetarian options too, but the dedicated experience could see a starter such as pigeon carpaccio with horseradish remoulade and beetroot purée, followed by roast grouse with game and barley broth plus sausage made of grouse leg.

★ The Witchery by the Castle

352 Castlehill (225 5613, www.thewitchery.com). Bus 23, 27, 41, 42, 67. **Open** noon-11.30pm daily. **Main courses** £17-£42. **Map** p48 C4 ㉒ **Modern European**

The Witchery has been one of the most unmissable restaurants in the city since it opened in 1979, in authentic late 16th-century premises very close to the Castle Esplanade. With its red leather banquettes, candles, oak panelling and ornate ceiling, it's almost like stepping back four centuries into a Gothic version of Edinburgh's past. In 1989, a second dining room was added to the premises: the Secret Garden, a lighter space with the feel of an antique conservatory. To complete the picture, there are also nine extraordinary suites offering perhaps the most extravagant B&B in Scotland (*see p245*). That said, the Witchery pre-dates the 1990s modernisation of Edinburgh's dining scene, it's at the geographical ground zero of Scottish tourism and its

Crafter's Barn.

longstanding reputation creates expectations that are far from consistently fulfilled. From the à la carte, you might have wood pigeon with grilled potato terrine, gem lettuce and burnt onion mayo to start; roast venison loin with black quinoa, ironbark pumpkin and pickled pear to follow; and mascarpone and passionfruit trifle with pistachio biscotti for dessert.

Pubs & Bars

City Café
19 Blair Street (220 0125, www.citycafeedinburgh. co.uk). Nicolson Street–North Bridge buses. **Open** 9am-1am daily. *Food served* 9am-10pm daily. **Main courses** £6.50-£9.50. **Map** p49 E4 ㉓
This was one of the first modern café-bars on the scene in the 1980s; these days, it feels as much a part of Edinburgh as the castle. There is a pool table and fill-you-up meals, to be devoured in the booths at the back or the more open seating at the front. With its retro American diner styling, it's packed with pre-clubbers at weekends; the late mornings bring a clientele in search of a serious breakfast.

Crafter's Barn
9 North Bank Street (226 1178, www.crafters-barn. co.uk). Bus 23, 27, 41, 42, 67. **Open** 11am-1am Mon-Fri; 9am-1am Sat, Sun. *Food served* 11am-9pm Mon-Wed; 9am-10pm Thur-Sun. **Main courses** £8-£26. **Map** p48 C3 ㉔
This was a 2014 debutante in the heart of tourist country, offering Belgian beer – plus mussels, pizza, steak and more – in a large modern room looking out to the Mound. The beer even leaks into the cooking, with Leffe butter in the seared fish with mussels, while the house fish stew features a creamy Hoegaarden sauce. You can, of course, just pop in for a couple of Tripel Karmeliets.

Ecco Vino
19 Cockburn Street (225 1441, www.eccovino edinburgh.com). Nicolson Street–North Bridge buses. **Open** *Bar* 11am-11pm Mon-Thur, Sun; 11am-midnight Fri, Sat. *Food served* 8am-10pm Mon-Thur, Sun; 8am-11pm Fri, Sat. **Main courses** £10-£11. **Map** p48 D3 ㉕
The formula at this perennial Old Town wine bar has a simplicity bordering on genius. Create a basic Italian menu (snacks, antipasti, salads, pasta) and throw in some specials. Devise an Italian-slanted wine list with affordable mini-flights. Store the bottles of wine along one wall as a design feature, run the bar along the other side of the room, light some candles, then watch the customers come back time after time. It even opens early for coffee these days.

Shops & Services

★ Coda Music
12 Bank Street (622 7246, www.codamusic.co.uk). Bus 23, 27, 41, 42, 67. **Open** 9.30am-5.30pm Mon-Sat; 11am-5pm Sun. **Map** p48 D4 ㉖
Books & music
Here you have a decent chance of popping in, rummaging through the racks and emerging with something new, vibrant and Scottish – in the folk idiom – rather than clichéd, tartan and mawkish, which is more than you can say for some of the other shops in the locale. It also sells blues, country, Scottish country dance, bagpipe, world music and jazz CDs.

Cookie
29 Cockburn Street (622 7260). Nicolson Street–North Bridge buses. **Open** 9.30am-6pm Mon-Sat; noon-5pm Sun. **Map** p48 D3 ㉗ **Fashion**
A youthful fashion outlet where you can buy dresses like the Lacey by Friday On My Mind (Mad

EXPLORE

Men-style), the Doris by Emily and Fin (wholesome with an ironic undertone), or the Daisy Wrap by Eucalyptus (your dad's gran going off to the movies). So not the best place for an understated LBD, but ideal for some retro lines and colours.

Ness

336-340 Lawnmarket (225 8815, www.ness.co.uk). Bus 23, 27, 41, 42, 67. **Open** 10am-6pm daily. **Map** p48 C4 ㉓ **Fashion**
You come here for dresses, skirts, tops, knitwear, footwear and accessories, although a tweed mini-kilt in Applecross tartan or a semi-fitted coat in North Sea blue would be typical purchases; Ness makes a virtue of its heritage. Since it launched in Edinburgh nearly 20 years ago, the formula has proved successful: there are another two branches in the city, as well as others in Scotland and England.
Other locations 367 High Street, Old Town (226 5227); Ocean Terminal, Leith (554 5231).

Royal Mile Whiskies

379 High Street (524 9380, www.royalmilewhiskies. com). Bus 23, 27, 41, 42, 67. **Open** 10am-6pm Mon-Wed, Sun; 10am-7pm Thur-Sat. **Map** p48 D4 ㉙ **Food & drink**
In the heart of tourist country, across the street from the High Kirk of St Giles, this whisky shop has an almost peerless selection of single malts from Scotland alongside blended and grain whiskies, as well as bottles from Ireland, Japan, the USA and elsewhere. There are other spirits and a good beer selection too, and staff have an enormous depth of expertise. A brilliant place to browse or buy.

Underground Solu'shn

9 Cockburn Street (226 2242, www.underground solushn.com). Nicolson Street–North Bridge buses. **Open** 10am-6pm Mon-Wed, Fri, Sat; 10am-7pm Thur; noon-6pm Sun. **Map** p48 D3 ㉚ **Books & music**
This shop's original premises were in a basement – hence the name – but it eventually emerged blinking into the daylight at this address. It's been the city's leading light for independent dance, electronic and specialist music for nearly 20 years. It's where you come for house, techno, dubstep, deep house and drum 'n' bass releases; there's loads on vinyl, plus CDs and DJ gear. It even sells rock and pop.

THE ROYAL MILE – EAST

Beyond the traffic lights, just behind the Inn on the Mile Hotel, Niddry Street dips steeply down towards Cowgate. Behind its shabby walls, which have been done up to represent 19th-century Edinburgh in at least one BBC drama, is a warren of cellars built into the arches of South Bridge, atmospheric places that are brought to life through a number of guided tours (*see p26*). The **Banshee Labyrinth** (29-35 Niddry Street, 556 8642, www.thebansheelabyrinth.com) is touted as the city's most haunted pub and has a shabby appeal, even on karaoke Tuesdays.

A little way down the High Street, past the self-consciously historical exterior of the modern **Radisson Blu Hotel** (*see p217*) and hostel-heavy Blackfriars Street, sits the glass façade of the **Museum of Childhood**, the first museum of its type in the world. A little further down, and across the street, is the **Scottish Storytelling Centre**, which incorporates **John Knox House**.

If you've time, nip down Trunk's Close, just beside the Storytelling Centre. If you're in luck, the gate will be open and you can access a small landscaped garden, one of the many little treasures that lurk behind the unassuming entrances of the city's closes. Another of them, nearby Tweeddale Court, holds one of only a few surviving stretches of the old city wall. The Netherbow, the eastern city gate, used to stand at roughly this point on the High Street.

The Canongate takes its name from the route used by Augustinian canons, who arrived at Holyrood Abbey in 1141, to reach the gates of Edinburgh. Situated outside the old city walls, the Canongate was separate from Edinburgh as recently as 1856; the Netherbow Port marks the spot where one burgh ended and the other began. Now mostly residential, it's dotted with museums

EXPLORE

Coda Music.

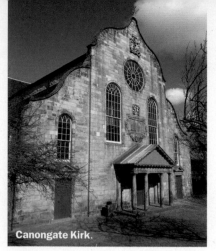

Canongate Kirk.

and shops offering everything from old-fashioned fudge to hog roast rolls.

Despite the area's history, not everything here is old – the Canongate is no stranger to modern-day architectural controversies (*see p231* **Welcome to New Waverley**). Nearby, along East Market Street, a block north, sit the City of Edinburgh Council's modern administrative offices. At the door is *Everyman* by the German sculptor Stephan Balkenhol, which resembles a small oil derrick with a full-colour bronze man on top. Locals were not amused by its reported £100,000 price tag.

By contrast, the Canongate Tolbooth was built back in 1592 and is instantly recognisable by its clock, bell tower and external stairway. The building has served at various points as a council chamber, a police station and a prison; it now houses the **People's Story**, a social history museum.

Opposite sits the **Museum of Edinburgh**, which offers fascinating insights into the other end of the social spectrum. The museum is contained within several adjoining timber-framed buildings dating from the 16th century that were later extended; the frontage is surmounted by three overhanging white-painted gables of a kind that were once common in the Old Town. One of the buildings is called Huntly House; under it, Bakehouse Close leads to the offices of Architecture+Design Scotland (formerly the Royal Fine Art Commission for Scotland), a national body responsible for promoting high standards of planning and architectural design. An old building made new, it's worth a peek.

The bell-shaped Dutch design of the **Canongate Kirk** stands out from the tenement buildings on the Royal Mile. Its construction was ordered in 1688 to house the displaced congregation of Holyrood Abbey, after James II & VII turned the abbey into a royal chapel for

the use of the Knights of the Thistle. It remains Edinburgh's official military church; the royal family often worship here when they're staying in the Palace of Holyroodhouse, and the Queen's granddaughter, Zara Phillips, had her wedding here in 2011. Outside on the pavement is a statue (*see p57* **Statuesque**) to local poet Robert Fergusson, cited by Burns as an inspiration. Fergusson's mental health was fragile and he died in the city asylum in 1774 at the age of just 24.

The Canongate Kirkyard has excellent views over Calton Hill, and is the resting place for some well-known figures. Among them are Burns' muse Clarinda, further commemorated in the name of a nearby tearoom. Just the right side of twee, it's a sedate place to relax over a nice cup of tea and a cake. Works by Burns and many other poets through the ages, including Fergusson, can be found at the **Scottish Poetry Library**, down Crichton's Close. Nearby is **Dunbar's Close Garden**, at the end of an unassuming close on the north side of the Canongate; it's a lovely place to linger for a few moments.

Further east along the Canongate sit some hugely attractive houses, most obviously the two well-kept, gleaming white edifices of the **Canongate Manse** and **Whitefoord House**. The latter is now a residence for Scottish war veterans. White Horse Close is also very pretty; the gabled building at the end was once a coaching inn, and the departure point for the stage coach to London. It was called into service in 1745 as the officers' quarters of Prince Charles Edward Stuart's army.

At the bottom of the Royal Mile is the **Palace of Holyroodhouse**, the British monarch's official residence in Scotland. From a distance, the building appears perfectly symmetrical, but on closer inspection, it becomes clear that the left tower is much older than the right. The palace was damaged by Cromwell's forces, who accidentally burned down the south wing. However, it was restored in the reign of Charles II and lovingly decorated by Queen Victoria, whose influence can still be seen today during the guided audio tours that run when the royals aren't in residence. The purpose of the strange, squat building straddling the fence along Abbeyhill, dubbed Queen Mary's Bath House, is unknown: it might have been a bathhouse, or perhaps a doocot (where 'doos', or pigeons, nest).

The grounds of the palace contain the ruins of **Holyrood Abbey**, founded by David I in 1128 and irrevocably linked with the Scottish monarchy. James II, born in the abbey lodgings, was married there, as were James III, James IV and Mary, Queen of Scots. James V and Charles I were crowned there; David II, James II, James V and Lord Darnley were all buried within its walls. The abbey suffered extensive attacks throughout the centuries: it was sacked by Edward II in

IN THE KNOW
HORNY HORNY HORNY

The national animal of Scotland is, bafflingly, the unicorn. Two featured on the royal coat of arms before the Union of the Crowns (1603) and a single unicorn remained on the British royal coat of arms thereafter – although the unicorn appears on the left in the Scottish version (a dexter supporter), on the right in the more generic UK version (a sinister supporter). In all cases, the unicorns have been tethered with a chain because, according to folklore, a free unicorn is a dangerous creature.

1322, damaged in 1544 and again in 1570 (with the loss of the choir and transepts) and violated yet further by a mob of Presbyterian zealots in 1688. It was finally abandoned in 1768.

Close by Holyrood sits Enric Miralles' **Scottish Parliament**, controversial from the outset but now settled into the fabric of the city. The complex features an array of distinctive buildings made from glass, concrete and wood, but also takes in the restored and refurbished original buildings, among them the 17th-century, Grade A-listed **Queensberry House**. The second Duke of Queensberry returned home here in 1707 after attempting to placate crowds opposed to the dissolution of the original Scottish parliament. After a hard day at the office, he was probably none too pleased to find that his lunatic son had escaped his guards and spit-roasted the kitchen boy. The oven he used still exists, though the building is now used as parliamentary offices.

The Scottish Parliament is just one of many new constructions and refurbishments around Holyrood Road, which now has student residences, hotel accommodation, offices and more. The *Scotsman* newspaper moved here in 1999, into the modern office building Barclay House, but the paper's decline saw it relocate once more, in 2014, to more modest premises elsewhere in the city. Barclay House has been taken over by Rockstar North, the company behind computer game Grand Theft Auto; a sign of the times. Just next to the Parliament buildings sits the interactive, child-oriented earth sciences museum **Our Dynamic Earth**.

Sights & Museums

FREE Museum of Childhood
42 High Street (529 4142, www.edinburgh museums.org.uk). Bus 35 or Nicolson Street–North Bridge buses. **Open** 10am-5pm Mon-Sat; noon-5pm Sun. **Admission** free. **Map** p49 E3 ③
This popular attraction was founded in 1955 by local councillor Patrick Murray, who made sure that visitors understood the difference between a museum of childhood and a museum for children. While grins of recognition are usually spread wide across the faces of kids-at-heart of all ages, older generations may enjoy their trip down memory lane more than pre-teens. While there are hundreds of bygone toys on display – some collectors' items displaying a level of craftsmanship absent on today's shelves – there's a lack of the kind of interactivity that 21st-century kids tend to expect from museums. That said, a few early automaton boxes – the Haunted House, Sweeney Todd et al – do provide entertainment. There's a small shop at the entrance. Check online for details of guest exhibitions.

FREE Museum of Edinburgh
Huntly House, 142-146 Canongate (529 4143, www.edinburghmuseums.org.uk). Bus 35.

Palace of Holyroodhouse. *See p66.*

EXPLORE

Open *Aug* 10am-5pm Mon-Sat; noon-5pm Sun. *Sept-July* 10am-5pm Mon-Sat. **Admission** free. **Map** p49 G3 ③②

This council-run museum is looking considerably smarter since a 2012 facelift: it now boasts a striking red and yellow frontage, a freshened-up interior and an audio-visual floor display in one room that relates Edinburgh's history in 15 minutes; the courtyard is now an attractive external space. The displays remain fabulously eclectic: glass, silver, pottery, the collar and bowl of Greyfriars Bobby, architect James Craig's original plans for the New Town, a specs case that belonged to John Knox and – the most historically important – the National Covenant of 1638, with some of the signatures written in blood. The museum is based in a 16th-century townhouse and adjoining properties, so history is woven into its fabric as well as its artefacts and curios.

Our Dynamic Earth

112-116 Holyrood Gait (550 7800, www.dynamic earth.co.uk). Bus 6, 35, 60. **Open** *July, Aug* 10am-6pm daily. *Apr-June, Sept, Oct* 10am-5.30pm daily. *Nov-Mar* 10am-5.30pm Wed-Sun. **Admission** £12.50; £7.95-£10.50 reductions; free under-5s. **Map** p49 H3 ③③

Our Dynamic Earth is near the former home of Edinburgh-born James Hutton, the father of geology. It's anyone's guess what he'd make of its modern, tent-like exterior, but he'd surely approve of its educational aims: to take visitors back to the creation of the universe 14 billion years ago, then bring them forward to the present day. Aimed primarily at school-age children, it's a science museum that takes a gee-whizz approach to the Big Bang, the formation of the planet, tectonics and vulcanism, weather, evolution, ecology, climate and more. There's only one way through the galleries, stuffed with interactive displays, so it's worth taking your time. Kids can touch a real mini-iceberg, gaze quizzically at bubbling primordial 'soup' and squeal during an earthquake simulation; staff enter into the spirit of things. Accompanying scientific explanations are simplistic without being patronising. Book online in advance to save money on ticket prices.

IN THE KNOW
NO KIDS, NO KIDDING

Every year the City of Edinburgh Council publishes a statistical snapshot of the city and its demographics: 'Edinburgh by Numbers'. A recent edition showed that just over 80 per cent of households had no children at all. There are single people, couples and student flats galore, but a traditional set-up with 'a couple and one or two kids' accounts for only 12 per cent of homes in the Scottish capital these days.

★ Palace of Holyroodhouse & Queen's Gallery

Holyrood Road (556 5100, www.royalcollection. org.uk). Bus 6, 35, 60. **Open** *Apr-Oct* 9.30am-6pm daily (last entry 4.30pm). *Nov-Mar* 9.30am-4.30pm daily (last entry 3.15pm). Regularly closed in season for royal visits and other special events; call or check online. **Admission** *Palace* £11.60; £7-£10.60 reductions; free under-5s; £30.20 family. *Palace & Gallery* £16; £9-£14.50 reductions; free under-5s; £41 family. **Map** p49 H2 ③④

The Palace of Holyroodhouse owes its existence to adjacent Holyrood Abbey (now picturesque ruins), established in 1128 by David I. When Edinburgh was confirmed as the nation's capital, royal quarters were built next to the abbey and have been gradually upgraded and renovated over the years. The building is still used by the Queen as an official residence in Scotland. When there is no royal in residence, parts of the palace are open to the public, with an audio tour detailing the history of a series of plush bedrooms, galleries and dining rooms.

The tour takes you back to 1566 when, six months pregnant, Mary, Queen of Scots watched as four Scottish noblemen murdered her secretary David Rizzio here, with the consent of her husband, Lord Darnley. Some say Darnley wanted to kill the baby she was carrying (the future James I & VI), believing it not to be his. Darnley died soon after in deeply suspicious circumstances.

After Queen Victoria acquired Balmoral on Deeside in 1848, west of Aberdeen, she began to use Holyroodhouse as a stop-off point on the long journey north. It was Victoria who extensively redecorated the Edinburgh building's then-drab walls with the paintings and tapestries that remain on view today, just a small part of the extensive Royal Collection housed here.

The intricate and ornate entrance to the Queen's Gallery next to the palace leads most visitors to expect a grand and old-fashioned room; in fact, the interior is surprisingly contemporary. Made up of a series of flexible spaces, the gallery hosts exhibitions from the Royal Collection, with a focus on works from the Royal Library at Windsor Castle. There's also computer access to an e-Gallery, with interactive online exhibition catalogues and details of other works from the collection. *Photo p65.*

FREE People's Story

Canongate Tolbooth, 163 Canongate (529 4057, www.edinburghmuseums.org.uk). Bus 35. **Open** *Aug* 10am-5pm Mon-Sat; noon-5pm Sun. *Sept-July* 10am-5pm Mon-Sat. **Admission** free. **Map** p49 F3 ③⑤

The Canongate Tolbooth, which houses this museum, is one of the most emotionally resonant buildings in the Old Town. Back in the day justice was dispensed here and prisoners were detained to meet their fate: beheading, branding, burning, hanging or transportation. No surprise that a jail display

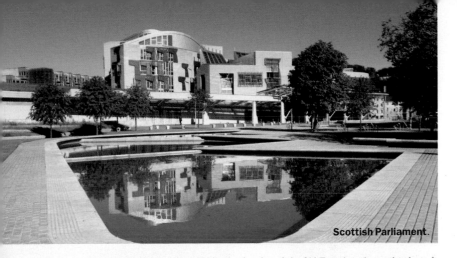

Scottish Parliament.

forms a part of the People's Story, but this local authority-run venue is more specifically dedicated to the social history of the city's working classes over the last four centuries or so. The exploits of the feared Edinburgh mobs are recorded, but most displays concern everyday life – friendly societies, guilds and unions are prominent, with various trades represented. Exhibits go up to the 1980s, even covering punk and football. The museum also offers a glimpse into the grinding poverty that some citizens endured in the past – something that continues into the present. Dwell on that and a picture emerges of an Edinburgh that's very different to the glamorous Festival City.

★ FREE Scottish Parliament

Canongate (348 5200, www.scottish.parliament.
uk). Bus 6, 35. **Open** 10am-5pm Mon, Fri, Sat,
9am-6.30pm Tue-Thur. Times can vary; always
call or check website before you visit. **Admission
& guided tours** free. **Map** p49 G3 ㉞

The people of Scotland had a long wait to see the building that houses their new parliament. When the scaffolding and coverings were finally removed, a confident, dynamic and innovative complex was revealed, different from any other parliamentary building in the UK. If you've time, take the one-hour tour, which explores areas not normally accessible to the casual visitor. If you just want to drop in, there's an exhibition about the building, plus a café, a shop and crèche facilities. On business days, tickets are available for the public gallery in the debating chamber for those who book ahead. Seating is limited.

If you don't have the time or inclination to venture inside, the building's exterior, along with the garden areas and water features, provide plenty of points of interest. The parliament's dedicated arts strategy is reflected by design components and art installations. Among them is the Canongate Wall, which is covered with quotations from centuries of Scottish writers engraved into blocks of different types of Scottish stone. At the end of the wall is a line drawing of the Old Town based on a sketch made by the building's architect Enric Miralles, who died before the project's completion.

FREE Scottish Poetry Library

5 Crichton's Close, Canongate (557 2876, www.
spl.org.uk). Bus 35. **Open** 10am-5pm Tue, Wed, Fri;
10am-7pm Thur; 10am-4pm Sat. **Admission** free;
donations welcome. **Map** p49 G3 ㉟

Founded in 1984, the Scottish Poetry Library has been housed since 1999 in this award-winning building – one of the unlikely architectural surprises concealed down the closes of the Royal Mile. Robert Burns is well represented on the shelves, but so are scores of other poets, writing in English, Gaelic, Scots or perhaps a language of their own invention. The library also has a small selection of books for sale, including works by local poets. Staff are happy to help with recommendations and will even try to help you track down a poem, even if you can only remember a line or two.

Scottish Storytelling Centre & John Knox House

43 High Street (556 9579, www.scottishstorytelling
centre.co.uk). Bus 35 or Nicolson Street–North
Bridge buses. **Open** *July, Aug* 10am-6pm Mon-
Sat; noon-6pm Sun. *Sept-June* 10am-6pm Mon-
Sat. **Admission** *Scottish Storytelling Centre*
free. *John Knox House* £5; £1-£4 reductions.
Map p49 E3 ㊳

The Scottish Storytelling Centre, which opened in 2006 in a space that once housed the Netherbow Theatre, offers a programme of storytelling events, small-scale theatre shows, music and more. Visitors are ushered through the unassuming doorway and past the bright café into a light, airy space that holds a permanent exhibition entitled Scotland's Stories. Aimed at all ages, it has an interactive wall that introduces all kinds of Scottish tradition and literature (Finn MacCuill to Morag), and is also full of mini tableaux behind doors, and touchy-feely

Blackfriars.

boxes for youngsters. A recess contains a sound and vision display on Robert Louis Stevenson. The fully refurbished theatre has been acoustically designed to meet the needs of the unamplified human voice, and a wall can be swung out into the exhibition space to provide a more intimate storytelling room.

John Knox House forms part of the same warren of rooms and spaces. The house is a genuine 15th-century property that was saved from demolition in the 19th century in the belief that the founder of Scottish Presbyterianism had lived there. Actually, it was home to one James Mosman, a goldsmith who was a fascinating character in his own right; it's his initials, along with those of his wife Mariota Arres, that can be seen on the external wall below a first-floor window. The museum says that Knox was in residence for a short period before his death in 1572, so the association with one of the key figures of the Reformation still stands. You can enjoy a visit because Knox passed this way, or just because it's an authentic slice of old Edinburgh.

Restaurants

Blackfriars
57-61 Blackfriars Street (558 8684, www. blackfriarsedinburgh.co.uk). Bus 35 or Nicolson Street–North Bridge buses. **Open** *Restaurant* noon-2pm, 6-10pm Wed-Sat, 12.30-2pm, 6-10pm Sun. *Bar* noon-midnight Wed-Thur; noon-1am Fri, Sat; 12.30pm-midnight Sun. **Main courses** £15-£24. **Map** p49 E4 ③⑨ **Modern British**
In premises once occupied by the renowned Black Bo's, this venture has split premises, with a restaurant space in one half, bar in the other. It's small and modern, with decent bistro-style cooking offering the likes of dover sole or steak and chips. Excellent snacks are available in the bar (spiced almonds, jamón ibérico de recebo and more).

£ Clarinda's
69 Canongate (557 1888, www.darindastearoom. co.uk). Bus 35. **Open** 8.30am-4.45pm Mon-Sat; 9.30am-4.45pm Sun. **No credit cards. Map** p49 G3 ④⓪ **Tearoom**
Hanging baskets of flowers outside, politesse inside: nothing changes at this classic Royal Mile tearoom. Along with your pot of tea, sample a chocolate crispy, a sherry trifle or a melting moment; the menu also runs to breakfasts, sandwiches and lunches. It's all very sweet and home-made, with cakes and biscuits that your Scottish granny might have baked. It's named in honour of Robert Burns' muse, Agnes Maclehose, who is buried in the nearby Canongate Kirkyard. The poet called her Clarinda.

★ David Bann
56-58 St Mary's Street (556 5888, www.davidbann. com). Bus 35 or Nicolson Street–North Bridge buses. **Open** noon-10pm Mon-Fri; 11am-10.30pm Sat, Sun. **Main courses** £11-£13. **Map** p49 F4 ④① **Vegetarian**
David Bann took vegetarian cooking by the scruff of the neck in Edinburgh some years ago, and remains peerless in the craft and inventiveness he brings to his range of eclectic dishes; the cooking is as good as this genre gets in Scotland. Three courses might bring olive polenta with roast vegetables and goat cheese curd to start; chilli pancake with grilled sweet potato, courgette and chocolate sauce as a main; apple, cinnamon and calvados tart for dessert; bottle of biodynamic Austrian pinot noir on the side. Superior all-day brunches at the weekends too.

Dubh Prais
123B High Street (557 5732, www.dubhprais restaurant.com). Bus 35 or Nicolson Street–North Bridge buses. **Open** 5-10.30pm Tue-Sat. **Main courses** £16.50-£25. **Map** p49 E3 ④② **Scottish**

EXPLORE

This discreet cellar has served Scottish cuisine for more than 25 years and locals tend to take it for granted. Pity: chef James McWilliams has a great CV, and his cooking has been praised by the likes of Raymond Blanc. His kitchen delivers nicely executed food with robust flavours, as demonstrated by dishes such as medallions of venison in thyme sauce or fillet of Scotch beef with mustard and whisky sauce. It's a small place, though, so book ahead.

£ Edinburgh Larder

15 Blackfriars Street (556 6922, www.edinburgh larder.co.uk). Bus 35 or Nicolson Street–North Bridge buses. **Open** 8am-5pm Mon-Fri; 9am-5pm Sat, Sun. **Main courses** £4-£8.50. **Map** p49 E3 **㊸ Café**

Just a few yards from the Royal Mile, this small, modern café has been a great stop for good, simple food since its 2009 launch. It does breakfasts, soup, sandwiches and plates (cheese, fish, salad), hot drinks and a limited selection of wine and beer. With lots of Old Town venues screaming for your attention, this discreet side-street option is a pleasant alternative. The owners also have a bistro (*see p101*) in the West End part of the New Town.

Empires

24 St Mary's Street (466 0100, www.empirescafe. wordpress.com). Bus 35. **Open** 5-11pm Mon-Thur; noon-11pm Fri-Sun. **Main courses** £14-£15. **Map** p49 F3 **㊹ Turkish**

Empires is a small Turkish meze venue where most people order a selection of small dishes, although there are a few main course alternatives such as kofte or moussaka. It's not expensive and what it lacks in finesse it makes up for with an enduring charm; the decor transports you to an obscure corner of Istanbul's Grand Bazaar. If you're really lucky, there might be a belly dancer.

★ La Garrigue

31 Jeffrey Street (557 3032, www.lagarrigue.co.uk). Bus 35 or Nicolson Street–North Bridge buses. **Open** noon-2.30pm, 6-9.30pm daily. **Set meal** £35. **Map** p49 E3 **㊺ French**

The designer-rustic decor of Edinburgh's specialist Languedoc restaurant – tables and chairs are by the late, great furniture-maker Tim Stead – is in keeping with chef Jean-Michel Gauffre's philosophy. The food is top-quality regional cooking (duck galantine to start, then Castelnaudary cassoulet, perhaps) and there are some predictably good wines. One of Edinburgh's best-loved venues, since 2001.

Wedgwood

267 Canongate (558 8737, www.wedgwoodthe restaurant.co.uk). Bus 35. **Open** noon-3pm, 6-11pm Mon-Sat; 12.30-3pm, 6-11pm Sun. **Main courses** £15-£28. **Map** p49 F3 **㊻ Modern European**

Royal Mile restaurants can tend to the gimmicky, to attract attention, but the gimmick here is simple.

This is a very good restaurant with quality cooking. Small and contemporary, it's somewhere to pop in for some sesame and soy-glazed trout or mustard-crusted lamb; a haven in tourist country. If you're not sure what to order, try the Deciding Time pre-starter (champagne plus assorted amuse-bouches for £9.95) as you browse the menu.

Pubs & Bars

Canons' Gait

232 Canongate (556 4481, www.gait.bar). Bus 35. **Open** noon-11.30pm Mon-Thur; noon-1am Fri, Sat; 12.30-11pm Sun. **Map** p49 F3 **㊲**

Roomy and comfortable, with a decent pub-grub menu and music on some evenings, the Canons' Gait offers a modern take on what a Royal Mile bar should be. Despite its location in the middle of tourist country, the place still attracts its share of locals, and cask and craft ale aficionados.

Waverley

1 St Mary's Street (557 1050). Bus 35 or Nicolson Street–North Bridge buses. **Open** 5-11pm Mon-Thur; 5pm-midnight Fri; 7pm-midnight Sat. **Map** p49 E3 **㊽**

Not so much a pub as a manifestation from another universe that drifts in from time to time and opens its door at random. Young people come in, carry items up the stairs then never come down again. The barmaid is ethereal. There is hardly any choice of beer. Crisps are sometimes free. This is the type of pub you visit once, then dream about years later, wondering if it ever existed at all. If you drop by for a pint, it will probably be closed.

Whiski

119 High Street (556 3095, www.whiskibar.co.uk). Bus 35 or Nicolson Street–North Bridge buses. **Open** 11am-1am daily. *Food served* 11am-10pm Mon-Thur; 10am-10.30pm Fri-Sun. **Main courses** £10-£24. **Map** p49 E3 **㊾**

With self-consciously traditional decor, folk music nightly, plus haggis, salmon, steak and venison on the bar menu, Whiski doesn't skimp on the Scottishness. Then again, it has a remarkable selection of whiskies – one of the very best in the city – and even whisky cocktails. There's a sister venue elsewhere in the Old Town, which incorporates a bar, restaurant, tasting room and whisky shop, but the High Street bar has the better atmosphere. **Other location** Whiski Rooms, 4-7 North Bank Street, Old Town (225 7224).

Whistlebinkies

7 Niddry Street (557 5114, www.whistlebinkies. com). Bus 35 or Nicolson Street–North Bridge buses. **Open** 5pm-3am Mon-Thur; 1pm-3am Fri-Sun. **Map** p49 E3 **㊿**

Edinburgh's late-night, live music dive par excellence, Whistlebinkies takes in rock and pop acts,

EXPLORE

indie kids, singer-songwriters, troubadours and open mics. There's a cover charge from midnight on Fridays and Saturdays; if it's 1am and you're looking for beer and music, you're not going to quibble.

Shops & Services

Cadenhead's
172 Canongate (556 5864). Bus 35. **Open** 10.30am-5.30pm Mon-Sat. **Map** p49 F3 ⑤
Food & drink
In one form or another, Cadenhead's has been around since 1842, surviving as one of the very few remaining independent bottlers of decent Scotch – nothing chill-filtered, no caramel colouring. Its Canongate shop in Edinburgh is a magnet for whisky aficionados; some of the whiskies sold here are nothing short of magnificent; it also does good rum and gin. The parent company website is www.wmcadenhead. com, while the Edinburgh shop is on Facebook; search for 'Cadenhead's Whisky Shop Edinburgh'.

★ Corniche
2 Jeffrey Street (556 3707, www.corniche.org.uk). Bus 35 or Nicolson Street–North Bridge buses. **Open** 10.30am-5.30pm Mon-Sat. **Map** p49 E3 ⑤
Fashion
A byword for serious fashion labels with a history stretching from 1976, Corniche is where you go for Vivienne Westwood Melissa shoes, a Sandrine Philippe dress or a Comme des Garçons men's jacket. If you tentatively inquire about 'a nice hat', you might find that it's by A Child of the Jago, in burgundy or brown, and costs £165. Some gorgeous stuff, though.

Geoffrey the Tailor
57-61 High Street (557 0256, www.geoffreykilts. co.uk). Bus 35 or Nicolson Street–North Bridge buses. **Open** 9.30am-6pm Mon-Sat; 11am-6pm Sun. **Map** p49 E3 ⑤ **Kilts**
One of the most acclaimed businesses in the city when it comes to traditional made-to-measure kilts, Highland dress and accessories for gents, ladies and children. If you're getting kitted out for graduation, a wedding, or a big night out, all you need is here. The company also has the expertise to design and register a tartan just for you.

Kleen Cleaners
10 St Mary's Street (556 4337, www.kleen-cleaners. co.uk). Bus 35 or Nicolson Street–North Bridge buses. **Open** 8.30am-5.30pm Mon-Fri; 10am-4pm Sat. **Map** p49 F3 ⑤ **Dry cleaners**
Instantly recognisable by the display of wedding dresses in the window, this store offers dry cleaning and laundry (specialising in delicate fabrics such as cashmere or silk) as well as alterations and repairs. Relatively close to Holyrood, Kleen Cleaners also provides services 'by appointment to Her Majesty the Queen'.

Psychomoda
22 St Mary's Street (557 6777, www.psychomoda. com). Bus 35 or Nicolson Street–North Bridge buses. **Open** 11am-6pm Mon-Sat. **Map** p49 F3 ⑤
Fashion
Womenswear here ranges from the impeccably formal, such as made-to-measure red carpet gowns or graduation dresses, to recycled and upcycled garments, Baby Moda frocks for pre-school girls and more. Once visited, and trusted, customers tend to go back, which is testament to how founder and designer Alison Harm runs her business.

Ragamuffin
278 Canongate (557 6007). Bus 35 or Nicolson Street–North Bridge buses. **Open** 10am-6pm Mon-Sat; noon-5pm Sun. **Map** p49 F3 ⑤ **Fashion**
This shop is packed full of clothes and knitwear from the British Isles, Ireland and elsewhere. As well as its own Angels Don't Trudge brand, Ragamuffin stocks many other designers, including Privatsachen, Flax, Grizas and Oska. Expect flowing dresses, quirky knits and a relaxed style in natural fibres, plus a wide range of accessories. For updates, check the blog (www.ragamuffinloves.blogspot.co.uk).

COWGATE & SOUTH

At the junction of Cowgate and Holyrood Road, turn off and head south along the Pleasance. A short distance uphill is the 16th-century boundary of the Old Town, marked by a remaining corner of the Flodden Wall. Turning right from here on to Drummond Street provides a neat shortcut to the National Museum of Scotland on Chambers Street. Alternatively, turn north where Cowgate and Holyrood Road meet, up shop-packed St Mary's Street towards the Royal Mile. This crossroads was at one time an entrance to the Old Town, known as the Cowgate Port.

Cowgate was originally used by cows and herdsmen passing to and from the fields. However, little of its history is tangible today. By the middle of the 19th century, it had become one of the most densely populated areas of the city, crammed with impoverished inhabitants and such a centre for Irish immigrants it was known as Little Ireland (*see p71* **In the Know**). Today, it's a popular centre for alcohol-fuelled hedonism. Come the weekend, the huge **Three Sisters** pub (139 Cowgate, 622 6802, www.thethreesistersbar. co.uk) takes the lion's share of the street's punters, while assorted nightclubs and bars provide bibulous competition.

On the corner of Blackfriars Street sits **St Cecilia's Hall**, which houses the Russell and Mirrey collections of early keyboard instruments. Owned by the University of Edinburgh, the hall also hosts musical performances and recitals (*see p186*). High School Wynd, opposite Blackfriars Street, then leads up to the **Old High School**

IN THE KNOW LITTLE IRELAND

Cowgate, running east from the Grassmarket under George IV Bridge and South Bridge, was once a focal point for the city's Irish population. The Great Famine in Ireland in the 1840s and its aftermath saw huge numbers leave the country, with Scotland as a significant destination. Many came to Edinburgh, settling in what was then a poor part of town. St Patrick's Church on South Gray's Close, on the Cowgate, served as the local Catholic place of worship from 1856, although the building dates from the 18th century.

(1777, now the Edinburgh Centre for Carbon Innovation), via Infirmary Street and **Dovecot Studios**, back to South Bridge and the University of Edinburgh's Old College.

Beyond the towering backs of the Edinburgh Sheriff Court buildings, Cowgate passes under George IV Bridge. The eagle-eyed will see a plaque to James Connolly. Born in the Irish immigrant community in the Cowgate in 1868, Connolly was a republican and socialist, executed for his role in the Easter Rising in Dublin in 1916. Between here and the Grassmarket, the **Magdalen Chapel** is dwarfed by surrounding buildings. In the bloody days of the 17th century, the chapel served as a mortuary for executed Covenanters, whose bodies were to be buried around the corner in **Greyfriars Kirkyard**.

There's been a market of some sort at the **Grassmarket** since at least 1477, when the area received its charter from James III; look out for a plaque laid on the 500th anniversary. Later, the Grassmarket developed a darker history as a venue for executions; these days, though, it's a little more approachable. Indeed, thanks to a £6 million makeover completed in 2008, it can even look continental. There's ample space for drinkers and diners to go alfresco (weather permitting) at pubs such as the **White Hart Inn** (34 Grassmarket, 226 2806, www.whitehart-edinburgh.co.uk) and many others. Robert Burns is said to have written 'Ae Fond Kiss' at the White Hart. These days, there are more events than ever at the Grassmarket, not to mention its Saturday market (10am-5pm) featuring street food, gifts, fruit and veg, confectionery and more.

At the west end of the Grassmarket are Granny's Green Steps, which lead up to Johnston Terrace and a walk around the base of Castle Rock. If you don't fancy it, head up the West Bow at the east end of the Grassmarket, which quickly turns into steep, bending and quirky Victoria Street. There's an eclectic range of shops here, selling everything from classy clothes or whisky to greetings cards and comedy wigs.

Nearby Candlemaker Row, off Cowgate, leads up to **Greyfriars Kirk** and George IV Bridge. At the top of Candlemaker Row stands one of Edinburgh's more curious attractions, a small statue of a dog named **Greyfriars Bobby**. When a man named John Gray was buried in the kirk's graveyard, so the story goes, his loyal Skye terrier Bobby kept constant watch over his grave

EXPLORE

Grassmarket.

for 14 years, until his own death in 1872. It's not uncommon to see passers-by touch the dog's nose for luck. Bobby's long vigil is far from the strangest occurrence reputed to have happened at Greyfriars Kirk, said to be one of the most haunted burial grounds in Britain. Sightings of ghosts and ghouls have been recorded at regular intervals; some night-time visitors have even reported receiving scratches and bruises. The 'violence' is often attributed to the ghost of the 17th-century judge and Lord Advocate Sir George Mackenzie. Commonly known as 'Bluidy' Mackenzie, he was the scourge of the Covenanters, many of whom were buried here after being executed. Perhaps it's the undulating ground or the result of centuries of burials, but even hardened sceptics admit to finding the place spooky.

Greyfriars Kirk and its kirkyard have played a pivotal role in the history of Scotland. The National Covenant was signed before the pulpit in 1638; later, the survivors of the Battle of Bothwell Brig (1679) were kept in the south-west corner of the yard in the Covenanters' Prison for five months under desperate living conditions. Not all of them survived their incarceration; the **Martyrs' Monument**, with its chilling inscription ('Halt passenger, take heed of what

BATHHOUSE ART

Adding to the great tapestry of existence.

Opened in 1887, the Infirmary Street Baths were the product of Victorian public health improvements. The bathhouse gave locals somewhere to swim and, more importantly, to wash, making a stand against cholera and other infectious diseases; modern-style bathrooms were not a feature in working-class households at the time. The baths remained in use for more than a century, eventually closing in 1995. They lay derelict for more than a decade, but eventually found an unlikely new purpose.

The roots of **Dovecot Studios** (see p73) go back to 1912, when the fourth Marquess of Bute created a tapestry studio in Corstorphine with the help of two craftsmen who had previously worked with William Morris. The enterprise collaborated regularly

with high-calibre artists, among them Henry Moore, Stanley Spencer and Graham Sutherland. At the end of the 1990s, the seventh Marquess of Bute no longer felt able to support the studios, which were threatened with closure. But with assistance from Alastair and Elizabeth Salvesen (of the family behind transport and logistics firm Christian Salvesen), the studios were reborn in 2001. In 2008, Dovecot moved into its new home – the old Infirmary Street Baths, transformed by Malcolm Fraser Architects with £8 million of the Salvesens' money.

At entrance level, there's a reception, café and exhibition space, in which Dovecot stages a rotating programme of temporary exhibitions dedicated to art, crafts and design (not solely tapestry). Further inside, the old swimming pool has been converted into an impressive weaving studio, complete with its own mezzanine viewing gallery (open noon-3pm Mon-Sat). All told, it's a fresh, attractive space, certainly one of the more engaging Edinburgh galleries to have opened in the last decade.

you do see, This tomb doth shew for what some men did die…'), is their memorial, and stands in the north-eastern part of the yard.

Opposite the entrance to Greyfriars, the bold, impressive lines of the modern **National Museum of Scotland** (1998) mask a warren of corridors that opens up on to spectacular drops and huge spaces. The museum's roof terrace and its **Tower** restaurant boast fantastic views of Arthur's Seat and the castle. The adjacent **Royal Museum** (1888) was designed by Captain Francis Fowke along grand Victorian lines. A major refurbishment project, completed in 2011, married the two in sympathetic style. Both now operate under the National Museum of Scotland name, but with clearly contrasting architectural styles.

Bordered on its northern edge by Chambers Street, **Old College** is the oldest of the city's university buildings. Architect Robert Adam began work on it in 1789, only to be interrupted by the Napoleonic Wars. William Playfair then finished it, and Rowand Anderson added the landmark dome in 1883. Entrance to the main quadrangle is either through the small entrance of the **Talbot Rice Gallery**, up West College Street, or through the monumental arch on South Bridge. For years, the Old College quad was a rather dowdy space, but a major revamp in 2011 smartened it up considerably and added a rather splendid lawn. Some areas of Old College are open to the public. Most notable are the Playfair Library, with its superb classical interior – used as an events venue – and the old Upper Museum, now part of the Talbot Rice Gallery. The museum features a table from Napoleon's lodgings on St Helena, complete with a cigar burn that was allegedly made by the Corsican.

Two of the university's student unions are nearby, on either side of Bristo Square. Teviot Row House (the world's oldest purpose-built student union) is the grander of the two and a major Fringe venue every August; Potterrow is easily identifiable by its modern, domed roof. Next to Teviot Row House is the **Reid Concert Hall** (see p187), which hosts classical concerts and houses the University of Edinburgh Museum of Instruments.

George Heriot's School, the impressive renaissance pile off nearby Lauriston Place, was established in 1628 by the goldsmith and jeweller to the late James I & VI. It was used in its early days as a military hospital for Cromwell's troops. Now it is just one of the city's many private, fee-paying schools.

Sights & Museums

FREE Dovecot Studios
10 Infirmary Street (550 3660, www.dovecot studios.com). Nicolson Street–North Bridge buses.

Open *Galleries, shop & studios* 10.30am-5.30pm Mon-Sat. *Café* 8.30am-5pm Mon-Fri; 10am-5pm Sat. **Admission** free. Map p49 E4 ⑰
See p72 Bathhouse Art.

FREE Greyfriars Kirk & Kirkyard
1 Greyfriars, Candlemaker Row (225 1900, www. greyfriarskirk.com). Bus 2, 23, 27, 35, 41, 42, 47, 60, 67. **Open** *Kirk* Apr-Oct 10am-4.30pm Mon-Sat. Nov-Mar Tours by appointment only; see website. *Kirkyard* 24hrs daily. **Admission** *Kirk* free; donations welcome. *Kirkyard* free. Map p48 D5 ⑱
Located on the site of a Franciscan friary, Greyfriars dates from 1620. The west end of the church was reduced to ruins in 1718 after the local council's gunpowder store exploded; 127 years later, much of the kirk was then gutted by a fire. After renovations, both the exterior and the interior are impressive once more, with some sympathetic, traditional harling on the outer walls and elegantly sparse spaces within. The small exhibition on the church's history contains a display about the National Covenant, but most people go to see the portrait of Greyfriars Bobby, painted by John MacLeod in 1887. Check online for regular musical recitals held here. The administratively separate kirkyard, meanwhile, with its 16th-century roots, is even older than the kirk: atmospheric and allegedly haunted. It played a key role in 17th-century Scottish history and the travails of the Covenanters; it's also one of the Old Town sites where you can really believe you have hopped back in time to a far more Gothic and unforgiving Edinburgh.

FREE Magdalen Chapel
41 Cowgate (220 1450, www.scottishreformation society.org). Bus 2. **Open** 10am-2pm Tue, Thur, Fri. **Admission** free. Map p48 D4 ⑲
Built between 1541 and 1547 (the steeple was added in 1626), the Magdalen Chapel is the headquarters of the Scottish Reformation Society. The chapel held the first congregation of the Church of Scotland in December 1560, which included John Knox. Its walls are lined with 'brods' (receipts for money or goods donated to

EXPLORE

the chapel) from the 16th to the 19th centuries, wrapping round the walls like a frieze. The chapel also contains the only surviving pre-Reformation stained-glass window in Scotland – a modest venue in its own way but historically significant.

★ FREE National Museum of Scotland

Chambers Street (0300 123 6789, www.nms. ac.uk). Bus 2, 23, 27, 35, 41, 42, 67. **Open** 10am-5pm daily. **Admission** free. **Map** p48 D5 ⑥

Designed by Benson & Forsyth, the NMS was judged to be the Scottish Building of the Year after opening in 1998. The huge, airy complex is full of stairways and windows that lead to or look out on other levels, reminiscent of the city's architecture of centuries gone by. It was plonked down next to the old Royal Museum, a beautiful Victorian space dating from 1888, and the two existed side by side until a major revamp saw them properly bonded together, relaunching under the NMS banner in 2011. The modern part of the complex covers everything from Scotland's geological origins to Neolithic artefacts, Roman silver, Dark Ages stone carvings, medieval reliquaries and the like, through the industrial period and right up to the modern day. Thousands of items are on display, from everyday objects to a whisky still. Grim relics of the darker side of Edinburgh's past are also on show, among them the Maiden of Edinburgh guillotine and an iron gaud used to restrain prisoners on the old Tolbooth. It's all you need to know about Scotland under one roof.

The former Royal Museum, with its soaring atrium and side galleries, entertains and informs with technology, natural history, anthropology and more. Kids are enthralled by the animal gallery; the exhibition gallery, as its name suggests, hosts major travelling exhibitions.

FREE Talbot Rice Gallery

Old College, South Bridge (650 2210, www.trg. ed.ac.uk). Nicolson Street–North Bridge buses. **Open** 10am-5pm Tue-Fri; noon-5pm Sat. **Admission** free. **Map** p49 E4 ⑥

Situated off William Playfair's stately, grand Old Quad in Old College, the Talbot Rice Gallery is named after the Watson Gordon Professor of Fine Art, David Talbot Rice (1903-72), famed for his writings on Islamic art. There are two main exhibition spaces: the contemporary-style White Gallery, which stages a changing programme of shows throughout the year, and the Georgian Gallery, a neoclassical space designed by Playfair. The latter has academic, experimental and historic shows, while its upper level has the university's Torrie Collection, mostly 17th-century Dutch paintings and bronzes.

Restaurants

Divino Enoteca

5 Merchant Street (225 1770, www.divino edinburgh.com). Bus 2, 23, 27, 41, 42, 67.

National Museum of Scotland.

Open 4pm-midnight Mon-Thur; noon-1am Fri, Sat. **Main courses** £10-£20. **Map** p48 D4 ⑥ Italian

This is the upmarket restaurant and wine bar from the Vittoria Group that owns a number of decent pasta 'n' pizzas venues in the city. Divino Enoteca occupies an Old Town basement with just the right balance of modernity and shade, the wine is good and the food ranges from tapas-style nibbles to Italian mains featuring duck, monkfish or veal.

Grain Store

30 Victoria Street (225 7635, www.grainstore-restaurant.co.uk). Bus 2, 23, 27, 41, 42, 67. **Open** noon-2pm, 6-10pm Mon-Sat; 6-10pm Sun. **Main courses** £17-£29. **Map** p48 C4 ⑥ French-Scottish

Looking out from the first-floor level down a curving, cobbled brae, the Grain Store has been a comforting presence for more than 20 years. The cooking takes Scottish ingredients and treats them in the French style: oysters with spinach and hollandaise, for example, followed by roast partridge with tarte tatin, raisins and baby carrots, perhaps. It's particularly good for a lazy and good-value lunch (£12.50 for two courses, £15 for three). A stone-walled, wooden-floored, candlelit local favourite.

Maison Bleue

36-38 Victoria Street (226 1900, www.maison bleuerestaurant.com). Bus 2, 23, 27, 41, 42, 67. **Open** noon-10pm daily. **Main courses** £9-£28.50. **Map** p48 C4 ⑥ French/North African

Maison Bleue is a warren of a restaurant over three floors, French with North African and other influences, using Scottish ingredients. Despite the slightly cluttered nature of that description, it works. The menu consists of bouchées (one as a starter, two or three as a main), brochettes (char-grilled skewers) and other mains, so it's flexible; you could have anything from steak and chips to canard confit à la paysanne, lamb tagine to a merguez brochette. It's fun, it's quirky and a refurbishment in early 2015 means it looks that bit more presentable.

★ Mother India Café
3-5 Infirmary Street (524 9801, www.motherindia. co.uk). Nicolson Street–North Bridge buses. **Open** noon-2pm, 5-10.30pm Mon-Wed; noon-10.30pm Thur; noon-11pm Fri, Sat; noon-10pm Sun. **Dishes** £3.50-£5.95. **Map** p49 E4 ⑥⑤ **Indian**
Mother India's empire in Glasgow expanded east in 2008 and set up here, offering an entire menu of starter-sized, tapas-style Indian dishes. The decision proved a very shrewd one: the restaurant has been a great success, and with good reason. The decor is simple and the cooking impressive, with dishes that range from simple chickpeas and potatoes in yoghurt-based sauce to the livelier likes of king prawns and spiced haddock.

£ Oink
34 Victoria Street (07771 968233 mobile, www. oinkhogroast.co.uk). Bus 2, 23, 27, 41, 42, 67. **Open** 10.30am-6pm daily. **Hog roast rolls** £2.95-£4.95. **Map** p48 C4 ⑥⑥ **Takeaway**
Oink started life as a market stall selling hog roast rolls, with pork sliced directly from a roast pig. These days, it has this takeaway, another in the Canongate and runs a stall at Edinburgh Farmers' Market (*see p109*) every Saturday. There are a couple of tables for rainy days, but you generally buy your hog roast roll (small, medium or large) then go. *Photo p77.*

£ Ting Thai Caravan
8-9 Teviot Place (225 9801). Bus 2, 23, 35, 41, 42, 47, 60, 67. **Open** 11.30am-10pm Mon-Thur, Sun; 11.30am-11pm Fri, Sat. **Main courses** £6.80-£9.40. **Map** p48 D5 ⑥⑦ **Thai**
Ting Thai had 'popped up' as an eatery before settling into these permanent premises with exposed ducting, brick and benches in 2013. Food comes in bowls and takeaway cartons; the venue has an almost provisional feel. The cooking is good, however, these are familiar Thai dishes and the prices won't traumatise your credit card.

Tower
National Museum of Scotland, Chambers Street (225 3003, www.tower-restaurant.com). Bus 2, 23, 27, 35, 41, 42, 67. **Open** 10am-11pm daily. **Main courses** £16.50-£41.50. **Map** p48 D4 ⑥⑧ **Modern European**
Located on level five of the National Museum of Scotland, the Tower is one of local restaurateur James Thomson's trio of upmarket venues (along

with Rhubarb and the Witchery). It's a contemporary space with a terrace and a great Old Town outlook, where menus include brunch, a two-course lunch, afternoon tea, a table d'hôte and à la carte. If you're choosing hand-dived scallops or the 200g Buccleuch fillet steak from the latter, don't expect economy dining.

£ Union of Genius
8 Forrest Road (226 4436, www.unionofgenius. com). Bus 2, 23, 35, 41, 42, 47, 60, 67. **Open** 8.30am-4pm Mon-Fri; noon-4pm Sat. **Soups** £3.60-£4.20. **Map** p48 D5 ⑥⑨ **Takeaway**
It does brilliant soup such as caldo verde, Moroccan harira or puy lentil and cumberland sausage; it doesn't do much else. Most people buy to go, but there are a couple of tables for sitting in. The owners also have a takeaway van in nearby George Square. The simplicity and quality of the offering justify the venue's name.

Pubs & Bars

★ Bow Bar
80 West Bow (226 7667). Bus 2, 23, 27, 35, 41, 42, 47, 60, 67. **Open** noon-midnight Mon-Sat; 12.30-11pm Sun. **Map** p48 C4 ⑦⓪
This simple one-room pub – with traditional-style wooden fixtures and fittings, brewery mirrors and leather benches – has one of the largest and most interesting ranges of single malt whisky, not just in Edinburgh but in Scotland. The award-winning Bow also offers an ever-changing choice of cask ales from Scotland and down south. On any given day you might find beers from Orkney, Edinburgh, Northumbria, North Yorkshire, London or elsewhere. The bar menu is limited to pies, albeit pretty good ones. You don't come here to eat, though; you come for a pint of cask ale and an unchill-filtered, oloroso-finished single malt, or similar.

BrewDog
143 Cowgate (220 6517, www.brewdog.com). Bus 2 or Nicolson Street–North Bridge buses. **Open** noon-1am Mon-Sat; 12.30pm-1am Sun. **Map** p48 D4 ⑦①
You can rely on this modern, bustling bar – the Edinburgh outpost of the expanding Brewdog empire – for keg beer that actually tastes of something. Not only do you get Brewdog's own products (Dead Pony, This Is Lager, Punk IPA, Libertine Black, among others), but also guests on draught from all over the world: Canada, Germany or the US, for example. There are numerous bottles, too, and the bar does pizza for peckish moments.

Royal Oak
1 Infirmary Street (557 2976, www.royal-oak-folk.com). Nicolson Street–North Bridge buses. **Open** 11.30am-2am Mon-Sat; 12.30pm-2am Sun. **No credit cards. Map** p49 E4 ⑦②

EXPLORE

Fall through the doors of this tiny, two-floor pub and you're virtually guaranteed to be regaled with a flurry of fiddles, squeezeboxes and guitars. There are regular folk sessions, with the Wee Folk Club (£5 entry) taking over the lounge bar on Sundays with guest artists. No frills: just beer and tunes.

★ Sandy Bell's
25 Forrest Road (225 2751, www.sandybells edinburgh.co.uk). Bus 2, 23, 35, 41, 42, 47, 60, 67. **Open** noon-1am Mon-Sat; 12.30pm-midnight Sun. **Map** p48 D5 ⑳
Like the Royal Oak, this is a folkies' hangout: there are nightly sessions with regulars, guests and anyone who has the chops to join in. Also like the Royal Oak, there are no frills – but if you're looking for a traditional Scottish pub that draws everyone from bus drivers to philosophers (the University of Edinburgh is nearby), this is the place.

Shops & Services

Aha Ha Ha
99 West Bow (220 5252, www.hijinks.me). Bus 2, 23, 27, 35, 41, 42, 47, 60, 67. **Open** 10am-6pm Mon-Sat; 11am-5pm Sun. **Map** p48 C4 ⑭ **Gifts & souvenirs**
If you want to buy magic tricks, accessories for a hen or stag night, garish wigs or scary rubber masks, even fake poo, this is the place to come. With a massive comedy nose and specs above the door, Ah Ha Ha is hard to miss. Its sister business, Hijinks, round the corner in the Grassmarket, sells and hires out fancy dress for when you absolutely must turn into Wonder Woman.
Other location Hijinks, 88 Grassmarket, Old Town (225 3388).

★ Armstrongs
83 Grassmarket (220 5557, www.armstrongs vintage.co.uk). Bus 2. **Open** 10am-5.30pm Mon-Thur; 10am-6pm Fri, Sat; noon-6pm Sun. **Map** p48 C4 ⑮ **Fashion**
There are three branches of Armstrongs around the city; each is jam-packed, but the Grassmarket outlet was the first and remains the biggest and most alluring. You'll find old suits, uniforms, kilts, shirts and coats for men; women can browse flapper chic, tea dresses from the 1940s, prom dresses from the 1950s and much more – including accessories.
Other locations 14 Teviot Place, Old Town (226 4634); 64-66 Clerk Street, South Edinburgh (667 3056).

Deadhead Comics
27 Candlemaker Row (226 2774, www.deadhead comics.com). Bus 2, 23, 27, 35, 41, 42, 47, 60, 67. **Open** 10am-6pm Mon-Sat; noon-3pm Sun. **Map** p48 D4 ⑯ **Books & music**
If you like your comic shops in a Gothic location, very close to an atmospheric church graveyard and – in the nicest possible sense – with authentic backstreet cachet, then Deadhead is the shop for you. Small, stacked with titles and boasting staff who really know their stuff, it's a comics aficionado's delight.

Demijohn
32 Victoria Street (225 4090, www.demijohn. co.uk). Bus 2, 23, 27, 41, 42, 67. **Open** 10am-6pm Mon-Sat; 11.30am-5pm Sun. **Map** p48 C4 ⑰
Food & drink
Founded in Edinburgh in 2004 but such a success it now has branches in Glasgow, Oxford and York, Demijohn is a deli for runny things. You chat with staff, you sample what's on offer, then you buy a demijohn and the shop fills it with the chosen product: liqueur, wine, whisky or another spirit (elderflower vodka or damson gin, say), oil and vinegar. It also sells chutneys and jams.

Fabhatrix
13 Cowgatehead (225 9222, www.fabhatrix.com). Bus 2. **Open** 10.30am-6pm Mon-Fri; 10.30am-5.30pm Sat; noon-5pm Sun. **Map** p48 D4 ⑱
Accessories
With its roots in the craft stalls that pop up in Edinburgh during the festival month of August, Fabhatrix launched its shop in 2002 and has been covering heads in the city ever since. There are styles for men and women, from a Harris Tweed trilby to a soft felt Raeburn that looks like something worn by a model in a Colourist painting. A fun place to shop.

Hawico
71 Grassmarket (225 8634, www.hawickcashmere. com). Bus 2. **Open** 10am-6pm Mon-Fri; 10am-5pm Sat. **Map** p48 C4 ⑲ **Fashion**
From its origins in the Hawick Hosiery Company in the Scottish Borders in 1874 to its current globe-bestriding, luxury cachet, Hawick Cashmere produces chic, classic cardigans and jumpers for ladies and gents, scarves and other knits – all still created in Hawick, under the Hawico trademark. Warm, functional and light, your purchase will have tradition and design contemporaneity at the very same time.

★ IJ Mellis
30A Victoria Street (226 6215, www.mellischeese. co.uk). Bus 2, 23, 27, 41, 42, 67. **Open** 9.30am-6pm Mon-Sat; 11am-5pm Sun. **Map** p48 C4 ⑳
Food & drink
Dating back more than 20 years, this branch of IJ Mellis was the original and certainly carried the most impact when it opened. Like a small, cool cave of tile and brick, it radiates the aroma of artisan cheeses from all over the UK, Ireland and Europe. For cheese fans, it's a veritable wonderland. If you're in South Edinburgh or Stockbridge, there are branches there too.
Other locations 6 Bakers Place, Stockbridge (225 6566); 330 Morningside Road, South Edinburgh (447 8889).

Oink. See p75.

other practitioners. The shop is well stocked with homeopathic and herbal medicines and also carries natural cosmetics.

Old Town Bookshop
8 Victoria Street (225 9237, http://oldtown bookshop-edinburgh.co.uk). Bus 2, 23, 27, 41, 42, 67. **Open** 10.30am-6pm Mon-Sat. **Map** p48 D4 **Books & music**
Specialists in antiquarian titles, in art and architecture, Scottish history and topographical studies. There are also good collections of maps and prints.

Ripping Records
91 South Bridge (226 7010, www.rippingrecords. com). Bus 35 or Nicolson Street–North Bridge buses. **Open** 9.30am-6.30pm Mon-Sat; 12.30-5.30pm Sun. **Map** p49 E4 **Books & music**
Ripping doubles up as a music store and a ticket vendor of some repute for gigs in Edinburgh, Glasgow and elsewhere. You can buy tickets for a gig tomorrow, next week, next month – or in the case of really huge events – next year. Alternatively, have a browse through the CD racks.

SImplyFixIt
1 Forrest Road (549 8820, www.simplyfixit.co.uk). Bus 2, 23, 35, 41, 42, 47, 60, 67. **Open** 9am-6pm daily. **Map** p48 D5 **Computers**
SimplyFixIt fixes iPhones, iPads, Macs and other laptops in need of TLC. It has another six branches in and around Edinburgh; see the website for details of your most convenient.
Other locations throughout the city.

Swish
22-24 Victoria Street (225 7180, www.swishlife. co.uk). Bus 2, 23, 27, 41, 42, 67. **Open** 10am-6pm Mon-Sat; 11am-5pm Sun. **Map** p48 D4 **Fashion**
A small family-run boutique, trading for more than 20 years, Swish sells brands for men, women and children. That means Bellfield jackets and jeans for gents, Yumi dresses for ladies, Barts hats for the bairns and much more besides. Its sister business, SwishPrint, specialises in unusual T-shirts; you can even print your own.
Other locations SwishPrint, 50 Cockburn Street, Old Town (226 7020, www.swishprint.co.uk).

Transreal Fiction
46 Candlemaker Row (226 6266, www.transreal. wordpress.com). Bus 2, 23, 35, 41, 42, 47, 60, 67. **Open** 11am-6pm Mon-Fri; 10am-6pm Sat. **Map** p48 D5 **Books & music**
Formerly at Cowgatehead, Transreal made the short hop to Candlemaker Row in 2011 and remains the city's specialist sci-fi and fantasy bookshop. Beloved of fans and authors, it sells not just high street SF, but also imports you won't find anywhere else but the internet.

Medusa
6-7 Teviot Place (225 6627, www.medusahair. co.uk). Bus 2, 23, 35, 41, 42, 47, 60, 67. **Open** 9am-6pm Mon-Wed, Fri; 9am-8.30pm Thur; 9am-4.30pm Sat. **Map** p48 D5 **Health & beauty**
Prices at Medusa start at £12.50 for a graduate stylist to cut the hair of your pre-school child and rise to £85.50 for a creative director to do woven, full-head highlights. The business has another three branches in the city, plus one just outside, in Musselburgh, East Lothian.
Other locations 63 Grassmarket, Old Town (225 6700); 34 South Clerk Street, South Edinburgh (667 7766); 26 Bread Street, South Edinburgh (622 7277).

★ Mr Wood's Fossils
5 Cowgatehead (220 1344, www.mrwoods fossils.co.uk). Bus 2. **Open** 10am-5.30pm Mon-Sat. *July, Aug, Dec* also open Sun; call for hrs. **Map** p48 C4 **Gifts & souvenirs**
Vending genuine and ancient remains of animals, plants and other organisms to locals and visitors since 1987, this place has fossils of all sizes and to suit all budgets. There are also minerals (polished and unpolished), some meteorites (au naturel or sliced) and a range of jewellery incorporating fossils and more affordable gemstones.

Napiers
18 Bristo Place (225 5542, www.napiers.net). Bus 2, 23, 27, 41, 42, 67. **Open** 9.30am-6pm Mon-Fri; 9.30am-5.30pm Sat; noon 5pm Sun. **Map** p48 D5 **Health & beauty**
Napiers offers consultations, treatments and therapies from herbalists, acupuncturists and various

EXPLORE

The New Town

Edinburgh's New Town arrived thanks to a succession of construction projects that started in 1767. Their combined legacy is some of the finest neoclassical architecture in the entire world. From the West End to the edge of Broughton in the east, Princes Street in the south to the fringes of Stockbridge in the north, the entire area merits its status as a UNESCO World Heritage Site. Despite the presence of commerce at its southern reaches, which is Edinburgh's modern-day city centre, the New Town is still chiefly residential, home to some of the Scottish capital's wealthiest residents. The richest have townhouses, which often come with keys to exclusive communal gardens; those with lesser budgets inhabit the still-expensive, elegant tenement blocks that populate the area.

EXPLORE

Scottish National Portrait Gallery.

Don't Miss

1 Scottish National Gallery From the Renaissance via rococo to Rembrandt and Rodin (p88).

2 Scott Monument The height of early Victorian Gothic (p89).

3 Scottish National Portrait Gallery Social history told through selfies (p88).

4 Joseph Bonnar Gorgeous jewellery at grandiose prices (p97).

5 Georgian House New Town life as lived from the 1790s (p87).

INTRODUCING THE NEW TOWN

As the Old Town became more and more congested in the 18th century, the city fathers realised that they needed to provide alternative housing. However, this wasn't social housing as we understand the term today. The scheme was motivated not by concern for the general populace's health and welfare, but to allow the city's monied classes a means of escape from the tumbling tenements and questionable hygiene in what then constituted the town centre.

Salubrious districts had already been erected to the south of the Old Town, most notably Brown Square in 1763 and George Square three years later, but more were needed. A competition was announced in April 1766, inviting architects to submit plans for an area north of the castle that was more accessible thanks to the draining of the Nor' Loch and the construction of the original North Bridge. The contest was won by a then-unknown 22-year-old architect named James Craig, whose genteel, regimented vision can still be seen today. His original New Town was defined by **Princes Street** in the south, **Queen Street** in the north, **Charlotte Square** in the west and **St Andrew Square** in the east.

While the warren-like Old Town was Scottish through and through, this New Town's identity centred around wider British influences. The street names honoured the House of Hanover (Hanover Street), George III and his family (Charlotte Square, Frederick Street, George Street, Princes Street) and the 1707 Act of Union (Rose and Thistle streets).

The new quarter was initially shunned by Edinburghers, who considered it too exposed to winds blowing in from the Forth. But they soon saw the error of their ways, abandoning the chaotic jumble of the Old Town for the new refined streets and mansions. Into the 19th century, the New Town grew far beyond Craig's initial plan, spilling towards Broughton in the east and the old village of Stockbridge in the north. The area immediately west of Princes Street, now known as the city's **West End**, is also seen as part of that expansive vision.

THE ORIGINAL NEW TOWN

Although **Princes Street** was part of the first phase of New Town construction, it's suffered more than any other thoroughfare since it was built. Today, it's an unsympathetic jumble of Edwardian aesthetics, functional blocks and glass-fronted edifices, all of which presumably seemed a good idea at the time. All the same, its mix of high-street department stores, chains and cheap souvenir shops continues to attract sizeable crowds. The massive St James development by the east end of Princes Street, on the site of an old

Princes Street.

civil service building and the unlovely St James Shopping mall, could be transformative – but with an estimated completion date of 2020 at the earliest, any positive effects are some way off.

Also at this end of Princes Street, **Scotland's People Centre** encompasses both the 18th-century General Register House and the adjacent, 19th-century New Register House. Designed by Robert Adam and planned while North Bridge was being built, General Register House first opened in 1789, despite being only half complete. Under Robert Reid's supervision, it was finished in the 1820s; the Corinthian pillars, beautifully balanced front elevation and restrained frieze work are magnificent. Completed in 1863, the New Register House sits on West Register Street, which leads through to St Andrew Square past Victorian-era pubs the **Guildford** and the **Café Royal**. The interior of the latter is particularly spectacular, its central bar offset by a succession of ceramic murals depicting inventors such as Faraday and Watt.

On Princes Street itself, outside General Register House, is Sir John Steell's dramatic statue of the **Duke of Wellington** on his horse Copenhagen. Most equestrian statues have four feet on the ground, but Wellington's horse rears up on its hind legs as his master points symbolically towards Waterloo Place, the start (or finish) of the A1 trunk road leading to St Paul's Cathedral in London.

Back along Princes Street at no.48 is **Jenners** (0344 800 3725, www.houseoffraser.co.uk), the department store that stands as the city's grande dame of retail therapy. Founded in 1838 by two Leith drapers, the shop was rebuilt in 1893 after a fire. An estimated 25,000 people gathered for the unveiling of its elaborately carved, statue-encrusted frontage, inspired by the façade of Oxford's Bodleian Library. It remained independent until 2005 when it was bought by the House of Fraser chain, but it still retains a certain cachet.

If this stretch of the street hints towards the city's 19th-century extravagance, no single structure exemplifies it better than George Meikle Kemp's colossal **Scott Monument**, which reaches 200 feet into the sky. It was originally meant to have been sited in Charlotte Square, but instead dominates the skyline on the corner of Waverley Bridge (named for Scott's *Waverley* novels) and rather overshadows the nearby statue of Lanarkshire-born explorer **David Livingstone**. Also close by are statues of **Adam Black** (twice Lord Provost of Edinburgh and founder of the *Edinburgh Review*) and **John Wilson** (a professor of moral philosophy at the University of Edinburgh).

For more comemmoration, look to the the west side of the intersection of the Mound and Princes Street, where you find a handsome 1903 statue of wigmaker-turned-poet **Allan Ramsay**. However, this area is dominated by the twin neoclassical temples of the **Royal Scottish Academy** building and the **National Gallery** building, two 19th-century piles designed by prolific architect William Playfair. The plainer and more refined National Gallery, set back from the street, was completed in 1859; the more florid Academy opened in 1826 but was substantially remodelled in the 1830s. The latter is topped by sphinxes and an incongruous statue of the young Queen Victoria. Completed by John Steell in 1844, the statue was originally displayed at street level, but it's said that Victoria, displeased by her chubby appearance, demanded that it was elevated to a rooftop location in order that she should avoid close scrutiny by her subjects.

Into the new millennium, extensive work took place in and around these galleries. Both appear traditional from the outside, but refurbishments mean that the gallery spaces within are state of the art, with temperature and humidity controls, air conditioning and specialised lighting. While the two buildings seem to stand in splendid isolation from each other, they have been connected at basement level since 2004 and there is access in Princes Street Gardens. The Gardens Entrance, as it was officially dubbed, has an IT gallery, an education centre (with a 200-seat lecture theatre and cinema), a restaurant, a café and a shop. Taken together, the National Gallery and Academy buildings, plus the Gardens Entrance with its facilities, are referred to as the **Scottish National Gallery** complex.

The verdant **Princes Street Gardens**, meanwhile, protected by an Act of Parliament since 1816, act as a buffer between the New Town and Old Town. Indeed, the first thing that strikes most visitors about Princes Street is how lopsided it appears, with solid rows of buildings fronting

Princes Street Gardens.

EXPLORE

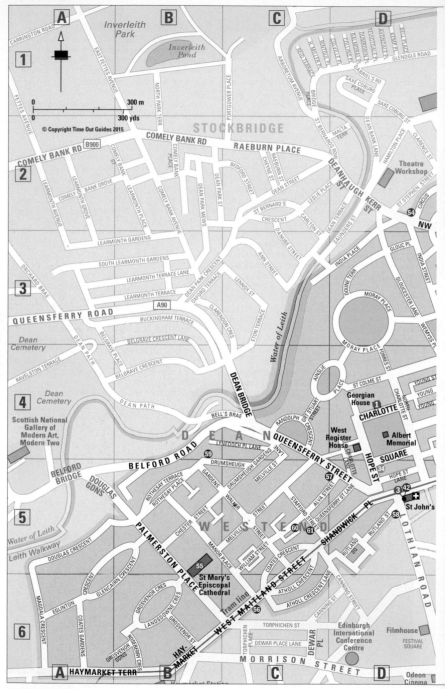

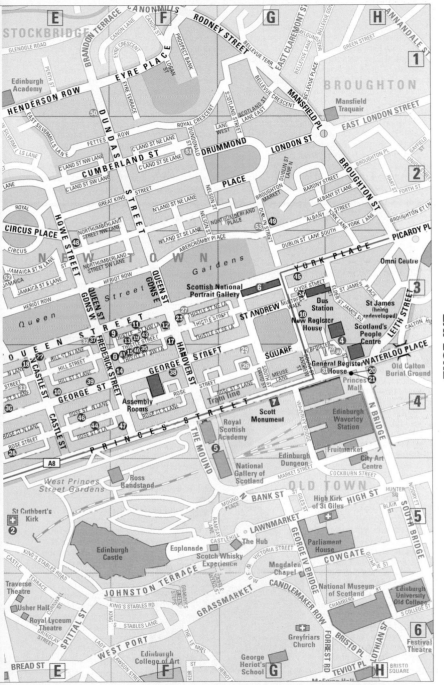

on to this green expanse. Yet more statues are dotted around the gardens, and the far western end is home to two churches: **St John's Episcopal Church** and, at the foot of Edinburgh Castle, the Presbyterian **St Cuthbert's Parish Church**. In summer, the western part of the gardens is often awash with people listening to music emanating from the Ross Bandstand. During the winter, a Christmas fair is set up by the Scott Monument in the eastern part of the gardens, with an ice rink, big wheel and other fairground attractions.

A final word on Princes Street: its traffic management schemes are legendary and the introduction of trams in 2014 enlivened matters no end. Cyclists, meanwhile, have been getting their wheels trapped in tram tracks and falling over. Public transport or walking makes a lot more sense than trying to drive or use a bicycle.

From the Scott Monument, South St David Street leads north to **St Andrew Square**, named for Scotland's patron saint. It sits at the eastern end of George Street, a grassy haven adding punctuation to one of the city's more upmarket thoroughfares. In the centre stands the Melville Monument, a 140-foot column topped by a statue of Henry Dundas. For more on the redevelopment of the square and on Dundas, *see below* **Tower of Power**.

For years, St Andrew Square and environs was the heart of Edinburgh's financial industries, until regeneration programmes encouraged key

TOWER OF POWER

Who was Harry the Ninth?

Although St Andrew Square Garden, at the eastern end of George Street, formed part of the original 18th-century plan for the New Town, its gates remained locked to the hoi polloi for more than two centuries. In 2008, things changed. The local authority and the local enterprise agency forked out £2.6 million between them to renovate and open up the garden, which became an instant hit with office workers, shoppers and tourists.

The garden is still dominated by a towering monument to a now largely forgotten figure from the city's past, Henry Dundas. Modelled on Trajan's Column in Rome, and designed by William Burn, the Melville Monument was completed in 1823, with the statue on top added in 1828.

Born in 1742 to an aristocratic family in Dalkeith, Midlothian, then educated in Edinburgh, Dundas rose to become a senior politician and lawyer, holding a string of important offices in Edinburgh and London. While Great Britain was getting to grips with an independent United States, the French Revolution and the Napoleonic Wars, he served variously as Treasurer of the Navy, Home Secretary, Secretary of State for War and First Lord of the Admiralty. He was rewarded with the title Viscount Melville in 1802, nine years before he died.

An interesting aspect of all this is that no one in Scotland liked him much. For a great deal of his career, Dundas ran things north of the border by manipulating elections through influence and patronage; he was popularly known as 'Harry the Ninth, the Uncrowned King of Scotland'. When the French Revolution inspired dissent in Scotland, effigies of Dundas were burned.

In 1792, he impeded Wilberforce's efforts to abolish the slave trade, which did little for his current image, but the final nail in the coffin of his reputation was his impeachment in 1806 for the misappropriation of public funds. He was acquitted, but it was the end of his career.

Historian Michael Fry's revisionist take on his life and times (*The Dundas Despotism*, 1992) describes him as a man whose ideals of good government stemmed in some regard from the Scottish Enlightenment, but whose thinking was firmly rooted in its time. Not corrupt in other words, just misunderstood. Whether or not you accept this interpretation, you can now sit in St Andrew Square Garden and look up at someone who was once more important than all contemporary Scottish Westminster MPs put together.

Assembly Rooms.

institutions to move west into glass and brick new-builds. Among them was the Royal Bank of Scotland, now based in Gogarburn on Edinburgh's western edge, but the bank's registered office remains a former mansion on the square. Set behind a lawn, this Palladian-style edifice was built in 1772 for Sir Lawrence Dundas on a site that, in Craig's plan for the New Town, had been reserved for a church; it's a mark of Sir Lawrence's political muscle that he was able to overrule the city fathers. The building is still a working branch of the bank; the sumptuously decorated iron dome of the Telling Room, added by J Dick Peddie in 1857, is open during banking hours. Next door sit more more outlandish bank premises: a branch of Bank of Scotland in an 1851 pseudo-palazzo complete with rooftop statues.

St Andrew Square has been the subject of multi-million-pound redevelopment programmes in the last decade and more. First to be completed was Harvey Nichols, which provided a shot in the arm for the town's shopping scene when it opened in 2002. The town's new bus station opened next, albeit without attracting the same levels of media coverage. Between them sits Multrees Walk, a tidy pedestrianised road lined with smart designer boutiques. Trams began running through the square in 2014, while the latest development, at the south-western corner, is scheduled to bring yet more office, retail and residential space by 2016. With the anchor tenant being discount clothing store TK Maxx, it's a long way from James Craig's 18th-century concept of elegant living.

The **Scottish National Portrait Gallery** sits just north of the square, at the eastern end of Queen Street. Designed by Sir Robert Rowand Anderson and completed in the late 19th century, it's a confident building, its Gothic style representing a departure from the Georgian neoclassical constraint of much of the New Town. The interior was sympathically refurbished at the end of 2011, while its red sandstone façade is best seen aglow in the late evening summer sun.

Named in honour of King George III – and connecting St Andrew Square with Charlotte Square – **George Street** was conceived as the New Town's main thoroughfare. It was part of Edinburgh's old financial district, but many of the institutional buildings here have been transformed to meet more modern needs. The street is now a major shopping drag, with its fair share of bars and other good-times venues, but the ambience is different from that of nearby Princes Street: the pace is more leisurely and the price tags are higher. One constant throughout the street's history has been the excellent views offered from here of the steep descent down to north Edinburgh and, on clear days, the hills of Fife, particularly down Dundas Street.

Through all the changes, George Street has retained its dignity even though national fashion chains aplenty have crept in behind the street's toned-down façades and few of the city's traditional names remain; the estimable jewellers **Hamilton & Inches** is one notable survivor. The buildings themselves, however, lend a suitably stylish austerity to the parade of bars, hotels, restaurants and shops.

The **Assembly Rooms**, built by public subscription, has been a feature of the street since 1787. The rooms became a favoured haunt of Edinburgh's Regency partying set: it was here, in 1827, that local resident Sir Walter Scott revealed that he was the author of the anonymously published *Waverley* novels. Though it briefly served as a labour exchange

IN THE KNOW TO THE MAX

The influential and gifted scientist James Clerk Maxwell was born in Edinburgh in 1831. A crater on the Moon and a mountain range on Venus have been named in his honour, but it took until 2008 for his home city to unveil an appropriate statue. It's at the east end of George Street.

EXPLORE

St Andrew's & St George's Church.

Sir Lawrence Dundas saw it moved here. It was later the site of what became known as 'the Disruption' of the Church of Scotland: in 1843, 472 ministers marched from here to the Tanfield Hall at Canonmills and established the Free Church of Scotland.

At the west end of George Street is **Charlotte Square**, always part of James Craig's original New Town plan, but initially called St George's Square. Some say the name change was to avoid confusion as there was already a new George Square in the south of the city; others say it was named for the wife of George III, Charlotte; yet others say it was named for their daughter, also Charlotte, born in 1766, or for the queen and the princess jointly. In any event, it remains one of the most pleasant spaces in the city centre.

The actual building design in the square was down to Kirkcaldy-born architect Robert Adam in 1791, the year before his death. He was responsible for the palatial frontages, discreetly ornamented with sphinxes and pediments, but each house was then built separately by the owners of the individual plots, creating an effect of harmony in diversity. Many illustrious types have lived here or hereabouts down the years; among them was Alexander Graham Bell, the inventor of the telephone, who was born at 16 South Charlotte Street in 1847.

The best-preserved façades are on the north side of the square, an excellent example of Adam's famous 'palace-front' design. At no.7 is the **Georgian House**, run by the National Trust for Scotland. Next door, at no.6, is **Bute House**, the official residence of Scotland's First Minister.

The imposing, domed building on the west side of the square is **West Register House**, originally built as St George's Church in 1814. It is now home to part of the National Archives of Scotland. In the grassy centre of the square, which hosts the **Edinburgh International Book Festival** (*see p37*) in August, there's a monument to Prince Albert (1819-61), husband of Queen Victoria.

Reached via the north-eastern corner of Charlotte Square and tucked away from the grander streets, Young Street was once home to the New Town's less financially blessed residents. In recent years, though, it's become a favoured haunt of Detective Inspector John Rebus, who drinks at the **Oxford Bar** in Ian Rankin's best-selling novels. To the north of Young Street, parallel to George Street, lie Queen Street and Queen Street Gardens. Like many of the green spaces that punctuate the New Town, they were created as retreats for the residents of the grand squares and terraces to enjoy, and are still open only to residents. Further along Queen Street, at no.8, is a townhouse built by Robert Adam; at no.9 stands Thomas Hamilton's neoclassical **Royal College of Physicians**.

and recruiting centre during World War I, it's remained a popular venue for concerts, plays and performances of all kinds, and is busy throughout August. Although quite the grande dame, the interior had a serious revamp in 2011-12 and the venue now combines Georgian splendour with 21st-century facilities. It even plays host to branches of chain restaurant Jamie's Italian (202 5452, www.jamieoliver.com), Kiehl's skincare (220 1731, www.kiehls.co.uk) and boutique jeweller Rox (541 2209, www.rox.co.uk).

George Street was once a popular quarter for Edinburgh's literary types. The poet Shelley and his first wife Harriet Westbrook honeymooned at no.84 (currently home to the Northern Lighthouse Board) and Sir Walter Scott lived just off George Street at 39 North Castle Street. Meanwhile, 45 George Street was the headquarters of the influential literary journal, *Blackwood's Magazine*, which counted Henry James and Oscar Wilde among its contributors. (The premises are now home to a branch of the fashion chain LK Bennett.)

Towards the east end of the street, **St Andrew's & St George's Church** (1787) was originally intended for a plot of land on St Andrew Square but the intervention of

Sights & Museums

Georgian House

7 Charlotte Square (0844 493 2117, www.nts. org.uk). Princes Street buses. **Open** *July, Aug* 10am-6pm daily. *Apr-June, Sept, Oct* 10am-5pm daily. *Mar* 11am-4pm daily. *Nov* 11am-3pm daily. **Admission** £6.50; £5 reductions; £11.50-£16.50 family. **Map** p82 D4 **①**

When John Lamont, the 18th Chief of Clan Lamont, bought this house in 1796, it cost him the princely sum of £1,600. It's now run by the National Trust for Scotland (NTS), with excellent reconstructions that open a window on to how the upper classes lived during the late 18th and early 19th centuries. The rooms are packed with period furnishings and details, right down to sugar cones, locked tea caddies and newspapers. The basement contains an informative video presentation; well-informed staff and volunteers are happy to answer questions.
▶ *Other NTS attractions in the city include Gladstone's Land (see p56) in the Old Town.*

FREE St Cuthbert's Parish Church

5 Lothian Road (229 1142, www.st-cuthberts.net). Princes Street buses. **Open** 10am-4pm Tue-Wed, Thur. *Services* see website. **Admission** free. **Map** p83 E5 **②**

St Cuthbert's has claim to being one of the oldest places of worship in or around the city; it's said that its heritage dates to a mud and wattle church built by the Northumbrian St Cuthbert in the seventh century. There is documentary evidence for a church from the 12th century, and the steeple dates from 1789, but the current fabric of the main building is as recent as 1894. Its kirkyard is the resting place of some notable names: writer Thomas de Quincey, mathematician John Napier and the Reverend David Williamson, the church's covenanting minister celebrated in song as 'Dainty Davie'. When Williamson was buried here in 1706, no stone was erected by his widow, presumably because she would have had to list on it the names of her six predecessors. Inside the church, there's an impressive frieze behind the apse modelled on Leonardo's *Last Supper*, while the stained-glass window of David and Goliath is by Tiffany of New York.

FREE St John's Episcopal Church

3 Lothian Road (229 7565, www.stjohns-edinburgh.org.uk). Princes Street buses. **Open** *Apr-Sept* 8am-4.45pm Mon-Fri; 8am-4pm Sat; 8am-7pm Sun. *Oct-Mar* 8am-4.45pm Mon-Fri; 8am-12.30pm Sat; 8am-2pm, 5-7pm Sun. **Admission** free. **Map** p82 D5 **③**

Designed in the perpendicular Gothic style by William Burn, who also built the Melville Monument in St Andrew Square, St John's began life in 1816, just before the Act of Parliament that outlawed any further building on Princes Street Gardens; as such, the church's view of Edinburgh Castle is legally protected. The collection of stained glass is said to be the finest in Scotland, while the building's glitziness does make it stand out among the austerity of the country's more prevalent Presbyterian interiors. At the north-eastern end of the church, the external mural area often features thought-provoking, politicised images while the basement terrace houses a branch of Henderson's, Edinburgh's original vegetarian café, and the One World Shop.

EXPLORE

Scottish National Gallery.
See *p88.*

EXPLORE

★ Scotland's People Centre

2 Princes Street (314 4300, www.nrscotland. gov.uk). Princes Street buses. **Open** 9am-4.30pm Mon-Fri. **Map** p83 H3 ❹

Opened in 2008, Scotland's People Centre is the country's leading facility for family research, describing itself as the 'official government source of genealogical data for Scotland'. It brings together the resources of the General Register Office for Scotland, the National Archive of Scotland and the Court of the Lord Lyon; they hold material on everything from births, deaths and marriages to parish registers, census results, wills and coats of arms. Anyone can visit the shop or the café (no admission charge) or attend taster sessions on family history (£7). Access for research costs £15 a day, £490 for three months or £1,450 a year for individuals (with additional costs for assisted searches). Some of the research rooms are stunning, particularly the Matheson Dome and Adam Dome.

★ FREE Scottish National Gallery

The Mound (624 6200, www.nationalgalleries.org). Princes Street buses. **Open** *Aug* 10am-6pm Mon-Wed, Fri-Sun; 10am-7pm Thur. *Sept-July* 10am-5pm Mon-Wed, Fri-Sun; 10am-7pm Thur. **Admission** free; charges for special exhibitions vary. **Map** p83 F4 ❺

Scott Monument.

A degree of confusion surrounds the galleries at the Mound as there are demonstrably two of them, the official website says there are three buildings, Edinburgh locals talk about them as quite separate, while the official name for some years has been the Scottish National Gallery.

To make it as simple as possible, both neoclassical edifices were designed by William Playfair and both are 19th-century creations. The one set back from the road is the National Gallery, housing Scotland's national art collection; the one sitting on Princes Street is the Royal Scottish Academy (RSA) building, home to the distinguished arts organisation of the same name. The third building that the official website talks about is the Gardens Entrance, with café, IT gallery, lecture theatre, restaurant and shop. This opened in 2004, linking with the National Gallery and the RSA building at basement level, and has its own entrance in Princes Street Gardens. It's the overall complex that is titled the Scottish National Gallery.

All that said, the wealth of great works in the National Gallery building is undeniable, from Byzantine-like Madonnas through the Northern Renaissance and High Renaissance (highlights include Botticelli's *Virgin Adoring the Sleeping Christ Child*, and a handful of Titians) and on to the early 20th century. Impressionist and Post-Impressionist work includes *Olive Trees* by Van Gogh, *A Seascape, Shipping by Moonlight* by Monet, and *A Group of Dancers* by Edgar Degas. The permanent collection of Scottish art encompasses works by artists such as Ramsay, Wilkie and McTaggart, with the latter's semi-abstract seascapes taking you away from the New Town to the Argyll coast on a wild day – you can almost feel the salt spray on your face. It's one of the gallery's strengths that it can draw visitors deeper into Scotland but also far beyond its confines, to the wider world, all during a visit to one venue.

Meanwhile, the RSA building adjacent was occupied by the RSA (225 6671, www.royalscottishacademy.org) in the early years of the 20th century. It is now effectively a large-scale space for temporary exhibitions, big-ticket blockbusters as well as shows devoted to less well-known artists. A number of annual events focus on Scottish art, others are specifically run by the RSA itself. Chief among the latter are the RSA Annual Exhibition (usually May/June) and RSA New Contemporaries, focusing on recent graduates (usually March/April). *Photo p87.*

Scottish National Portrait Gallery

1 Queen Street (624 6200, www.nationalgalleries. org). Bus 10, 11, 12, 16, 26, 41, 44, 44A. **Open** 10am-5pm Mon-Wed, Fri-Sun; 10am-7pm Thur. **Admission** free. **Map** p83 G3 ❻

One of the flagship venues run by the National Galleries of Scotland, this incredible Gothic-style building opened for business in 1889. A recent major refurbishment saw a fresh curatorial eye applied

Scottish National Portrait Gallery.

to its collection and a revamp of the interior. Since its 2011 relaunch, the impressive atrium, with its star-spangled ceiling and frieze of figures from Scottish history, feels more vibrant than ever. The art tells the tale of Scotland since the 16th century through the medium of portraiture, the gallery also hosts photography exhibitions, both contemporary and historical. The airy, expansive café is a good stop for a coffee or a light meal.

★ Scott Monument

East Princes Street Gardens (529 4068, www. edinburghmuseums.org.uk). Princes Street buses. **Open** *Apr-Sept* 10am-7pm Mon-Sat; 10am-6pm Sun. *Oct-Mar* 9am-4pm Mon-Sat; 10am-4pm Sun. **Admission** £3. **No credit cards.** **Map** p83 G4 **7**

Travellers emerging from Waverley Station expecting to see austere classical architecture may gawp in disbelief at the Victorian Gothic exuberance of the Scott Monument. Designed by the self-taught architect George Meikle Kemp, the monument houses a vast white marble statue of Sir Walter Scott (by

---

IN THE KNOW MEIKLE CALAMITY

The **Scott Monument** (*see above*) was designed by George Meikle Kemp and officially opened in 1846. Kemp was not there to see the pomp and celebration, however, as he fell into the Union Canal, in West Edinburgh, and drowned in March 1844 after having a meeting with the building contractor.

John Steell) as well as 64 statuettes, mostly of Scott's characters but with a few notable figures from Scottish history thrown in for good measure. It was completed in 1846, 14 years after Scott's death, using funds raised from public donations, which shows how dearly Edinburghers felt about the famous author. The views from the top are superb, but the final flight of steps (there are 287, in all) up to the pinnacle is a squeeze.

Restaurants

Café Marlayne

76 Thistle Street (226 2230, www.cafemarlayne. com). Bus 6, 10, 11, 12, 16, 23, 24, 27, 29, 41, 42, 61, 100. **Open** noon-10pm Mon,Tue; 8am-10pm Wed-Fri; 9am-10pm Sat; 10am-10pm Sun. **Main courses** £8-£18. **Map** p83 F4 **8** **French**

Wicker chairs and wooden tables create a relaxed environment at this small restaurant, tucked away just off Frederick Street, where you can count on finding solid cuisine – with a hint of France – based around such ingredients as duck, boudin noir, rabbit or sea bass. Puddings are straightforward (apricot sponge pudding with orange sauce, say) and the wine list won't break the bank.

Other location 13 Antigua Street, Calton Hill & Broughton (558 8244).

Café St Honoré

34 Thistle Street North West Lane (226 2211, www.cafesthonore.com). Bus 6, 10, 11, 12, 16, 23, 24, 27, 29, 41, 42, 61, 100. **Open** noon-?pm, 5.15-10pm Mon-Fri; noon-2pm, 6-10pm Sat, Sun. **Main courses** £18-£22. **Map** p83 F3 **9** **Modern British**

Café St Honoré. *See p89.*

With an excellent chef (Neil Forbes), a contemporary menu (confit duck leg with braised lentils and spinach as a main, sea buckthorn posset to follow), a long-running reputation as one of Edinburgh's better and more discreet restaurants (hidden away on a New Town lane) and its French fin de siècle look, this venue continues to please a discerning clientele.

Contini Ristorante

103 George Street (225 1550, www.contini.com). Bus 10, 11, 12, 16, 24, 29, 41, 42, 61 or Princes Street buses. **Open** 7.30am-midnight Mon-Sat; 10am-10pm Sun. **Main courses** £9-£32. **Map** p83 E4 ⑩ **Italian**

In 2004, Victor and Carina Contini, part of the family behind the celebrated Valvona & Crolla delicatessen and café, decided to set up on their own in this former bank, a grandiose space done out with modern fixtures and fittings. The aim is simple: to produce high-quality Italian food, from pizza and pasta to Milanese-style veal and other more involved dishes. At the front of the premises, the bar area caters to coffee fiends, wine-sippers and snackers. After a decade as Centotre, the venue was rebranded as Contini Ristorante in 2014, but its ownership and standards remain the same.

▶ *The Continis also run Contini Cannonball (see p59) in the Old Town, and the Scottish Café & Restaurant in the Gardens Entrance of the Scottish National Gallery complex (see p88) in the New Town.*

Dining Room at 28 Queen Street

Scotch Malt Whisky Society, 28 Queen Street (220 2044, www.thediningroomedinburgh). Bus 23, 24, 29, 42, 61. **Open** noon-3pm Mon, Tue, Sun; noon-3pm, 5-9pm Wed-Sat. **Set meal** £35. **Map** p83 F3 ⑪ **Modern European**

Complementing the Leith original, the Scotch Malt Whisky Society has this second members' room in a restored Georgian townhouse with modern interior design touches. The ground floor holds the Dining Room, open to members and non-members alike; food has a pronounced contemporary flavour (roast venison with barley, smoked malt sauce, burnt honey and carrots is a typical main). The whisky flight with dinner is £15.50.

▶ *For the Leith branch, see p163.*

★ Dogs

110 Hanover Street (220 1208, www.thedogs online.co.uk). Bus 6, 10, 11, 12, 16, 23, 27, 61. **Open** noon-4pm, 5-10pm daily. **Main courses** £10.50-£15. **Map** p83 F3 ⑫ **Modern British**

Dave Ramsden ran two of the most celebrated Edinburgh restaurants of modern times (FitzHenry and Rogue, both now gone), but in 2008 he went down the value route with the Dogs. The decor is simple (huge dog picture excepted), and the food is both good and good value, bringing, say, marinated mullet and sardine stew to start, followed by mutton,

FIT FOR THE 21ST CENTURY

Three famous venues bring themselves up to date

Heritage is a double-edged sword. Visitors come to Edinburgh in huge numbers for its many attractions, but buildings that date from the 18th or 19th centuries, for example, don't always adapt well to the technical demands of modern life. They also tend to have suffered a little wear and tear over their impressive lifetimes. The people who run some of Edinburgh's most iconic venues are very well aware of this, and the last few years have seen serious money spent bringing them up to scratch once more.

This started at the **National Museum of Scotland** (see p74) in 2008. The modern museum at the end of Chambers Street in the Old Town opened in 1998, next door to the old Royal Museum, completed in 1888. The former is a cutting edge maze of levels with displays that cover the story of Scotland from geological origins to the modern era. This contrasted with the older museum, with its soaring grand gallery and smaller side galleries covering natural history, technology, social anthropology and more.

It took three years and cost more than £47 million, but in 2011 the new National Museum of Scotland complex, properly incorporating both spaces, was opened to the public. Aside from adding sparkle to the galleries of what was the Royal Museum, a whole new floor was added there at ground level with lifts, restaurant and more. The complex is the country's most popular visitor attraction.

In the New Town, meanwhile, the **Scottish National Portrait Gallery** (see p88) was closed for two and a half years, 2009-11, to bring its Victorian fabric and spaces up to standard. More floor space was created for works of art that had long lain in storage, while the entire collection was subjected to a fresh curatorial eye. What resulted was far from a tepid gallery of portraiture and more a social history of Scotland through the representation of its people and their pastimes. The central hall with its ambitious mezzanine-level frieze of figures from Scottish history was freshened up, while the museum retained the ability to host major touring exhibitions of art and photography. The revamp cost £17.6 million, but few would argue that it wasn't money well spent.

The next well-known name to close temporarily for refurbishment was the **Assembly Rooms** (see p85) in George Street. Opened during the reign of George III, it has long served the city as a venue for shows and entertainments, thanks to its grand ballroom, music hall and suites. Over a period of 18 months, and at a cost of £9.3 million, delicate plasterwork was restored, chandeliers were reconditioned and the original Georgian features were overhauled. Behind the scenes, modern infrastructure was put in place. Inaugurated in January 1787 for the Caledonian Hunt Ball, it reopened after refurbishment in time to serve as a venue for the 2012 Edinburgh Festival Fringe.

EXPLORE

Assembly Rooms.

Honours.

apricot and root vegetable pie. It's a straightforward spot, but clearly enduring.

Dusit
49A Thistle Street (220 6846, www.dusit.co.uk). Bus 6, 10, 11, 12, 16, 23, 24, 27, 29, 41, 42, 61, 100. **Open** noon-3pm, 6-11pm Mon-Sat; noon-11pm Sun. **Main courses** £10-£20.50. **Map** p83 F3 ⑬ **Thai**

Dusit has been among the city's leading Thai restaurants for more than a decade, offering a modern dining space with subtle Thai touches, alongside a menu that covers all the expected Thai dishes and much more besides. The signature Dusit curry, for example, involves a marinated and chargrilled 8oz sirloin in 'mysterious red sauce'.

£ Eteaket
41 Frederick Street (226 2982, www.eteaket.co.uk). Bus 10, 11, 12, 16, 24, 29, 41, 42, 61, 100. **Open** 9am-6pm Mon-Wed, Fri-Sun; 9am-9pm Thur. **Main courses** £6-£8.50. **Map** p83 F4 ⑭ **Tearoom**

Eteaket is the tea boutique face of an online tea vending business, where customers can choose from all kinds of quality varieties. These premises act as a lush but polite, modern café where food options include anything from breakfast to soup, sandwiches or light meals but also afternoon tea (£14.95), tea cocktails and cakes. The tea range runs from a basic breakfast blend to delicate white teas, rooibos teas and more.

Fishers in the City
58 Thistle Street (225 5109, www.fishersbistros. co.uk). Bus 6, 10, 11, 12, 16, 23, 24, 27, 29, 41, 42, 61, 100. **Open** noon-10.30pm daily. **Main courses** £16-£21. **Map** p83 F4 ⑮ **Fish & seafood**

The owners of the long-established Fishers in Leith (*see p159*) ventured into the city centre in 2001 with this smart, modern seafood eaterie. Hardy perennials include excellent creamy fish soup, oysters and a seafood platter; alternative mains might be scallops with smoked pancetta butter, edamame salad and teriyaki dressing. A long-term dependable.

★ Harvey Nichols Forth Floor
Harvey Nichols, 30-34 St Andrew Square (524 8350, www.harveynichols.com). Princes Street buses. **Open** *Brasserie* 10am-5pm Mon; 10am-10pm Tue-Sat; 11am-5pm Sun. *Restaurant* noon-3pm Mon; noon-3pm, 6-10pm Tue-Fri; noon-3.30pm, 6-10pm Sat; noon-3.30pm Sun. **Main courses** *Brasserie* £11-£19. *Restaurant* £16-£25. **Map** p83 G3 ⑯ **Modern European**

The fourth floor of Harvey Nichols is food and drink central, with many options to tempt the casual shopper. The Forth Floor Brasserie & Restaurant – punnily named but with great views – provide the more formal alternatives to in-store branches of Chocolate Lounge or Yo! Sushi. Both are spacious and contemporary, discreetly partitioned from each other, but the latter is the more accomplished with its white linen and elaborate Mod Euro menu. Here, a typical main could be seared halibut with corn and cod brandade fritter, radish, curried butter nage, toasted cumin potato crisp, baby courgette and pattypan. In summer, eating on the terrace is sheer joy.

£ Henderson's
94 Hanover Street (225 2131, www.hendersons ofedinburgh.co.uk). Bus 6, 10, 11, 12, 16, 23, 27, 61. **Open** 8am-10pm Mon-Wed; 8am-11pm Thur-Sat. **Main courses** £11-£12.50. **Map** p83 F3 ⑰ **Vegetarian**

pig cheek with lentils and kohlrabi, then sea bass en papillote with ginger and lemongrass prawns.

★ Number One
Balmoral Hotel, 1 Princes Street (557 6727, www.restaurantnumberone.com). Princes Street buses. **Open** 6.30-10pm daily. **Set meal** £70. **Map** p83 H4 **⑳ Modern European**
An enviable address, a keen reputation, Jeff Bland as executive chef, Brian Grigor as head chef and a Michelin star held for some years. It all adds up to one of the best dining experiences in Scotland in a modern, haute style that could bring Borders roe deer with pistachio, pomegranate and bulgur as a typical main course.
▶ *For the Balmoral Hotel itself, see p249; for its afternoon teas, see below.*

★ Palm Court at the Balmoral Hotel
Balmoral Hotel, 1 Princes Street (556 2414, www.thebalmoralhotel.com). Princes Street buses. **Open** 8am-8pm Mon-Thur, Sun; 8am-9pm Fri, Sat. *Afternoon tea served* noon-5.30pm daily. **Afternoon tea** £29; £45 with a glass of champagne. **Map** p83 H4 **㉑ Café**
Tea at the Balmoral has an honourable pedigree. Served in the elegant Palm Court, it comes with an amuse-bouche, sandwiches, savouries, scones, pastries and tea or coffee. Kick back and feel like an old-school aristocrat to the accompaniment of live harp music. Add a glass of Bollinger, though, and the price goes up considerably. You can also pop in to the Palm Court for breakfast or elevenses (8am-noon daily) or tea and sandwiches (11am-5.30pm daily)

Restaurant Mark Greenaway
69 North Castle Street (226 1155, www.mark greenaway.com). Bus 16, 24, 29, 41, 42, 61, 100. **Open** noon-2.30pm, 5.30-9.30pm Tue-Sat. **Main courses** £22-£32. **Map** p83 E4 **㉒ Modern European**
Chef Mark Greenaway brought his talent and sense of theatricality here in 2013, after a couple of years elsewhere in the city. The interior is very New Town with its arched windows and views to Queen Street Gardens; the predominantly blue decor pleasantly smart. A typical main is pork belly, slow roasted for 11 hours, with pork cheek pie, blackened fillet, sweetcorn and toffee apple jus. Expect dishes that smoke, broth that must be poured and edible constructions you have to ask the staff to explain. You'll either love it or run away somewhere else to demand steak and chips. Further north in New Town is Bistro Moderne, Greenaway's more informal venue. *Photo p94.*
Other location Bistro Moderne, 15 North West Circus Place, New Town (225 4431, www.bistro moderne.co.uk).

Urban Angel
121 Hanover Street (225 6215, www.urban-angel. co.uk). Bus 6, 10, 11, 12, 16, 23, 27, 61. **Open**

This basement vegetarian restaurant was a true pioneer back in the 1960s and is still going strong. Food – hearty soups, stews, salads and desserts – is served canteen-style: you queue, you pay, you find a table. The Henderson's empire has expanded over time; there's now a deli upstairs, plus a bistro with table service round the corner, and a café in the basement of St John's Episcopal Church.
Other locations Bistro, 25C Thistle Street, New Town (225 2605); Henderson's @ St John's, St John's Episcopal Church, 3 Lothian Road, New Town (229 0212).

Honours
58A North Castle Street (220 2513, www.the honours.co.uk). Bus 16, 24, 29, 41, 42, 61, 100. **Open** noon-2.30pm, 6-10pm Tue-Sat. **Main courses** £16.75-£32. **Map** p83 E4 **⑱ British**
A sister venue to Restaurant Martin Wishart (*see p161*) in Leith, the Honours is Wishart's take on upmarket brasserie dining. It's fairly formal, the decor is classy and modern and the menu ranges from a crab cappuccino starter to mains such as rabbit à la moutarde or steaks of impeccable provenance. You can spend serious money here, although there is a prix fixe for most of the year.

Iris
47A Thistle Street (220 2111, www.irisedinburgh. co.uk). Bus 6, 10, 11, 12, 16, 23, 24, 27, 29, 41, 42, 61, 100. **Open** noon-11pm daily. **Main courses** £10-£18. **Map** p83 F3 **⑲ Modern European**
Iris opened just as the credit crunch began to bite, so its endurance tells you something about its popularity. A small, modern, informal, bistro-style venue with understated decor and a Modern European approach, two courses here might involve ibérico

EXPLORE

8am-5pm Mon-Fri; 9am-5pm Sat, Sun. **Main courses** £8-£13.50. **Map** p83 F3 ㉓ **Café**
Urban Angel has a smart country-city crossover look, with wooden floors and white walls. It maintains high ethical standards, using organic, free range and Fairtrade ingredients. In the daytime, it's known for its high-quality takeaways and snacking menu (all-day brunch, cakes, coffee, tea), but the kitchen can also pull off some accomplished lunch dishes, such as salmon fillet with quinoa and avocado, or venison burger with juniper mayo. There are decent wines and beers too.

Valvona & Crolla Ristorante & VinCaffè

Multrees Walk, St Andrew Square (557 0088, www.valvonacrolla.com). Bus 10, 11, 12, 16, 26, 41, 44, 44A, 67. **Open** 8.30am-9.30pm Mon-Fri; 9am-10pm Sat; 8.30am-8pm Sun. **Main courses** £10-£22. **Map** p83 G3 ㉔ **Italian**
Valvona & Crolla is Edinburgh's distinguished Italian deli (*see p122*) on Elm Row, dating from 1934. In 2004, the proprietors added this operation on Multrees Walk. Breakfast, cakes, coffee and light meals are available in the ground-floor VinCaffè, a full menu in the Ristorante upstairs. Two courses in the latter might involve smoked trout with lime, rocket and dill to start, then roast lamb marinated in olive oil and rosemary, with roast vegetables.

Wildfire

192 Rose Street (225 3636, www.wildfire restaurant.co.uk). Princes Street buses. **Open** 5.15-10pm Mon; noon-2.30pm, 5.15-10pm Tue-Thur; noon-3pm, 5.15-10pm Fri-Sun. **Main courses** £15-£27. **Map** p83 E4 ㉕ **Scottish**
A small steak and seafood bistro, established for more than a decade, Wildfire is simply but smartly furnished and, given the sheer retail noise and bar-hopping on Rose Street, stands as a real oasis of sanity and good cooking. It doesn't make a great fuss of its steaks the way other local boutiques de boeuf do, but they're all aged for a minimum of 21 days and sourced from the Scottish Borders. You can have fillet, rib-eye or sirloin; add king prawns sautéed in garlic butter for surf 'n' turf. Non-steak fans have the likes of casserole of the day or fisherman's pie.

Restaurant Mark Greenway. *See p93.*

Pubs & Bars

Abbotsford

3 Rose Street (225 5276, www.theabbotsford. com).
Princes Street buses. **Open** 11am-11pm Mon-Thur,
Sun; 11am-midnight Fri, Sat. *Food served* noon-
10pm Mon-Sat; 12.30-10pm Sun. **Main courses**
£9-£17. **Map** p83 G4 ㉖
The Abbotsford is owned by DM Stewart, a long-established Edinburgh publican with a small stable of
very good pubs across the city. This one has genuine Edwardian roots, an attractive island bar and a
particularly ornate ceiling. The tables are big – you'll
probably end up sharing – and, given its location, the
pub can get very busy at peak times. It offers five,
ever-rotating guest ales as well as other beers on
tap and an extensive range of whiskies. The menu
includes a selection of Scottish classics plus a few
more adventurous dishes; if you don't fancy the bustle of the bar, the first-floor restaurant, called Above,
is more genteel.

★ Bramble

*16A Queen Street (226 6343, www.bramblebar.
co.uk). Bus 10, 11, 12, 16, 23, 24, 26, 27, 29,
41, 42, 44, 44A, 61.* **Open** 4pm-1am daily.
Map p83 F3 ㉗
In a discreet sub-basement on Queen Street, this
venue has everything you could ever want from
a contemporary cocktail bar. It's crepuscular,
modestly labyrinthine, there are DJs at the weekend, and the cocktails are among the very best in
Scotland, served by enthusiasts who know their
mixology. If you're looking to pep up your day, a
Mint 500 usually does the trick: Hendrick's gin, elderflower, apple juice, lime juice, mint, basil, peach bit-
ters and vanilla gomme syrup.
▶ *Bramble has two sister bars: the Lucky Liquor Co
(see below), also on Queen Street; and the Last Word
Saloon (see p110) in Stockbridge.*

★ Café Royal Circle Bar

*19 West Register Street (556 1884, www.cafe
royaledinburgh.co.uk). Princes Street buses.*
Open 11am-11pm Mon-Wed; 11am-midnight
Thur; 11am-1am Fri, Sat; 12.30-11pm Sun.
Food served 11am-10pm Mon-Sat; 12.30-10pm
Sun. **Main courses** £13-£35. **Map** p83 H4 ㉘
In situ since the 1860s, an island bar dominates this
attractive and elegant pub, where the walls are decorated with Royal Doulton tiles of famous inventors.
You find all sorts here, from tourists to people dropping in after work. It gets very busy on Saturdays
as people seek respite from shopping, and it's still a
favoured haunt of the rugby crowd on days when
Scotland are playing at Murrayfield. The good food
majors on fish and seafood, with oysters, Shetland
mussels and scallops, but also beef pie and burgers.
▶ *For a more formal dining experience, try
the beautiful Café Royal Oyster Bar at no.19A
(556 4124).*

Dome

*14 George Street (624 8624, www.thedome
edinburgh.com). Princes Street buses.* **Open** *Grill
Room* noon-11pm Mon-Wed; noon-midnight Thur;
noon-1am Fri-Sun. *Food served* noon-9.45pm Mon-
Fri; noon-10.45pm Sat, Sun. **Main courses** £13.50-
£32. **Map** p83 G4 ㉙
In the 1840s, the Commercial Bank of Scotland built
this elaborate old pile as its head office. There's a
striking classical frontage, and the whole thing is
crowned by a dome, now housing a grand bar and
restaurant called the Grill Room. The building also
has a separate and more discreet but very ornate dining space called the Club Room (call for details), and
other nooks and crannies. If nothing else, it's worth
popping into the Grill Room for a quick drink, just to
look at the ostentation.

★ Guildford

*1-5 West Register Street (556 4312, www.guildford
arms.com). Princes Street buses.* **Open** 11am-11pm
Mon-Thur; 11am-11.30pm Fri, Sat; 12.30-11pm Sun.
Food served noon-2.30pm, 5.30-9.30pm Mon-Thur;
noon-2.30pm, 5.30 10pm Fri; noon-10pm Sat; 12.30-
3pm, 5.30-9.30pm Sun. **Map** p83 H4 ㉚
This has been a pub since 1898 – a very grand pub
at that, which might even justify the word opulent
with its ornate ceiling, decorative touches, gallery
and attractive bar. The cask ales come from all over
the British Isles, so you're as likely to find something
local to Edinburgh as you are a well travelled guest
ale from Devon. Good menu, wonderful signature
sausages and very handy for Waverley Station.
▶ *Scotland's People Centre, the genealogical
research facility, is next door; see p88.*

Juniper

*Hotel Indigo, 20 Princes Street (652 7370, www.
juniperedinburgh.co.uk). Princes Street buses.*
Open noon-1am daily. *Food served* noon-10pm
daily. **Map** p83 G4 ㉛
The Cairn Group, which owns and runs hotels across
the UK, acquired these premises on Princes Street
in 1997, but it took until 2012 for the company to
announce some serious investment in the property.
The popular and award-winning Juniper subsequently arrived in 2013, a first-floor cocktail bar with
great views, inventive cocktails and a snacky 'street
food' menu. The hotel then promptly morphed into a
branch of Hotel Indigo in 2015 but Juniper remained
part of the plan to seduce new customers.

Lucky Liquor Co

*39A Queen Street (226 3976, www.luckyliquorco.
com). Bus 10, 11, 12, 16, 23, 24, 27, 29, 41, 42, 61.*
Open 4pm-1am daily. **Map** p83 E3 ㉜
Sister venue to Bramble (*see above*), the Lucky
Liquor Co has 13 bottles of base spirits behind the
bar and uses them to conjure up 13 cocktails, the list
changing every three months or so. The approach is
creative and experimental, the decor is simple and

EXPLORE

NEW TOWN SHOPPING

What's where.

The New Town is the heart of Edinburgh's shopping culture, especially along Princes Street. You'll find many big names from the British high street here: department stores such as Debenhams and House of Fraser, fashion favourites including H&M and Gap, a huge branch of Waterstones bookstore and newer arrivals such as Apple. The street has suffered from a certain malaise in recent years, and the Scottish souvenir stores pumping out electro-bagpipe music haven't helped; however, it still draws the crowds.

Stroll north from here up Frederick or Hanover streets, into George Street, and the landscape changes immediately. In Georgian and Victorian buildings that once housed banking institutions and offices, you'll find the likes of Hollister, Fat Face, Jack Wills, Karen Millen, Phase Eight and much, much more. Running parallel to Princes Street, George Street has any number of these designer brands, beauty stores and other enterprises; scattered in between them are a decent selection of bars and cafés, as well as the ubiquitous coffee franchises.

Keep heading north from George Street, down Dundas Street, and the New Town offers an array of independent galleries for those interested in taking home some art. By contrast, beyond the east end of George Street, through St Andrew Square, stands Harvey Nichols. It's the gateway to Multrees Walk: one of Edinburgh's newer streets and lined with upscale designer stores such as Michael Kors, Burberry and Sandro.

West of George Street, meanwhile, sits William Street in the West End, home to some gorgeous independent boutiques offering stylish women's fashions.

Across the New Town are shops you will know and shops that exist in other cities, but also one-offs that could only be found in the Scottish capital. For anyone interested in retail, it's a rewarding area to browse.

---

stripped-down, and the staff are dedicated to mixing great drinks. A typical cocktail would be Our Lady: Laphroaig Quarter Cask single malt, home-made fig brandy, lemon juice and Lady Grey tea syrup.

Oxford Bar

8 Young Street (539 7119, www.oxfordbar.co.uk). Bus 61. **Open** 11am-midnight Mon-Thur; 11am-1am Fri, Sat; 12.30-11pm Sun. **Map** p83 E4 ③
Cramped, dowdy and clannish, the Oxford's bar area offers no space whatsoever; although there's a rather plain room adjacent with seating, it's all pretty basic, and it sometimes feels as if you have to be a member to drink here. Why bother? Because it ploughs its own furrow, and doesn't give a monkey's about the white noise of contemporary style. It also enjoys minor celebrity status as a favoured haunt of Ian Rankin's Detective Inspector Rebus (and, for that matter, Rankin himself).

Whigham's Wine Cellars

13 Hope Street, Charlotte Square (225 8674, www.whighams.com). Princes Street buses. **Open** noon-midnight Mon-Thur; noon-1am Fri, Sat; 12.30pm-midnight Sun. *Food served* noon-9.45pm daily. **Main courses** £9-£19. **Map** p82 D5 ③
In business for more than 30 years, Whigham's is divided into two interlinked areas: there's an actual old cellar with atmospheric, candlelit alcoves, as well as a more modern, open space. Eat or drink in either. Snacks include tapas such as squid with garlic mayo, nachos and similar, and even snackier canapés. The main menu might bring dishes like a smoked salmon starter, then roast beef salad with new potatoes as a main.

Shops & Services

★ 21st Century Kilts

48 Thistle Street (220 9450, www.21stcentury kilts.com). Bus 6, 10, 11, 12, 16, 23, 24, 27, 29, 41, 42, 61, 100. **Open** 10am-6pm Tue-Sat. **Map** p83 F4 ③ **Kilts**
Howie Nicholson's take on the kilt has helped alter the garment's image. Here, you can find ready-to-wear versions for men, such as the Cheviot (chic black), a denim kilt or one in Harris Tweed. Women's kilts can be short enough for clubbing, long enough for a formal occasion; they do kids' versions too.

Anta

117-119 George Street (225 9096, www.anta. co.uk). Bus 10, 11, 12, 16, 24, 29, 41, 42, 61 or Princes Street buses. **Open** 10am-6pm Mon-Wed, Fri, Sat; 10am-7pm Thur; 11am-5pm Sun. **Map** p83 E4 ③ **Homewares**
Run by designers Annie and Lachlan Stewart, Anta sells ceramics, fabrics, furniture, household items and rugs. Clearly Scottish, it's restrained and tasteful and the materials are sourced locally: wool from the Western Isles, woven in the Borders; oak furniture made in the Highlands.

Archipelago Bakery
39B Queen Street (07932 462715 mobile, www.
archipelagobakery.co.uk). Bus 10, 11, 12, 16,
23, 24, 27, 29, 41, 42, 61. **Open** *9am-2.30pm*
Mon-Sat. **Map** p83 E3 🟢 **Food & drink**
A tiny basement bakery where you would least
expect to find one, which offers great organic breads,
cakes, savouries, soup and quiche to take away.

Arran Aromatics
46A George Street (220 1012, www.arran
aromatics.com). Bus 10, 11, 12, 16, 24, 29, 41,
42, 61 or Princes Street buses. **Open** *9am-5.30pm*
Mon-Sat; noon-5pm Sun. **Map** p83 F4 🟢 **Health &**
beauty
Arran Aromatics started life on the island of Arran
off the west coast in the late 1980s and has since
grown into a business with outlets across Scotland.
With bath and body products, fragrances and
scented candles it has been one of the country's more
successful brands for more than 25 years.

Hamilton & Inches
87 George Street (225 4898, www.hamilton
andinches.com). Bus 10, 11, 12, 16, 24, 29, 41,
42, 61 or Princes Street buses. **Open** *9.30am-*
5.30pm Mon-Fri; 9am-5pm Sat. **Map** p83 E4 🟢
Accessories
Jewellers and silversmiths since 1866, holding a
royal warrant from Her Maj since 1955, smart George
Street premises (and another store in Knightsbridge,
London): Hamilton & Inches is not where you buy
a ten-quid friendship bangle. It's more for brands
such as Wellendorff or Stephen Webster, diamond
set wedding bands costing the best part of £3,000

or a Bremont Victory gentleman's watch at nearly
£12,000. That includes VAT, though…

Jane Davidson
52 Thistle Street (225 3280, www.janedavidson.
co.uk). Bus 6, 10, 11, 12, 16, 23, 24, 27, 29, 41,
42, 61, 100. **Open** *9.30am-6pm Mon-Wed, Fri, Sat;*
9am-7pm Thur; noon-5pm Sun. **Map** p83 F4 🟢
Fashion
Founded by Jane Davidson as an Old Town boutique
more than 40 years ago, this shop is now run by her
daughter Sarah Murray and has developed into
a chic New Town store occupying four floors of a
townhouse. It sells top fashions for women including Diana von Furstenberg gowns, Queene and Belle
knitwear, Escada coats, Roksanda Ilincic skirts and
a range of accessories besides.

Joseph Bonnar
72 Thistle Street (226 2811, www.josephbonnar.
com). Bus 6, 10, 11, 12, 16, 23, 24, 27, 29, 41,
42, 61, 100. **Open** *10.30am-5pm Tue-Sat.*
Map p83 F4 🟢 **Accessories**
With beautiful pieces of antique and period jewellery, this shop is praised by both customers and
experts. Pause to admire a 14-carat gold ring with
a large cabochon amethyst, for instance, or an
Etruscan Revival-style bangle in 15-carat frosted
gold. It all looks good enough for a design museum,
and even the prices, drifting through four into five
figures, seem worth it. *Photo p98.*

One World Shop
St John's Episcopal Church, Princes Street
(229 4541, www.oneworldshop.co.uk), Princes

21st Century Kilts.

EXPLORE

Street buses. **Open** 10am-5.30pm Mon-Sat; 11am-5.30pm Sun. **Map** p82 D5 **42** **Gifts & souvenirs**

In the basement of St John's Episcopal Church at the corner of Princes Street and Lothian Road, this shop sells ethical brands such as Divine chocolate, Braintree bamboo socks or Carishea geranium and lemongrass body balm, as well as Indonesian earrings and Indian incense. It's a grotto of ideologically sound consumption and great for buying gifts.

★ Pam Jenkins

41 Thistle Street (225 3242, www.pamjenkins. co.uk). Bus 6, 10, 11, 12, 16, 23, 24, 27, 29, 41, 42, 61, 100. **Open** 10am-5.30pm Mon-Sat; noon-4pm Sun. **Map** p83 F3 **43** **Accessories**

Jimmy Choo, Christian Louboutin, Miu Miu, Gianvito Rossi. Sold on the idea of this place yet? It's designer heaven for serious shoe shoppers, both men or women – although, given the range, the bags and other accessories, women get more choice. If you're taken with the idea of splashing out the best part of £500 on a pair of burgundy metallic pumps, this is where to come.

Rogerson Fine Footwear

126-128 Rose Street (220 1775, www.rogerson shoes.com). Princes Street buses. **Open** 9.30am-6pm Mon-Wed, Fri, Sat; 9.30am-7pm Thur; noon-5pm Sun. **Map** p83 E4 **44** **Accessories**

This Scottish family business caters to both ladies and gents, selling the likes of Accatino shoes or Peter Kaiser ankle boots for women, and Crockett & Jones brogues or Ecco boots for men. Whether you're looking for smart, casual or a particularly toasty slipper, you'll find it here – the shop also has a range of designer handbags.

Stringers

13 York Place (557 5432, www.stringersmusic. com). Bus 10, 11, 12, 16, 26, 41, 44, 44A. **Open** 9.30am-5.30pm Mon-Sat. **Map** p83 G3 **45** **Musical instruments**

There are people who've lived in Edinburgh for decades and never heard of this shop. Given that it sells cellos, double basses, violas and violins, also bows and accessories – and it's on York Place, where retail outlets hardly clutter the streetscape – that can be no surprise. The staff are experts, and all customers are welcomed, from beginners to professionals.

★ Tiso

123-125 Rose Street (225 9486, www.tiso.com). Princes Street buses. **Open** 9.30am-5.30pm Mon-Tue, Fri, Sat; 10am-5.30pm Wed; 9.30am-7.30pm Thur; 11am-5pm Sun. **Map** p83 E4 **46** **Sport & outdoors**

Tiso has been trading as long as some middle-aged Scottish hillwalkers have been alive. For decades, it's been the first port of call for decent boots, rucksacks, tents, sleeping bags, waterproofs, ski gear, watersports equipment and more. It has branches across Scotland; this outlet has four floors of anything you could conceivably need to venture out on to the hills, lochs or snow.

Other location 41 Commercial Street, Leith (554 0804).

Xile

12 Frederick Street (225 3390, www.xileclothing. com). Princes Street buses. **Open** 9am-6pm Mon-Wed; 9am-8pm Thur; 9am-7pm Fri, Sat; 11am-6pm Sun. **Map** p83 F4 **47** **Fashion**

Xile is a Scottish business that's been going for more than 30 years, providing label clothes for the young

Joseph Bonnar. *See p97.*

EXPLORE

gent about town. It's where to find gear by Adidas, Barbour, Carhartt, Replay or Weekend Offender: jeans, jackets, knitwear, sweats, hoodies, T-shirts. **Other location** Ocean Terminal, Leith (561 4496).

THE SECOND NEW TOWN

Running north, and downhill from Queen Street, **Dundas Street** is the backbone of what was known as 'the Second New Town', a further stage of development in the early 19th century, built as landowners cashed in on the city's need for upmarket dwellings. Scottish property laws allowed them to stipulate the architectural style, so the New Town's cohesive neoclassical formality is played out with little interruption north of Queen Street Gardens to Stockbridge and Broughton.

Resolutely residential and exclusive, these new areas were designed to deter outsiders or Old Town riff-raff. Churches aside, there were originally no public buildings, squares or markets. Two centuries later, shops and restaurants have long been open along the main roads and Dundas Street itself has more than half a dozen small, commercial art galleries, but this area remains the New Town's residential heart. The best way to explore is simply follow your feet.

The immediate locale has a literary heritage. At **17 Heriot Row**, look for the stone-carved inscription commemorating the fact that author Robert Louis Stevenson lived here as a child. It's said the adjacent gardens influenced his novel *Treasure Island*, and the gas light that once stood outside the house inspired his poem 'The Lamplighter'. Nearby, JM Barrie, author of *Peter Pan*, lodged at **3 Great King Street** when he was a student. Just beyond lies **Drummond Place**, a crescent-shaped street once called home by *Whisky Galore* author Compton Mackenzie (no.31). And just over the intersection is **Scotland Street**, the setting for best-selling Edinburgh writer Alexander McCall Smith's novel *44 Scotland Street*.

Further west is the Moray Estate. Designed by James Gillespie Graham in 1822 at the behest of the Earl of Moray, it's one of the grandest of the New Town's residential quarters. Its focus is **Moray Place**, less than 100 yards from the north-western corner of Queen Street Gardens. This is a remarkable circus, with handsome neoclassical frontages incorporating dozens of columns, and a private, central garden providing the splash of green that offsets the greying sandstone and diluted honey colour of the residences. It may look like a film set, but people really do live here.

Restaurants

£ Leo's Beanery

23A Howe Street (556 8403, www.leosbeanery. co.uk). Bus 24, 29, 42, 61. **Open** 8.30am-5pm Mon-Fri; 9am-5pm Sat; 10am-5pm Sun. **Main courses** £6.50-£8. **Map** p83 E3 ④ **Café**

Sometimes all you want is a small and friendly basement café where you can have a decent coffee, a light meal, sandwich or bowl of soup. Leo's Beanery is then your destination of choice in this stretch of the New Town if you're out exploring the architecture and general splendour. Good breakfasts too.

Stac Polly

29-33 Dublin Street (556 2231, www.stacpolly. com). Bus 10, 11, 12, 16, 26, 41, 44, 44A. **Open** 6-9.30pm Mon-Thur, Sun; 6-10pm Fri, Sat. **Main courses** £18-£27. **Map** p83 G2 ④ **Franco-Scots**

For over 25 years, Stac Polly has made a virtue of its Scottishness with a traditional French twist, offering the likes of haggis in filo with sweet plum and red wine sauce as a starter; main courses are built around Borders lamb, pheasant or Aberdeen Angus beef. In culinary terms it may not be cutting edge, but the New Town basement premises, the atmosphere and the resonance do impress visitors. **Other location** 38 St Mary's Street, Old Town (557 5754).

Pubs & Bars

Clark's Bar

142 Dundas Street (556 1067). Bus 23, 27, 36, 61. **Open** 11am-11.30pm Mon-Thur, Sun; 11am-12.30am Fri, Sat. **Map** p83 E1 ⑤

Sparse and traditional, this old howf (read: cosy pub) opened in 1899 and hasn't changed in years: it still features red leather seats, shiny brass table tops and a dark red ceiling. You'll find a reasonable malt whisky selection, cask ales, football on the telly and live music some nights: an antidote to the excesses of George Street up the hill.

★ Cumberland

1-3 Cumberland Street (558 3134, www. cumberlandbar.co.uk). Bus 23, 27, 61. **Open** noon-midnight Mon-Wed; noon-1am Thur-Sat; 11am-midnight Sun. *Food served* noon-3pm

EXPLORE

Mon, Tue; noon-9pm Wed-Sat; 12.30-6pm Sun.
Main courses £9-£13. **Map** p83 F2 🟡

In the residential heart of the New Town, the Cumberland almost looks out of place among the surrounding townhouses. But with its attractive wooden interior, good cask ale, superior pub food (including sharing platters, sandwiches and mains such as smoked haddock with black pudding, duck egg and butter sauce) plus the adjacent sunken beer garden for summer evenings or warm afternoons, it remains one of the city's finer pubs.

Kay's

39 Jamaica Street (225 1858, www.kaysbar.co.uk). Bus 24, 29, 42, 61. **Open** 11am-midnight Mon-Thur; 11am-1am Fri, Sat; 12.30-11pm Sun. *Food served* noon-2.30pm Mon-Sat. **Map** p83 E3 🟡

Drink has been a mainstay here for nearly two centuries: before morphing into a pub in 1976, these premises housed a wine merchant. Now, along with a reputation as a patrician New Town howf, this historic spot – with predominantly red decor – offers an excellent choice of single malts and a perfect environment in which to sample them. Typical bar food choices would be pie, chips and beans; curry and rice; mince and tatties – all at budget prices.

Spit/Fire

26B Dublin Street (556 5967, www.spitfirebars. com). Bus 10, 11, 12, 16, 26, 41, 44, 44A. **Open** noon-1am Mon-Thur; noon-2am Fri; 10.30am-2am Sat; 10.30am-1am Sun. *Food served* noon-10pm Mon-Fri; 10.30am-10pm Sat, Sun. **Main courses** £6.50-£13.50. **Map** p83 G2 🟡

From the makers of the estimable craft beer bar Hanging Bat in South Edinburgh, Spit/Fire appeared at the end of 2014 with two bars in one venue. Upstairs is Spit, more of a bar-restaurant, with craft beer, wines by the glass and a menu where use of an actual rotisserie, big enough for a suckling pig, looms large. The design flourish is tiled bench seating that's heated, while beer casks and kegs make a guest appearance in the toilets. Downstairs, Fire has a snackier menu, a more bar-like demeanour and also craft beer.

▶ *Find the Hanging Bat at 133 Lothian Road, 229 0759, www.thehangingbat.com.*

Shops & Services

Eden

18 North West Circus Place (225 5222, www.eden retail.co.uk). Bus 24, 29, 42, 61. **Open** 10am-4pm Mon; 10am-5.30pm Tue-Sat. **Map** p82 D2 🟡

Fashion

Another of Edinburgh's accessible boutiques, Eden reflects the sensibilities of proprietor Ruth George (in charge since 2006). With a small and smart interior, it stocks a decent range of Spanish labels such as Compañía Fantastica, Kling and Malahierba, as well as styles from Denmark, Italy and the UK.

THE WEST END

The western end of the New Town, popularly known as the West End, is a quiet and pleasant place for a wander. Shandwick Place, Queensferry Street and the women's fashion hub of William Street have shoppers, but much of the rest of the area is almost entirely bereft of activity. Most of the buildings are either high-rent offices or apartments, but one street in particular is astonishing in its outlook. Looking south-west along **Melville Street** to the triple spires of **St Mary's Episcopal Cathedral** reveals a view that would make Baron Haussman, 19th-century renovator of Paris, shed tears of envy.

Nearby, at the west end of Princes Street, the **Waldorf Astoria Edinburgh – The Caledonian** (*see p250*) is a colossal red sandstone and brick edifice built in 1903 as a railway hotel, although the adjacent station is long gone. It was originally the subject of complaints, with locals suggesting that such a vulgarian effort was better suited to Glasgow, but more than a century of respectability has helped it blend in. The unwieldy name comes from its ownership by Hilton (since 2000), the application of Hilton's Waldorf Astoria branding (since 2012) and the building's heritage as the Edinburgh hotel of the long-defunct Caledonian Railway.

Sights & Museums

🆓 St Mary's Episcopal Cathedral

Palmerston Place (225 6293, www.cathedral.net). Bus 2, 3, 4, 12, 25, 26, 31, 33, 44, 44A. **Open** varies daily; check website for services. **Admission** free. **Map** p82 B5 🟡

The foundation stone at St Mary's was laid in 1874, complete with what would now be called a time capsule. The main building was finished and consecrated five years later, but the spires, which form an integral part of Edinburgh's skyline, were not finished until World War I. With its 270-foot central spire, the rose window of the south transept, and the fossils in the granite steps of the high altar, the cathedral is filled with artistic and ecclesiastical points of interest; Edinburgh has many beautiful churches, but perhaps none more beautiful than this. It's open daily for services, so there is access for architecturally curious visitors; choral evensong (5.30pm Mon-Wed, Fri; 3.30pm Sun) can be a haunting, contemplative time.

Restaurants

China Town

3 Atholl Place (228 3333, www.chinatownedinburgh. com). Bus 2, 3, 4, 12, 25, 26, 31, 33, 44, 44A. **Open** noon-2pm, 5.30-11pm Mon, Wed-Sun. **Main courses** £7.50-£19.50. **Map** p82 C6 🟡 **Chinese**

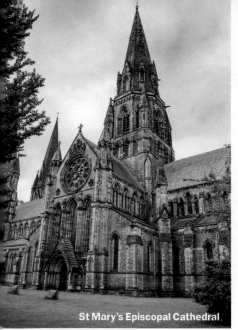

St Mary's Episcopal Cathedral

over the dining rooms, notably the rococo-styled Pompadour, dating from the 1920s. Although the Galvins have their name on the website, the top man behind the kitchen door is the talented Craig Sandle. If the aesthetics of your dining environment would make Louis XV happy, it follows that the menu has classic French overtones too; a typical main course would be roast monkfish with lobster ravioli, caramelised cep, cauliflower and shellfish bisque.

▶ *For less formal dining in the hotel, try the Galvin Brasserie de Luxe (222 8988, www. galvinbrasseriedeluxe.com).*

Restaurant at the Bonham
Bonham Hotel, 35 Drumsheugh Gardens (623 9319, www.thebonham.com). Bus 19, 37, 41, 43, 47, 113. **Open** 6.30-9.30pm Mon, Tue; noon-2.30pm, 6.30-9.30pm Wed-Sat, 12.30-3pm, 6.30-9pm Sun. **Main courses** £16-£18. **Map** p82 B4 ❺❾ **Modern French**
This discreet eatery is part of the Bonham Hotel. The space is wood-panelled and attractive with contemporary touches, while the range of menus includes the Express Lunch (one course plus coffee), the Boozy Snoozy (tables of four only, three courses, two bottles of wine per table), standard à la carte and others. The approach is generally Modern European; a typical main would be beef cheeks with mash, curly kale and bordelaise sauce.
▶ *For the hotel, see p251.*

Shops & Services

Arkangel & Felon
4 William Street (226 4466, www.arkangel andfelon.com). Bus 3, 4, 12, 25, 26, 31, 33, 44, 44A. **Open** 10am-5.30pm Mon-Wed, Fri, Sat; 10am-6.30pm Thur. **Map** p82 C5 ❻⓪
Fashion
One of the city's best-loved boutiques, independently owned Arkangel & Felon is home to a seductive range of chic clothing. The pieces – from names such as Almost Famous, Lilith, Sarah Pacini and others – will wow anyone looking for a more individual look. There's also a great range of designer jewellery, plus bags, shoes and other accessories. It's for those moments when only a beige flocked skater dress will do.

★ Studio One
10-16 Stafford Street (226 5812, www.studio one. co.uk). Bus 3, 4, 12, 25, 26, 31, 33, 44, 44A. **Open** 10am-6pm Mon-Wed, Fri, Sat; 10am-7pm Thur; 11am-5pm Sun. **Map** p82 C5 ❻❶
Accessories/ Gifts & souvenirs/Homewares
Studio One sells homewares from clocks to picture frames, kitchen gadgets, glassware, designer dinner plates, jewellery and candles, as well as assorted toys, and fashion accessories for today's stylish, urban lady. It's a genre-defying gift shop and browsing heaven. People even buy things occasionally.

China Town first opened back in the mid 1980s in Newington in South Edinburgh, but upped sticks and moved to the West End in 2008. It's developed into a local favourite with its Cantonese-focused cooking, avoidance of MSG in dishes and aspirational decor. The full evening menu offers everything from the simplest haddock with ginger and spring onion to a pretty rare and premium-priced dish of abalone with Chinese vegetables in oyster sauce. It's handy for Haymarket Station too.

Edinburgh Larder Bistro
1A Alva Street (225 4599, www.edinburgh larder.co.uk). Bus 19, 36, 37, 41, 43, 47. **Open** 12.30-2.30pm, 5.30-10pm Tue-Sat. **Main courses** £12.50-£20. **Map** p82 D5 ❺❼ **Modern British**
Sister venue to the Edinburgh Larder café (*see p69*) in the Old Town, this bistro opened in summer 2012. People seriously fell in love with it after a 2013 redesign. There are lobster creel lights, rough wooden furniture made from reclaimed scaffolding boards and a menu that delivers such dishes as a hefty steak and chips to halibut with scurvy grass pesto.

Pompadour by Galvin
Waldorf Astoria Edinburgh – The Caledonian, Princes Street (222 8975, www.thepompadour bygalvin.com). Princes Street buses. **Set meal** £58. **Open** 6.30-10pm Wed, Thur; 6-10pm Fri, Sat. **Map** p82 D5 ❺❽ **Modern European**
A massive investment in this distinguished old hotel saw it rebranded in 2012 and the accomplished London restaurateurs the Galvin brothers take

EXPLORE

Stockbridge

Once a small village to the north of Edinburgh by the banks of the Water of Leith, Stockbridge was incorporated into the city thanks to the early 19th-century expansion of the New Town. For a time in the 1970s, its epicentre had the air of a bohemian enclave, although the legacy of that period involves bric-a-brac emporia, charity shops and delicatessens rather than cultural revolution.

The neighbourhood's architecture is as varied as the shopping. The handsome John Hope Gateway at the Royal Botanic Garden Edinburgh opened only in 2009, while the sense of Georgian gentility associated with the area is best reflected in the elegant curve of Royal Circus, a William Playfair creation of 1823, or the small and exclusive Ann Street, co-designed by artist Sir Henry Raeburn in 1814. Conversely, the path along the Water of Leith to Dean Village is an unexpectedly bucolic escape from the city, while the Royal Botanic Garden Edinburgh itself is not only scientifically important but a truly beautiful and contemplative space.

EXPLORE

Scottish National Gallery of Modern Art.

Don't Miss

1 **Royal Botanic Garden Edinburgh** Cultivating beauty since 1820 (p105).

2 **Scottish National Gallery of Modern Art** Paolozzi to Picasso in two imposing buildings (p113).

3 **Stockbridge Market** The most fun you can have here on a Sunday (p109).

4 **Scran & Scallie** A top-notch gastropub (p110).

5 **Dean Village** Your local mill village analogue by the Water of Leith (p112).

FROM NEW TOWN TO THE BOTANICS

You can take the well-trodden New Town path to Stockbridge via Howe Street and South East Circus Place, but there is a short and worthwhile detour. If you continue north along Howe Street where it turns into St Vincent Street (just after Great King Street), you'll get a close-up look at Thomas Playfair's impressive **St Stephen's Church** (1828), which dominates the view all the way downhill from the city centre. The huge arched entrance was Playfair's answer to the awkward shape and sloping ground of the site.

Bear left by the church and you'll be at the less interesting end of St Stephen Street. Continue along it, though, and things soon get more engaging. Along St Stephen Street, round the corner into Kerr Street, Deanhaugh Street and into **Raeburn Place** (Stockbridge's main drag), there's a huge concentration of independent shops selling everything from artisan cheeses to chandeliers, CDs to children's clothes, designer jewellery to interior design items. Around this stretch there are also more than a dozen charity shops, including specific book and music outlets run by Oxfam (25 and 64 Raeburn Place, 332 9632 and 332 7593, www.oxfam.org.uk), as well as any number of bars, cafés, delicatessens, pubs and restaurants.

There is more to Stockbridge than eating and shopping, however, as you'll see if you wander up to the corner of India Place and Gloucester Street (opposite the west end of St Stephen Street). Built in the 1790s using stones recovered from buildings in the Lawnmarket that had been demolished during construction of the Mound, **Duncan's Land** was the birthplace of artist David Roberts (1796-1864), who specialised in visions of souks, monuments of Egypt, vistas of the Holy Land, and who was known for dressing in exotic clothing. Look out for the higgledy-piggledy doors and windows, and the lovely lintel that reads 'Fear God Onlye' (now above the front door of a Thai restaurant).

Another memento of the area's past can be found in the form of the gateway that once led to Stockbridge's meat and vegetable market. It now stands in attractive isolation off St Stephen Street, at St Stephen Place, but still announces the availability of 'Butcher Meat, Fruits, Fish and Poultry'. Built in 1826 after a public campaign, the market was a poke in the eye for city officials who hoped that Stockbridge would leave such undignified practices behind and remain market-free, much like the neighbouring New Town. Although that market is long gone, the good news is that another started locally in 2011 (see p109 **Don't Quit Your Stalling**).

Where Deanhaugh Street crosses the Water of Leith, down the stairs next to a branch of Pizza

Express, you can join the waterside path that comes out at Bridge Place and takes you along to the **Stockbridge Colonies** (off Glenogle Road), more evidence of the area's less patrician past. Following the curve of the river, this group of 'one-ended' streets were conceived by the Edinburgh Co-operative Building Company and named after its members and supporters. The design of the buildings is two-storey but double-sided: the entrance to the upper dwelling is accessed from stairs on one street, with access to the ground floor from the street on the opposite side. Built between 1861 and 1911 as homes for artisans and other workers, their cutesiness now makes them highly desirable.

At the bottom of one of the colony streets, Bell Street, you can pick up the Water of Leith Walkway again, heading towards **Canonmills**, where Stockbridge, Broughton and the New Town meet. There are a cluster of eateries here, not far from the East Gate of the **Royal Botanic Garden Edinburgh**, among them the **Bluebird Café** and the **Water of Leith Café Bistro**.

On the other side of the river from the colonies, Arboretum Avenue takes you towards Arboretum Place and the West Gate of the Royal Botanic Garden Edinburgh, home to **Inverleith House**.

Sights & Museums

FREE Inverleith House
Royal Botanic Garden, Inverleith Row (248 2971, www.rbge.org.uk). Bus 8, 23, 27 to the East Gate. **Open** *Mar-Sept* 10am-5.30pm Tue-Sun. *Feb, Oct* 10am-4.30pm Tue-Sun. *Nov-Jan* 10am-3.30pm Tue-Sun. **Admission** free. **Map** p106 B4 ❶
Set bang in the middle of the botanical gardens, this impressive, four-square Georgian building was designed in 1774 by David Henderson for James Rocheid, whose family owned the Inverleith estate. The house eventually became the home of the garden's Regius Keeper and then, from 1960 until 1984, the Scottish National Gallery of Modern Art, but it's now owned and run as an art gallery by the Royal

Botanic Garden Edinburgh. Exhibitions are largely contemporary; a highlight in 2013 was Mostly West, with pieces by the late Austrian artist Franz West made in collaboration with the likes of Douglas Gordon, Sarah Lucas and others.

★ FREE Royal Botanic Garden Edinburgh

Inverleith Row (248 2909, www.rbge.org.uk).
Bus 8, 23, 27 to the East Gate. **Open** *Gardens &*
John Hope Gateway Mar-Sept 10am-6pm daily.
Feb, Oct 10am-5pm daily. Nov-Jan 10am-4pm daily.
Glasshouses Mar-Sept 10am-5.30pm daily. Feb, Oct
10am-4.30pm daily. Nov-Jan 10am-3.30pm daily.
Admission *Gardens & John Hope Gateway* free.
Glasshouses £5; £4 reductions; free under-15s.
Map p106 B4 ❷

The Royal Botanic Garden Edinburgh, aka the 'Botanics', has delighted both plant-lovers and casual strollers since it was sited here in 1820, although the history of botanical gardens elsewhere in the city goes back to the 17th century. Edinburgh's most peaceful tourist attraction, the garden is also a noted centre for botanical and horticultural research, and houses the oldest botanical library in the British Isles. Access to its grand Victorian palm house is free, but it costs to delve deeper into the themed zones of the 1960s-era glasshouses. Still, it's worth the modest fee for the privilege of viewing all kinds of orchids, ferns, tropical plant life and rainforest species. The Plants & People glasshouse, for example, displays bananas, cocoa, rice and sugar as vegetation; the pond in the middle is covered with freakishly large water lily leaves every summer.

There's also the John Hope Gateway at the West Gate, a carbon-neutral biodiversity and information centre that includes a restaurant (552 2674, www. gatewayrestaurant.net), a shop, a plant nursery and education rooms. With a wood and slate exterior and a rooftop wind turbine, the Gateway is the Botanics' leap into the 21st century. Another highlight is the gorgeous Aeolian harp created by harp maker Mark Norris. Made from wych elm felled here in 2003, it was on display in 2009, then returned permanently in 2012 in its own pavilion, 'to be played by the wind'.

The Botanics is a controlled space (no ball games, no bikes, no dogs, no jogging). However, during the Edinburgh International Festival and Festival Fringe, it often hosts outdoor art installations. And year-round there's plenty that appeals to children. The pond, with its waterfowl, and the waterfall in the rock garden area, where a heron sometimes fishes, are particularly popular.

The Terrace Café (552 0606), by Inverleith House, is good for a quick coffee and a snack; if you choose to sit outside, you may be pestered by pigeons or squirrels or even seagulls. But even the mighty gulls are intimidated by Edinburgh's middle-class mothers of young children, many of whom stroll here in all weathers.

Restaurants

Bell's Diner

7 St Stephen Street (225 8116). Bus 24, 29, 36,
42, 61. **Open** 6-10pm Mon-Fri, Sun; noon-10.15pm
Sat. **Main courses** £8-£15. **Map** p107 E4/5 ❸
North American

Royal Botanic Garden.

EXPLORE

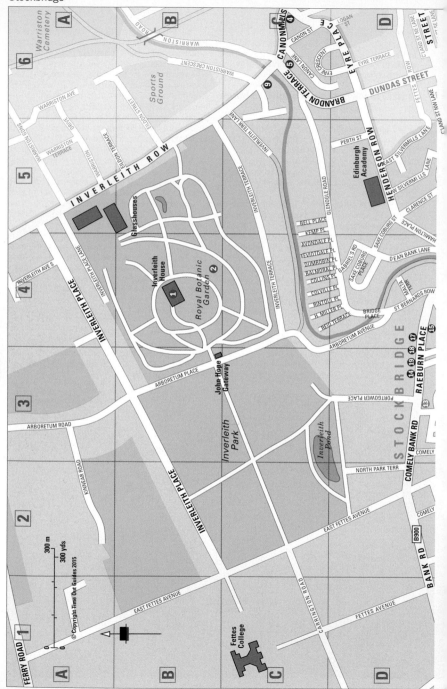

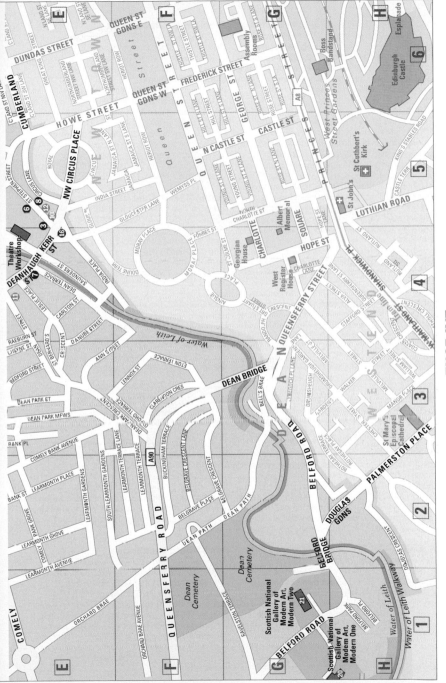

EXPLORE

Bell's Diner has been pursuing its simple formula on St Stephen Street since 1972, which should tell potential customers something about the effectiveness of the approach. It's a small, simple room that sells fabulous burgers (including vegetarian ones), good steaks and hearty desserts to delight your inner child. No frills, no messing.

Bluebird Café

5 Canonmills (07749 971847 mobile, www.the bluebird cafe-edinburgh.com). Bus 8, 23, 27, 36. **Open** 6am-4pm Mon-Fri; 8am-5pm Sat, Sun. **Main courses** around £6. **Map** p106 C6 ❹ **Café**
Tiny, approachable and less than ten minutes' walk from the East Gate of the Royal Botanic Garden Edinburgh, the Bluebird opened in 2014 and built an immediate following thanks to its enthusiastic staff and great home cooking. The sandwiches, scones and soups are all worth sampling; you can sit in or take away. The friendly atmosphere was inspired by the owner's motorbike travels around the US.

Circle by Di Giorgio

1 Brandon Terrace (624 4666, www.thecircdecafe. com). Bus 8, 23, 27, 36. **Open** 8.30am-5pm Mon-Thur, Sun; 8.30am-10pm Fri, Sat. **Main courses** £9-£12. **Map** p106 C6 ❺ **Café**
There's been a small, smart café here for many years, but it was taken over in 2013 by Nadia and Ivan Di Giorgio. It remains a daytime café (coffees, breakfast, sandwiches, sharing plates, hot dishes), but also opens on Friday and Saturday evenings. More substantial meals might have a Mediterranean flavour: lasagne or marinated lamb kebab, for example.

Purslane

33A St Stephen Street (226 3500, www.purslane restaurant.co.uk). Bus 24, 29, 36, 42, 61. **Open** noon-2pm, 6-11.30pm Tue-Sun. **Set meal** £24.95, £29.95. **Map** p107 E5 ❻ **Modern European**
A small basement restaurant where overhearing the next table's conversation is almost inevitable; fortunately, the thoughtful and well-presented food focuses attention back to the matter in hand. A sculpture of guinea fowl with roast salsify, broccoli, carrot purée and jus gras is a typical main course. The two- or three-course set meals are good value, and there's a tasting menu.

Rafael's

2 Deanhaugh Street (332 1469, www.rafaels-bistro.wikidot.com). Bus 24, 29, 36, 42, 61. **Open** 6.30-10pm Tue-Sat. **Set meal** £22.50, £26.50. **Map** p107 E4 ❼ **Spanish**
Ever get that feeling where you just want to go somewhere cosy and welcoming, slump at a table and order up some Spanish-flavoured food and wine? Rafael's offers precisely that, and has done for more than a decade. It's not one of the city's higher-profile eateries, but the affability, the blackboard specials and the ambience often hit the spot.

Stockbridge Restaurant

54 St Stephen Street (226 6766, www.the stockbridgerestaurant.com). Bus 24, 29, 36, 42, 61. **Open** 7-9.30pm Tue-Sat; 7-9pm Sun. **Main courses** £21-£25. **Map** p107 E5 ❽ **Modern European**
Another small basement venue, though very nicely appointed with bare stone and much art, this well-established restaurant pulls off classic haute cooking with a contemporary twist. This could bring herb-crusted cod with warm potato salad and white wine tartar sauce as a starter; roast grouse with bacon, bread sauce, game chips and watercress as a main. You pay for this kind of quality, but there is a set menu too (two courses £20.95, three £24.95).

Water of Leith Café Bistro

1 Howard Street (556 6887, www.thewater ofleithcafebistro.com). Bus 8, 23, 27, 36. **Open** 9.30am-5pm Tue-Fri; 10am-5pm Sat; 10am-4pm Sun. **Set meal** £12, £14.50. **Map** p106 C6 ❾ **Café**
Down by the Water of Leith at Canonmills, this café-bistro has the expertise of Mickael Mesle in the kitchen, and the ever-friendly Ana Mesle running front of house. It functions as a café from breakfast through to afternoon coffee and cake, but the cooking is seriously accomplished: two courses at lunchtime brings dishes such as salmon and smoked mackerel roulade to start, much-accessorised slow-roast pork belly as a main. Formerly in Leith, they moved here in 2014; the clients kept coming.

Pubs & Bars

Bailie

2 St Stephen Street (225 4673, www.thebailie bar.co.uk). Bus 24, 29, 36, 42, 61. **Open** 11am-midnight Mon-Thur; 11am-1am Fri, Sat; 12.30pm-midnight Sun. *Food served* 11am-10pm Mon-Sat; 12.30-10pm Sun. **Main courses** £9-£17. **Map** p107 E5 ❿
This old-style basement pub somehow combines New Town money with Stockbridge bohemia. The beer includes mainstream cask ales and some offbeat keg, while food ranges from cajun chicken breast, steak and teriyaki pork chops to more typical pub grub.

Bon Vivant Stockbridge

4-6 Dean Street (315 3311, www.bonvivant-stockbridge.co.uk). Bus 24, 29, 36, 42, 61. **Open** noon-1am daily. *Food served* noon-10pm daily. **Main courses** £14-£17.50. **Map** p107 E4 ⓫
Part of the Bon Vivant's Edinburgh empire (which includes another bar-restaurant, a bar-kitchen, a bottle shop and a café-takeaway), this place is many things to many people. You could pop in for brunch, for a snack and a glass of wine, for a full-blown dinner or just for a drink. Tucked away up a side street, it's easy to miss.

DON'T QUIT YOUR STALLING

Weekly street markets are back in business.

Edinburgh Farmers' Market.

You can blame the inexorable, post-war rise of supermarkets, alienation from the countryside or the weather. Whatever the cause, the tradition of outdoor food markets faded a long time ago in most parts of Scotland, including Edinburgh.

Towards the end of the 1990s, however, the local council felt that reviving some sort of Saturday market could kill two birds with one stone: bring fresh, healthy food into the heart of the city on the busiest shopping day of the week, and give small Scottish farmers and producers another outlet for their wares. The irresistible synergy of economic development and 'eating local' saw the birth of **Edinburgh Farmers' Market** (www.edinburghfarmersmarket.co.uk) in 2000. It's now held every Saturday (9am-2pm), just south-west of Castle Rock, at Castle Terrace, South Edinburgh. There are usually around 50 stalls selling everything from mini-kegs of cask ale to organic chickens, fresh herbs to superior cheese on toast for eating on the spot. Anyone seeking an iconic Scottish purchase won't have to look too hard for smoked salmon or venison, but the market also springs some endearing surprises: buffalo steaks from Fife, for instance.

The good news, for locals and visitors, is that its success kicked off an enthusiasm for weekly markets elsewhere. This is most evident at **Stockbridge Market** (10am-5pm Sun, Jubilee Gardens, Stockbridge, www.stockbridgemarket.com). Launched in 2011,

it regularly has more than 40 stalls featuring great street food from the likes of **Union of Genius** (*see p75*) or Glasgow's **Babu Bombay** (*see p198*), as well as fresh produce and gifts.

Also in 2011, back in the city centre, a small food market was launched in the pedestrianised area between Leith Street and York Place: the **Tram Stop Market** (11am-5pm Sat, outside St Mary's Catholic Cathedral, New Town, www.edinburgh markets.co.uk). In the Old Town, the **Grassmarket Weekly Market** (10am-5pm Sat, Grassmarket, www.greatergrassmarket.co.uk) is useful if you fancy some more browsing after the nearby Edinburgh Farmers' Market closes.

Even Waverley Station got in on the act in 2014, with **Waverley Market @ Platform 2** (11am-7pm Fri, www.localmotivemarkets.co.uk). Shop for street food, gifts and crafts while you're waiting for a train.

Grassmarket Weekly Market.

Henri's of Edinburgh.

Other locations Bon Vivant, 55 Thistle Street, West Edinburgh (225 3275); The Devil's Advocate, 9 Advocates Close, Old Town (225 4465); Pep & Fodder, 11 Waterloo Place, Calton Hill (556 5119).

Last Word Saloon

44 St Stephen Street (225 9009, www.lastword saloon.com). Bus 24, 29, 36, 42, 61. **Open** 4pm-1am daily. **Map** p107 E5 ⑫

These premises have pre-history as a bar, but the Last Word arrived in 2012 bringing a sense of fun to cocktails and mixology. Here you can have Black Bottle blended whisky with ice cream soda or go for a Last Word itself, a Detroit classic from the 1920s: Beefeater Gin, Green Chartreuse, Maraschino liqueur and fresh lime juice. This is a basement bar that feels like a speakeasy, but more irreverently creative.

▶ *The Last Word is owned by the same people as Bramble and the Lucky Liquor Co, both in New Town; see p95.*

Scran & Scallie

1 Comely Bank Road (332 6281, www.scran andscallie.com). Bus 24, 29, 42. **Open** 11am-1am daily. *Food served* noon-3pm, 6-10pm Mon-Fri; noon-10pm Sat, Sun. **Main courses** £9.50-£22. **Map** p106 D3 ⑬

Billing itself as 'a public house with dining', S&S is basically a small bar with a much larger dining area attached. The beer is good and the venue's pedigree

is impeccable; it comes from the same stable as local Michelin-starred restaurants Castle Terrace (*see p132*) and the Kitchin (*see p160*). Consequently, the pub-style menu offers everything from sausage and mash to whole roast partridge. Once you've had dinner, you can still hang around for a drink or two.

Shops & Services

Armstrong's

80 Raeburn Place (315 2033, www.armstrongsof stockbridge.co.uk). Bus 24, 29, 42. **Open** 7am-5.30pm Tue-Fri; 7am-5pm Sat. **Map** p106 D3 ⑭
Food & drink

A traditional Scottish fishmonger trading since the 1940s, Armstrong's endures thanks to the quality of its produce, its service, the expertise of staff and its speciality smoked salmon. You can get all the expected North Atlantic and North Sea species, and also more exotic, foreign-landed species from the Indian Ocean or the South Atlantic. It's all fresh, though; that crab in the window is still moving.

Bliss

5 Raeburn Place (332 4605, www.bliss-stockbridge.co.uk). Bus 24, 29, 42, 61. **Open** 9.45am-6pm Mon; 9.30am-6pm Tue-Fri; 10am-6pm Sat; 11am-6pm Sun. **Map** p106 D4 ⑮
Gifts & souvenirs

This is a small, tidy shop full of cute things. There are handmade greetings cards, soft toys, baby shoes, soaps, candles, handbags by Ness, jewellery by Watch This Space, children's nightwear by Powell Craft and much else besides.

Edinburgh Floatarium
29 North West Circus Place (225 3350, www. edinburghfloatarium.co.uk). Bus 24, 29, 36, 42, 61. **Open** 10am-3pm Mon; 10am-8pm Tue-Fri; 10am-6pm Sat; 10am-5pm Sun. **Map** p107 E4 ⑯
Health & beauty
Edinburgh isn't short of spa facilities, but the Floatarium enjoys something of a unique reputation as it was the first – and remains the only – local spa to offer a floatation experience. Aside from the get-in-a-tank business, it also offers massages, facials, reflexology, waxing and all the other usual spa treatments.

★ Galerie Mirages
46A Raeburn Place (315 2603). Bus 24, 29, 42. **Open** 10am-5.30pm Mon-Sat; noon-4.30pm Sun. **Map** p106 D4 ⑰ **Accessories**
To describe this small spot as a jewellery shop doesn't quite do it justice. Set back from the main street, under a little arch that reads 'Specialists in ethnic art', it's a treasure chest of jewellery, scarves, objets and more. You'll find pieces using amber, silver and semi-precious stones; although some items are premium-priced, there's a great deal that is eminently affordable.

★ Henri's of Edinburgh
48 Raeburn Place (332 8963, www.henrisof edinburgh.co.uk). Bus 24, 29, 42. **Open** 9am-7pm Mon-Thur, Sat; 9am-11pm Fri; 9am-6pm Sun. **Map** p106 D3 ⑱**Food & drink**
Henri's moved here from Morningside in 2011, to provide some French-slanted competition to Herbie

a few doors along, particularly on the cheese and charcuterie front. It also incorporates a café: opt for a deli platter or decent sandwich, and a glass of wine, and you'll be a happy snacker.

★ Herbie of Edinburgh
66 Raeburn Place (332 9888, www.herbie ofedinburgh.co.uk). Bus 24, 29, 42. **Open** 9am-7pm Mon-Fri; 9am-6pm Sat. **Map** p106 D3 ⑲
Food & drink
Ploughing the good food furrow for more than 20 years, Herbie's prides itself on its breads, brie de meaux, chicken liver pâté and smoked ham. It also offers all the usual deli standards, including wine, oil, vinegar, fine jam, chutney, cold meats, nibbles and more besides. Not much space, lots to look at.

AROUND STOCKBRIDGE

Back in Stockbridge proper, at the top of Leslie Place (off Deanhaugh Street), sits **St Bernard's Crescent**. It's all that remains of the Raeburn Estate, a grandiose property development financed by artist and Stockbridge native Sir Henry Raeburn (*see p112* **In the Know**). Although St Bernard's House, the central focus of the development, is long gone, the Georgian crescent remains a fine example of neoclassical elegance. The central section, with Doric columns, is unexpectedly impressive. It's a regular star of period dramas.

Nearby **Ann Street**, thought to have been designed jointly by architect James Milne and Sir Henry Raeburn, and named after Raeburn's wife, may be the prettiest street in the city. It's unusual in Edinburgh for the fact that each of its Georgian houses has its own front garden.

EXPLORE

St Bernard's Crescent.

IN THE KNOW
HE'S BEEN FRAMED

Born in Stockbridge, portrait painter Sir Henry Raeburn (1756-1823) created some iconic Scottish images, including *The Reverend Robert Walker Skating on Duddingston Loch* (also known as *The Skating Minister*). The picture – which can be seen at the **Scottish National Gallery** (*see p88*) – even inspired some design elements in the Scottish Parliament at Holyrood. Raeburn owned estates around Stockbridge, and the main road through the area is still called Raeburn Place. He is buried at **St Cuthbert's** (*see p87*) on the southern edge of the New Town.

Thomas de Quincey, the author of *Confessions of an English Opium-Eater*, lived here for a time.

To reach **Dean Village** from Stockbridge, simply follow the Water of Leith upstream by taking the path at the south-west end of Saunders Street, off Kerr Street by the bridge. Within minutes, you'll find yourself in a very different kind of environment, with the shops and the bustle replaced by trees. You won't get to enjoy the rural idyll for long, as it's less than half a mile to Dean Village, but you will pass **St Bernard's Well** en route. Around 30 years after a mineral spring was discovered here in 1760, the landowner commissioned a pump house modelled on the Temple of Sybil at Tivoli, complete with a statue of Hygeia, the goddess of hygiene and good

health. The entire monument was refurbished in 2013; its decorative pump room interior is open to the public on **Doors Open Day** (*see p30*).

Dean Village itself was named for the dean, or deep valley, in which it sits. It's best seen from the surprisingly high **Dean Bridge**, designed by Thomas Telford, which crosses the Water of Leith between Stockbridge and the West End. From at least the time of David I (who reigned from 1124 to 1153), flour was milled for Edinburgh and the surrounding area here. The old buildings were largely replaced in the 19th century by breweries and distilleries; they now form part of a conservation area that's resolutely residential and has bags of character. The cobbled streets, varied architecture and steep braes form an attraction all by themselves, although mementoes of the milling trade remain.

At the top of Bell's Brae stands quirky **Kirkbrae House**. Eagle-eyed visitors will spot a panel taken from the ruins of a granary named Jericho, which was built for the Incorporation of Baxters (bakers) in 1619 and once stood in the dean immediately below the house. The ornate panel depicts the sun and cherubic heads with an inscription reading 'In the sweat of thy face shalt thou eat bread, Gen.3 verse 19'. Down the hill by Bell's Brae Bridge is another old panel, carved with two crossed bakers' peels (used for taking hot loaves out of ovens). The nearby window lintels with inscriptions are now getting too weathered to be read easily.

From Dean Village, you can follow the Water of Leith further upstream to the **Scottish National Gallery of Modern**

Dean Village.

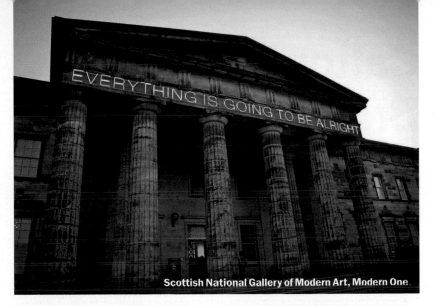

Scottish National Gallery of Modern Art, Modern One

Art's two buildings: cross Bell's Brae Bridge into Damside and the Water of Leith Walkway continues, on the north side of the river. In less than half a mile, some steps lead up to Belford Road; the galleries are around the corner to the right. Alternatively, continue along the walkway under the Belford Bridge; you'll soon reach a gate and some steps behind the Scottish National Gallery of Modern Art, Modern One. Don't tarry, though: the gate is locked at 6pm from April to September, and at dusk from October to March.

Sights & Museums

★ FREE Scottish National Gallery of Modern Art, Modern One

Belford Road, west side (624 6200, www.national galleries.org). Edinburgh Coach Lines bus 13 or National Galleries of Scotland shuttle bus from the Mound (£1). **Open** 10am-5pm daily. **Admission** free; charges vary for special exhibitions. **Map** p107 H1 ⑳

IN THE KNOW
HOME ON THE GRANGE

The **Grange Club** (www.thegrangeclub.com) in Portgower Place, Stockbridge, offers hockey, squash and tennis to members, but is also a key venue for Scottish cricket. Scotland's international team has played matches here in recent years against the likes of Australia, England and Pakistan, drawing respectable four-figure crowds.

Since 1984, Scotland's national collection of modern art has been housed in this neoclassical structure, designed by William Burn in the 1820s as an institution for fatherless children, lately dubbed Modern One. There are regular changing exhibitions, as well as a strong permanent collection with pieces by Freud, Mondrian, Hepworth and a few Damien Hirsts; the most beautiful work may be Picasso's Blue Period *Mère et Enfant* (1902).

As you walk downstairs to the café with its pleasant terrace, look across to the supporting wall on the stairwell. It's been given over to the artist Douglas Gordon, who has neatly printed on it the names of every person he has ever met and whose name he can remember. The fun continues outside the main entrance, where the lawn has been remodelled by American landscape architect Charles Jencks into an extensive, spiral sculpture entitled *Landform*.

★ FREE Scottish National Gallery of Modern Art, Modern Two

Belford Road, east side (624 6200, www.national galleries.org). Edinburgh Coach Lines bus 13, or National Galleries of Scotland shuttle bus from the Mound (£1). **Open** 10am-5pm daily. **Admission** free; charges vary for special exhibitions. **Map** p107 G1 ㉑

Housed in an impressive 1831 building that was originally an orphanage, Modern Two (formerly known as the Dean Gallery) also offers an impressive range of modern art. It hosts special and visiting exhibitions, as well as a permanent mock-up of the London studio of Leith-born sculptor Sir Eduardo Paolozzi (1924-2005). His 15ft metal sculpture, *Vulcan*, looms above the café. Modern Two is directly across the road from Modern One; you can comfortably visit both in an afternoon.

Calton Hill & Broughton

Robert Louis Stevenson was not popularly known as a comedian, but his description of Calton Hill as the city's best vantage point does have a certain witty merit. In *Edinburgh: Picturesque Notes*, published in 1878, Stevenson wryly points out that you can't see the castle from the castle, and you can't see Arthur's Seat from Arthur's Seat, but you can see both from Calton Hill. The views are supplemented by an array of historic monuments and a contemporary art gallery in an old observatory.

Back at the bottom of the hill, north of the modern Omni Centre, stands the adjacent district of Broughton. One of Edinburgh's more fashionable neighbourhoods, it's home to the city's small but cultured gay scene as well as a wide selection of bars and restaurants.

EXPLORE

Old Calton Burial Ground

Don't Miss

1 **Calton Hill** A view with a thrill: the city, the Forth, the Pentlands and beyond (p119).

2 **Beltane** Pagan shenanigans every spring, up Calton Hill (p29).

3 **Old Calton Burial Ground** Final resting place of philosopher David Hume, and others (p116).

4 **Old Royal High School** Built 1829, searching for a new role in life since 1968 (p118).

5 **Broughton Street** Interesting eats, drinks and shops (p123).

Old Royal High School. *See p118.*

INTO THE NEIGHBOURHOOD

Calton Hill was formed in the same geological upheaval that created the Castle Rock and Arthur's Seat. The area may have been occupied during the Bronze Age, up to 4,000 years ago, but there's far more documentary evidence covering the last 550 years. In the middle of the 15th century, for instance, James II gave an amphitheatre to the city on Greenside, modern-day Greenside Place. It remains a focus for entertainment today, thanks chiefly to the Edinburgh Playhouse and the Omni Centre.

This confection of history and fun, ancient and modern, is typical of the area. On Waterloo Place, the Old Calton Burial Ground dates from 1718, but sits opposite a hotel that opened as recently as 2009; not far away, on Royal Terrace, one of Edinburgh's most creative restaurants, **Paul Kitching 21212**, offers 21st-century dining in a 19th-century townhouse. And in Broughton, new bars and cafés appear regularly in a district that was first documented – as 'Bruch toun' – as far back as the early 12th century.

In practical terms, the area known as Calton Hill is bordered by Calton Road to the south (Waterloo Place and Regent Road running more obviously above), and London Road and its environs in the north. Leith Street and the neighbourhood of Broughton are to the north-west, the latter eventually linking to the genteel suburb of Canonmills at the edge of Stockbridge.

EAST OF PRINCES STREET

Waterloo Place effectively forms an eastern entrance to the modern city centre, an impression aided by the austere façades on either side of the street and the twin ceremonial archways of Regent Bridge that carries the thoroughfare over Calton Road below. Both the bridge and the original Waterloo Hotel building on the north side (now the **Apex Waterloo Place**; *see p256*) opened in 1819, just four years after the celebrated battle at which the Prussian Marshal Blucher, the Anglo-Irish Duke of Wellington and their allies saw off Napoleon once and for all.

As the **Old Calton Burial Ground** had already been in situ on this side of Calton Hill for a century or so, the creation of Waterloo Place in the years immediately after the Battle of Waterloo effectively sliced the cemetery in two. The greater part is still accessible on the south side up some obvious steps, and serves as the final home for a number of figures from the Scottish Enlightenment; David Hume (*see p240*) is the most celebrated, with a monument by Robert Adam that befits his status.

Elsewhere in the cemetery stand a couple of other memorials that resonate with more modern sensibilities. One of them, a towering obelisk that dates from 1844, is a tribute to a group of late 18th-century political reformers led by Thomas Muir, who were transported for sedition after having had the audacity to demand parliamentary reform. The other, an 1893 monument that features Abraham Lincoln and a freed slave, was designed to commemorate the Scots who died during the American Civil War. Elsewhere, up against the cemetery's east wall, sits the **Governor's House** (1817), from the now-demolished Calton Gaol. Described by Robert Louis Stevenson as 'castellated to the point of

> ### IN THE KNOW
> ### STONES OF SCOTLAND
>
> Regent Road Park (off Regent Road) is home to the Stones of Scotland, an art installation of 32 stones. Each represents one of the country's local authority areas, from Shetland in the north to Dumfries and Galloway in the south. The stones have a great geological diversity and the installation overlooks the scenic eastern part of the Old Town and Holyrood Park. It's less than five minutes' walk from the old Royal High School.

EXPLORE

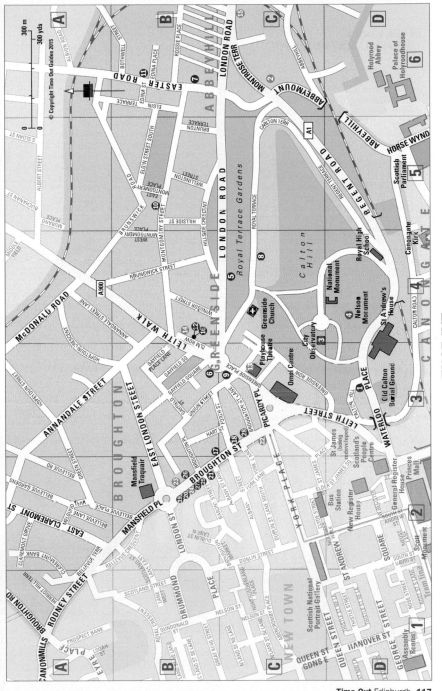

EXPLORE

IN THE KNOW
TO HUME IT MAY CONCERN

The first proper footpath around Calton Hill was created after philosopher David Hume presented a petition to the town council in 1775, claiming that such a path would contribute 'not only to the pleasure and amusement, but also to the health of the inhabitants of this crowded city'.

folly', it was largely swept away during the construction of nearby St Andrew's House.

Back on the north side of Waterloo Place, at the corner by **Howies** restaurant, is a lane called Calton Hill. Down here, you'll find the **Parliament House** hotel (*see p257*), and the old **Rock House** at the end of the residential tenements opposite. This building was home to a succession of photographers from the 1830s, among them David Octavius Hill. Working in collaboration with Robert Adamson, Hill helped to give photography credibility as a modern art form; 5,000 of the pair's calotypes and negatives now form a major part of the photography collection held by the National Galleries of Scotland at the **Scottish National Portrait Gallery** (*see p88*).

Waterloo Place also offers one of the most obvious routes up Calton Hill. Take the steps to your left as the street morphs into Regent Road.

One of the two truly imposing buildings along Regent Road, **St Andrew's House** on the south side was created as a home for Scottish civil servants and is now part of the Scottish Government apparatus. It may be historically and aesthetically important, but it does smack of authoritarianism, all the more unsettling since construction was completed in 1939.

By contrast, the **old Royal High School** building a little further along on the north side is perhaps the finest example of neo-classical architecture in the city. Completed in 1829, it was designed by Glasgow-born architect Thomas Hamilton, a renowned Greek revivalist, and modelled on the Temple of Theseus in Athens (since reassigned to the god of fire and called the Hephaisteion). Because of the old pile's monumental size, it's difficult to get a proper perspective on it, even from the other side of the road. It's best to focus on the detail, then contemplate its grandeur from one of the closes at the lower (eastern) end of the Royal Mile.

The school was the site of a round-the-clock, 1,980-day vigil in the 1990s by protestors campaigning for Scottish devolution. This was triggered by the 1992 general election, which once again returned a Conservative government to Westminster, at odds with public opinion in Scotland. The Royal High building seemed an auspicious location for a protest, as it had been made ready to house a parliamentary chamber before a referendum for a Scottish Assembly back in 1979. At that time, the vote didn't carry. But at a second referendum in 1997, the Scottish people finally said 'yes' in sufficient numbers; two years later, a parliament was convened. A cairn on Calton Hill commemorates the vigil.

The old Royal High building was vacated by its namesake school in 1968. Since then, various

National Monument.

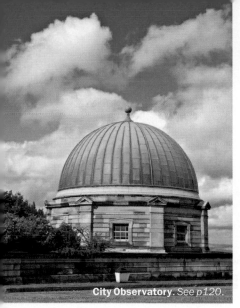

City Observatory. *See p120.*

plans have been mooted to make use of such a remarkable building, but none has come to fruition. The latest, in early 2015, involved a £55 million hotel development, which, even at the consultation stage, provoked 'concerns'. Edinburgh is waiting to see what happens.

Across the road from the old school is another Hamilton effort: the **Robert Burns Memorial**, a small Greek temple that seems completely out of sync with its purpose. However, it's worth straying this far to take in the fantastic view up to the castle, and down to the environs of Holyrood. Alternatively, head along **Regent Terrace** (closed to through traffic since shortly after 9/11 to protect the resident US Consulate) for a stroll by the elegant townhouses around that side of the hill, before heading around to Royal Terrace and back to the top of Leith Walk. The **Regent** pub would be your pit stop for a beer or some food.

Restaurants

Howies

29 Waterloo Place (556 5766, www.howies. uk.com). Playhouse or Princes Street buses. **Open** 5.30-9.30pm daily. **Main courses** £6.50-£15. **Map** p117 D3 ❶ **Modern European**
In classy Georgian premises just yards from the east end of Princes Street, Howies is a good option for an economical meal (chicken liver parfait to start, then baked salmon with herb crust, perhaps). You'll only stretch your budget if you choose rib-eye steak. There are Scottish comfort food options too: burgers, haggis, macaroni cheese.
Other location 10-14 Victoria Street, Old Town (225 1721).

Pubs & Bars

★ Regent

2 Montrose Terrace (661 8198, www.theregentbar. co.uk). Bus 15, 35, 104, 113. **Open** noon-1am Mon-Sat; 12.30pm-1am Sun. *Food served* noon-10pm Mon-Sat; 12.30-10pm Sun. **Main courses** £5.50-£8. **Map** p117 C6 ❷
When people visit Calton Hill, they tend to head back to the city centre afterwards for eats or drinks, unaware that there's a friendly bar just a short walk along Regent Road – in the other direction, it's not far from the Palace of Holyroodhouse or the Scottish Parliament either. The Regent serves cask ale and comforting pub grub, so if you want to sit quietly with a plate of nachos and a decent pint, it's perfect; it also fits the bill for an evening out with a few friends in a relaxed environment. The pink union flag that flutters outside might tell you this is a gay venue, but everyone is welcome here.
▶ *The owners also run Café Nom de Plume (see p124) in Broughton Street.*

CALTON HILL & NORTH

Calton Hill yields a multitude of handsome views. Looking west, you'll see the clock tower of the Balmoral Hotel, with the castle behind. To the south, you can take in the diagonal slash of Salisbury Crags, Holyrood Park behind to the south-east. On the eastern horizon, look out for the volcanic pimple of North Berwick Law. But for the most pressing sense of space, look north-east, up to where the land falls away to Leith and the Firth of Forth beyond it opens up towards the North Sea. Catch it at the right time, when a chilly sea mist blankets the Forth, encroaching on the city, and it appears quite magical. Locals call this mist a 'haar'.

Although the views from it are terrific, Calton Hill is most famous for its array of bizarre architecture, and no single structure is more folly-like than the 12 Doric columns that form the **National Monument** to the dead of the Napoleonic Wars. Designed by William Playfair, the architect responsible for some of Edinburgh's finest buildings, this was designed to be a replica of the Parthenon in Athens. Sadly, funds ran out in 1829 before it was completed. The structure came to be known as 'Scotland's Disgrace', although locals are far from offended by its presence nearly two centuries later; it's actually held in some affection these days.

Meanwhile, the enclosed precinct on Calton Hill referred to in the singular as 'the Observatory' contains a number of buildings. The oldest is the rather Gothic 1776 creation at the western side, designed by New Town architect James Craig but never quite completed nor used for stargazing; instead, it served as lodgings for

EXPLORE

the man who ran the first observatory here. In the middle of the precinct sits Playfair's elegant Royal Observatory, based on the Temple of the Winds at Athens. Construction began in 1818; four years later, it was given a regal imprimatur, although it wasn't fully functional until 1831. As an institution, the Royal Observatory moved away to Blackford Hill in South Edinburgh in 1896; the buildings became known simply as the **City Observatory**, and another dome was added at the north-east. Also on the site, at the south-eastern corner, is a neo-classical monument that stands as a memorial to Playfair, who died in 1823.

The Astronomical Society of Edinburgh began using the observatory buildings in 1938, but in the modern era persistent vandalism and theft from what is an insecure site caused it to think again. In 2009, the society held what members described as a 'quiet closing event', and then vacated the buildings forever. It was a sad ending to 233 years of astronomy on Calton Hill. Where there was once science, there is now art, however: visual arts outfit **Collective** began a staged move to the precinct in 2013.

Other buildings on the hill include the towering **Nelson Monument** and, nearby, the Playfair-designed **Monument to Dugald Stewart** (1753-1828), professor of moral philosophy at the University of Edinburgh. They all provide a dramatic backdrop to the annual **Beltane** celebration (*see p29*), which draws thousands of people on the night of 30 April every year for a modern interpretation of the ancient rites of spring. The festival pre-dates the monuments: Beltane, or something like it, has been held on Calton Hill since at least the 18th century, and probably much earlier.

WHERE THE WILD THINGS ARE

From one hill to another…

EXPLORE

Looking south-east from Calton Hill, Holyrood Park takes up the bulk of the skyline, with Arthur's Seat sitting high in the centre. Managed by Historic Scotland, and designated as a Site of Special Scientific Interest, the park has deep undergrowth, lochs, marsh, rocky cliffs and unimproved grassland. The environment is great for visitors who want to get away from the city streets, but it's even better for the species of animals and plants that live in it.

The story behind the 12th-century creation of Holyrood Abbey at the park's northern edge involves a deer (*see p214*), but you won't find mammals quite so large grazing here these days. There were sheep until as late as 1977, but these too have gone. Look out, instead, for brown hares and grey squirrels. At quieter times, you might catch a hedgehog scuttling about (or squashed on the road), a fox or even a weasel. At Duddingston Loch, an important site for a variety of birds (swans, geese, ducks, coots, grebes and, in the surrounding trees, herons), keep your eyes peeled for water voles and otters. Common in Central Scotland in the middle of the last century, otters were almost wiped out by agricultural pesticides. However, they have gradually been moving back into suitable habitats in recent years; you may even see one or two on the Water of Leith in the city.

Elsewhere in the park, St Mary's Loch and Dunsapie Loch are both frequented by waterfowl, and there's more bird life in the shape of finches, skylarks, summer visitors such as sedge warblers, and tits. The unsettling squawk of corbies (carrion crows) is common – you might see one perching nearby as you walk on Arthur's Seat or Calton Hill – but the glamour birds are the raptors. You can often see kestrels hovering above Holyrood Park, as well as sparrowhawks, short-eared owls and the occasional buzzard. Frogs, smooth newts and toads also call the park home, as do all kinds of bats, butterflies, damselflies and moths. It's estimated that there are around 350 plant species in the park, with as many as one-sixth of them quite rare in Scotland. Although lots of people come here for fresh air, not so many get down on their knees to check whether that dark pink flower is really a sticky catchfly. But for the enthusiast, the various micro-environments provide all kinds of interest.

▶ *For more on the park's nature, visit the Holyrood Lodge Information Centre; see p129.*

Paul Kitching 21212

Sights & Museums

FREE Collective

City Observatory & City Dome, 38 Calton Hill (556 1264, www.collectivegallery.net). Princes Street or Playhouse buses. **Open** *Aug* 10am-6pm daily. *Apr-July, Sept* 10am-5pm Tue-Sat. *Oct-Mar* 10am-4pm Tue-Sat. **Admission** free. **Map** p117 C4 ❸
Established more than 30 years ago to support new and emerging artists, Collective moved in 2013 to take over Calton Hill's former observatory buildings. Expect challenging art and an open-air café provided by Milk (*see p149*). Check the website for Observers' Walks, artist-made audio guides designed to be downloaded and listened to as you wander the hill.

Nelson Monument

Calton Hill (556 2716, www.edinburghmuseums. org.uk). Princes Street or Playhouse buses. **Open** *Apr-Sept* 10am-7pm Mon-Sat; noon-5pm Sun. *Oct-Mar* 10am-3pm Mon-Sat. **Admission** £4. **No credit cards. Map** p117 D4 ❹
If the views from Calton Hill aren't grand enough for your liking, you can get an even better all-round vantage point from the top of the Nelson Monument. Designed to mimic the shape of Nelson's telescope, to commemorate his victory (and death) at Trafalgar in 1805, it was completed in 1815. A time-ball, weighing a mighty 1,680lb, was added at the top in 1852. The ball and the One O'Clock Gun at the castle were synchronised, helping ship captains in Leith Harbour to reset their chronometers. The ball is still dropped daily (except Sunday) at 1pm.

Restaurants

Gardener's Cottage

1 Royal Terrace Gardens (558 1221, www. thegardenerscottage.co). Playhouse buses. **Open** noon-2.30pm, 5-10pm Mon, Thur, Fri; 10am-2.30pm, 5-10pm Sat, Sun. **Set dinner** £35. **Map** p117 C4 ❺ **Modern British**

Based in the the historic gardener's cottage (designed by William Playfair) on the south side of the gardens along London Road, this restaurant (opened 2012) takes its commitment to local, seasonal food very seriously; the premises include a kitchen garden where they grow produce. Decor is simple, tables are communal. Lunch is à la carte – a typical main would be halibut with toasted spelt, savoy cabbage, bacon, capers, braised celery and potato – while dinner is a six-course set menu. There's brunch at weekends.

Khushi's

10 Antigua Street (558 1947, www.khushis.com). Playhouse buses. **Open** noon-10pm Mon-Thur, Sun; noon-11pm Fri, Sat. **Main courses** £8-£16. **Map** p117 B34 ❻ **Indian**
The daddy of Edinburgh's Indian restaurants. The original was set up in 1947 by Kushi Mohammed from Jalandhar in Punjab, who'd moved to the city before the war. Mr Mohammed passed away in 1977, but the family kept the business going. It's been here since 2011. Modern interior, brilliant fish pakora, BYO (but note the bottle recycling charge).

★ £ Manna House

22-24 Easter Road (652 2349, www.themanna housebakery.co.uk). Bus 1, 35. **Open** 8am-6pm Mon-Sat; 9am-5pm Sun. **Main courses** £2.50-£5. **No credit cards. Map** p117 B6 ❼ **Café**
A long-established, quite brilliant pâtisserie offering coffee, cake, elaborate tarts, salads and savouries. It arrived in Easter Road as the area was becoming partially gentrified during the property boom; its sheer quality meant it survived the ensuing downturn. Tables are few, so many customers take away.

★ Paul Kitching 21212

3 Royal Terrace (0845 222 1212, www.21212 restaurant.co.uk). Playhouse buses. **Open** noon-1.45pm, 6.45-9.30pm Tue-Sat. **Set lunch** £22-£55. **Set dinner** £49-£69. **Map** p117 C4 ❽ **Modern European**

Chef Paul Kitching and manager Katie O'Brien made their names at Juniper in Manchester; when they moved north in 2009, their new venue instantly joined Edinburgh's top tier and a Michelin star followed within a year. The interior is lush, the cooking among the most creative in Scotland. Lunch is a two- to five-course affair; dinner brings three to five courses – there's no carte as such. What arrives, plated, sometimes takes some working out, both visually and from its title: Lamb Curry CCCC? Pac-Man? However, it's very good indeed; eating here is an adventure and Kitching is an excellent guide.
▶ *There are four rooms above the restaurant for anyone seeking a very swish B&B; see p256.*

Pomegranate
1 Antigua Street (556 8337, www.pomegranates restaurant.com). Playhouse buses. **Open** noon-11.30pm Mon-Fri; 11am-11.30pm Sat, Sun. **Main courses** £8-£15. **Map** p117 C3 ❾ **Middle Eastern**
The ground-floor restaurant has some decorative concessions to the aesthetics of the Middle East; it's where you come for very good meze, kebabs, shawarma dishes or mains such as marinated chargrilled lamb with rice. Downstairs is the shisha space and a particularly attractive and atmospheric Arabian dining room for groups.

£ Renroc
91 Montgomery Street (556 0432, www.cafe renroc.co.uk). Playhouse buses. **Open** 9am-5pm Mon; 9am-6pm Tue, Wed; 9am-8pm Thur; 9am-11pm Fri; 10am-5pm Sat, Sun. **Main courses** £6-£8. **Map** p117 B5 ❿ **Café**
If Manna House (*see p121*) demonstrates the gentrification of the southern end of Easter Road, nearby Renroc does the same in the tenement hinterland between Easter Road and Leith Walk. There's a café menu (breakfast, soups, sandwiches, light meals, sharing boards), wine and a couple of craft beers on tap. The tables outside are a real suntrap.

Rivage
126-130 Easter Road (661 6888, www.rivage restaurant.co.uk). Bus 1, 35. **Open** noon-2pm, 5.30-11pm daily. **Main courses** £7-£11. **Map** p117 B6 ⓫ **Indian**
With chunky decor and bare brickwork, the family-run Rivage is a neighbourhood winner. It goes beyond the basics to provide some elaborate starters and inventive mains (such as the Hyderabadi Biryani Pots), while substantial side dishes such as baingan bharta (aubergine with chill, garlic and coriander) weigh in at under £4.

★ Valvona & Crolla Caffè Bar
19 Elm Row, Leith Walk (556 6066, www. valvonacrolla.com). Playhouse buses. **Open** 8.30am-6pm Mon-Sat; 10.30am-4pm Sun. **Main courses** £8-£19. **Map** p117 B4 ⓬ **Italian**

Still the most celebrated delicatessen (*see below*) in Scotland, V&C also features this simple, tasteful space at the rear, a converted stables. Food includes breakfast, toasted paninis and more substantial dishes at lunch. One big bonus is the wine list: pick any bottle from the award-winning selection in the deli and you can enjoy it in the caffè for the retail price plus £6 corkage.

Pubs & Bars

★ CC Blooms
23-24 Greenside Lane (556 9331, www.ccblooms edinburgh.com). Playhouse buses. **Open** 11am-3am daily. *Food served* 11am-9pm daily. **Main courses** £7.50-£9.50. **Map** p117 C3 ⓭
A legend on the Edinburgh gay scene, open for more than 20 years, this venue can be all things to all people: somewhere to eat (tapas, flatbreads, pub grub-style mains), to drink, to chill out and listen to DJs or to dance the night away on Fridays and Saturdays. It enjoyed a sorely needed revamp in 2012 and now looks all set for another couple of fun decades.

Joseph Pearce's
23 Elm Row, Leith Walk (556 4140, www.bodabar. com/joseph-pearces). Playhouse buses. **Open** 11am-midnight Mon-Thur, Sun; 11am-1am Fri, Sat. *Food served* 11am-9.30pm Mon-Thur, Sun; 11am-9pm Fri, Sat. **Main courses** £6-£10.50. **Map** p117 B4 ⓮
Owned by a Swedish couple who run a number of offbeat, contemporary bars, Joseph Pearce's was converted from a typical old geezers' drinking den to its current hipster-friendly incarnation. It's child-friendly during the day, and serves everything from breakfast to Swedish meatballs. Things liven up later, especially on weekend evenings.

Safari Lounge
21 Cadzow Place (661 4741, www.thesafarilounge. co.uk). Bus 4, 5, 15, 19, 26, 34, 44, 44A, 45, 104, 113. **Open** noon-midnight Mon-Thur, Sun; noon-1am Fri, Sat. *Food served* noon-9pm Mon-Thur; noon-10pm Fri, Sat; noon-8pm Sun. **Main courses** £5-£10. **Map** p117 C6 ⓯
'An interesting, modern bar? With food? Here?' Locals were quite surprised when the Safari Lounge arrived in 2013, as Cadzow Place, which forms part of London Road, was not exactly known for quality drinks or cuisine. A decent addition to the neighbourhood, it offers brunch, an eclectic range of snacks, hot dogs, wraps and curries – even a beer or two – in a traditional bar space, smartened up by neon, animal prints and quirky design touches.

Shops & Services

★ Valvona & Crolla
19 Elm Row, Leith Walk (556 6066, www. valvonacrolla.com). Playhouse buses. **Open**

8.30am-6pm Mon-Thur; 8am-6.30pm Fri, Sat; 10.30am-4pm Sun. **Map** p117 B4 ⓰ **Food & drink** Edinburgh's best-known stop-off for Italian deli goodies, gourmet treats and wine has been trading at this address since 1934. The range of comestibles is wide and the quality top notch. There's a well-known caffè (*see p122*) at the back; in August, part of the premises morphs into a Fringe venue.

BROUGHTON

Broughton has long been a cultured and unconventional part of the city. Once notorious for witchcraft, it's been home to Edinburgh's gay community for many years, and the area around Broughton Street, Picardy Place and the Edinburgh Playhouse has been nicknamed 'the pink triangle'. However, there's more to the neighbourhood, which has steadily become one of the city's most fashionable corners.

The approach down **Leith Street** from the east end of Princes Street has altered hugely in recent years with a row of modern offices and, to the south side, the huge **Omni Centre**, filled with chain eateries, a health centre, a multiscreen cinema and even a hotel (the **Glasshouse**; *see p256*). The north side of Leith Street is also changing thanks to the massive St James development (*see p233*). One of the few constants is the **Playhouse** (*see p190*), one of the city's major venues for touring musicals, gigs and Edinburgh International Festival performances. In situ since 1929, and with a capacity of 3,000, it's not going anywhere.

Flanking the huge roundabout at the top of Leith Walk, **Picardy Place** was named

for a colony of cambric weavers who came to Edinburgh from Picardy, France, in 1732. Today, assuming you're not popping into one of the local bars or eateries, the main reason to visit is for a peek at the statue of **Sherlock Holmes**; his creator, Sir Arthur Conan Doyle, was born at a now-demolished property at no.11. The statue was put into storage in 2009 because of the Edinburgh Trams project – a line was supposed to come down Leith Walk, and Sherlock was in the way. But the line was never built and the detective was reinstated in 2012.

Nearby, in front of St Mary's Catholic Cathedral, are a set of outsize sculptures by the late Sir Eduardo Paolozzi, born in Leith and a favourite son of the city; there's more of his work in the **Scottish National Gallery of Modern Art, Modern Two** (*see p113*). This site is also home to the small **Tram Stop Market** (*see p109* **Don't Quit Your Stalling**). Outside the Omni Centre, opposite the Paolozzi works, stand a more populist pair of scrap-metal giraffes, created in 2005 by Helen Denerley; *see p57* **Statuesque**.

Heading downhill (north-west) from Picardy Place, **Broughton Street** is the heart of the neighbourhood and lined with several decent shops, cafés and restaurants. You can stroll down here with no intention to stop or buy anything and then, hours later, return to your hotel well fed, well watered and armed with several carrier bags containing anything from upcycled clothes to artisan chocolate.

At the bottom of the street, across the roundabout, is **Mansfield Church**, the walls of which are covered in wonderful murals created in the 1890s by leading Arts and Crafts artist

EXPLORE

Broughton Street.

Phoebe Traquair. The church was bought by the Mansfield Traquair Trust (www.mansfield traquair.org.uk) in the 1990s and houses offices of voluntary organisations. The murals (restored in 2005) are occasionally open to the public, usually on the second Sunday of the month.

Continuing down the hill, a look down **Bellevue Crescent** on the left reveals more elegant New Town façades. Eventually, via Rodney Street, you come to Canonmills (*see p104*).

Restaurants

£ Artisan Roast

57 Broughton Street (07858 884756 mobile, www.artisanroast.co.uk). Bus 8 or Playhouse buses. **Open** 8am-7.30pm Mon-Fri; 10am-6.30pm Sat, Sun. **Map** p117 B3 ⑰ **Café**

Artisan Roast supplies quality coffee to cafés and restaurants, but in Broughton Street it has its very own premises. It's a laid-back spot, far from po-faced, although staff take their coffee very seriously. There are some cakes and pastries, but the main attraction is the fresh-roast arabica. The success of the venture has seen it expand in Edinburgh and Glasgow. **Other location** 138 Bruntsfield Place, South Edinburgh (07858 884771 mobile).

Burger Meats Bun

1 Forth Street (556 7023, www.burger-meats-bun. co.uk). Bus 8 or Playhouse buses. **Open** noon-9.30pm Tue-Sun. **Burgers** £7.50-£8.50. **Map** p117 C3 ⑲ **Burgers**

Edinburgh eventually caught up with the burger fad; this venue opened in 2014. The interior is fun if basic, the focus very much on the meat. Burgers range from the Fiery (beef) or Hot Chic (chicken) to a vegetarian version. Fries cost extra, but come in basic, cheese, or Thai chilli cheese varieties.

Café Nom de Plume

60 Broughton Street (478 1372). Bus 8 or Playhouse buses. **Open** 11am-11pm Mon-Sat; noon-11pm Sun. **Main courses** £6.50-£8.50. **Map** p117 B2 ⑲ **Café**

Run by the same people as the Regent (*see p119*), the predominantly gay Nom de Plume provides a relaxed café environment. One room faces Broughton Street; the other overlooks a garden at the back. Food is simple (sandwiches, toasted bagels or larger dishes such as fish pie or macaroni cheese), or you can just drop by for a coffee, wine or beer. A useful place to pick up information on the city's gay scene.

L'Escargot Bleu

56 Broughton Street (557 1600, www.lescargot bleu.co.uk). Bus 8 or Playhouse buses. **Open** noon-2.30pm, 5.30-10pm Mon-Thur; noon-3pm, 5.30-10.30pm Fri, Sat. **Main courses** £16-£20. **Map** p117 B2 ⑳ **French**

With wood floors, white walls and an elevated bistro vibe, this place feels completely authentic. As for the food, Barra snails with crab bisque, camembert in filo, and rib-eye steak with roquefort sauce all feature. Accompany that last dish with some vin rouge and you'll be a contented diner.

Other location L'Escargot Blanc, 17 Queensferry Street, New Town (226 1890, www.lescargotblanc. co.uk).

Port of Siam & the Bangkok Bar

1 Barony Street (478 7720, www.portofsiam. com). Bus 8 or Playhouse buses. **Open** *Restaurant* noon-2pm, 6-10pm Mon-Fri; noon-10pm Sat; 12.30-10pm Sun. *Bar* noon-10pm Mon-Sat; 12.30-10pm Sun. **Main courses** £12-£18. **Map** p117 B2 ㉑ **Thai**

With entrances on both Broughton Street (Bangkok Bar) and Barony Street (Port of Siam), this is

L'Escargot Bleu.

The cluster of gay venues around the top of Broughton Street and by the Playhouse includes the legendary **CC Bloom's** (see p122), **Café Habana** next door (DJs, drink, food, karaoke) and **Planet** (a more basic bar) a few doors further down. At the corner of Broughton Street and Picardy Place are the **Street** (café-bar with dancefloor) and **Chalky's** (bar and nightclub), while down Broughton Street itself is **Café Nom de Plume** (see p124).

simple menu ranges from nibbles to pâté with bread and pickles to sharing plates (cheese, charcuterie or a mix of the two). Launched in 2013, it's a thoroughly civilised addition to Broughton Street.

Shops & Services

★ Concrete Wardrobe
50A Broughton Street (558 7130, www.concrete wardrobe.com). Bus 8 or Playhouse buses. **Open** 11am-7pm Mon-Sat; noon-5pm Sun. **Map** p117 B2 ㉕ **Fashion/Accessories**
This boutique specialises in contemporary Scottish craft, covering fashion, jewellery, ceramics, homewares, soft furnishings and more. It's always worth a look for nicely priced clothing from designers both local and further afield. Staff are savvy.

Crombies
97-101 Broughton Street (557 0111, www. sausages.co.uk). Bus 8 or Playhouse buses. **Open** 8am-5.30pm Mon-Fri; 7.30am-5pm Sat. **Map** p117 B2 ㉖ **Food & drink**
Award-winning and family-run, Crombies is simply a very good butcher's. Beef is locally sourced from West Lothian and pork from the Scottish Borders, while the selection of black pudding, haggis, pies and pastries is excellent. The signature product is sausages: more than 40 types ranging from the traditional (pork) to the downright creative (whisky, hog and wild thyme).

Joey D
54 Broughton Street (557 6672, www.joey-d.co.uk). Bus 8. **Open** 10.30am-6pm daily. **Map** p117 B2 ㉗ **Fashion**
Almost defying description, Joey D takes vintage fabric, deconstructs it, reconstructs it according to an idiosyncratic, dystopian aesthetic and turns it into a contemporary fashion desirable. There are items for men and women, as well as excursions into accessories and furniture. Belted miniskirt made from an old curtain? Armchair covered in mismatched patching? Bags apparently made from Harris Tweed samples? This is the place to come.

Real Foods
37 Broughton Street (557 1911, www.realfoods. co.uk). Bus 8. **Open** 8am-9pm Mon-Fri; 9am-6.30pm Sat; 10am-6pm Sun. **Map** p117 C3 ㉘ **Food & drink**
Spices, grains, oils, vinegars, nuts, pulses, muesli, sugar-free jam, fresh fruit and veg, beer, wine, bread, eco toiletries and household cleaners, vitamin supplements, beauty products – the list goes on and on. Real Foods sells it all and has done since 1975. You could buy a bag of organic porridge oats or stock up on supplies that would feed a gluten-intolerant family of eight for three months.
Other location 8 Brougham Street, South Edinburgh (228 1201).

effectively one venue split into two. You get chatty staff and Thai street food in the smaller bar, a full Thai restaurant menu in Port of Siam. That means pork skewers or tom yum soup and a beer in the former; duck and apple red curry, coconut rice and a glass of white wine in the latter.

Pubs & Bars

Basement
10A-12A Broughton Street (557 0097, www. basement-bar-edinburgh.co.uk). Bus 8 or Playhouse buses. **Open** noon-1am daily. *Food served* noon-10pm Mon-Thur, Sun; noon-11pm Fri, Sat. **Main courses** £8.50-£12.50. **Map** p117 C3 ㉒
Broughton Street's original style bar has a history going back over 20 years, although it was acquired by a local pub group and given a serious makeover in 2014. Older customers shook their heads, new ones loved the Mexican food menu, smarter environment and pot plant wall theme. Times change.

★ Cask & Barrel
115 Broughton Street (556 3132). Bus 8 or Playhouse buses. **Open** 11am-12.30am Mon-Thur, Sun; 11am-1am Fri, Sat. *Food served* noon-2pm daily. **Dishes** £3.50-£4.50. **Map** p117 B2 ㉓
The choice of real ales at this traditional pub is impressive, with examples from across Britain, from Orkney's Highland Brewing to West Sussex's Dark Star. It's the kind of place where, if you ask bar staff to remove the sparkler from the tap before they pour your pint, they know what you're talking about. Basic bar meals (chilli, baked tatties, nachos) are available. Popular for football and rugby.

Pickles of Broughton Street
56A Broughton Street (557 5005, www.getpickled. co.uk). Bus 8 or Playhouse buses. **Open** 4.30pm-midnight Mon-Thur, Sun; 4pm-1am Fri, Sat. *Food served* 4.30-11.30pm Mon-Thur, Sun; 4pm-12.30am Fri, Sat. **Dishes** £6-£12. **Map** p117 B2 ㉔
Pickles inhabits a basement, looks like a cross between a wine bar and an upmarket deli and has a reasonable selection of wine and bottled beers. The

Arthur's Seat & Duddingston

Those who make it to the top of Arthur's Seat, the central feature of Holyrood Park, will be rewarded with fantastic views. The glorious 360-degree panorama encompasses the Bass Rock and the hills of North Berwick Law and Traprain Law to the east, the Lammermuir Hills and Moorfoot Hills to the south-east, the Pentland Hills to the south, and the Firth of Forth and Fife to the north. On clear days, you can see up to 40 miles, to the fringe of the southern Highlands.

The hill's deep geological history of vulcanism adds to its lustre, as does the variety of wildlife scattered across its green expanses. Edinburghers walk to the top of the hill, run and cycle round its flanks or picnic by its lochs; some end their visit by wandering to the charming old village of Duddingston, tucked away under the hill's south-east flank.

The summit.

Don't Miss

1 **The summit** The top of Arthur's Seat, 823 feet above sea level (p128).

2 **Salisbury Crags** Dramatic sandstone cliffs overlooking Edinburgh (p128).

3 **Duddingston Loch** How can a city be this scenic? (p129)

4 **Duddingston Kirk** From the 12th century (p129).

5 **Raptors** Keep your eyes on the skies (p129).

ARTHUR'S SEAT

Although Arthur's Seat is commonly referred to as Edinburgh's volcano, it hasn't erupted for a while. Its last explosion was roughly 350 million years ago, when it was on a different part of the Earth's crust and standing in a shallow sea. What remains today is not the volcano itself, but a basalt lava plug that choked its neck. The intervening epochs have seen it subject to all kinds of geological stresses; more recent erosion and glaciation created the shape that now stands.

The earliest evidence of human activity on Arthur's Seat is the series of linear bumps in the grass just above Dunsapie Loch, the remains of cultivation terraces that may date to the Bronze Age. Further below, Duddingston Loch has more solid evidence of Bronze Age occupation. And above Dunsapie Loch on Dunsapie Hill, there are traces of a later Iron Age hill fort.

No one really knows why the main hill is called Arthur's Seat. Some say it's a corruption of the Gaelic 'ard-na-saighead', 'the height of arrows', or even of Archer's Seat. The latter gains currency from both the Iron Age fortifications and the fact that, during the 12th century, the surrounding parkland was a royal hunting ground.

Climbing Arthur's Seat

Several paths lead to the summit, standing 823 feet high. Some are simple and others tougher, but all are within the ability of any moderately fit adult or child, and can be taken at an easy pace to enjoy the view. A word of warning: while it's basically safe, **Holyrood Park** has cliffs, precipices and sudden drops. Wear strong shoes or boots that won't slip on the grass or slopes, and take appropriate clothing; there is no shelter from the elements.

The easiest route is from **Dunsapie Loch** on the eastern part of **Queen's Drive**. The loch is almost halfway to the top anyway; from here it's a straight pull up a grass slope to the summit. No matter which approach you take up, this is also the best way back down to the road.

If you're after a more challenging hike, enter the park through the **Holyrood Park Road** entrance. Once inside, go right at the second, upper roundabout and walk uphill a short way on Queen's Drive. An obvious track soon cuts in to your left, between the south end of Salisbury Crags and the massif of Arthur's Seat itself. The track leads down into a small valley known as Hunter's Bog. At the lip of the valley, immediately up to the right, you'll see a zig-zagging and rough-cut stone pathway leading directly up the gully between Arthur's Seat and Nether Hill. It's arduous, and you'll certainly need a head for heights but it's a great way to access the peak. Alternatively, **Hunter's Bog** itself is a placid

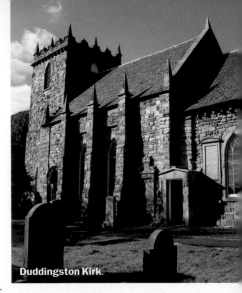

Duddingston Kirk.

space, enclosed by Arthur's Seat on one side and the slopes behind Salisbury Crags on the other.

Another route to the top is via the northern approach. Enter the park via the **Holyrood Lodge Information Centre**, then head along Queen's Drive in the direction of the man-made **St Margaret's Loch**. The loch itself, created in 1856 as part of a series of works drawn up by Prince Albert, is a picturesque spot and good for those who can't manage a more strenuous hike. Walking in its direction takes you past a grille set into the wall on the right; this is **St Margaret's Well**, which, during plague years, was relied upon as a source of clean drinking water. Before reaching the loch, turn right by the ruins of the 15th-century **St Anthony's Chapel**, obvious on its crag. Here, an easy rising glen, the Dry Dam, leads uphill and the path to the summit is obvious.

Elsewhere in Holyrood Park

Skirting the foot of Salisbury Crags, high above Queen's Drive, the **Radical Road** was built by unemployed weavers from the west of Scotland in the wake of the Radical War of 1820, also known as the Scottish Insurrection. They were put to work on the project at the suggestion of Sir Walter Scott.

Part of the rock face alongside the road is known as **Hutton's Section**. In the late 18th century, it was used by James Hutton, popularly known as the father of modern geology, to show that the sill of Salisbury Crags was formed by lava forcing its way into older sedimentary rock layers. In later years, it was said to be a favourite spot of George VI, father of Elizabeth II, who took early-morning strolls here when he was staying at the Palace of Holyroodhouse. Today, it's the best place from which to make sense of the architectural complexity of the **Scottish Parliament** (*see p67*) or just to look at the city.

The park offers more than just constitutionals, of course. Every May Day, faithful souls take a pre-dawn jaunt to follow the pagan tradition of washing their faces in the dew at sunrise. The origins of the ritual are unknown. Also unknown is the identity of whoever left 17 miniature coffins, each containing a wooden doll, in a hillside cave at some point in the 18th or 19th centuries. The coffins were found in 1836 by boys out looking for rabbits; they can be seen in the **National Museum of Scotland** (*see p74*).

Sights & Museums

FREE **Holyrood Lodge Information Centre**
Horse Wynd, by the Scottish Parliament at Holyrood, Old Town (no phone, www.historic-scotland.gov.uk). **Open** 9.30am-3pm daily. **Map** p281 L6.

Housed in an old stone lodge, this unstaffed centre has displays on the history, geology and archaeological features of Holyrood Park, along with useful leaflets and up-to-date information about wildlife. You could see hares, grey squirrels, lots of waterfowl, herons, voles and even otters, toads, frogs, newts and rare plants. Look up and there could be a kestrel, or other raptors, out hunting

This facility shouldn't be confused with the Holyrood Park Education Centre – a modern building just east of the Palace of Holyroodhouse perimeter wall. It serves as a base for the park rangers and police, and is usually open only for school parties, not the public. Rangers do run guided walks and other activities, however (details on 652 8150, hs.rangers@scotland.gsi.gov.uk).

DUDDINGSTON

People have lived in Duddingston for thousands of years. The shores of Duddingston Loch were settled during the Bronze Age, perhaps as far back as 800 BC; a cache of broken Bronze Age weapons and tools was dredged up here in 1778. The first documentary evidence of the area concerns David I, who gifted lands here to Kelso Abbey in the early 12th century.

For years, the villagers made a living weaving a coarse linen cloth known as 'Duddingston hardings'. Parishioners also made use of the benign environment: farming was a common occupation, as the land surrounding Arthur's Seat is very fertile, and there was also a small salt industry. No longer, though. Designated an Outstanding Conservation Area in 1975, the village today is smart, residential and tranquil It effectively comprises two parallel streets: Old Church Lane has the church, while the Causeway has the pub.

If you're walking around Queen's Drive, the road that rings Holyrood Park, you can descend into the village from the steps to the south of man-made Dunsapie Loch. Otherwise, the easiest stroll is from the park's Holyrood Park Road entrance, taking a right at the first roundabout and following the 'low road' that skirts under the cliffs of the Lion's Haunch and then along to the north side of **Duddingston Loch**. This does miss the views of the loch itself, and the adjacent nature reserve, afforded by the 'high road' above.

Another alternative is to look for the blue cycle route sign pointing left off Holyrood Park Road as you approach the park entrance. This leads into a small, modern residential area and eventually to an old disused railway line, the **Innocent Railway**, which goes through a tunnel, along to Duddingston Loch's south side, then to Duddingston Road West, just over half a mile from the centre of Duddingston village.

The aforementioned church is **Duddingston Kirk** (661 4240, www.duddingstonkirk.co.uk), built in 1124 and one of the oldest churches in Scotland still used for regular worship. There are services on Sundays at 10am and 11.30am; the church is only open to the public during weekend afternoons in August (call or check online before setting out). Next to the church's gates is a two storey tower, once used as a graveyard lookout point and now called the Session House. Such features were common in Edinburgh kirkyards in the 19th century, when grave-robbers such as Burke and Hare were on the prowl for fresh cadavers to sell to the medical schools. Look out too for the 'loupin'-on-stane' (jumping-on stone), used for mounting horses, and a punishment collar known as 'the jougs'.

The small, octagonal building at the foot of the Duddingston Kirk manse, and the edge of the loch, is **Thomson's Tower**. The structure was named for the Reverend John Thomson (1778-1840), one of the parish's best-known ministers and also a landscape painter. It was Thomson who coined the Scots phrase, 'We're a' Jock Tamson's bairns' (meaning, all men are equal in the eyes of God). Thomson's visitors included the artist JMW Turner and the writer Sir Walter Scott, who wrote part of *The Heart of Midlothian* in the tower. In the mid 1790s, the loch also provided the setting for one of Scotland's most famous paintings: Sir Henry Raeburn's *Reverend Robert Walker Skating on Duddingston Loch*. You can see the original in the **National Gallery** (*see p88*).

Back in the village, most visitors end up in the **Sheep Heid Inn** (Causeway, 661 7974, www.thesheepheidedinburgh.co.uk), a country pub far removed from urban life. There's been an inn on this spot since 1360; one of its incarnations was a favourite of James VI, who presented the innkeeper with a snuffbox decorated with a ram's head, which inspired the name.

South Edinburgh

From the fringes of the Old Town to the Pentlands, South Edinburgh encompasses the urban, the suburban and the positively pastoral. Close to the heart of the city, Lothian Road boasts what is almost an arts village, while the grassy expanse of the Meadows offers accessible green space. Attractions such as Surgeons' Hall Museums and the Edinburgh College of Art are also held tight to Edinburgh's bosom, but single items of interest become rarer in the great residential sweep from Merchiston in the west to Newington in the east. Among the southern suburbs, the commanding views from Blackford Hill deconstruct the city's geography. And then, just where you least expect it, you'll find Edinburgh's other castle.

Timberyard.

Don't Miss

1 George Square in August A focal point for fun on the Fringe, with food stalls and bars (p137).

2 Craigmillar Castle Mary, Queen of Scots was here, twice, in the 1560s (p141).

3 Surgeons' Hall Museums Jars of pickled people parts and more (p138).

4 Timberyard Modern British food at its best (p133).

5 Arthouse movies Find them at the Filmhouse and the Cameo (p174).

EXPLORE

Usher Hall.

LOTHIAN ROAD & TOLLCROSS

Running south from the western end of Princes Street is **Lothian Road**. The thoroughfare has long had a reputation for carousing: when Robert Louis Stevenson was a student, more than 140 years ago, he came here with his friends in search of beer and bawdiness. Later, in **St Cuthbert's Churchyard**, Stevenson was said to have found a gravestone bearing the name 'Jekyll', which he went on to use in his most famous book.

Over the years, a degree of respectability has crept in to the area. The construction of the **Royal Lyceum Theatre** (*see p190*) in 1883 and the elegant **Usher Hall** (*see p187*), three decades on, helped; much later, the arrival of the **Filmhouse** (*see p174*) and the **Traverse Theatre** (*see p191*) created an artistic cluster. However, seediness remains. As darkness falls, the area's pubs and clubs become magnets for young drinkers. Within the space of a block or two, you could watch traditional ballet or lap dancing, get into a discussion about Lars von Trier or get into an argument in a taxi rank. Jekyll and Hyde, indeed.

The **Union Canal** (*see p146*) starts just west of Lothian Road, while at the road's south end sits **Tollcross**, a major junction now dominated by the Princes Exchange. Built in 2001 to house financial-sector offices, it may be big but it's hardly beautiful. For something with a little more soul, keep walking up Home Street and Leven Street until the **Barclay Viewforth Church** looms out of nowhere. With one of the tallest spires in Edinburgh, this Franco-Venetian, Gothic extravaganza was created by FT Pilkington in 1864, and first known as the Barclay Church.

Sights & Museums

FREE Edinburgh College of Art
Hunter Building, 74 Lauriston Place (651 5800, www.eca.ed.ac.uk). Bus 2, 23, 27, 35, 45, 47. **Map p134 C3 ①**
With 18th-century roots, the Edinburgh College of Art (ECA) took its current name in 1907; in 2011, it merged with the University of Edinburgh, but it still retains a separate identity and character. Students learn architecture, art, design and music; its main annual events are the runway fashion show (late April) and the week-long degree art show (late May/June – see website for details). At these jamborees, the public gets to see work from the hopelessly quotidian to the really rather special: the 2009 Turner Prize-winner, Richard Wright, is an ECA graduate.

Restaurants

★ Castle Terrace
33-35 Castle Terrace (229 1222, www.castle terracerestaurant.com). Bus 1, 2, 10, 11, 15, 15A, 16, 22, 24, 34, 35, 36, 47. **Open** noon-2pm, 6.30-10pm Tue-Sat. **Main courses** £27-£42. **Map p134 B3 ② Modern British**
From the same stable as the Michelin-starred Kitchin in Leith (*see p160*), and sharing its 'From Nature to Plate' ethos, Castle Terrace, under the guiding hand of chef-patron Dominic Jack, also has a Michelin star. It was awarded for its exemplary tasting menus – including a vegetarian version – and for main courses such as saddle of roe deer with celery, celeriac, apple and caramelised walnuts. The understated modern decor does not distract from the food or able service.

Passorn

23-23A Brougham Place (229 1537, www.passorn thai.com). Bus 10, 11, 15, 15A, 16, 23, 24, 27, 36, 45, 47. **Open** 5.30-11pm Mon; noon-2.30pm, 5.30-11pm Tue-Fri; 12.30-3pm, 5.30-11pm Sat. **Main courses** £11-£20. **Map** p134 B4 ❸ **Thai**
Passorn has built an enviable reputation for the quality of its cooking; no MSG is used and the kitchen has a commitment to quality Scottish and Thai produce. Vegetarian spring rolls are recommended, as is the tom yum soup. Mains are as standard as pad thai, as deft as pla samun pri (monkfish in turmeric and coconut sauce with lemongrass, garlic and chilli).

El Quijote

13A Brougham Street (478 2856, www.quijote tapas.co.uk). Bus 10, 11, 15, 15A, 16, 23, 24, 27, 36, 45, 47. **Open** 5-10pm Mon-Thur; 12.30-10pm Fri-Sun. **Main courses** £15-£22. **Map** p134 B4 ❹ **Spanish**
Small, rustic and genuinely Iberian, what you get here is inventive tapas of some quality, bigger versions to share, and main course-sized portions along the lines of flamenquin cordobes (pork loin in breadcrumb with serrano ham, manchego cheese and peppers, served with salad and chips).

Shebeen

8 Morrison Street (629 0261, www.shebeenbar. co.uk). Bus 1, 2, 10, 11, 15, 15A, 16, 24, 34, 35, 36, 47. **Open** 5-10pm Mon-Wed; noon-2pm, 5-10pm Thur, Fri; 3-10pm Sat; 4-10pm Sun. **Main courses** £10-£26. **Map** p134 A3 ❺ **South African**
Just off Lothian Road, simply furnished but with a touch of Africa in the decor, Shebeen's menu runs from starters such as sosatie (beef, pork or prawn on a skewer), burgers such as the Soweto Mess and the almost legendary Bunny Chow (curry in a hollowed-out loaf). Steaks, meanwhile, come in two absolutely insane sizes: cheetah (under 480g) and lion (500g or over).
Other location 103 Dalry Road, West Edinburgh (629 3030).

★ Timberyard

10 Lady Lawson Street (221 1222, www. timberyard.co). Bus 1, 2, 10, 11, 15, 15A, 16, 22, 24, 34, 35, 36, 47. **Open** noon-2pm, 5.30-9.30pm Tue-Sat. **Main courses** £21-£26. **Map** p134 B3 ❻ **Modern British**
A leading restaurateur in Edinburgh for more than 20 years, Andrew Radford opened his latest fine-dining venture in a former timber yard in 2012. It's a Radford family affair (son Ben is in charge of the kitchen, other Radfords elsewhere), raw materials are carefully sourced and the food is excellent, both in flavour and presentation. A dish of smoked beef, cauliflower, kohlrabi, mushroom, kale, radish and ramps (wild garlic) sounds like an assemblage, but looks and tastes like art.

Tuk Tuk

1 Leven Street (228 3322, www.tuktukonline. com). Bus 10, 11, 15, 15A, 16, 23, 27, 36, 45. **Open** noon-10.30pm daily. **Dishes** £3.95-£5.25. **Map** p134 B5 ❼ **Indian**
Given the Scottish weather, it's a blessing to find Indian street food indoors, although Tuk Tuk is clearly a restaurant rather than an errant market stall. The informing ethos is 'rustic, roadside and railway station', while the menu offers a range of tapas-sized dishes from chicken curry on the bone or tarka dal, to more fun creations such as the Mumbai burger and chicken lollipops. No licence, BYOB.

EXPLORE

Fashion show, Edinburgh College of Art.

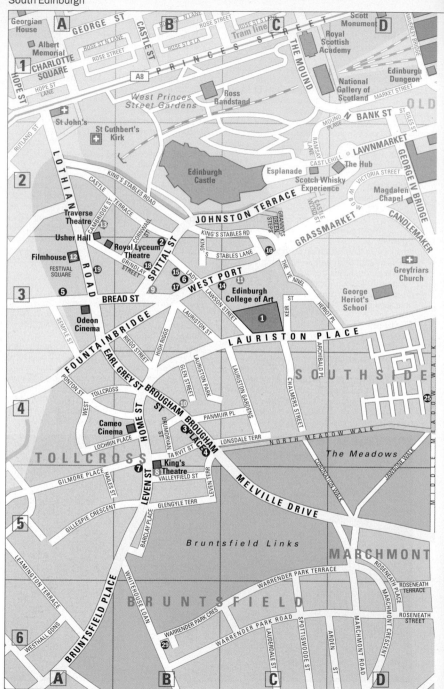

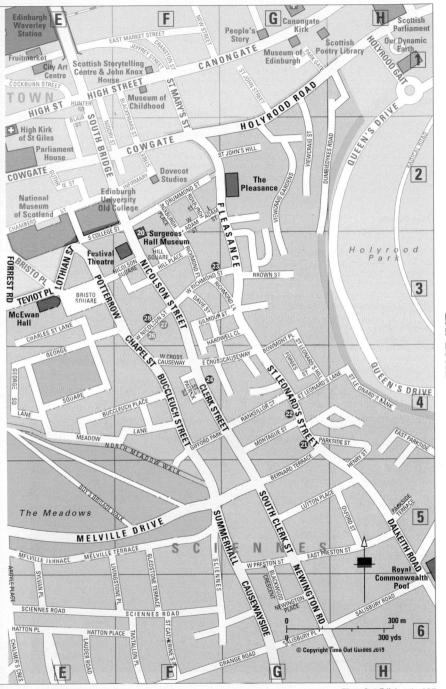

Pubs & Bars

Bennet's

*8 Leven Street (229 5143, www.bennetsbar.co.uk).
Bus 10, 11, 15, 15A, 16, 23, 27, 36, 45.* **Open**
11am-1am Mon-Sat; noon-1am Sun. *Food served*
11am-10pm daily. **Main courses** £10-£14.
Map p134 B5 ⑧
A marvel of Victorian design. A long wooden bar
occupies one side of the room, with alcoves along the
top of the gantry accommodating a huge selection
of single malts, while the opposite wall has fitted red
leather seats and more wooden fittings. Enjoy the
whisky selection and hearty pub grub in the bar or
in the Skean Dhu restaurant space through the back.
▶ *An ideal place for a drink before or after a show at
the King's Theatre next door; see p190.*

Blue Blazer

*2 Spittal Street (229 5030). Bus 1, 2, 10, 11, 15,
15A, 16, 22, 24, 34, 35, 36, 47.* **Open** 11am-1am
Mon-Sat; 12.30pm-1am Sun. **Map** p134 B3 ⑨
Sandwiched between the lads-night-out chaos of
Lothian Road and the lap-dancing bars at the top of
the West Port, the Blue Blazer is a cosy place to hide
away and chat, with a decent pint of cask ale, especially in the wee room through the back.

★ Cloisters

*26 Brougham Street (221 9997, www.cloistersbar.
com). Bus 10, 11, 15, 15A, 16, 23, 24, 27, 36, 45,
47.* **Open** noon-midnight Mon-Thur; noon-1am
Fri, Sat; 12.30pm-midnight Sun. *Food served*
noon-2.30pm, 5-9pm Tue-Fri; noon-9pm Sat;
12.30-5.30pm Sun. **Main courses** £7.50-£12.50.
Map p134 B4 ⑩

Ploughing its quality beer furrow for nearly 20
years, Cloisters is based in a former parsonage
beside a church, which accounts for the sparse,
ecclesiastical interior. There's good cask and keg
beer from all over Britain, but primarily Scotland.
Food includes superior burgers, seafood options and
snacky light bites.

Dragonfly

*52 West Port (228 4543, www.dragonflycocktail
bar.com). Bus 2, 35.* **Open** 4pm-1am daily.
Food served 4-10pm daily. **Map** p134 C3 ⑪
Now more than a decade old and a little shabby
around the edges, Dragonfly endures as a handy
place for a cocktail whether you ask the staff to mix
you something classic or opt for a drink from their
menu such as Sage Against The Machine (tequila,
sage, pineapple, lime and sugar syrup).

Filmhouse Café Bar

*88 Lothian Road (229 5932, www.filmhouse
cinema.com). Bus 1, 2, 10, 11, 15, 15A, 16, 24,
34, 35, 36, 47.* **Open** 8am-11.30pm Mon-Thur;
8am-12.30am Fri; 10am-12.30am Sat; 10am-
11.30pm Sun. *Food served* 8am-10pm Mon-Fri;
10am-10pm Sat, Sun. **Main courses** £7-£9.
Map p134 A3 ⑫
The city's independent arthouse cinema has been
around since the 1980s and continues to draw a loyal
crowd of cinephiles. With coffee, snacks, light meals
(chickpea and coconut curry, chilli) and a couple of
good beers on tap, the Filmhouse's café-bar – currently getting an upgrade – is the perfect place to
meet friends before or after a movie. These days, you
can even have breakfast here.
▶ *For the cinema itself, see p174.*

Cloisters.

Traverse Theatre Bar

10 Cambridge Street (228 5383, www.traverse. co.uk). Bus 1, 2, 10, 11, 15, 15A, 16, 24, 34, 35, 36, 47. **Open** 11am-midnight Mon-Wed; 11am-1am Thur-Sat. *Food served* 11am-8pm Mon-Sat. **Main courses** £6.50-£9. **Map** p134 A2 ⑬

The basement bar at the Trav gets very busy pre-and post-performance, especially during the Fringe. The modern, roomy establishment is mainly open plan, but there's a more closed-off dining space in one corner. It attracts a typical café-bar crowd: you can pick up anything from a coffee or a glass of wine to puy lentil chilli with potato wedges.

▶ *For the theatre, see p191.*

Shops & Services

Armchair Books

72-74 West Port (229 5927, www.armchairbooks. co.uk). Bus 2, 35. **Open** 10.30am-6.30pm daily. **Map** p134 B/C3 ⑭ **Books & music**

If the interior of a second hand and antiquarian bookshop ought to feel like a wander through a novel by Borges or Eco, then this is a winner. Small, cramped, untidy, appealing and with a droll Twitter feed, it offers everything you want from an establishment of this nature. Stock covers subjects from architecture to travel, and there's fiction too.

Boardwise

4 Lady Lawson Street (229 5887, www.boardwise. com). Bus 1, 2, 10, 11, 15, 15A, 16, 22, 24, 34, 35, 36, 47. **Open** 10am-6pm Mon-Sat; noon-5pm Sun. **Map** p134 B3 ⑮ **Sports**

Skateboards, longboards, decks, trucks, wheels, bearings and shoes; snowboards, boots, bindings and bags; surfboards, paddleboards, wax, leashes, fins, wetsuits; assorted clothing. If you want to stand on a plank and move, Boardwise has the expertise to sell you that plank, what you need to maintain that plank and a Burton hoody as well.

Godiva

9 West Port (221 9212, www. godivaboutique. co.uk). Bus 2, 35. **Open** 10.30am-6.30pm Mon-Sat; 10.30am-6pm Sat; 11.30am-5.30pm Sun. **Map** p134 C3 ⑯ **Fashion**

Making a reasonable claim to be the hippest boutique in the city, Godiva sells items by independent designers, as well as vintage clothing and accessories. This covers everything from women's printed T shirts by Naromode via mega pleat skirts by Rowanjoy to recycled leather shopper bags by Paperthinks, or perhaps just a chic, second-hand frock.

Herman Brown

151 West Port (228 2589, www.hermanbrown. co.uk). Bus 2, 35. **Open** 1-6pm Mon-Sat. **Map** p134 B3 ⑰ **Fashion**

With roots going way back to 1980s market stalls, Herman Brown is now a well-established vintage

fashion store, where you can find anything from bodystockings or leotards to 1940s dresses, plus jewellery, accessories and a small range of items for men. With plenty of classy finds at good prices, it's definitely worth a look.

McAlister Matheson Music

1 Grindlay Street (228 3827, www.mmmusic. co.uk). Bus 1, 2, 10, 11, 15, 15A, 16, 22, 24, 34, 35, 36, 47. **Open** 9.30am-6pm Mon-Fri; 9am-5.15pm Sat. **Map** p134 B3 ⑱ **Books & music**

At the other end of Grindlay Street from the Usher Hall concert venue, this shop is your first stop in Edinburgh for classical music expertise. There's a huge number of discs on display, both new and back catalogue, and the staff know precisely what they're talking about and you can pick up magazines such as *Gramophone* here too.

Wonderland

97, 101 & 103 Lothian Road (229 6428, www.wonderlandmodels.com). Bus 1, 2, 10, 11, 15, 15A, 16, 24, 34, 35, 36, 47. **Open** 9.30am-6pm Mon-Sat. **Map** p134 A3 ⑲ **Toys & games**

This is far more than a toyshop. It does have model trains, gorgeous dolls' houses, kites and more, but how many children prefer playing Minecraft or Grand Theft Auto these days? Consequently, Wonderland bills itself as a model shop and hobby store, attracting grown-ups as well as kids, and catering to the former with camera-mountable quadcopters, radio-controlled airplanes and boats, or Scalextric sets for over two hundred quid.

GEORGE SQUARE & AROUND

Located between Bristo Square and the Meadows, **George Square** should cause the city fathers of yore to hang their heads in shame. Built around handsome central gardens, the square was once completely lined by elegant houses dating from the 1760s; no.25 was home to a young Walter Scott. Some survive along the west side of the square, but many were levelled in the 1960s for the University of Edinburgh. The central gardens are open during the week (until 7pm in summer, 4pm in winter), but they really come into their own every August when temporary Fringe venues, food stalls and bars move in and the whole place operates like a fair until late at night, seven days a week.

Around the gardens, the squat **University Library** (1967) by Sir Basil Spence on the south-west corner was constructed to replace the more elegant library in Old College (in the Old Town); the south-east corner commemorates the city's most famous philosopher son in the shape of the **David Hume Tower** (1963). The **Appleton Tower** (1966), set back from the east side of the square on Crichton Street, is widely regarded as one of the city's more objectionable buildings.

EXPLORE

EXPLORE

Refurbishment work in the last decade and more hasn't improved its aesthetics to any great degree, although the university's latest crack at architecture on Crichton Street, known as the **Informatics Forum** (2008), serves to obscure it from some angles.

To the north, **Bristo Square** is also home to some modern-era university buildings, but also squeezes in a riot of architectural Victoriana. Highlights include **Teviot Row House** (1889), Britain's oldest, purpose-built student union; the **Reid Concert Hall** (1859; *see p187*); the **McEwan Hall** (1897), designed by Sir Robert Rowand Anderson and funded by brewing magnate Sir William McEwan; and, next door on Teviot Place, the old **Medical School** (1888).

A little further west on Lauriston Place is the extensive **Quartermile** development, a whole new neighbourhood on the site of the old Royal Infirmary. Its mix of residential and commercial space blends Victorian and contemporary, with some of the original hospital buildings standing alongside sleek, dark cuboids constructed since 2006.

Head in the opposite direction from Bristo Square, and you'll find another building by the celebrated architect Sir William Playfair. Completed in 1832, **Surgeons' Hall** was where Burke and Hare brought their freshly deceased victims. It remains home to the Royal College of Surgeons of Edinburgh, a body that can trace its roots to 1505, and to the celebrated **Surgeons' Hall Museums**. In contrast, the site directly opposite has housed a place of entertainment since the 1820s. The Empire Palace Theatre was built here in 1892 and it was this that was radically refurbished, and given a modern façade, in 1994 to create the **Festival Theatre** (*see p188*).

Sights & Museums

★ Surgeons' Hall Museums

Nicolson Street (527 1711, www.museum. rcsed.ac.uk). Nicolson Street–North Bridge buses. **Open & admission** see website. **Map** p135 F2/3 ⓴

Currently closed for refurbishment but scheduled to reopen in autumn 2015, the collections at Surgeons' Hall Museums allow visitors to trace the history of medicine in the city from 1505, when the Barber Surgeons of Edinburgh were incorporated, to the development of modern surgical techniques. The main attraction is the celebrated and sometimes quite disturbing pathology collection – pickled remains of body parts to the fore – although the revamp is adding an 18th-century anatomical theatre to the premises, and further exhibition space. The dental collection, meanwhile, will make you very glad that you live in an era of 21st-century technology when it comes to getting that wobbly tooth fixed.
▶ *The Royal College of Surgeons of Edinburgh also owns a hotel on site, Ten Hill Place; see p259.*

Restaurants

★ Aizle

107 St Leonard's Street (662 9349, www.aizle. co.uk). Bus 14 or Nicolson Street-North Bridge buses. **Open** 6-9.30pm Wed-Sun. **Set meal** £45. **Map** p135 G4 ㉑ **Neo-bistro**
Scottish-American couple Stuart Ralston (chef) and Krystal Goff (mixologist) were working in elevated establishments in the US and the Caribbean, but moved across the Atlantic to introduce 'bistronomie' to Edinburgh: small dishes of high gastronomic standard served bistro style. There is no menu as such; what you get are five courses fashioned from a monthly changing ingredients list. Dishes could include ricotta gnudi with burnt leeks, mussels and purslane; perhaps torched mackerel with ras el hanout, cauliflower and golden raisins. Aizle is fun, creative and was the most significant restaurant opening in Scotland in 2014.

Blonde

75 St Leonard's Street (668 2917, www.blonde restaurant.co.uk). Nicolson Street–North Bridge buses. **Open** 6-9pm Mon; noon-2.30pm, 6-10pm Tue-Sat; noon-2.30pm, 6-9pm Sun. **Main courses** £11-£19. **Map** p135 G4 ㉒ **Modern European**
This modern neighbourhood restaurant, named for its pale wood interior, has long been a real asset to the area. The menu takes an eclectic approach: dishes such as tofu and mint koftas as a starter, venison casserole with red wine, root vegetables and chocolate as a main, perhaps.

Bonsai

46 West Richmond Street (668 3847, www.bonsaibarbistro.co.uk). Nicolson Street– North Bridge buses. **Open** noon-10pm Mon-Fri; noon-10.30pm Sat, Sun. **Main courses** £4-£9. **Map** p135 F3 ㉓ **Japanese**
The team in this small establishment produces a range of dishes, many of which are designed not to spook the local palate: cheese gyoza (cheddar in

Aizle.

pastry with soy-chilli dip), prawn tempura, vegetable tempura, beef teriyaki or California gaijin-zushi (featuring avocado and crabstick). If you want to go with actual sushi, however, you can.
Other location 14 Broughton Street, Broughton (557 5093).

Kalpna
2-3 St Patrick Square (667 9890, www.kalpna restaurant.com). Nicolson Street–North Bridge buses. **Open** noon-2pm, 5.30-10.30pm Mon-Sat; 6-10.30pm Sun (Oct-Apr closed Sun). **Main courses** £6.50-£15.50. **Map** p135 F4 ㉔ **Indian/ Vegetarian**
A purveyor of Indian vegetarian food for more than three decades, Kalpna has stayed much the same over the years, although the main dining area has been well maintained. The lunchtime buffet is cheap, but you need to catch the dishes when they're fresh. At dinner, try a thali, a dosa or dam aloo kashmeri, the restaurant's signature dish: potato 'barrels' filled with vegetables, nuts and paneer in a complex sauce.

£ Peter's Yard
27 Simpson Loan, off Middle Meadow Walk (228 5876, www.petersyard.com). Bus 2, 23, 27, 35, 41, 42, 45, 47, 60, 67. **Open** 7.30am-7pm Mon-Fri; 9am-7pm Sat, Sun. **Main courses** £6. **Map** p134 D4 ㉕ **Café**
Part of the Quartermile development that's brought hundreds of new luxury apartments to the site of the former Royal Infirmary, Peter's Yard is a bright, airy café with a Swedish slant, offering superior cakes and pastries, platters, soup, hot drinks and more. The breads are very good too. Since opening in 2007, it's spawned a mini empire with bakery-shops and another café.
Other location 3 Deanhaugh Street, Stockbridge (332 2901).

Pubs & Bars

Pear Tree House
38 West Nicolson Street (667 7533, www.pear-tree-house.co.uk). Nicolson Street–North Bridge buses. **Open** 11am-midnight Mon-Thur; 11am-1am Fri, Sat; 12.30pm-midnight Sun. **Map** p135 F3 ㉖
Just over the road from the University of Edinburgh's George Square campus, the Pear Tree's cobbled beer garden has played host to generations of thirsty students and Fringe-goers looking for an alfresco beer on a sunny day. Inside, the decor is classic trad Scots pub, with the addition of a big screen for the football. Outside term-time and August, the beer garden is fairly placid.

Usher's of Edinburgh
32B West Nicolson Street (662 1757, www. ushersofedinburgh.co.uk). Nicolson Street– North Bridge buses. **Open** noon-midnight Mon-Thur, Sun; noon-1am Fri, Sat. *Food served* noon-10pm daily. **Main courses** £8.50-£15.50. **Map** p135 F3 ㉗
This 2014 debutante was an instant hit, getting the balance right between its modern basement bar atmosphere, food menu and undoubted commitment to beer. You can eat well here (all-day breakfasts, steak and chips, curried lamb in barley ragout), but when it comes to draught beer it has 15 keg taps, another five on cask – all rotating – and a 'beerbot' on Twitter to help you keep track.

Shops & Services

Word Power
43 West Nicolson Street (662 9112, www.word-power.co.uk). Nicolson Street–North Bridge buses. **Open** 10am-6pm Mon-Sat; noon-5pm Sun. **Map** p135 F3 ㉘ **Books & music**

EXPLORE

Launched more than 20 years ago with Scottish writer James Kelman performing at the opening ceremony, Word Power is the exact opposite of Waterstones. Small, independent and radical, it was described by another Scottish writer, Alan Bissett, as 'a hive-mind of the anti-mainstream and the left-field'. It's also a very good bookshop.

THE SOUTHERN SUBURBS

The immediate centrepiece of South Edinburgh is the **Meadows**, stretching east to west from Newington to Tollcross and north to south from Quartermile to Marchmont. Formerly the site of the shallow Burgh Loch, it once provided the city's drinking water. It was drained in the 17th and 18th centuries, leaving the distinctive flat, grassy area known today. Tree-lined paths, complete with cycle lanes, cut across in every direction; it's a popular spot for joggers and amateur sports. A lack of lighting means caution should be taken late at night, but it's generally a safe place.

South of the Meadows, **Marchmont** is composed almost entirely of grand tenements, built between 1876 and 1914. The cobbled sweep of Warrender Park Road gives a flavour of the district, which has long had the atmosphere of a student ghetto thanks to its proximity to various colleges. The nearby **Warrender Swim Centre** (55 Thirlestane Road, 447 0052, www.edinburghleisure.co.uk), housed in a beautiful Victorian building dating from 1887, is an impressive place for a dip.

To the south-west, the Meadows opens out into **Bruntsfield Links** – all that's left of the old Burgh Muir ('town heath') gifted to the city by David I in 1128. It's long been associated with golf: there are records mentioning golf at Bruntsfield in the 17th century and there was an official Bruntsfield Links Golfing Society from 1761, although it eventually moved elsewhere in the city. A short-hole course was set up in 1895; it still exists, effectively a 36-hole pitch-and-putt (see http://golf.tollcross.org). The claim that Bruntsfield Links is the oldest golf course in the world may have some merit.

Bruntsfield itself is a thriving locale, mostly residential away from its main thoroughfare and the retail-minded Bruntsfield Place. Further south sits **Morningside**, once notoriously snooty but now merely exclusive.

Beyond these suburbs is **Blackford Hill** with its astonishing views back towards the city and Arthur's Seat. At its top stands the twin-teacake structure of the **Royal Observatory**, moved here in 1896 after the facilities on Calton Hill (*see p119*) were deemed no longer fit for purpose. The Blackford Hill observatory has a visitor's centre, but it's only open for pre-booked group visits and public astronomy evenings (see www.roe.ac.uk). East of Blackford Hill is **Craigmillar Castle**. Further south still lie the **Braid Hills**; these also offer amazing views back towards central Edinburgh.

Sights & Museums

Craigmillar Castle

Craigmillar Castle Road (668 8600, www. historic-scotland.gov.uk). Bus 21, 24, 33, 38, 49. **Open** *Apr-Sept* 9.30am-5.30pm daily. *Oct-Mar* 9.30am-4.30pm Mon-Wed, Sat, Sun. **Admission** £5.50; £3.30-£3.40 reductions. *See p141* **Edinburgh's Other Castle**.

Restaurants

£ Falko Konditorei

185 Bruntsfield Place (656 0763, www.falko. co.uk). Bus 11, 15, 15A, 16, 23, 36, 45. **Open** 10am-6pm Wed-Fri; 8.30am-6pm Sat; 9.30am-6pm Sun. **Main courses** £5-£7.50. **Café**
This authentic German bakery with some café space is just the place to sit down with a generous lump of black forest gateau and a strong coffee. Cakes are made fresh every day; by late afternoon, the best ones might be gone. Also on offer: great bread, creative sandwiches and very good teas and coffees.

Nonna's Kitchen

45 Morningside Road (466 6767, www.nonnas-kitchen.co.uk). Bus 5, 11, 15, 15A, 23, 36. **Open** 10am-2.45pm, 5-10pm Tue-Sun. **Main courses** £8-£25. **Italian**
Friendly, family-run, with a simple fresh look and capable of anything from a decent pizza margherita or penne all'arrabbiata to veal escalope, fillet steak or king prawn thermidor, Nonna's Kitchen is exactly the kind of relaxed Italian restaurant you'd want in the neighbourhood.

★ Rhubarb

Prestonfield, Priestfield Road (225 7800, www. prestonfield.com). Bus 2, 14, 30. **Open** noon-2pm, 6.30-10pm Mon-Thur; noon-2pm, 6-11pm Fri, Sat; 12.30-3pm, 6.30-10pm Sun. **Main courses** £18-£35. **Modern European**

IN THE KNOW SEVEN HILLS

Every June, hundreds of runners take part in the Seven Hills of Edinburgh Race & Challenge (www.seven-hills.org.uk). It starts on Calton Hill, then goes via Edinburgh Castle, Corstorphine Hill and Craiglockhart Hill East, on to Braid Hill and Blackford Hill (both in South Edinburgh), then to Arthur's Seat and back to Calton Hill for the finish. The distance is just over 14 miles with 2,200 feet of ascent.

EXPLORE

EDINBURGH'S OTHER CASTLE

All of the history, none of the crowds.

A fabulous and largely intact ruin with parts dating back around six centuries, **Craigmillar Castle** (*see p140*) is a little out of the way; two and a half miles south of the city centre, give or take. However, there's an advantage to its relative isolation: you can get a taste of a little history with none of the Old Town queues. A couple of hundred visitors during a peak summer Sunday constitutes 'busy'. Sometimes, on a weekday in February, you could be here completely on your own.

The L-shaped tower house at the castle's core dates from around 1400, and was built by a notable local family called the Prestons. They later added a mighty curtain wall and, in the early 16th century, yet more layers of defence and protection. After it was captured by the English in 1544, the castle got another makeover, and it was in this condition that it greeted its most famous guest, Mary, Queen of Scots, who stayed here for a few days in September 1563 and again for more than a fortnight in November/ December 1566, when she discussed high politics and low deeds with leading nobles

of the day. She was not quite 24 years old at the time. By the following summer, Mary had been imprisoned and forced to relinquish the Scottish throne in favour of her infant son, James VI. By 1568 she was in custody in England; she led an eventful if tragic life.

Craigmillar Castle passed out of the hands of the Prestons and was acquired by another family of notables, the Gilmours, in 1660. They moved out in the 18th century to more modern accommodation and left the castle to moulder, romantically, as part of their estate. It eventually fell into ruin and persisted in this condition until the government stepped in just after World War II.

Today, the castle is the responsibility of Historic Scotland. It's a shell, with none of the museum ambience that defines Edinburgh Castle. However, the atmosphere and the views make a visit worthwhile. What's more, if you have children, they can happily stage mock fights here with plastic swords, which isn't something that would go down too well in more confined and security-conscious parts of Edinburgh Castle.

EXPLORE

View of Arthur's Seat from Blackford Hill. *See p140.*

James Thomson, the brains behind the Witchery (*see p61*) and the Tower, took over the old Prestonfield House Hotel, gave it a sumptuous makeover and reopened it as Prestonfield over a decade ago. Rhubarb, the hotel's gorgeous restaurant, offers the likes of white and green asparagus (in season), morels and squat lobster as a starter; guinea fowl with confit leg pastilla, date and red onion purée, spiced couscous and preserved lemon as a main. You can also have afternoon tea (£20) or a light two-course lunch (also £20) if you want to experience the luxury without splashing out on the à la carte.
► *For the hotel, see p258.*

£ S Luca

16 Morningside Road (446 0233, www.s-luca. co.uk). Bus 11, 15, 15A, 16, 23, 36, 45. **Open** 9am-10pm daily. **Main courses** £5-£8. **Café**
S Luca is famed for the ice-cream served at its legendary Musselburgh café (*see p203*). This branch – located near Holy Corner, so named because it has more churches per square foot than Vatican City – also offers sandwiches and panini melts, burgers, pasta and pizza. But who needs them when there are snowballs and fudge sundaes to be scoffed?

Secret Herb Garden

32A Old Pentland Road (07525 069773 mobile, http://secretherbgarden.co.uk). Bus 15, 15A. **Open** 10am-4pm daily. **Café**
Just on the other side of the Edinburgh City Bypass not far from Fairmilehead, the Secret Herb Garden is around 15 minutes' walk from the main road where you get off the bus – it's easier to get here by bike or car. What you find is a herb garden, shop and café in assorted greenhouse and farm buildings that are undeniably charming. The café has excellent soup, quiche and cakes, and every full moon the venue hosts a Full Moon Dinner in one of the glasshouses. As far as the Edinburgh dining scene goes, it's a unique experience, and the wines are good too.

Toast

146 Marchmont Road (446 9873, www.toast edinburgh.co.uk). Bus 5, 24, 41. **Open** 10am-10pm Mon-Sat; 10am-4pm Sun. **Main courses** £11.50-£14. **Café-bistro**
With art on the walls and lots of blond wood, Toast looks smart and contemporary. The menu changes as the day progresses, kicking off with some excellent breakfast dishes (french toast with bacon and maple syrup, say), moving through to sandwiches, salads and light meals, then becoming more of a bistro in the evenings. It's one of the best places for a relaxed weekend brunch – if you can get a table.

£ Zulu Lounge

366 Morningside Road (466 8337, www.the zululounge.com). Bus 5, 11, 15, 15A, 23, 36. **Open** 7.30am-5pm Mon-Fri; 8am-4pm Sat, Sun. **Main courses** £2.75-£7.25. **Café**
A café and takeaway for those hankering after a boerewors roll or some pap and wors (sausage and chakalaka sauce on maize meal, with a side salad), the Zulu Lounge may be cramped, but it offers a taste of South Africa, along with more familiar lunch snacks and ice-cream floats for your inner child.

Pubs & Bars

★ Canny Man's

237 Morningside Road (447 1484, www.canny mans.co.uk). Bus 5, 11, 15, 15A, 16, 36. **Open** 11am-11pm Mon-Wed, Sun; 11am-midnight Thur, Sat; 11am-1am Fri. *Food served* noon-3pm, 6.30-9pm Mon-Thur, Sat; noon-3pm Fri, Sun. **Main courses** £7.50-£30.
A gastropub before the term was invented, a venue of great character, a warren of rooms and knick-knacks, a genuine survivor from the Victorian era and a specialist in smørrebrød, the Canny Man's has a great deal to recommend it. There are cocktails, champagnes, single-malt whiskies and good beer.

The main menu offers fresh seafood (dressed crab, oysters, cocktail de crevettes, lobster), as well as those smørrebrød options. But. The venue's froideur to new faces is legendary and even its own website quotes the line, 'Loved by those who know it and hated by those who do not understand it.'

Montpeliers
159-161 Bruntsfield Place (229 3115, www.montpeliersedinburgh.co.uk). Bus 11, 15, 15A, 16, 23, 36, 45. **Open** 9am-1am daily. *Food served* 9am-10pm daily. **Main courses** £8-£17.

Although it's been trading since 1992, the odd refurb has ensured that Montpeliers hasn't become dated or dowdy: it's currently rocking the arty-banquette-Ikea-studio-kitchen-bar look. The menu runs from a very good breakfast to coffee and cakes; sandwiches, salads, burgers and more at lunch; sea bass, pork belly or lamb shank in the evenings. There are seats outside for summer evenings on Bruntsfield Place.

Shops & Services

Coco
174 Bruntsfield Place (228 4526, www.coco chocolate.co.uk). Bus 11, 15, 15A, 16, 23, 36, 45. **Open** 10am-6pm Mon-Sat; noon-4pm Sun. **Food & drink**

Bizarre as it may sound, some people don't eat chocolate – or not until they visit Coco. The range on sale (dark chocolate with chilli or raspberry, for example) tempts the most recalcitrant denier, and service is unerringly friendly. If you want to learn about handling chocolate, you can sign up for classes at the Coco Chocolate Kitchen.

Other location 71 Broughton Street, Broughton (558 2777).

Earthy Market Café & Store
33-41 Ratdiffe Terrace (667 2967, www.earthy. uk.com). Bus 42, 67. **Open** 9am-7pm Mon-Fri; 9am-6pm Sat; 10am-6pm Sun. **Food & drink**

An attractive, comprehensively stocked wholefood store, with a rustic-looking café offering everything from breakfast to a hearty moussaka.

Other locations 1-6 Canonmills Bridge, Stockbridge (shop 556 9699, café 556 9696); 19 Windsor Place, Portobello (shop 344 7930).

Edinburgh Bicycle Cooperative
8 Alvanley Terrace (228 3565, www.edinburgh bicycle.com). Bus 11, 15, 15A, 16, 23, 36, 45. **Open** 9am-6pm Mon-Wed, Fri; 9am-7pm Thur; 10am-6pm Sat, Sun. **Map** p134 B6 ⑭ **Sports**

Since 1977, this has been a centre of expertise for bike sales (its own Revolution brand, Cannondale, Marin, Specialized, Whyte and many others), bike parts and repairs. EBC also offers classes on bike maintenance or cycling skills.

Other location 30 Rodney Street, Broughton (557 2801).

Coco.

West Edinburgh

Aside from a few big attractions (Edinburgh Zoo, Jupiter Artland, Murrayfield Stadium), West Edinburgh doesn't generally detain or divert visitors. But it's here that layers of Edinburgh's economic history fit together like a topological puzzle, just as a few old buildings hidden away in residential warrens hint at a less urban past. Big hitters of the modern-day business world are clustered around the western side of Lothian Road, and Morrison Street, in an area known as the Exchange, not far from the ghosts of Victorian brewing at Fountainbridge. Meanwhile, picturesque Cramond with its Roman associations, Corstorphine Hill and the Craiglockhart Hills offer a break from urban life; as too, in parts, does the surprisingly sylvan Union Canal, its terminus tucked discreetly away just off Lothian Road.

EXPLORE

Edinburgh Zoo

Don't Miss

1 Edinburgh Zoo Pandas, penguins, pygmy hippos and much more (p170).

2 Jupiter Artland Site-specific sculpture by some big names from the art world (p151).

3 Murrayfield The national rugby stadium; also hosts music gigs (p151).

4 Union Canal Your alternative, waterside route out of the city (p146).

5 Edinburgh Trams The smoothest airport link in the British Isles (p147).

Edinburgh Quay, Union Canal.

WEST OF LOTHIAN ROAD

The first recognised name to put down roots in what is now the **Exchange** business district wasn't a bank or a life insurance company, but the **Sheraton Grand Hotel** (*see p257*) on Festival Square, built in 1985. It wasn't until three years later that the local authority decided to promote the locale as a dedicated financial quarter. So began a construction boom, the most important in the city for many years, although very much superseded since.

At first, development centred around the **Edinburgh International Conference Centre**, which arrived on Morrison Street in 1995. Ambitious, albeit flat-topped and functional, it was designed by Sir Terry Farrell. Other new buildings followed: in 1997, the **Standard Life** head office at the corner of Lothian Road and the West Approach Road; four years later, the sweeping crescent roof of **Scottish Widows** on Morrison Street. Also in 2001, the Sheraton Grand redeemed itself for its brutalist façade with the addition, at the rear, of the colourful, Farrell-designed **One Spa**.

Today, every scrap of land around the Exchange seems to be taken up with identikit offices. Such paucity of imagination is only too evident in comparison to the tenements at the corner of Fountainbridge and Grove Street nearby. Built in the mid 1860s by maverick Victorian architect FT Pilkington, they offer a hint of fantasy that's a welcome relief from the area's orthogonal tedium.

South-west of the Exchange and sliced in two by the West Approach Road, the **Fountainbridge** neighbourhood doesn't win many beauty contests. The arrival of the **Fountain Park** (www.fountainpark.co.uk)

complex in 1999 added amenities (it now has a casino, cinema, gym and ten-pin bowling), but it didn't improve the look of the place. There used to be a massive brewery at Fountainbridge, dating from the Victorian era when Edinburgh was a world centre for beer. It brewed its final pint in 2005 and the land was given over to development. That said, it remains the only district within a mile or two of the castle with living memories of an industry that once drove the local economy. Economic obsolescence is just another facet of modern capitalism, but one that holds special resonance for West Edinburgh; this kind of shift has been played out in the area before, with a superhighway that pre-dates the Victorian era.

Completed by Irish navvies and French stonemasons in 1822, the **Union Canal** once ran all the way from Lothian Road (where the Odeon cinema now stands) to Camelon, near Falkirk, linking with the Forth & Clyde Canal that went on to Glasgow. Coal, building materials and passengers came in to Edinburgh; merchants' goods, horse manure and more passengers went out. However, the waterway's heyday proved short-lived, and the canal was bought out by a competitor railway company in 1848. Traffic died out after the 1860s; the Lothian Road terminus was built over in 1922, and the waterway was mothballed in the 1960s.

It took a while for the idea of reopening the canal as a civic amenity and a potentially green transport network to gain support, but it eventually achieved critical mass. The Union Canal was reopened in 2001 with a new city-centre terminus established at Lochrin Basin, off Fountainbridge, as part of the revitalisation. Residential developments, offices, restaurants and cafés arrived, lending it the air of a prosperous little marina. Formally known as

EXPLORE

Edinburgh Quay, it won an award for the best regeneration project in Scotland in 2005.

Popular with walkers, joggers and cyclists, the towpath runs from Lochrin Basin right through the city, out to the countryside. If you leave the towpath as it goes through the suburb of Craiglockhart, just over a mile from Lochrin Basin, you'll be within a stroll of the **Craiglockhart Hills**; among them is the **Easter Craiglockhart Hill**, a local nature reserve with great views and a welcome diversion from the necessarily flat canal. Also in the vicinity is the Business School of Edinburgh Napier University, off Colinton Road, home to the **War Poets Collection**.

Back on the canal, you can keep going as far as your stamina will take you. If you make it to the village of Ratho, around eight miles from the city centre, reward yourself with a meal and a beer at the splendid **Bridge Inn** (27 Baird Road, 333 1320, www.bridgeinn.com), which has outdoor seating overlooking the waterway.

Since 2014, the latest transport wheeze in the area is, of course, the **Edinburgh Trams** line (see p264). Although it has a New Town terminus the vast majority of the line runs through West Edinburgh from Haymarket Station to Murrayfield Stadium and eventually

on to Edinburgh Airport. The project was controversial, badly managed, late and over budget – but it does provide a pretty slick way to get to the city centre from the airport.

Sights & Museums

FREE War Poets Collection

Edinburgh Napier University, Business School, Craiglockhart Campus, 219 Colinton Road (455 4260, www2.napier.ac.uk/warpoets). Bus 4, 10, 27, 36, 45. **Open** 8.45am-9pm Mon-Thur; 8.45am-8pm Fri; 10am-4pm Sat, Sun. Summer hours vary; check website for details. **Admission** free.

This Victorian building at Craiglockhart was used as a sanatorium for shell-shocked soldiers in World War I. Among them were soldier-poets Wilfred Owen and Siegfried Sassoon, who met here in 1917 and wrote some of their best work while recovering from their frontline experiences. The place gives a hint of the world of a century ago, when Craiglockhart offered the men a brief respite before they were sent back into battle. Sassoon survived, but Owen was killed a mere week before the war ended.

▶ *For a fictional account of Sassoon and Owen's meeting at Craiglockhart, see Pat Barker's novel Regeneration (1991) or the film adaptation directed by Gillies MacKinnon (1997).*

EXPLORE

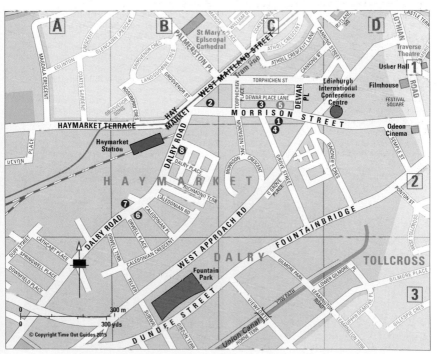

© Copyright Time Out Guides 2015

Restaurants

Atelier

*159-161 Morrison Street (629 1344, www.the
atelierrestaurant.co.uk). Bus 2, 3, 4, 25, 33, 44,
44A.* **Open** noon-2.30pm, 6-9.30pm Mon-Thur;
noon-2.30pm, 5.30-10.30pm Fri, Sat; noon-2.30pm,
6-9pm Sun. **Main courses** £13.50-£26. **Map**
p147 C1 ❶ **Modern European**

With exposed stone walls and simple modern
decor, this venue certainly looked the part when
it opened in 2013 – and the kitchen delivers on the
promise of its menu. You'll find friendly service,
relaxed surroundings, great cooking and intense
flavours. Typical main courses are a 12oz sirloin,
chips, smoked tomato, chimichurri and mushroom
ragout; or venison with smoked parsnip, pickled red
cabbage and cocoa nibs jus.

£ Chop Chop

*248 Morrison Street (221 1155, www.chop-chop.
co.uk). Bus 2, 3, 4, 25, 33, 44, 44A.* **Open** noon-
2pm, 5.30-10pm Mon-Fri; noon-2pm, 5-10pm Sat;
12.30-2.30pm, 5-10pm Sun. **Main courses** £7.50-
£10.50. **Map** p147 B1 ❷ **Chinese**

Founded in 2006 by foodie entrepreneur Jian
Wang from north-east China, Chop Chop remains
Edinburgh's go-to eatery for dumplings. The set
meals are fun; hot and sour soup followed by stir-
fried spicy beef, fried rice, and spinach and peanuts
makes for a decent two-course dinner. But don't miss
the dumplings: boiled or fried, with assorted fillings.

A FESTIVAL ON THE HOOF

The Royal Highland Show: putting the cult into agriculture.

EXPLORE

The Royal Highland & Agricultural Society
of Scotland was founded in 1784 to promote
regeneration of rural areas. In 1822, it
staged a little event in Edinburgh: the first
ever **Royal Highland Show** (*see p29*). Nearly
two centuries later, it's one of the biggest
events on the local calendar.

The Royal Highland Centre at Ingliston,
next to Edinburgh Airport, has been its
home since 1960. For four days each year,
towards the end of June, it's the centre of the
universe for all things agricultural. Farmers
show their animals in competition; trade
exhibitors sell everything from tractors to
slurry-handling equipment; and there are
displays on equestrianism, falconry, drystane
dyke building and many other rural crafts.

If the show was just a bunch of farmers
talking about milk yields, the public would
not have taken it to heart. In reality, it's both
a serious agricultural event and, for the
outsider, a lot of fun. Getting close enough
to touch thoroughbred beasts – Aberdeen
Angus, Highland and other cattle breeds you
can't even name – is surprisingly enjoyable.
The rare breeds of sheep and goats are
even more beguiling to uneducated urban
eyes. The food and drink element almost
constitutes a show-within-a-show; the world's
biggest haggis appeared in 2014, courtesy
of Scottish haggis and sausage manufacturer
Hall's. Verified by the Guinness Book of
Records, it weighed a mighty 2,226lb, 10oz:
as much as a small car.

The show has plenty for kids too, another
reason why it's so popular. Some 178,000
people turned up in 2014: that makes it
about as popular as Glastonbury music
festival, except with fewer drumming
collectives and much more livestock.

£ Milk

232 Morrison Street (629 6022, www.cafemilk. co.uk). Bus 2, 3, 4, 25, 33, 44, 44A. **Open** 7.30am-4pm Mon-Fri; 8am-4pm Sat; 8am-3pm Sun. **Main courses** £4.50-£7. **Map** p147 C1 ❸ **Café**

A small venue with tiled walls, wooden flooring and seating on benches or stools, the factor that makes Milk stand out is its commitment to good food. There are breakfasts (such as mushroom and rosemary burrito), sandwiches (pastrami, Thai chicken) and hot dishes (barley, beef and beer stew), as well as soups and salads.

▶ *Milk also runs the open-air café at the Collective gallery on Calton Hill; see p121.*

Pho Vietnam House

3 Grove Street (228 3383, www.vietnamhouse. co.uk). Bus 2, 3, 4, 25, 33, 44, 44A. **Open** 5-10pm daily. **Main courses** £8-£12 **Map** p147 C2 ❹ **Vietnamese**

Small, characterful, welcoming – and rare: Scotland is not known for its surplus of Vietnamese restaurants. The speciality is pho, the iconic noodle soup available in half a dozen different styles, but there are other distinctive dishes too, such as braised pork in caramel sauce or roll-your-own rice pancakes with beef or pork. Unlicensed, so BYOB.

Pubs & Bars

Thomson's

182 Morrison Street (228 5700, www.thomsons baredinburgh.co.uk). Bus 2, 3, 4, 25, 33, 44, 44A. **Open** noon-11.30pm Mon-Wed; noon-midnight Thur, Sun; noon-1am Fri; 4-11.30pm Sun. **Map** p147 C1 ❺

Set up like a traditional pub, but in the wood-panelled style of Scottish architect Alexander 'Greek' Thomson (1817-75), this venue has an attractive interior, good cask ales, a decent choice of whiskies and legendary steak pies that come with beans and gravy. It's a simple, old-school formula that has seen the bar prosper. If a PowerPoint presentation at the nearby Edinburgh International Conference Centre has driven you to drink, this is where to come.

DALRY & GORGIE

West of the castle, Edinburgh was once a bucolic stretch of farmland and small hamlets. Given the riot of tenements and modern developments crammed into the area, such tranquillity hardly seems credible today. But seek out some of the older buildings, and a different picture emerges. Down Distillery Lane, for instance, off Dalry Road by Haymarket Station, sits a simple 18th-century mansion called **Easter Dalry House**. Further along Dalry Road at Orwell Terrace is **Dalry House** (1661), originally the country seat of the Chiesley family. Hemmed in by tenements, it was restored as an old folks'

Milk.

home in the 1960s, then transformed into flats in 2006. Back in 1689, John Chiesley, son of the original owner, was found guilty of murder, had his hand hacked off before being hanged, and was then (allegedly) buried in the back garden. The building is said to be haunted by Chiesley's ghost: 'Johnny One-Arm'.

Continue along Dalry Road and you'll reach Gorgie, where Heart of Midlothian (or, simply, Hearts) play their football at **Tynecastle Stadium** (0871 663 1874, www.heartsfc.co.uk). Nearby is the child-pleasing **Gorgie City Farm** (*see p170*). And at the junction of Gorgie Road and Balgreen Road, look out for **Saughton Park**. Complete with winter gardens, a sunken Italian garden and a celebrated rose garden, it's a good point to access the Water of Leith Walkway. The park may be in line for some serious upgrading between 2016 and 2018.

Close to Saughton, and even more anomalous than its fellow country houses in Dalry, is **Stenhouse Mansion**, a solid 16th-century pile, extended in 1623, on a meander of the Water of Leith. Used by Historic Scotland as a conservation centre, it's not open to the public. The motto above the door reads 'Blisit be God for all his giftis', which, around 400 years ago, would have included an uninterrupted view of Edinburgh Castle, just over two and a half miles away, over open land.

Restaurants

First Coast

97-101 Dalry Road (313 4404, www.first-coast. co.uk). Bus 2, 3, 4, 25, 33, 44, 44A. **Open** noon-2pm, 5-10pm Mon-Sat. **Main courses** £10-£20. **Map** p147 B2 ❻ **Modern European**
A success for more than a decade, this bistro has a familiar, comfortable feel and a menu that could bring something as straightforward as smoked haddock fish cakes as a starter, but then follow that up with sea bass with coconut and red pepper sauce as a main, or maybe some Thai chicken salad.

Locanda de Gusti

102 Dalry Road (346 8800, www.locandadegusti. com). Bus 2, 3, 4, 25, 33, 44, 44A. **Open** 5-10.30pm Mon; noon-2.30pm, 5-10.30pm Tue-Fri; noon-10.30pm Sat. **Main courses** £8-£18.50. **Map** p147 B2 ❼ **Italian**
This venue operated at different premises across the city for several years, moving here in 2014. With its farmhouse kitchen decor, Neapolitan chef and dishes drawing on the cuisine of south Italy, it has been a fantastic addition to the neighbourhood, a beacon of Italian cooking in West Edinburgh.

★ Sushiya

19 Dalry Road (313 3222, www.sushiya.co.uk). Bus 2, 3, 4, 25, 33, 44, 44A. **Open** noon-2.30pm, 5-10.30pm Tue-Thur, Sun; noon-2.30pm, 5-11pm Fri; noon-3pm, 5-11pm Sat. **Main courses** £8-£14. **Map** p147 B2 ❽ **Japanese**
Blink and you could miss this Japanese restaurant: if the outside hardly announces itself, the inside is tiny. Food is excellent – there is real talent at work in the kitchen – and the menu goes beyond deservedly acclaimed sashimi and sushi to dumplings, ramen noodles, tempura, teriyaki and more.

Pubs & Bars

Caley Sample Room

42-58 Angle Park Terrace (337 7204, www.the caleysampleroom.co.uk). Bus 2, 3, 3A, 4, 25, 33, 44, 44A. **Open** noon-midnight Mon-Thur; noon-1am Fri; 10am-1am Sat; 10am-midnight Sun. *Food served* noon-10pm Mon-Fri; 10am-10pm Sat, Sun. **Main courses** £10-£17.50.
Red brick on the outside and roomy on the inside, the Caley has wooden seating and simple, functional decor. With good cask ales on offer, it invariably gets packed before Hearts games at nearby Tynecastle. The menu includes a decent all-day brunch at weekends, and mains such as tempura haddock and chips or pork belly with sweet potato mash.

THE REST OF THE WEST

A mile or so west of Haymarket Station, **Murrayfield Stadium** is the home of Scottish rugby union. An extraordinary place during any Six Nations match, it's one of the UK's largest sporting arenas. Since its 1990s redevelopment it has also played host to big rock and pop concerts; boy band One Direction drew an audience of more than 60,000 in 2014.

Further west still is **Edinburgh Zoo** (*see p170*). While it's principally a family-oriented attraction, its penguin parade is one of the city's most bizarre sights, whatever your age. Legend has it that the parade began in the 1950s, when a keeper accidentally left the door to the penguin enclosure open and they followed him out.

Behind the zoo, **Corstorphine Hill** is a favourite with joggers and walkers. Trees hide many of the best panoramas, but there is one fabulous lookout point, east of the higher reaches of the zoo. Known as 'Rest-and-Be-Thankful', it's where travellers journeying in from the west once got their first real glance at Edinburgh. The hill's quirkiest claim to fame, however, is the Cold War nuclear bunker that sits here. Built in 1952, it remained secret for a decade until protestors made its existence public in 1963. The local council eventually inherited the complex and sold it to a developer, who struggled to do anything worthwhile or lucrative with it. In 1992, a fire caused major damage; the bunker has lain derelict ever since, although there are long-term plans to turn it into a museum (for updates, see www.barntonquarry.org.uk).

Near the city limits by the airport, **Ingliston** is the long-time home of the annual **Royal Highland Show** (*see p148* **A Festival on the Hoof**). If you swing back towards the Firth of Forth instead, however, the road to Cramond in the north-west of the city passes **Lauriston Castle**. The magnificent lawns at the front belong to Edinburgh Croquet Club (www. edinburghcroquetclub.com).

Go all the way to **Cramond** and you reach the earliest known settlement in the Lothians. Waste has been found from the camps of Mesolithic people who inhabited the area in 8500 BC. The Romans set up camp around AD 140 – you can see the floorplan remains of their old fortification next to the church. Finds here have included the Cramond Lioness, a remarkable sandstone statue of a lioness that was dug out of the River Almond and is now on show in the **National Museum of Scotland** (*see p74*).

During the 18th century, the water power available from the Almond proved irresistible to industrialists, who built iron mills along its banks; the village exported nails around the world. The small **Cramond Heritage Trust** exhibition in the Maltings (Apr-Sept 2-5pm Sat,

Murrayfield Stadium.

Sun only, www.cramondassociation.org.uk) offers an intriguing slant on local history. These days, Cramond is a handsome and desirable suburb of Edinburgh.

Finally, almost beyond the city altogether, at the fringes of West Lothian, **Jupiter Artland** has been a major addition to the Scottish arts scene since it opened in 2009.

Sights & Museums

Jupiter Artland

Bonnington House, Wilkieston (01506 889900, www.jupiterartland.org). First Bus 27, X27. **Open** *mid May-June, Sept* 10am-5pm Thur-Sun. *July, Aug* 10am-5pm daily. **Admission** £8.50; £4.50-£6 reductions.

This sculpture park sits in the grounds of Bonnington House, a 17th-century mansion. Owners Robert and Nicky Wilson undertook a labour of love to create a wonderland of environmental and site-specific art from the likes of Andy Goldsworthy, Anish Kapoor, Tania Kovats and more. Some works are permanent, others come and go; there's usually something new to see each year. It takes around two hours to wander round, and is certainly worth the trek from town. There is parking, a café and a shop.

Lauriston Castle

2 Cramond Road South, Davidsons Mains (336 2060, www.edinburghmuseums.org.uk). Bus 16, 21, 27, 29, 37, 41, 42. **Open** *House tours* Apr-Oct 2pm Mon-Thur, Sat, Sun. Nov-Mar 2pm Sat, Sun. *Grounds* dawn-dusk daily. **Admission** *House* £5; £3 reductions. *Grounds* free.

Set in large grounds looking over the Forth, and now in the care of the local authority, this neo-Jacobean fortified property was built in the 1590s (as a tower

house for Sir Archibald Napier), extended during the 1820s and last used as a private home in 1926, when owner William Reid left it to the nation. Reid and his wife were enthusiastic antiques collectors who furnished the house with their finds; the property's Edwardian interiors have been carefully preserved. Access is by tour only; it lasts around an hour.

Murrayfield Stadium

Murrayfield (tours 346 5160, tickets 0844 335 3933, www.scottishrugby.org). Bus 12, 26, 31. **Open** *Tours* 11am Mon-Fri. *Stadium & shop* 9am-5pm Mon-Sat. Hours vary on match days; see website. **Admission** *Tours* £6; £3 reductions.

There's been a stadium here since 1925. The current version, Scotland's national rugby stadium – all-seated, all-covered and with a capacity in excess of 67,000 – is the biggest sporting arena in Scotland. It plays host to Scotland home games in the Six Nations and the autumn tests, and major music gigs, and occasional football matches. The stadium even has its own tram stop.

Restaurants

★ Redwood Bistro

1 Meadow Place Road (281 2576, www.redwoodbistro.co.uk). Bus 12, 26, 31, 100. **Open** 5.30-9pm Tue-Sat. **Main courses** £11-£22. **North American**

Chef-patron Annette Sprague closed her bistro in Stockbridge in 2011 to look for bigger premises. It finally re-emerged at the end of 2014, out west beyond the zoo (it's by the roundabout where St John's Road turns into the Glasgow Road). Sprague is Californian, so her menu reflects that West Coast eclecticism with Italian, North African and Pacific Rim influences.

Leith & the Coast

Despite its proximity to the centre of Edinburgh (it's only about two miles north-east of the Old Town), Leith is very much a separate place, with a history and a modern-day atmosphere all of its own. It has been a medieval fishing settlement, a port, a shipbuilding centre, resolutely working class, then largely but not completely de-industrialised. Since the 1980s, however, revitalisation has taken hold, which has seen the formerly depressed docks area and environs transformed by apartments, bars, offices, restaurants and shops.

Of course, Leith is just one part of the city's long coastline, which stretches all the way from Queensferry in the west via leafy Cramond, the old harbours at Granton and Newhaven, through Leith, then on to the long beach at Portobello in the east. It's a reminder that, although most of Edinburgh's attractions are inland, this is a coastal city.

Plumed Horse.

Don't Miss

1 **Royal Yacht Britannia**. Sleek outside, suburban on the inside (p158).

2 **Dinner** Leith has an enviable selection of seriously good eateries.

3 **The pubs** Rare whisky or robust chat, it's all here.

4 **Portobello Beach** A mile of sands for winter walks and summer days (p164).

5 **22 bus** From Leith to Princes Street in 20 minutes; the next one will be along very soon.

Oil industry vessel, Albert Dock.

EXPLORE

FROM THE CITY TO THE SHORE

When David I founded Holyrood Abbey in 1128, he endowed it with various patches of land. Among them was a small area around the Water of Leith, which included a sparse fishing settlement. David himself built another fishing village nearby, roughly around the site of what's now the Shore. It grew neatly enough and became a conduit for international trade. But in 1329, Robert the Bruce, in one of his last acts as King of Scotland, granted Leith Harbour to the burgesses of Edinburgh, a move that was to have dramatic repercussions.

Edinburgh tried to keep Leith under its thumb for the next 500 years – despite the fact that for much of that time, it was Scotland's premier port. When cargoes were landed at Leith, for instance, someone had to walk or ride all the way to the Edinburgh Tolbooth, up on the High Street, to pay duty before the cargo could be unloaded. This absurd situation sums up the relationship between the two: Leith the doorway to the wider world, the commanding capital skimming a share of its trade. Mutual mistrust simmered down the years, with Leith always maintaining a sense of distance.

IN THE KNOW CRUISE LIFE

Leith is a working port, but the maritime traffic is not all grain-carrying cargo ships or oil industry supply vessels. Its Western Harbour, in particular, is an important stopping-off point for cruise ships. There are dozens each year and some are pretty big, pushing the port's size limit for vessels – 50,000dwt – right to its maximum.

Leith was often caught in the middle of the conflicts between Scotland and England. But locals viewed such distractions as a minor impediment to business as usual. By the second half of the 18th century, strong trade links had been forged with the Baltic, the Low Countries, Scandinavia and further afield, and the harbour had capacity for hundreds of ships. Brandy, citrus fruits and wine arrived from France, Portugal and Spain, with rice, rum and timber coming from America and the West Indies. Business boomed.

The tensions between Leith and Edinburgh never went away, and the pair divorced in 1833 over mismanagement of the docks, corruption and other financial issues. But then, in 1920, the expansion of Edinburgh saw Leith swallowed up again, before it was left to crumble in the post-war doldrums. The port's struggles were compounded into the 1970s by drugs, prostitution, ill-advised public-sector housing developments and unemployment.

In the 1980s, encouraged by cheap rents, a few intrepid entrepreneurs set up in Leith, leavening the social mix while also encouraging others to follow. Things began to improve, but Leith really took off after the docks were privatised (in 1992), and the newly created Forth Ports Plc looked at alternative uses for its land. Flagship projects followed: the first ever Malmaison hotel opened here in 1994, the Scottish Government building arrived at Victoria Quay in 1996, the Royal Yacht *Britannia* berthed permanently in the Western Harbour in 1998 and the Ocean Terminal shopping mall was completed by 2001, all joined in the pre-credit crunch years by a frenzy of new residential construction. But despite all the changes, Leith retains an element of its disreputable past. It's still a working port, after all, where you can see

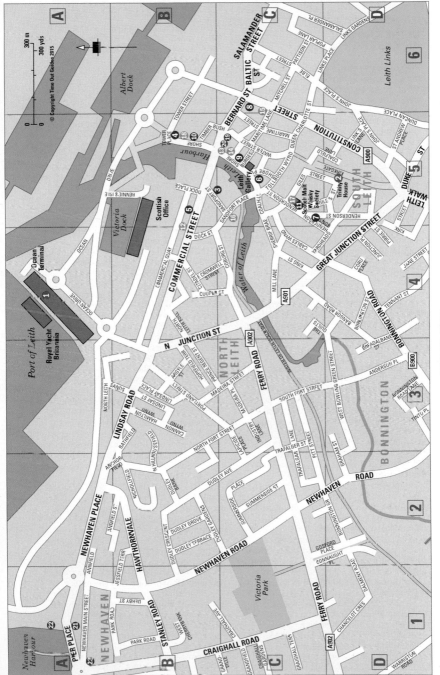

EXPLORE

EXPLORE

cargo ships, oil industry vessels, survey ships and Scottish marine protection vessels, as well as visiting cruise ships in summer. Crews come ashore for the pubs and saltier entertainments, while the new gentility of recent decades sits cheek-by-jowl with an older deprivation.

One other novelty in recent years has been the number of Polish voices. Following the 2004 EU expansion, thousands of young Poles headed for Scotland to find work, and many gravitated to Leith. The relationship between Poles and Leithers seems to be an economic symbiosis, the modern equivalent of the Baltic trade from centuries past. While the Leith of today is a different proposition to the Leith of more than 30 years ago, some things simply don't change.

LEITH

The most direct route to Leith from Edinburgh city centre, **Leith Walk** only came into its own after the construction of the first North Bridge, finally completed in 1772. The road was built on the line of an old earthwork running from Calton Hill to Leith, which had been thrown up as a defence against Cromwell's invading army in 1650. These days, it's flanked by tenements, with shops, pubs and cafés running all the way from the Edinburgh Playhouse at the city end to Leith's **New Kirkgate Shopping Centre**.

The pedestrianised **Kirkgate** behind the shopping centre, which used to link the foot of Leith Walk to the waterfront, was once Leith's main street. Its tenements were demolished and its memory obliterated by 1960s public housing, but two structures give a clue as to what the road was once like. The first is **Trinity House**; the other is the somewhat Gothic **South Leith Parish Kirk**, a Victorian structure erected on the site of a 15th-century church that was damaged when the English besieged Leith in 1560. Both are still handsome buildings, but they're overwhelmed by the crass modern architecture that surrounds them.

Before the development of Leith's docks and the addition of the various waves of urban housing in the 19th and 20th centuries, **Leith Links** was a flat, grassy stretch of land by the

Firth of Forth, stretching east. This was the site of Edinburgh's first racecourse; there are records of silver cups being presented as prizes as far back as 1655, although the course was eventually moved east to Musselburgh in 1816. The racecourse isn't Leith Links' sole claim to sporting fame. The Honourable Company of Edinburgh Golfers, the oldest golf club in the world, teed off here in 1744, and drew up the rules that were adopted by the Royal & Ancient Club at St Andrews ten years later.

What's left of the old links, now stretching for just over half a mile from the foot of Easter Road, is surrounded by houses and tenements, and is popular with footballers, cricketers and others simply out to enjoy the sunshine. At night, though, it can be a different story. For a time, Leith Links was a haunt for street prostitution, although a 2007 crackdown on kerb-crawling curbed the activity. But single women should still be wary of the area after dark.

The heart of Leith, however, is the **Shore**; it's where Leith really began, the site of the 12th-century fishing settlement that grew into a major international port. The Shore itself is actually the name of the street running down the south bank of the Water of Leith, which is home to an assortment of bars and restaurants that face modern apartments on the north bank. A stroll along it reveals just how dramatic the area's recent regeneration has been. However, it's by no means the whole story.

Just behind the Shore is a maze of narrow old streets that gives some indication of what the area was like before redevelopment. The oldest building is **Andrew Lamb's House** on Burgess Street, a warehouse with merchants' quarters recently refurbished to become an architect's office and a family home. The existence of a house on the site was first recorded in the 16th century, when it was visited by Mary, Queen of Scots, although the current building may date from the 17th century.

By the post-war period the building was in a state of decay; taken over by the National Trust for Scotland (NTS), it spent some years as a day-care centre for the elderly. The NTS was too strapped for cash to refurbish the place and it was sold to conservation architects Groves-Raines, who did a thoroughly sympathetic job on the restoration, mostly completed by 2012. It's private property, but there has been public access on **Doors Open Day** (*see p30*).

A short walk away, at the corner of Giles Street is the **Vaults**, a 1682 wine warehouse topped with an upper storey added in 1785. The ground floor has played host to various restaurants over the years; up the external stairway is the **Scotch Malt Whisky Society**. Signs of Leith's 19th-century prosperity are visible just a few minutes' walk from the Shore.

Look out for grand, confident edifices such as the old, domed **Leith Bank** in Bernard Street (1804); the **Custom House** at the east end of Commercial Street (1812); the **Exchange Buildings** and the **Assembly Rooms** in Constitution Street (1809); the **Municipal Buildings** (1827), also on Constitution Street and now a police station; and the **Corn Exchange** building (1862) at the corner of Baltic Street, which now houses the Creative Exchange (www.creativeexchangeleith.com), providing flexible workspace for the creative industries.

Back on the quayside, opposite 30 The Shore, a plaque commemorates the arrival here in 1822 of George IV, the first British monarch to set foot in Scotland for nearly two centuries. It reads, 'Geo IV Rex O Felicem Diem' or 'King George IV Oh Happy Day'.

Described by Robert Louis Stevenson as 'that dirty Water of Leith', Edinburgh's urban river meets the sea at Leith Docks – just yards from the plaque for George IV – and wasn't always the most bucolic of waterways. However, it's been cleaned up in recent years and can be a

LEARNING FROM HISTORY

Are parliament problems and tram traumas now a thing of the past?

If you looked only at newspaper headlines, you would get the impression that Edinburgh and large-scale project management simply don't go together very well.

Over a decade ago, the delays and cost overruns in the **Scottish Parliament** development at Holyrood were legendary. The initial budget estimate was £40 million, although that seemed to be plucked from thin air by a civil servant. Once a site and a design had been identified, there was a more realistic estimate of £109 million. When the building was completed in 2004, three years late, the cost had risen to £414.4 million. The city, indeed Scotland as a whole, thought nothing could be worse. There was a public inquiry and a typically Scottish crisis of self-confidence over the whole affair.

Then along came **Edinburgh Trams**. This was mooted as a city-wide network costing just over £0.5 billion, when deals were signed and work began in earnest in 2008. There was major disruption to the city centre for years, and the project was bedevilled by logistical problems and funding issues. One line eventually started running in 2014, from York Place in the New Town to

Edinburgh Airport – a fraction of the system as envisaged, which was originally supposed to encompass Leith and Newhaven. It was three years late and the estimated final cost went over £1 billion, including interest on money borrowed to get the slimmed-down project over the finishing line.

Fortunately, it's not all doom and gloom. When the Forth Road Bridge was identified as aged and creaky in the middle of the last decade, discussions started about a new road bridge immediately west of the existing one. Construction began in 2011, the bridge was dubbed the **Queensferry Crossing** and the cost was £1.6 billion. In 2013, it was reported that things were going so well that not only would the project be finished on schedule, by the end of 2016, but it would also be – gasp! – under-budget at around £1.45 billion.

Meanwhile, back in the city centre there have been some very successful refurbishments of major public facilities and attractions, benefiting both Edinburgh citizens and visitors (see p91 **Fit for the 21st Century**). After the Parliament and the trams, Scotland may be learning some lessons.

Royal Yacht Britannia.

good option for a daytime stroll. Start at the northern end of the Sandport Place Bridge and follow it upstream; it's a great way to see behind the façade of the city.

The **Water of Leith** also serves to divide Leith, north and south. It flows under Junction Bridge at the corner of Ferry Road and North Junction Street, overlooked by a gable-end mural depicting the history of Leith as a jigsaw. The final piece is a picture of a Sikh man reaching to take the outstretched hand of the community. Many of Edinburgh's Sikhs live in Leith; a Sikh temple now occupies a converted church, back over the bridge and down Sheriff Brae, back towards the Shore. To the left of the mural are two imposing but unfussy buildings, **Leith Library** and the **Leith Theatre** (28-30 Ferry Road). Both opened in 1932 and are curious examples of between-the-wars design; there are moves by a local charitable trust to restore the theatre to full working order in the coming years.

Along the otherwise unremarkable North Junction Street, the **Leith School of Art** (no.25) inhabits the oldest Norwegian seamen's church outside Norway, a small Lutheran kirk dating from 1868. Walk on a little way to the corner of Prince Regent Street and at its far end you'll discover the Doric portico and classical steeple of **North Leith Parish Kirk**, dating from 1816. A newer Leith, in the shape of the Sir Terence Conran-designed **Ocean Terminal** shopping mall and the **Royal Yacht Britannia**, is just a few minutes further along.

East of Conran's retail shed is **Commercial Quay**, a row of bonded warehouses that have been renovated and converted into apartments, upmarket shops, restaurants and offices. Facing them across the quayside is an implacable chunk of modern architecture housing the **Scottish Government**. The design may not be to everyone's taste, but its brashness epitomises modern-day Leith to a tee.

Sights & Museums

★ Royal Yacht Britannia
Ocean Terminal, Ocean Drive (555 5566, www.royalyachtbritannia.co.uk). Bus 11, 22, 34, 35. **Open** *Apr-Sept* 9.30am-4.30pm daily. *Oct* 9.30am-4pm daily. *Jan-Mar, Nov, Dec* 10am-3.30pm daily. **Admission** £14; £8.50-£12.50 reductions; free under-5s; £40 family. **Map** p155 A4 ❶
Launched in 1953, the year of Queen Elizabeth II's coronation, the Royal Yacht *Britannia* was used by the royal family for state visits, holidays and diplomatic functions for more than four decades. After it was decommissioned at the end of 1997, it was left in permanent residence in Leith, where it's consistently drawn big crowds. Since the 2009 addition of the Royal Deck Tea Room, you can even linger a while with tea and cake. Although the ship's exterior

has an art deco beauty, stepping on board – enter from the second floor of Ocean Terminal, through the Britannia Experience – is like regressing into a suburban 1950s aesthetic. You'll get to see the large state dining room, which has entertained everyone from Gandhi to Reagan, although the impeccable engine room will be more to the taste of the grease monkeys. Photographs of the royals in the Britannia Experience, dating back more than 50 years, give a top-down flavour of British social history in the second half of the 20th century as the empire faded away.

Trinity House
99 Kirkgate (554 3289, www.historic scotland. gov.uk). Bus 7, 10, 12, 14, 16, 21, 22, 34, 35, 49. **Open** Guided tours by appointment Mon-Fri; call for details. **Admission** free. **Map** p155 D5 ❷
The original Trinity House was built in 1555 as the Hospital for the Fraternity of Masters & Mariners of Leith. Completely rebuilt in neoclassical style in 1816 by Thomas Brown, these days it's hemmed in by local authority housing of a 1960s vintage. However, the interior is fabulous and now serves as a museum of Leith's seafaring history. Highlights include the war memorial window of 1933, added in remembrance of merchant sailors who died in the Great War, a number of portraits by Sir Henry Raeburn and the very impressive convening room upstairs. Trinity House is little visited and does smack of a bygone era, much like nearby Leith Docks with its surplus of derelict quayside, but it remains one of the most affecting museums in the city given the area's maritime heritage.

▶ *As part of its charitable work, Trinity House was a major contributor to the construction of both North Leith Parish Kirk on Madeira Street and the South Leith Parish Kirk.*

Restaurants

Café Domenico
30 Sandport Street (467 7266, www.cafedomenico. co.uk). Bus 16, 22, 35, 36. **Open** 10am-2.30pm, 5-10pm Mon-Thur; 10am-2.30pm, 5-10.30pm Fri; 10.30am-10.30pm Sat; 11am-10pm Sun. **Main courses** £9-£13. **Map** p155 C5 ❸ **Italian**
This tiny, informal Italian sits just around the corner from Commercial Street, and isn't immediately apparent to passers-by. The food combines heartiness with joie de vivre, and makes up in personality what it lacks in finesse. Once you've struggled through the generous antipasti (assorted charcuterie, cheeses and roast vegetables), the impending plate of pasta with spicy sausage, red pepper, cream and tomato might just finish you off – so bring an appetite. There's also a short breakfast menu.

Fishers in Leith
1 The Shore (554 5666, www.fishersbistros.co.uk). Bus 16, 22, 35, 36 **Open** noon-10.30pm daily. **Main courses** £13.50-£40. **Map** p155 B5 ❹ **Fish & seafood**
The nautical theme in the decor and fittings at Fishers, one of the city's leading seafood restaurants for more than two decades, is justified by the fact that the docks are on the doorstep; you can see working ships as you're walking down the Shore.

EXPLORE

Kitchin. See p160

Plumed Horse.

Eat in the bar or the raised area, choosing from starters such as brown shrimp rarebit, and mains such as Goan-style monkfish and king prawn curry.

▶ *There's a sister establishment in the New Town (see p92); the owners also operate adjacent bar-restaurant the Shore (see p163).*

★ Kitchin

78 Commercial Quay (555 1755, www.thekitchin. com). Bus 16, 22, 35, 36. **Open** *12.15-2.30pm, 6.30-10pm Tue-Thur; 12.15-2.30pm, 6.30-10.30pm Fri, Sat.* **Main courses** *£29-£36.* **Map** *p155 B5* ❺
Modern British

Tom and Michaela Kitchin set up in Leith in summer 2006, and were awarded a Michelin star within the year. Since then, they've operated at the apex of the city's restaurant scene, even expanding the premises in early 2015. The quality of raw materials is impeccable (Aberdeenshire veal, Borders game, Scrabster-landed white fish), while Kitchin's cooking has a little more edge than the classic French approach in evidence elsewhere; look out for dishes such as boned and rolled pig's head with langoustine and crispy ear salad. *Photo p159.*

£ Mimi's Bakehouse

63 The Shore (555 5908, www.mimisbakehouse. com). Bus 16, 22, 35, 36. **Open** *9am-5pm Mon-Fri; 9am-6pm Sat, Sun.* **Main courses** *£8-£10.* **Map** *p155 C5* ❻ **Café**

With its indulgent approach to afternoon tea, and slightly naughty decor (line drawings of topless ladies bearing cake while actually baring nothing immodest), Mimi's has been a massive hit since it opened in 2010; it's the spiritual heir of the late and much lamented Loopy Lorna's in Morningside. You can enjoy a good breakfast or lunch, but many folk favour the cakes, scones and traybakes.

★ Plumed Horse

50-54 Henderson Street (554 5556, www. plumedhorse.co.uk). Bus 16, 22, 35, 36. **Open** *12.30-1.30pm, 7-9pm Tue-Thur; 12.30-1.30pm, 6.30-9pm Fri, Sat.* **Set meal** *£55.* **Map** *p155 C5* ❼
Modern European

Tony Borthwick is a top chef and his restaurant has twice been recognised with a Michelin star: when it was in Crossmichael, Dumfries & Galloway, then after it moved here to Leith. Despite dropping off the edge of the Michelin table since, Borthwick has lost none of his talent. The Plumed Horse is small, politely decorated and very much embedded in the classic French style, which means main courses such as sautéed breast of young grouse with black pudding, caramelised apple and root vegetables.

£ Printworks Coffee

42 Constitution Street (555 7070). Bus 12, 16. **Open** *8am-5pm Mon-Fri; 9am-5pm Sat, Sun.* **Map** *p155 C6* ❽ **Café**

A small café that concentrates on doing the basics well. The coffee is very good, and the menu has soup, panini and wraps to eat in or take away. The moist falafel wrap is a thing of wonder, and the antipasti plate and Mediterranean platter are pretty decent too. For its online presence, search for 'Printworks Coffee' on Facebook.

£ Punjabi Junction

122-124 Leith Walk (07865 895022 mobile, www.punjabijunction.org). Bus 7, 10, 12, 14, 16, 22, 25, 49. **Open** 11am-4pm Tue-Thur; 11am-8pm Fri, Sat. **Main courses** £4-£8. Indian
This Punjabi café was launched in 2010 as Punjab'n De Rasoi, a community project to provide training and job opportunities for local Sikh women, and to help build up their skills and confidence. The menu is not elaborate; starters are mainly pakora and samosas, mains are chicken, fish or vegetable curries or thalis, dahl or lamb kofta, snacks and sides. But it's good, it's friendly and it's BYOB.

★ Restaurant Martin Wishart

54 The Shore (553 3557, www.martinwishart. co.uk). Bus 16, 22, 35, 36. **Open** noon-2pm, 7-10pm Tue-Fri; noon-1.30pm, 7-10pm Sat. **Set meal** £70. **Map** p155 C5 ⑨ Modern European
Located in the historical heart of Leith, Wishart's establishment has retained a Michelin star since 2001. The food is sublime; a meal here might take in Kilbrannan langoustine to start, Borders roe deer as a main and then calvados parfait to finish. Vegetarians are very well catered for, with a tasting menu all to themselves. The front-of-house staff are the slickest in the city, approachable and efficient.
▶ *Martin Wishart also has the Honours in New Town; see p93.*

Ship on the Shore

24-26 The Shore (555 0409, www.theshipon theshore.co.uk). Bus 16, 22, 35, 36. **Open** 9am-1am daily. *Food served* 9am-10pm daily. **Main courses** £21-£38. **Map** p155 B5 ⑩ Fish & seafood
Both parts (bar and restaurant) of the Ship have a nautical look, and its proximity to the docks adds yet more saltiness. Breakfast is served from 9am, the lunch menu kicks in at noon, dinner from 5pm. The latter brings dishes such as lemon sole meunière or chargrilled sea bass with crispy noodles and chilli-ginger stir fry. The crustacea and mollusc menu brings anything from half a dozen oysters to the fruits de mer royale that's suitable for two or three diners: a two-level platter containing pretty much everything you could think of, with chips.

£ VDeep

60 Henderson Street (563 5293, www.vdeep. co.uk). Bus 16, 22, 35, 36. **Open** noon-midnight Mon-Thur; noon-1am Fri; 11am-1am Sat;

11am-11pm Sun. **Main courses** £6-£7.50. **Map** p155 C5 ⑪ Indian
In recent years, this address has housed a tragic old pub, a fish restaurant and a craft beer bar-restaurant called the Vintage. In early 2015, the Vintage people relaunched with a craft beer and curry venue, in collaboration with TV presenter Hardeep Singh Kohli. Alongside draught Japiur IPA and bottled Brooklyn lager, you can order the expected beef bhuna or tandoori chicken, but also the unexpected 'bubble & sikh' and curried cauliflower cheese. The only surprise is that no one has tried this bar-restaurant-curry-craft-beer combo before.

Ship on the Shore.

EXPLORE

Pubs & Bars

★ Boda Bar

*229 Leith Walk (553 5900, www.bodabar.com).
Bus 7, 10, 12, 14, 16, 22, 25, 49.* **Open** 2pm-1am
Mon-Fri; noon-1am Sat; 1pm-midnight Sun.

Run by a Swedish couple with a friendly and laid-back
style, Boda operates like an inclusive neigh-bourhood
pub, albeit one with art, DJs, movies and
social events. Over the years, its owners have built
a mini-empire of great bars, all of which offer more
than the average pub.
Other locations Hemma, 73 Holyrood Road,
Old Town (629 3327); Joseph Pearce's (*see p122*);
Sofi's, 63 Henderson Street, Leith (555 7019);
Victoria, 265 Leith Walk, Leith (555 1638).

Bond No.9

*84 Commercial Street (555 5578, www.bondno9.
co.uk). Bus 16, 22, 35, 36.* **Open** 4pm-1am Mon-Thur;
noon-1am Fri-Sun. *Food served* 4-10pm
Mon-Thur; noon-10pm Fri-Sun. **Main courses**
£12.50-£19. **Map** p155 B5 ⑫

Housed in an old Leith warehouse, this spacious
bar and eatery has endured since 2008, which is a
lot longer than some of its neighbours managed.
The menu features lunch and weekend brunch, and
dishes such as steak and mash or vegetarian ravioli,
but the main event here is the bar, where the special-ity
is cocktails.

King's Wark

*36 The Shore (554 9260, www.thekingswark.com).
Bus 16, 22, 35, 36.* **Open** noon-11pm Mon-Thur;
noon-midnight Fri, Sat; 11am-11pm Sun. *Food
served* noon-10pm Mon-Sat; 11am-10pm Sun.
Main courses £8.50-£17. **Map** p155 C5 ⑬

The King's Wark has roots dating from the 15th
century (it was commissioned by James I to serve
as a royal residence, storehouse and armoury),
although there have been many rebuilds over the
years. Taverns have occupied the site since the 17th
century. The main room is a simple, well-worn, wel-coming
pub; there's also a smaller space to one side
for slightly more formal dining, for the likes of steak,
monkfish or duck. The beer-battered fish and chips
is always a winner.

Malt & Hops

*45 The Shore (555 0083, www.barcalisa.com).
Bus 16, 22, 35, 36.* **Open** noon-11pm Mon, Tue,
Sun; noon-midnight Wed, Thur; noon-1am Fri, Sat.
Food served noon-2pm Wed-Fri. **Main courses**
£4.50-£6.20. **Map** p155 C5 ⑭

This is a simple, traditional one-room pub with an
open fire, mildly decorated with dried hops, sitting
by the Water of Leith; lunch is offered Wednesday
to Friday. There are no distractions: no TVs, no gim-micks,
no hipsters, no attempt to be on trend. It just
sells good cask ale with lots of choice from Scottish
breweries, from Orkney in the north to the Borders in
the south. Ales from elsewhere in Britain feature too.

Port o' Leith

58 Constitution Street (554 3568). Bus 12, 16.
Open 11am-1am Mon-Fri; 9am-1am Sat; noon-1am
Sun. **Map** p155 C6 ⑮

Legendary proprietor Mary Moriarty may have
retired in 2009, but she was awarded a British
Empire Medal for her services to the community

EXPLORE

King's Wark.

in 2014 and the bar forever associated with her soldiers on. A small, neatly kept venue, it's patronised by everyone from merchant mariners to locals and students – it can get a little robust at times. Still, you could sit for hours looking at the details: ships' flags, lifebelts and so on. Essence of Leith.

★ Roseleaf
23-24 Sandport Place (476 5268, www.roseleaf. co.uk). Bus 16, 22, 35, 36. **Open** 10am-midnight Mon-Thur, Sun; 10am-1am Fri, Sat. *Food served* 10am-10pm daily. **Main courses** £6.50-£12. **Map** p155 C5 ⑯
This bar-café may look like a pub – because it once was - but it had a serious facelift a few years back, and has menus that would put many Edinburgh restaurants to shame. Old wooden fixtures and fittings are enhanced with the odd ornamental eccentricity (cocktails come in teapots, for example). The all-day brunch menu is among the city's best; main courses include pork belly, burgers, venison, and smoked macaroni cheese with a hint of truffle. It gets busy, so it's best to book, but you can always sit at the bar. Staff are a friendly lot.

★ Scotch Malt Whisky Society Members' Room
The Vaults, 87 Giles Street (555 2266, www.smws. com). Bus 16, 22, 35, 36. **Open** 11.30am-11pm Mon-Wed; 11.30am-midnight Thur-Sat; 12.30-10pm Sun. *Food served* noon-9pm Mon-Wed; noon-10pm Thur-Sat; 12.30-8pm Sun. **Main courses** £9.50-£19. **Map** p155 C5 ⑰
An Edinburgh institution, the society opened its doors here more than 30 years ago. Initially independent, it's now owned by Glenmorangie, itself part of luxury brands conglomerate LVMH. It's members only, with a fee of £122 for the first year; annual renewals thereafter are significantly cheaper. The venue itself is comforting and clubbish with tables and leather sofas, plus a small bar. Choose your cask strength whisky from the regularly updated list, where the tasting notes are always a source of great amusement, claiming 'cup cake', 'pineapple turnover' and 'jelly bean' aspects to a 24-year-old Islay, for example. The bar menu offers good lunches, snacks and evening meals.
▶ *For the New Town branch, see p90 Dining Room at 28 Queen Street.*

Shore
3-4 The Shore (553 5080, www.fishersbistros. co.uk). Bus 16, 22, 35, 36. **Open** noon-1am daily. **Food served** noon-10pm daily. **Main courses** *Bar* £9.50-£13. *Restaurant* £12.50-£21. **Map** p155 D5 ⑱
You'll find a small, bustling bar here and a couple of tables on the pavement outside; you can even take your drink across the street to the quayside. It's essentially a gastropub: the bar menu is fabulous (tiger prawns, fish and chips, steak and kidney pie),

and the adjoining wood-panelled room operates as a restaurant, serving the likes of lobster béarnaise, pigeon, steak or pork belly.
▶ *The Shore has the same owners as Fishers next door; see p159.*

Shops & Services

Beets Leith
49 Bernard Street (476 5086, www.beetsleith. co.uk). Bus 16, 22, 35, 36. **Open** noon-8pm Mon-Sat. **Map** p155 C5 ⑲ **Food & drink**
Providing an antidote to the drinks range next door at Sainsbury's Local, this bottle shop has a great selection of beers from Scottish brewers such as Black Isle, Brewdog, Cairngorm, Stewart and others, as well as beers from further afield. The wine choice isn't too shabby either, staff know their gins, there are other spirits, and even chocolate.

Flux
55 Bernard Street (554 1075, www.get2flux. co.uk). Bus 16, 22, 35, 36. **Open** 11am-6pm Mon-Sat; noon-5pm Sun. **Map** p155 C5 ⑳
Gifts & souvenirs
An Aladdin's cave of bits and bobs from British producers, or sourced from ethical companies overseas, Flux is the sort of shop where you might find rabbit-shaped LED nightlights, a recycled wool rug, happiness bangles, tea towels featuring the no.22 bus to Leith, or just a birthday card.

NEWHAVEN & GRANTON

Follow the coastline west along Lindsay Road, past the gleaming white silos of Chancelot Mill and the new housing developments on the side of the Western Harbour, to reach the old fishing village of **Newhaven**. Until the 20th century, this was an insular community whose residents were thought to have descended from the intermarriage of locals and the shipbuilding craftsmen brought over from France, Portugal, Spain and Scandinavia by James IV. It became famous for its colourfully dressed fishwives, who used to carry creels full of fresh fish up to Edinburgh to sell every morning. Much of the original village has been pulled down, but there are still some fishermen's cottages near the shore.

The former St Andrew's Kirk is now an indoor climbing centre, **Alien Rock** (8 Pier Place, 552 7211, www.alienrock.co.uk), but the major recreational centre in Newhaven is the huge **David Lloyd** health and sports facility (Newhaven Place, 0345 125 7016, www. davidlloyd.co.uk). When a sports and fitness centre first opened back in the 1990s, it was an anomaly on what was then a sparse stretch of waterfront, but with the number of new apartments that have been added since, it looks rather less lonely.

EXPLORE

Further west still, **Granton Harbour** was built by the fifth Duke of Buccleuch as part of his estate and opened in 1838. Engineering expertise for the project came courtesy of one of the lighthouse-building Stevenson family, relatives of Robert Louis. Like Newhaven, Granton was once the base for a fishing fleet, but only small leisure craft are found there today. It also boasted the world's first ferry train, offering the most direct route to Fife before the construction of the Forth Rail Bridge in 1890. And yes, there are new waterfront apartments here too.

Restaurants

C-Shack
3 Pier Place (467 8628, www.cshack.co.uk).
Bus 7, 10, 11, 16. **Open** 6-10pm Tue; noon-2pm, 6-10pm Wed-Fri; noon-5pm, 6-10pm Sat; 12.30-8pm Sun. **Main courses** £10-£19. **Map** p155 A1 ㉑
Fish & seafood
It's beside Newhaven Harbour, the interior looks like a well-appointed surf shack and there's ceviche, chowder, lobster, mussels and oysters on the menu. The specials board could bring scallops with black pudding and other simple, fresh dishes; given the owners' other restaurant – Thai bar-restaurant Port of Siam in Broughton (*see p124*) – there may be some Asian-inflected options at times.

Loch Fyne Seafood & Grill
25 Pier Place (559 3900, www.lochfyneseafood andgrill.co.uk). Bus 7, 10, 11, 16. **Open** 11.30am-10pm Mon-Thur; 11.30am-10.30pm Fri; 10am-11pm Sat; 10am-10pm Sun. **Main courses** £11-£25. **Map** p155 A1 ㉒ **Fish & seafood**
A well-established national chain with over 40 outlets, Loch Fyne actually started life in Argyll, but it took until 2007 for what was only its second Scottish outlet to open, at Newhaven Harbour. Housed in a late Victorian fishmarket, the restaurant is modern and spacious with sea views; the menu features the famous oysters along with other shellfish, salmon, lemon sole, haddock and more.

★ £ Porto & Fi
47 Newhaven Main Street (551 1900, www.portofi.com). Bus 7, 10, 11, 16. **Open** 8am-8pm Mon-Thur; 8am-10pm Fri, Sat; 10am-6pm Sun. **Main courses** £10-£14. **Map** p155 A1 ㉓ **Café**
This neat modern café (opened 2008) was originally aimed at residents of the new housing developments along the coast, but is now a well-known favourite well beyond the immediate neighbourhood. It offers seriously good breakfasts (eggs benedict, french toast with Ayrshire bacon and maple syrup), light meals (smoked chicken and mango salsa salad), coffee, tea and fabulous cakes (beetroot and chocolate). It gets very busy at peak times, especially for weekend brunch, so book ahead.

Newhaven Harbour lighthouse.
See p163.

Pubs & Bars

Old Chain Pier
32 Trinity Crescent (552 4960, www.oldchainpier. com). Bus 10, 11, 16, 32. **Open** noon-11pm Mon-Thur, Sun; noon-midnight Fri, Sat. **Main courses** £11-£17.
Formerly the booking office of an actual pier (which was destroyed by a storm in 1898), this place has been a pub since soon after World War II. Overlooking the Forth between Newhaven and Granton, it has great sea views from the picture windows. Typical main courses, to eat in the bar or conservatory, would be chicken supreme or fish and chips. Alternatively, sit with a pint (there are quality draught ales) and stare at the passing ships.

PORTOBELLO

The current focus of attention on Edinburgh's coastline is very much on Leith and points west. However, to the east, **Portobello** is worth a peek. It owes its suburban origins to a former sailor called George Hamilton, who built a cottage in the area in 1742. He'd served in the navy at the Battle of Porto Bello in Panama (1739) and gave the cottage that name. Other houses were built around it and Portobello duly became a coastal hamlet around three miles east of the Old Town.

Thanks largely to its handsome beach, it became a popular holiday resort by the Victorian

EXPLORE

period. Portobello followed other British seaside towns in the post-war years – its former vibrancy fading as people did other things with their leisure time. All the same, it has bars and cafés and, on sunny days, can still get packed. There's a mood of determined enjoyment whatever the weather that is peculiarly British; but if things get too inclement, the **Portobello Swim Centre** (57 Promenade, 669 6888, www.edinburghleisure. co.uk) with its pool and turkish bath offers an indoor alternative.

Restaurants

£ Beach House

57 Bath Street (657 2636, www.thebeachhousecafe. co.uk). Bus 15, 21, 26, 40, 42, 45, 49, 69, 113. **Open** 9am-5.30pm Mon-Fri; 9am-6pm Sat, Sun. **Main courses** £5-£9. **Café**

If you're down at the beach, you don't always want a beer, or takeaway fish and chips. Sometimes a decent café hits the spot and with its modern look, tables outside on the Promenade, coffee, cake, soup, sandwiches and light meals, this place is ideal.

Pubs & Bars

Dalriada

77 The Promenade (454 4500, www.dalriadabar. co.uk). Bus 15, 26, 40, 45, 113. **Open** 5-11pm Wed, Thur; 3pm-midnight Fri; noon-midnight Sat; 12.30-11pm Sun. *Food served* noon-3pm Sat; 12.30-3pm Sun. **Main courses** £4-£7.

Set directly on the Promenade, with beer gardens both front and back, this spacious old bar is the perfect pit stop if you've been strolling along the sands. There's good beer, coffee and at weekends basic bar food and live music. The opening hours are extended in spring and summer.

Espy

62-64 Bath Street (669 0082, www.the-espy.com). Bus 15, 21, 26, 40, 42, 45, 49, 69, 113. **Open** 10am-11pm Mon-Wed, Sun; 10am-1am Thur-Sat. *Food served* 10am-9pm Mon-Wed, Sun; 10am-10pm Thur-Sat. **Main courses** £9-£16.

Right by the beach, these premises have been home to several bars over the years. Locals appreciate the Espy's basic but comfortable decor and the family-friendly ambience, whether stopping in for a drink or some food (burgers, nachos, salads and more). It's a good stop-for-a-pint option.

Skylark

213 Portobello High Street (629 3037, www.the skylarkportobello.com). Bus 15, 26, 40, 45, 113. **Open** 10am-5pm Mon; 10am-midnight Wed-Sat; 11am-5pm Sun. *Food served* 10am-5pm Mon, Sun; 10am-9pm Wed-Sat. **Main courses** £5.50-£10.

While the main focus in Portobello is usually on the beach, the eastern reaches of the High Street saw the arrival of the Skylark in 2012, filling a gap in the market for a relaxed family-friendly café by day, bar by night. Food includes club sandwiches and croque monsieur, and the choice of beer, on draught or in bottles, is decent. Hours are longer in summer.

EXPLORE

Portobello beach.

Arts & Entertainment

Children

Come rain, shine or the dark, dreich days of winter, you'll want to keep your kids entertained. Happily, Edinburgh loves children, and children love it right back. The acres of parkland and the huge volcano thing (it is dormant, isn't it?) allow for plenty of outdoor romping when the weather holds. And when it doesn't, higgledy old houses and some Harry Potteresque buildings have their attractions, while the city's gloriously gory history should keep even the most jaded of teenagers switched on to the capital and its delights.

In August, your kids will enjoy whiling away many cash-free hours on the High Street watching Fringe performers drumming up trade for their shows. And if you're here at the end of December, don't miss the Torchlight Procession, as well as whatever other free spectacle the Hogmanay programme brings.

GETTING AROUND

The good news: most of the major sights in Edinburgh are confined within a few square miles. The bad news: many are separated by hills, some of which are pretty steep.

If it's too far or too high to walk, Edinburgh's bus network should come to the rescue. Under-fives travel free (up to two children per adult); children aged five to 15 pay 70p for a single fare; anyone aged 16 or over is classed as an adult and pays £1.50 for a single fare. Foldable buggies and pushchairs are permitted on Lothian Buses; they can be left unfolded if you're the only user, but common sense suggests that it's probably best to fold the buggy if there are others on board. Remember that wheelchair users always take priority. In the unlikely event that you, your buggy and child have to get off the bus to make way for a wheelchair user, the driver will give you a receipt so you can board the next available bus with no extra charge.

If you've got a very young baby and plan to make a number of journeys, it may be worth considering a sling as a space-saver. For more on buses, *see p263*.

Edinburgh's city centre is well served by black cabs. They can be costly but they do seat five, and you shouldn't need to fold your pushchair. Some cabs also have fold-down concealed toddler seats (ask the driver before getting in). For more on cabs, *see p264*.

SIGHTS
Around the Royal Mile

The cannons, creepier corners, ramparts and weapons at **Edinburgh Castle** (*see p55*) will doubtless engage the imagination of children,

IN THE KNOW ONE O'CLOCK FUN

The **One O'Clock Gun** has been fired from **Edinburgh Castle** (*see p55*) since 1861. Originally it helped sailors on the Forth set their chronometers, but now it's simply a city feature. At first the gun was a muzzle-loading cannon, but it has been periodically modernised; currently it's a 105mm howitzer, the British Army's L118 Light Gun, a type used on operations in the Falklands, the Balkans, Sierra Leone, Afghanistan and Iraq since 1982.

although they might struggle with its more formal and museum-like spaces. Meanwhile, the **One O'Clock Gun** (*see p168* **In the Know**) will scare younger kids and delight older ones: at 1pm daily except Sunday, an artillery piece is fired from the castle ramparts. To this day, locals on Princes Street below hear it and think, 'Oh, it's 1pm.' Tourists jump out of their skins. Yes, it fires a blank; it's still a sharp, percussive noise.

Down on Castlehill is the **Camera Obscura** (*see p55*), Edinburgh's oldest purpose-built tourist attraction. Again, kids may not go for the actual camera obscura, but you should have better luck with the other four floors of illusions, holographs and visual delights. The outside of the building has distorting mirrors that can entertain younger children for ages.

East along Market Street sits the **Edinburgh Dungeon** (*see p56*), run by the same company that operates the Blackpool, London and York Dungeons, focusing on the horrors of Scotland and Edinburgh's gory past. The dungeons claim to be family-friendly, but they're not recommended for under-eights; nor for anybody on the nervous side, for that matter.

Due south of here (go via George IV Bridge) sits the **National Museum of Scotland** (*see p74*), which has plenty to engage kids. Six new scientific discovery and invention galleries, with lots of hands-on opportunities, are due to open in 2016; the Discovery Zones teach history with games and dressing-up; and the Animal World gallery is always a winner. It's worth checking the 'What's on' section of the website for child-friendly activities.

Back up on the Royal Mile, continue east and you'll find the **Museum of Childhood** (*see p65*). It may be fascinating for adults or for any school-aged kids doing projects on the past, but it might not hold younger children's interest for long. In its favour, though, admission is free and the gift shop has some cute stock.

To learn about the Earth's deep history, continue eastwards to **Our Dynamic Earth** (*see p66*). This purpose-built venue charts the geological formation of the planet, its climate, evolution and much more, with dramatic presentations and ambitious displays that cover everything from volcanic fire to polar ice; there is also a full-dome cinema and a soft play area for under-tens.

Outside the centre

Respectively to the west and east of Edinburgh, both the **Almond Valley Heritage Centre** in Livingston (01506 414957, www.almondvalley. co.uk) and the **East Links Family Park** at Dunbar (01368 863607, www.eastlinks.co.uk) keep kids occupied with train rides, farm animals, play zones and soft play areas. It

Camera Obscura.

may be a good idea to pack a change of clothes as there can be wet and muddy areas at both.

Edinburgh's position on the Firth of Forth means there are some beach escapes within the city limits. There is beach and big open sky at **Cramond** (*see p150*), a walk to the tidal Cramond Island across a causeway for the adventurous – or take a trip to **Portobello** (*see p164*). Out of town heading east are the fantastic beaches and assorted attractions of **East Lothian** (*see p202*).

ACTIVITIES

Animal attractions

★ Craigie's
West Craigie Farm, by South Queensferry, West Edinburgh (319 1048, www.craigies.co.uk).
Open 9am-5pm daily.
At this working farm, you can pick your own soft fruit in season (also available by the punnet), get some fresh air or explore the Nature Detective trails on Craigie Hill nearby. There are free-range hens pecking around; kids can see the birds that have laid the eggs on sale in the farm shop and deli. There's a decent café too. Despite the postal address, Craigie's is within the City of Edinburgh boundary between the edge of the city itself and the village of Dalmeny. You'll need a car to get here, as there's no bus. Driving out of Edinburgh, leave the A90 around half a mile after Cramond Bridge, following the sign for Kirkliston; head west for just under a mile on the minor road until you see the Craigie's sign.

East Links Family Park. *See p169.*

★ Edinburgh Zoo

*Corstorphine Road, West Edinburgh (334 9171,
www.edinburghzoo.org.uk). Bus 12, 26, 31, 100.*
Open *Apr-Sept* 9am-6pm daily. *Mar, Oct* 9am-5pm
daily. *Nov-Feb* 9am-4.30pm daily. **Admission** £18;
£13.50-£15.50 reductions; free under-3s; £40.95-
£69.60 family.

Opened in 1913 on a large site occupying part of
Corstorphine Hill in the west of the city, the zoo has
long been a major Scottish visitor attraction – it's
in the country's top ten. The glamour animals of
late have been a pair of giant pandas (Tian Tian
and Yang Guang) who arrived from China in 2011,
although the penguin parade – where the birds leave
their enclosure and wander around – has been an
event since 1951 and still draws a crowd (daily at
2.15pm, if the penguins feel like it). Otherwise, there
are impressive, scary animals (leopards, lions), cute
ones (koalas, meerkats) and many others besides,
from armadillos to zebras. To get a flavour of the zoo
before your visit, check out the insanely cute Panda
Cam and Penguin Cam on its website.

FREE Gorgie City Farm

*51 Gorgie Road, West Edinburgh (337 4202,
www.gorgiecityfarm.org.uk). Bus 1, 2, 3, 21, 25,
33.* **Open** *Mar-Oct* 9.30am-4.30pm daily. *Nov-Feb*
9.30am-4pm daily. **Admission** free; donations
welcome.

This lovely, informal spot is a working farm so has
chickens (including some rare breeds), cows, ducks,
pigs, pygmy goats and sheep. There is also a pet
lodge, however, with hamsters, guinea pigs, rats,
rabbits, a snake called Dougal and three tortoises
(the oldest, Mrs Murdoch, was born in 1956). It's
educational, it's fun and it has a café with high chairs
(9.30am-4pm daily).

Scottish Seabird Centre

*The Harbour, North Berwick, East Lothian (01620
890202, www.seabird.org). North Berwick rail
then 15min walk.* **Open** *Apr-Aug* 10am-6pm
daily. *Feb, Mar, Sept, Oct* 10am-5pm Mon-Fri;
10am-5.30pm Sat, Sun. *Nov-Jan* 10am-4pm

Mon-Sat; 10am-5pm Sat, Sun. **Admission** £8.95;
£4.95-£6.95 reductions; under-3s free.

At heart, this is a conservation and education charity with a focus on seabirds, but it's developed into
a major attraction by knowing its audience. It has
a seafront location, live solar-powered webcams
showing birds on the islands offshore, and seasonal
boat trips around those islands (Apr-Oct, see website
for times and prices; also *see below* Seafari). There's
also a Wildlife Theatre (showing short films about
puffins, seals and more), the Kids' Zone soft play
area and the Kids' Corner quiet space (with marine-
themed storybooks). Amenities include a decent
café with a terrace overlooking the sea, plus good
nappy-changing facilities.

Seafari Adventures Forth

*Departures from the Harbour, North Berwick,
East Lothian (01620 890202) and Hawes Pier,
Queensferry, West Edinburgh (331 4857).* **Boat
trips** Easter-Oct. **Times & prices** see website
(www.seafari.co.uk).

From North Berwick, Seafari takes Scottish Seabird
Centre (*see left*) visitors and others out on a seabird
cruise on a small catamaran, or to the Bass Rock
or Isle of May on a rigid inflatable boat. From
Queensferry, right under the Forth Bridge, it offers
jaunts on the *Maid of the Forth* to Inchcolm Island
with its medieval abbey, and various other tours.
This latter vessel even has its own website (www.
maidoftheforth.co.uk).

Arts & culture

If your children are more into making art than
looking at it, there are a number of kid-friendly
studios in Edinburgh where they can learn some
new skills or just express their inner splurginess.

In Marchmont, South Edinburgh, sits **Doodles
Ceramic Workshop** (29 Marchmont Crescent,
229 1399, www.doodlesscotland.co.uk), a paint-
your-own-pottery enterprise. Kids and adults
can decorate a variety of white unglazed pottery
items with paints, stamps and stencils; staff then

glaze and fire them for you, to be collected three to four days later. A similar kind of activity is available at the **Ceramic Experience** (118 Ocean Drive, 554 4455, www.theceramic experience.com) near Ocean Terminal shopping centre in Leith. If you use acrylics, you can take your work away immediately; glazing and firing takes a day.

Finally, **Lauriston Castle** (*see p151*) in West Edinburgh has an annual craft workshop programme with sessions such as Springtime Flowerpots (green-fingered fun in the glasshouse), Pond Secrets (garden zoology) or A Sporting Chance (historical games).

Indoor play centres

The popularity of Edinburgh's various play centres seems to go in cycles. Sometimes it's simply a case of the newer ones boasting greater novelty value than their rivals. Below are a couple of the better options.

Maddie & Mark's Playtown
Ocean Terminal, Leith (555 1900). Bus 11, 22, 34, 35, 36. **Open** 9.30am-6pm Mon-Fri; 10am-6pm Sat, Sun. **Admission** £5.75-£6.75 4+; £4.75-£5.75 under 4s; £1 6-12mths; free under-6mths.
On the second floor of the Ocean Terminal shopping mall, Maddie & Mark's features soft play with mini-shops to explore (garage, shoe shop, supermarket),

Gorgie City Farm.

drop-in activities such as artwork or baking. There's also a café, and the Western Harbour and Royal Yacht *Britannia* (*see p158*) are directly outside. Standard weekday sessions are 90 minutes, but only 60 minutes at weekends and during school holidays.

★ Time Twisters
Unit 5, Catalyst Trade Park, 2B Bankhead Drive, Sighthill, West Edinburgh (308 2464, www.time twisters.co.uk). Bus 2, 18, 20, 25, 32, 34, 35, 63. **Open** 9.30am-6pm daily. **Admission** *1hr session* £4.25-£5.25 4+; £3.95-£4.65 under-4s; free babies.
This play centre has an Egyptian theme and is ideal for any child who's into crocodiles, mummies or pyramids. It's zoned for different age groups: four and over, three and under, plus a separate sports area. The venue also limits participation based on age and height: no one older than 12 allowed, no one taller than 4ft 9in. Admission is hourly, priced as standard or off-peak; if you think you're going to be in for longer, pay for the extra hour upfront as it works out cheaper. The owners' guiding principles are education, fun and health, so the café menu is quite commendable. But the best thing? Kids love it.

EATING & DRINKING

If self-styled 'family restaurants' aren't your scene, there are numerous places in the city that welcome children without making a song and dance about it.

All nine branches of **Pizza Express** (www.pizzaexpress.com) are family-friendly, but those in Leith (Waterview House, 38 The Shore, 554 4332) and in Stockbridge (1 Deanhaugh Street, 332 7229) are notably popular with kids. Alternatively, kids can make their own pizzas at **Giuliano's on the Shore** (1 Commercial Street, 554 5272, www.giulianos.co.uk) in Leith. For fish, try **Loch Fyne Seafood & Grill** (*see p164*) on the coast at Newhaven, which has a set-price two-course kids' menu. In the evening, you'll get lovely views of the sun setting over the Forth.

In the city centre, fans of elephants and Harry Potter should love the **Elephant House** (*see p59*) and, by the open green space of the Meadows, **Peter's Yard** (*see p139*) offers delicious snacks and meals to eat in or take away. Afterwards, the east end of the Meadows has perhaps the city's finest and most extensive public play area for kids, dubbed the Magnet. It has roundabouts, slides, swings and much more, all publicly funded and free to users.

The most child-friendly pub in the city, until 5pm, is **Joseph Pearce's** (*see p122*) on Elm Row, which has a raised area towards the back of the premises with high-chairs and toys. That said, the **Dalriada** (*see p165*) on the Promenade in Portobello always has lots of children in its garden on sunny weekends. The nearby **Skylark** (*see p165*) on Portobello High Street also attracts

ARTS & ENTERTAINMENT

a fair number of parents 'n' pushchairs during the day, as does the **Water of Leith Café Bistro** (*see p108*) in Canonmills.

ARTS & ENTERTAINMENT

While many of the festivals held throughout the year in Edinburgh have family-friendly aspects, several events cater purely for younger audiences. From late March into April, the Scotland-wide **Puppet Animation Festival** (www.puppetanimationfestival.org) brings a number of performances to Edinburgh. In the last week in May, **Imaginate** (www.imaginate. org.uk) is devoted entirely to performing arts for children, staged in various child-friendly venues and usually with a free crèche for under-fives. And in late October, the **Scottish International Storytelling Festival** at the **Scottish Storytelling Centre** (*see p67*) delivers plenty to entertain younger and older audiences.

Sticking with the theatre, several annual Christmas shows are usually worth a look. The **Royal Lyceum** (*see p190*) offers its own reinterpretation of a classic story, which generally holds the attention and imagination of children, even into their early teens. Alternatively, for a traditional panto, head to the **King's Theatre** (*see p190*).

Several local cinemas offer film screenings one morning a week for parents and carers with babies. The **Cameo** (*see p174*) has the Big Scream, **Cineworld** (www.cineworld.co.uk) has Cinebabies, the **Filmhouse** (*see p174*) has For Crying Out Loud, and the **Odeon** (www.odeon. co.uk) has Newbies. The Cameo also offers screenings that may be suitable for children and other moviegoers with autism. None of the above should be confused with special children's screenings of movies, such as the bargain Kids AM sessions at **Vue** at the Ocean Terminal mall in Leith (weekend mornings and school holidays).

EAT THE STREETS

Snacking on the hoof.

It's an old nugget of wisdom, usually worth repeating, that no one comes to Edinburgh for the weather. That said, the peak visitor months of the summer do see average high temperatures creep towards a balmy 20°C (68°F) and on most days from May to August it doesn't actually rain. That's when you and the kids are most likely to think, 'Maybe we can get some food to eat outside.'

The primary venues for street food are the city's markets (see p109 **Don't Quit Your Stalling**), while during August specifically, Festival Fringe venues such as George Square, the Pleasance, Summerhall and the Underbelly at Bristo Square all have an assortment of stalls selling everything from sparkly cupcakes to vegetarian curry.

There are also a number of hardy, intrepid and child-friendly street-food vendors that you can find on their usual pitches all year round. At the Grassmarket, the **French Connection** van has been a feature for a couple of decades, bringing crêpes to the masses. It offers savoury buckwheat galettes, hot dogs and drinks as well as sweet crêpes; the banana and chocolate version always goes down well. Leaving the Grassmarket, **Oink** (*see p75*) on Victoria Street is a shop with a few café chairs rather than a street stall, serving hearty hog roast rolls to take away in piglet, oink and grunter sizes.

Oink.

Further south, at the top of Middle Meadow Walk where it meets Lauriston Place, you'll find Brazilian crêpe stall **Tupiniquim** (07908 886184 mobile, www.tupiniquim.co.uk). Operating out of an old police box, its friendly owner provides gluten-free crêpes, in savoury and sweet varieties, as well as very good freshly pressed juices. Immediately south-east of Tupiniquim on George Square, look out for the **Union of Genius** soup van. Its sister café-takeaway (*see p75*) is on nearby Forrest Road, but the van, named Dumbo, offers excellent warming soup, which is especially welcome in the winter months. Open from 10.45am every weekday, it continues trading until the soup runs out.

Film

In Alfred Hitchcock's *The 39 Steps* (1935), the dashing hero flees London on the train for Edinburgh. At Waverley Station he reads a newspaper and sees that the authorities are searching for him; he alights, dramatically, on the Forth Bridge and heads further north. Edinburgh's unique cityscape lends itself to film-making and the city has a long-established regard for the movies. The first films were shown here at the end of the 19th century, the first purpose-built cinemas appeared in the Edwardian period, while the oldest surviving cinema opened in 1914. Meanwhile, the Scottish capital has played a key role in movies directed by talents as wide-ranging as Hitchcock, Hugh Hudson, Ron Howard and Danny Boyle. If you want a glimpse of Edinburgh's soul, however, try *The Illusionist* (2010), the French animation directed by Sylvain Chomet. The imagery alone speaks straight to the heart of residents, and exiles.

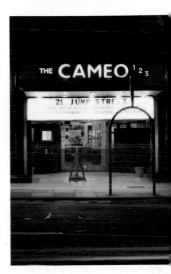

CINEMAS

In addition to multiplexes and a few independent operations, the city also has specialist enterprises such as the **Edinburgh Film Guild** (www.edinburghfilmguild.org.uk), the world's longest-running film society, in operation since 1930; the **Institut Français d'Ecosse** (www.ifecosse.org.uk), which promotes French culture and language, and shows French films; and the noteworthy **Scottish Documentary Institute** (www.scottishdocinstitute.com) based at the Edinburgh College of Art. Bars such as the **Brass Monkey** (14 Drummond Street, Old Town, 556 1961) and **Sofi's** (65 Henderson Street, Leith, 555 7019, www.bodabar.com) also show films in pretty relaxed settings. The big annual event is the **Edinburgh International Film Festival** (EIFF; *see p37*).

In the UK, films are classified according to the age of the audience. Universal (U) means anyone can watch it; Parental Guidance (PG) cautions that younger children might find the contents upsetting; 12A bars children under 12 unless accompanied by an adult; 12 bars under-12s completely; 15 bars under-15s; 18 bars under-18s.

The smallest screen at the Cameo and the front door at the Dominion (for both, *see p174*) have steps; wheelchair users should call ahead for access advice. Otherwise, all other screens in the city's cinemas are wheelchair-accessible.

IN THE KNOW
FESTIVALS GALORE

The **Filmhouse** (see p174) hosts an impressive number of specialist film festivals and seasons throughout the year. The **Edinburgh Mountain Film Festival** (www.edinburghmountainff.com, Feb) promotes adventure through film; the **Italian Film Festival** (www.italianfilmfestival.org.uk, Mar) shows contemporary and classic Italian cinema; **Dead by Dawn** (www.deadbydawn.co.uk, Apr) is Scotland's horror film festival; **Africa in Motion** (www.africa-in-motion.org.uk, Oct/Nov) brings the best of African cinema to Scotland – and this is just a taster. Check the Filmhouse website for up-to-date listings.

Edinburgh's multiplex options are Cineworld (www.cineworld.co.uk), Odeon (www.odeon.co.uk) and Vue (www.myvue.com).

★ Cameo Picturehouse

38 Home Street, South Edinburgh (0871 902 5723, www.picturehouses.com). Bus 10, 11, 15, 15A, 16, 23, 24, 27, 36, 45, 47. **Tickets** *£7-£10; £1.50-£9 reductions.* **Map** *p284 E9.*

This three-screen cinema is a real treat, offering around eight films a week from the edges of the mainstream. Chat up the friendly staff at the café-bar, then take your pint through when the film begins. Look out for various themed screenings, including Vintage Sundays shown at lunchtime (classics such as *Breathless*, *Duck Soup* or *The Sound of Music*) or Discover Tuesdays shown early evening (arthouse, cult and documentary). There are screenings for parents with babies under 12 months (Big Scream Club), autism-friendly screenings with low lights left on and the sound turned down, and more.

★ Dominion

18 Newbattle Terrace, Morningside, South Edinburgh (447 4771, www.dominioncinemas.net).

Bus 5, 11, 15, 15A, 16, 23, 36. **Tickets** *£9.50-£10.95; £7.50-£9.95 reductions.*

A beautiful reminder of what cinemas used to be like in the days before the multiplex, the Dominion's art deco interior dates from 1938. Family-run, it has four screens, impossibly comfortable seating and a small bar where you can get coffee or alcohol. It's more expensive than a multiplex, but far and away the most civilised cinema in the city.

★ Filmhouse

88 Lothian Road, South Edinburgh (information 228 2689, tickets 228 2688, www.filmhousecinema.com). Bus 1, 2, 10, 11, 15, 15A, 16, 24, 34, 35, 36, 47. **Tickets** *£5.50-£9; £4-£7.20 reductions.* **Map** *p284 E8.*

This three-screen indie opened its doors in 1978. Since then it's offered a mix of arty new films and classics as well as providing a headquarters for the Edinburgh International Film Festival (*see p37*). Its For Crying Out Loud screenings are for parents and babies only. Although the Cameo and the Glasgow Film Theatre might disagree, arguably this is Scotland's leading arthouse venue.

► *For the Filmhouse's long-established café-bar, see p136.*

CINEMA CITY

Edinburgh's relationship with the movies goes back a long way.

One of the first venues in Scotland's capital to show flickering images was the old Queen's Hall in Queen Street, way back in 1897. It later became the home of the BBC in Edinburgh and now houses the **Jam House** (see *p180*), an extensive bar-restaurant with live music.

The first cinemas as we would understand them today, however, arrived after 1910, when legislation stipulated that the audience space should be separate from the projection room, because of fire risk. In the few years before the Great War, these new-style cinemas sprang up all over the city: Dalry, Fountainbridge, Leith, Nicolson Street, several on Princes Street and elsewhere.

Their glory years were relatively brief. From the late Edwardian period until the aftermath of World War II, cinema had its heyday. It was gradually supplanted by television in the 1950s and 1960s. The Haymarket in Dalry Road, for example, opened in 1912, was renamed the Scotia in 1946, and showed its last movie in 1964. The Alhambra on Leith Walk appeared in 1914, closed in 1958 and was demolished in 1974 – a familiar story. Fortunately, not all the old stagers have disappeared.

The King's on Home Street, Tollcross, opened in 1914, became an arthouse cinema in 1949, was renamed the **Cameo** (see *above*) and still shows films today. It was a key venue for the fledgling Edinburgh International Film Festival (EIFF), which was founded in 1947 and provides more evidence of the city's enduring love affair with the medium. The other long-term survivors are the art deco **Dominion** (see *above*) in Morningside (opened 1938) and the **Odeon** on Lothian Road (also 1938, initially called the Regal).

In 1978, when the **Filmhouse** (see *above*) on Lothian Road launched to further test the water for arthouse cinema, it had one screen and room for an audience of 90. Now, nearly 40 years later, it has three screens, space for 450 or so, shows around 700 films a year and is a key Scottish venue for independent movies; it also provides a home for the EIFF.

Edinburgh isn't averse to mainstream movies that go bang and woo either. Lately, there have been five multiplexes with nearly 50 screens between them. In 2015, another Odeon opened at Kinnaird Park in the east of the city, adding to the total.

ESSENTIAL EDINBURGH FILMS

Six of the capital's star turns.

The Prime of Miss Jean Brodie.

GREYFRIARS BOBBY
DON CHAFFEY (1961)

Sentimental Disney tale about a Highland terrier (Bobby) who comes to Edinburgh with his old, unemployed owner. The man dies but then what happens to the dog? Set around Greyfriars Kirkyard, the cast is distinguished (Donald Crisp, Andrew Cruickshank, Gordon Jackson and more); the tears duly flow.

THE PRIME OF MISS JEAN BRODIE
RONALD NEAME (1969)

Female adolescent sexuality, a teacher–pupil affair, fatal politics and the peerless performance of Maggie Smith in the title role, all styled against an Edinburgh backdrop. It's very much a 1960s movie, portraying a story from the 1930s, but it remains la crème de la crème.

FESTIVAL
ANNIE GRIFFIN (2005)

An indie shot in Edinburgh during the 2004 festivals, this black comedy has an almost documentary verisimilitude when it comes to the city streets. It also features some very well-known faces, among them Amelia Bullmore, Stephen Mangan, Daniella Nardini, Chris O'Dowd and Clive Russell.

HALLAM FOE
DAVID MACKENZIE (2007)

A deeply confused young man from the Borders moves to Edinburgh, spots a woman who looks like his late mother, gets a job at the Balmoral Hotel where she works, then it starts to get very messy indeed. Jamie Bell stars; an iconic clock tower, abandonment and dysfunction take the supporting roles.

THE ILLUSIONIST
SYLVAIN CHOMET (2010)

A French animation depicting a magician who travels to Scotland, the audience for his talents in Paris dwindling in the face of more contemporary entertainments. The images of Edinburgh in the movie are haunting and beautiful; the plot will cut your heart apart because, ultimately, 'magicians do not exist'.

SUNSHINE ON LEITH
DEXTER FLETCHER (2013)

A musical featuring two Scottish soldiers, discharged from the British Army after service in Afghanistan, it involves love, heartache, secrets and reconciliation, lots of songs by the Proclaimers and scenes shot all around Edinburgh including, of course, in Leith. Unashamed, feel-good, sing-along fun.

Gay & Lesbian

Traditionally known as the Pink Triangle, Edinburgh's gay village is bounded on two sides by Broughton Street and the stretch of Leith Walk from Broughton Street to London Road. The triangle always lacked a third side, but, hey, it was a good slogan. Here, you find venues such as Planet, CC Blooms, Café Habana, Chalky's, the Street and Café Nom de Plume. Between them they offer everything from a cup of hot chocolate on a winter's afternoon to a dancefloor and a cheesy soundtrack to keep you moving until silly o'clock the next morning. If that isn't your idea of fun, then the LGBT Health & Wellbeing Centre website (see p266) can point you in the direction of any number of gay community groups, from sports teams to ramblers, book clubs to amateur dramatic societies. The website of Scotsgay magazine (www.scotsgay.co.uk) also has a good classified section covering community groups and venues across Scotland.

THE GAY SCENE

For more than 25 years, the Blue Moon café-bar on Broughton Street was an icon of gay Edinburgh. Operating at a couple of premises over its lifetime, it was a welcoming and sociable space, but it closed its doors in 2013 to be replaced by a Thai restaurant. The gay sex shop adjacent, Q Store, is still in business, but currently describes itself as pansexual, the proprietor observing that 'young people aren't labelling themselves these days'. The Newtown Bar on Dublin Street ceased trading in 2014 after more than 20 years as a gay venue and the address is now home to a bar-restaurant with a rotisserie and craft beer.

There is absolutely no doubt that there has been a shift in the city's gay scene in this second decade of the 21st century. In political and legislative terms, Edinburgh is the capital of a country where civil partnerships between same-sex couples have been possible since 2005; also where the Scottish Conservative Party has had an out lesbian leader since 2011, Ruth Davidson.

The Marriage and Civil Partnership (Scotland) Act was passed in February 2014 – seven months after similar legislation covering England and Wales – allowing civil partnerships to be converted into marriages, and for gay couples to marry. The law came into effect on 16 December 2014 and after the customary 15-day notice period, Scotland's first brace of gay weddings took place on the last day of the year; two couples tied the knot in Glasgow.

The witnesses at one celebration were First Minister Nicola Sturgeon and Scottish Green Party co-convenor Patrick Harvie MSP; at the other, the witnesses were Scots makar (national laureate) Liz Lochhead and Scottish government minister Marco Biagi MSP. It's a long way from the dark days of the 1980s when gay men and women across the British Isles were under siege from their own politicians.

Meanwhile, the hedonistic aspect of Edinburgh's gay scene has been dialled down lately, although it certainly hasn't disappeared altogether. Gay bars have closed, and there are fewer specifically queer club nights. In the venues that remain, the gay audience might be complemented by a straight hen party, seeking a good night out among some cute boys while getting in touch with their inner fag-hag. If there is karaoke, so much the better.

Even as the number of partying opportunities has declined, the civic side of gay life really does seem to have expanded. Aside from out politicians in the Scottish Parliament, there are sports clubs, support groups, arts groups and more. The **Edinburgh Gay Men's Chorus** (www.egmc. co.uk, pictured left) does a couple of charity fundraising gigs a year, for example, while **Loud & Proud** (www.loudandproudchoir.org), billing itself as Scotland's LGBT choir, performs four times a year. Both are worth a listen. Meanwhile, the idea of Scottish country dancing in Edinburgh with a community group called the **Gay Gordons** (www.gaygordonsedinburgh. com) is almost irresistible. It runs three terms of classes a year and holds occasional ceilidhs.

Perhaps the explanation for the shift in emphasis is simple. The party animals of the 1990s, and earlier, hit middle age at just the wrong time: the 2008-09 recession and its aftermath. Civil partnerships, stretched finances and the possibility of marriage came galloping over the hill in quick succession, presenting a generation of gay men and women with new choices.

Simultaneously, the straight world was coming to terms with the fact that being gay was just as much about the kids, choir practice and stress at work as it was about the poppers and house music stereotype. The bonus is a younger generation, gay, straight or whatever, who are much less likely to care about someone's sexuality at all.

Finally, the headline annual event in Edinburgh's gay calendar is the **Pride Scotia** festival (www.pride-scotia.org) every summer. Preceded by a couple of days of gigs and celebrations, the march itself goes through the Old Town from the Scottish Parliament to Bristo Square, where a stage is set up so various acts can entertain the crowd; it usually takes place on a Saturday in June.

RESTAURANTS

Café Nom de Plume
60 Broughton Street, Broughton (478 1372). Bus 8 or Playhouse buses. **Open** 11am-11pm Mon-Sat; noon-11pm Sun. **Main courses** £6.50-£8.50. **Map** p280 H4.

A popular café with a mainly gay clientele, under the same ownership as the Regent pub (*see p178*). For review, *see p124*.

PUBS & BARS

Café Habana
22 Greenside Lane, Leith Walk, Calton Hill (558 1270). Playhouse buses. **Open** 1pm-1am daily. **Map** p281 J5.

Next door to the Edinburgh Playhouse, this venue has been a fixture on the city's gay scene since the 1980s. It has undergone a couple of name changes in that time, but its Café Habana persona is very well established. Here, you find drag-queen hostesses leading themed nights, DJs, all-day karaoke on Sundays, and even, in summer, people sitting at the tables outside having a drink in the sunshine. There is no website, but you can find the venue on Facebook; search for 'Habana Edinburgh'.

★ CC Blooms
23-24 Greenside Lane, Calton Hill (556 9331, www.ccblooms edinburgh.com). Playhouse buses. **Open** 11am-3am daily. *Food served* 11am-9pm daily. **Main courses** £7.50-£9.50. **Map** p279 J5.

A mainstay of the city's gay scene for more than two decades. For review, *see p122. Photo p178.*

Café Nom de Plume.

Chalky's
*4 Picardy Place, Broughton (550 1780, www.
chalkysedinburgh.co.uk). Playhouse buses.* **Open**
11pm-3am Mon-Thur, Sun; 10pm-3am Fri, Sat.
Map p281 J5.
In premises that formerly housed the bar-nightclub
GHQ, Chalky's is another bar-nightclub, launched in
early 2014. It has a main room (contemporary chart
music and dance), a cheese room (retro-hits) plus a
VIP area, making it a go-to venue for late-night dancing, drinks and entertainment such as drag races.

Planet
*6 Baxter's Place, Leith Walk, Calton Hill
(556 1115). Playhouse buses.* **Open** 1pm-3am
daily. **Map** p281 J5.
A small bar with a reputation for keenly priced
drinks (shots, bombs, mini-cocktails), the emphasis
here is very much on fun, backed by a DJ soundtrack.
Planet doesn't have a dedicated website, but it is on
Facebook; search for 'Planet Bar Edinburgh'.

★ Regent
*2 Montrose Terrace, Calton Hill (661 8198,
www.theregentbar.co.uk).* **Open** noon-1am Mon-Sat; 12.30pm-1am Sun.
Food served noon-10pm Mon-Sat; 12.30-10pm Sun.
Main courses £5.50-£8. **Map** p281 M5.
Real ale and pub grub in a friendly gay bar. For
review, *see p119*.

Street
*2B Picardy Place, Broughton (556 4272, www.
thestreetbaredinburgh.co.uk). Playhouse buses.*
Open noon-1am Mon-Thur; noon-3am Fri, Sat;
12.30pm-1am Sun. *Food served* noon-9pm Mon-
Sat; 12.30-9pm Sun. **Main courses** £5.25-£7.50.
Map p281 J5.

In situ now for more than a decade, the Street is a
straight-friendly gay bar that offers everything from
burgers and chicken wraps to cocktails, afternoon
coffee to late-night DJs. The ground floor serves as
a café-bar, the basement as a bar-nightclub with
a dancefloor. The pub quiz every Wednesday is an
elaborate and amusing affair.

SHOPS & SERVICES

Number 18
*18 Albert Place, Leith Walk, Calton Hill (553 3222).
Bus 7, 10, 12, 14, 16, 22, 25, 49.* **Open** noon-
10pm daily. **Admission** £12; £7-£10 reductions.
Map p281 K3.
A long-established gay sauna with a discreet
entrance on the Albert Place stretch of Leith Walk.
Facilities include a steam room, sauna, shower and
more. Search for 'No.Eighteen' on Facebook.

Q Store
*5 Barony Street, New Town (477 4756). Bus 8
or Playhouse buses.* **Open** 11am-7pm Mon-Fri;
11am-6pm Sat; noon-5pm Sun. **Map** p280 H4.
Informed by its gay roots, Q Store now describes
itself as a pansexual shop selling DVDs, magazines,
underwear and toys to a metrosexual clientele, but
with a strong gay element.

Steamworks
*5 Broughton Market, Broughton (477 3567). Bus
8.* **Open** 11am-11pm daily. **Admission** £10-£15.
Map p280 H4.
Broughton Market is at the western end of Barony
Street and just two minutes' walk from Broughton
Street. Steamworks has a sauna room, spa pool,
steam room and other facilities. The £10 entrance fee
gets you a locker; a private cabin costs £15.

CC Blooms. *See p177.*

Nightlife

Nightclubs host gigs, gig venues run nightclubs and you're more likely to see King Creosote or Imelda May in a former church such as the Queen's Hall than you are in what might conventionally be described as a 'pop music auditorium'. Scotland's national rugby stadium has enormo-concerts, and the Esplanade at Edinburgh Castle, with the stands erected for the annual Tattoo, is yet another opportunity for Jessie J or Rod Stewart to do their thing in front of a few thousand people. The quaint idea that there is some categorical divide between a pub with a performance area, a nightclub, a traditional concert venue, a sports arena and even – in Edinburgh's case – a converted, Edwardian-era corn exchange simply doesn't stand up to scrutiny. From the scuzziest basement bar with a couple of decks and a wannabe DJ to a recently refurbished city-centre hall where you can see the Flaming Lips tear the house down, it's all nightlife.

ROCK & POP

Adaptability is the key word when thinking about Edinburgh's nightlife venues these days. At **Bannermans** in the Cowgate, you could sit happily on a Saturday afternoon in a pub environment with a pint and be completely unbothered by the high-decibel stylings of a sludgy noise combo from London in the adjacent alcove. Come back several hours later and they'll be hard to miss. Similarly, in the rather glossier surroundings of the **Voodoo Rooms** in West Register Street, you can finish your smoked chicken and goat's cheese salad, politely dab your mouth with a napkin, then head to the venue's Ballroom to hear a highly accomplished blues and soul singer start her set.

Precisely the same applies to establishments that host as many club nights as gigs, or vice versa, since it's no longer possible to draw a distinction between a nightclub and a similar sized live music venue. A typical week at the **Liquid Room** in Victoria Street, for instance, might see mid-evening gigs on four nights, late-evening clubs on three.

Once you've opened the doors and got the staff in, and the audience, you might as well maximise the fun – and the financial return – by providing something that people want. The live music fans can get the last bus home, the clubbers will keep going until 3am, although it should be noted that Edinburgh's jazzers keep late hours too, thanks to the **Jazz Bar** on Chambers Street. By comparison, the city's brace of folk clubs keep to a much more civilised schedule.

When it comes to bigger concert venues, these range from the ecclesiastical feel of the **Queen's Hall** on Clerk Street to **Murrayfield**, a fixture on the UK stadium tour circuit for visiting mega-acts. In the former, you could see a string quartet, a folk group and a rock band in the same week; in the latter, there may only be one or two gigs a year. Meanwhile, the **Usher Hall** on Lothian Road takes the same eclectic approach as the Queen's Hall, albeit on a much larger scale.

In short, whether you want to casually drop by a bar and see what's on, dance to hip hop into the next morning, or drool in anticipation over the stadium show you bought tickets for six months in advance, it's all possible in Edinburgh.

INFORMATION & TICKETS

To find out what's on, check the individual venue websites listed below, or keep your eyes peeled for posters around the more interesting areas of town. For a more comprehensive guide, try www.timeout.com/edinburgh. *The List* (online and in print) and freesheet the *Skinny* are also worth a look. Online, Jockrock (www.jockrock.org) also has up-to-date record and gig reviews.

The main shop outlets in Edinburgh for tickets are **Ripping Records** (*see p77*) and **Tickets Scotland** (127 Rose Street, New Town, 220 3234, www.tickets-scotland.com) while **Ticketmaster** (www.ticketmaster.co.uk) is a useful resource.

Bars with music

For Edinburgh's grungy, late-night music bar of note, **Whistlebinkies**, *see p69*.

Bannermans

212 Cowgate, Old Town (556 3254, www.bannermanslive.co.uk). Bus 2, 35 or Nicolson Street–North Bridge buses. **Open** noon-1am

Cabaret Voltaire.
See *p182*.

Mon-Sat; 12.30pm-1am Sun. **Admission** £4-£15. **Map** p285 J7.

Bannermans is at the bottom of Niddry Street, on the corner with the Cowgate, on the ground floor of a tenement that rises way over South Bridge above. The pub has a performance space in a low vaulted cellar and it's highly unlikely you'll have heard of the local bands that gig here (although someone occasionally pops up with some residual fairy dust). There's something on most nights; for a taster, check out the Under the Bridge compilations available to stream on the bar's website.

Banshee Labyrinth

29-35 Niddry Street, Old Town (558 8209, www.thebansheelabyrinth.com). Nicolson Street–North Bridge buses. **Open** 7pm-3am daily. **Admission** free-£7.50. **Map** p285 J7.

Another robust and subterranean Old Town pub, which really does feel like a labyrinth, the live music here fits into small spaces and could offer anything from an open mic night to Croatian crust punk, dark-wave bands to black metal. As an added bonus, the pub is haunted.

Henry's Cellar Bar

16 Morrison Street, South Edinburgh (629 4101, www.henryscellarbar.com). Bus 1, 2, 10, 11, 15, 15A, 16, 23, 24, 34, 35, 36, 47. **Open** 5pm-1am Mon-Thur, Sun; 5pm-3am Fri, Sat. **Admission** free-£6. **Map** p284 E8.

In a small basement bar just off Lothian Road, this is where gigs go off piste. Open mic night (Big Mouth Monday) has everything from music to spoken word; there is jazz on Thursdays, and an eclectic range of other performances and clubs including Click Clack, the monthly experimental live music night. An avant antidote to highly formatted television talent shows.

Jam House

5 Queen Street, New Town (220 2321, www.the jamhouse.com). Bus 10, 11, 12, 16, 26, 41, 44, 44A. **Open** 6pm-3am Fri, Sat. *Food served* 6.30-10.30pm Fri, Sat. **Admission** £3-£4. **Map** p280 G5.

Originally a church hall, then the BBC's Edinburgh headquarters, this cavernous venue became the Jam House in 2005. It has a bar, stage and dancefloor, and a balcony restaurant. Professional cover bands provide the entertainment. Smart-casual, over-21s only.

Voodoo Rooms

19A West Register Street, New Town (556 7060, www.thevoodoorooms.com). Princes Street buses. **Open** 4pm-1am Mon-Thur; noon-1am Fri-Sun. *Food served* 4-10pm Mon-Thur; noon-10pm Fri-Sun. **Admission** free-£15. **Map** p284 H6.

A bar-restaurant with an extravagantly gothic and shiny aesthetic, you expect this to be in New Orleans rather than upstairs from the Café Royal (*see p95*). It has two gig spaces: the Ballroom (capacity 200) and the smaller Speakeasy (80). Both are used for a range

ESSENTIAL EDINBURGH ALBUMS

Music of the city and the country.

THE HANGMAN'S BEAUTIFUL DAUGHTER
THE INCREDIBLE STRING BAND (1968)
With their roots in the 1960s Edinburgh folk scene, this group was built around Mike Heron and Robin Williamson, whose sometimes fey, psychedelic compositions came to be influential on performers as diverse as Led Zeppelin and Idlewild.

HANDFUL OF EARTH
DICK GAUGHAN (1981)
Gaughan, from Leith, covered everything on this album, from the Irish experience in Scotland to 17th-century English radicalism, songs of hope and beauty to acoustic guitar instrumentals. His talent as a musician and his political commitment inform every chord and lyric of a Scottish folk music classic.

GEOGADDI
BOARDS OF CANADA (2002)
Whenever Boards of Canada are discussed, it's obligatory to use words like 'disturbing', 'ethereal' and 'haunting'. A studio duo of some mystique rather than a live band, their electronica is probably best understood as the soundtrack to a novel that William Gibson hasn't written yet.

CAN'T STAND THE REZILLOS
THE REZILLOS (1978)
It takes real talent to appear cartoonish and trashy but crank out an album's worth of top-quality pop. A local art college band with frenetic energy and an ear for a tune, the Rezillos featured guitarist Jo Callis, who later joined the Human League and co-wrote some of their biggest hits.

SUNSHINE ON LEITH
THE PROCLAIMERS (1988)
With two albums in 1987 and 1988, the Proclaimers went from unknown to canonical status in just over a year. This was their second, including the hits 'I'm Gonna Be (500 Miles)' and the emotional title track that went on to become an anthem for both Hibernian FC and Leithers everywhere.

FOREVER 22
TV21 (2009)
After a few years of John Peel sessions, *Old Grey Whistle Test* appearances, gigs and one record, TV21 split in 1982, leaving fans thinking of what might have been. In 2009, they got together for a complete blast of a rock album, confirming that they were indeed the biggest band Edinburgh never had.

Sneaky Pete's.

of acts, from cover bands to unsigned local talent; there are also club nights, such as the long-running retro-jive of Vegas! (www.vegasscotland.co.uk), plus events like burlesque balls.

Club & gig venues

Edinburgh has a bundle of venues that provide clubs, gigs and other events. They tend to open around 10-11pm and keep going until 3am, although that varies depending on the night of the week and the club; check websites for specific nights. In smart George Street establishments such as **Lulu** (www.luluedinburgh.co.uk) in the basement of the Tigerlily hotel (*see p250*) and **Whynot** (www.wnclub.co.uk) in the Dome bar-restaurant (*see p95*), well-dressed hedonism rules. **Silk** (www.silknightclub.co.uk) on King's Stables Road is more youthful, and there is an ample selection of underground music too.

La Belle Angèle

11 Hastie's Close, Cowgate, Old Town (220 1161, www.labelleangele.com). Nicolson Street–North Bridge buses. **Admission** varies; see website. **Map** p285 J8.
Looking pretty sprightly for a venue that burned down in 2002, this much-loved old place was rebuilt and opened its doors again in 2014. A patrician presence on the local club scene, it attracts name DJs such as Kölsch, Oxide & Neutrino, Roni Size and more, as well as an eclectic range of bands. See its Facebook page for club and gig details and prices.

Bongo Club

66 Cowgate, Old Town (558 8844, www.thebongo club.co.uk). Bus 2, 35. **Admission** *Clubs* £2-£7. *Gigs* varies; see website. **Map** p285 K7.
Formerly in Moray House but occupying these Cowgate premises since 2013, the Bongo is a well-established presence on the city's alt-arts scene. Expect plenty of gigs, other events and regular club nights including Messenger (reggae), Mumbo Jumbo (funk and soul) or the self-explanatory I Love Hip Hop.

Cabaret Voltaire

36-38 Blair Street, Old Town (247 4704, www. thecabaretvoltaire.com). Nicolson Street–North Bridge buses. **Admission** *Clubs* free-£7. *Gigs* varies; see website. **Map** p285 J7.
More than just an underground club, Cabaret Voltaire offers a café-bar, gig space and acclaimed regular club nights ranging from Hector's House (house and its associates) to the more considered iAM (forward-thinking music for the masses). *Photo 180.*

Caves

12 Niddry Street South, Old Town (557 8989, www.thecavesedinburgh.com). Bus 35 or Nicolson Street–North Bridge buses. **Admission** varies; see website. **Map** p285 J7.
In the old vaults under South Bridge, the Caves complex hosts everything from weddings and private parties to Fringe shows in August. Throughout the year, there are gigs ranging from singer-songwriters to experimental ensembles, dance music duos to charity fundraisers.

Citrus Club

40-42 Grindlay Street, South Edinburgh (622 7086). Bus 1, 10, 11, 15, 15A, 16, 23, 24, 34, 35, 36, 47. **Admission** *Clubs* free-£5. *Gigs* varies. **Map** p284 E8.
A functional and far from glamorous venue that hosts the legendary retro club night Planet Earth, the Citrus has been bopping to Bowie, the Cure and the Smiths for well over 20 years. It also hosts live music in the shape of tribute bands and other acts that surprise you with their longevity, such as Bad Manners, the Members and TV Smith. Search for 'Citrus-Club Edinburgh' on Facebook for gig times and prices.

Electric Circus

36-39 Market Street, Old Town (226 4224, www.theelectriccircus.biz). Princes Street buses. **Admission** *Clubs* free-£6. *Gigs* varies; see website. **Map** p284 H7.
Good times reign at the Circus, where there are private karaoke rooms, gigs by up-and-coming indie hopefuls

and club nights ranging from retro and nostalgia to hip hop; regular sessions by guest DJs too. Every couple of months, Baby Loves Disco (www.babyloves. disco.co.uk) comes to town: a Saturday afternoon club for children up to six years old, and their parents.

Hive

15-17 Niddry Street, Old Town (556 0444, www. clubhive.co.uk). Bus 35 or Nicolson Street–North Bridge buses. **Admission** free-£4. **No credit cards. Map** p285 J7.
As the name suggests, this club is made up of a series of interlocking rooms, which host student-friendly nights across two dancefloors and several bars. Drinks promos make it a lively spot; check out the club's photos on Facebook to get a flavour (search for 'The Hive').

Liquid Room

9C Victoria Street, Old Town (225 2564, www.liquidroom.com). Bus 23, 27, 41, 42, 67. **Admission** varies; see website. **Map** p284 G7.
The Liquid Room was shut for 18 months due to water damage after a fire in an adjacent property, but finally made it back (in 2010) to reclaim its place among the city's leading live music and club venues. Here, you will see covers bands, but also the likes of Julian Cope, plus regular club nights and an additional space called the Annexe (entry from the Cowgate).

Mash House

37 Guthrie Street, Old Town (220 2514). Nicolson Street–North Bridge buses. **Admission** *Clubs* £3-£10. *Gigs* varies. **Map** p284 H8.
Tucked into a corner of Guthrie Street in the same SoCo development as La Belle Angèle (*see p182*),

the Mash House (launched at the end of 2013) is yet another venue looking to make best use of its space. It has club nights ranging from African and Balkan sounds to funk, house and techno, occasional exhibitions and around ten gigs a month (pop, psychedelia, punk, rock). Search for 'The Mash House' on Facebook.

Sneaky Pete's

73 Cowgate, Old Town (225 1757, www.sneaky petes.co.uk). Bus 2. **Admission** *Clubs* free-£5. *Gigs* varies; see website. **Map** p284 H8.
With some top Edinburgh club nights including In Deep and Wasabi Disco, plus gigs ranging from folk to Finnish singer-songwriters or slacker-indie rock, Sneaky Pete's is a brilliant asset to the city.

Studio 24

24-26 Calton Road, Calton Hill (558 3758, www. studio24club.co.uk). Bus 35. **Admission** *Clubs* £3-£5. *Gigs* varies; see website. **Map** p285 J6.
Hidden away from most of the other city clubs, down Calton Road, Studio 24 is where you might see Tantz doing Balkan tunes (with belly-dancer) or where the Alabama 3 might gig. Club nights cover everything from new-wave mutant disco to doo-wop and surf. Its audience really loves this place.

Wee Red Bar

Edinburgh College of Art, Lauriston Place, South Edinburgh (651 5859, www.weeredbar.co.uk). Bus 2, 23, 27, 35, 45, 47. **Admission** free-£5. **Map** p284 F9.
The spiritual home of Edinburgh's original art-punk scene (the Rezillos, the Fire Engines et al), the ECA's scarlet-painted students' union bar and club continues to provide a platform for happenings of

Studio 24.

ARTS & ENTERTAINMENT

every persuasion. There are various gigs and club nights, including the Egg, surely the city's longest-standing indie disco.

Large-scale venues

In addition to the venues listed here, some of the city's concert halls also stage regular rock and pop shows. The **Queen's Hall** (*see p187*) has hosted concerts by the likes of King Creosote, Richard Thompson and Imelda May; the **Usher Hall** (*see p187*) has featured Björk, the Flaming Lips and Robert Plant. The **Playhouse** (*see p190*) supplements its theatrical blockbusters with shows from big names; over the years, the Rolling Stones (1982), the Velvet Underground (1993), Tom Waits (2008) and Neil Young (2013) have all played there.

There are also two major open-air venues in the city. **Murrayfield Stadium** (*see p151*) has welcomed the likes of Madonna and One Direction in recent years; **Edinburgh Castle** (*see p55*) has hosted summer shows from acts including Jessie J, Rod Stewart and Paul Weller.

Corn Exchange

11 New Market Road, West Edinburgh (477 3500, www.ece.uk.com). Bus 4, 20, 34, 35, 44, 44A.
The main problem with this former corn exchange is its location, two and a half miles south-west of the city centre. The ambience at this oversized barn isn't that great, either. It has standing capacity for 3,000; big gigs are the order of the day. Blur were first when the venue opened in 1999 and since then Elbow, the Kaiser Chiefs and Kanye West have all appeared. The Corn Exchange also hosts awards ceremonies, conferences, corporate dinners, exhibitions and more.

FOLK & ROOTS

In the years since the 1960s Scottish folk revival and the founding of the **Edinburgh Folk Club** in 1973, a welter of bar-room activity has provided an informal focus on keeping the old traditions alive. The easiest way to discover the genuine article is to head to one of the various pub sessions. The best are held at **Sandy Bell's** (*see p76*) and the **Royal Oak** (*see p75*), the latter also home to Sunday evening's Wee Folk Club.

Venues

Edinburgh Folk Club

Pleasance Cabaret Bar, 60 Pleasance, South Edinburgh (650 2458). Nicolson Street–North Bridge buses. **Admission** £10; £7-£9 reductions. **Map** p285 K8.
The Pleasance Cabaret Bar hosts the Edinburgh Folk Club's weekly sessions (8pm Wednesday), continuing a folk tradition in the city that's lasted for more than 40 years. Some names you will have heard of, most you won't. Find it on Facebook; search for 'Edinburgh Folk Club'.

Leith Folk Club

Victoria Park House Hotel, 221 Ferry Road, Leith (07502 024852 mobile, www.leithfolk club.com). Bus 7, 11, 14, 21. **Admission** £7. **Map** p288 C2.
Leith's 2004 retort to the longer-established Edinburgh Folk Club up in town. There are weekly gigs here too (7.30pm Tuesday). It's the most fun you can have in Leith on a midweek night, albeit Hibernian FC and some Michelin-starred chefs might dispute that.

Jazz Bar.

Stand.

JAZZ

Edinburgh's jazz scene is effectively down to the one-man machine that is drummer and promoter Bill Kyle, who opened the Bridge Jazz Bar on South Bridge in 2002. It lasted all of a few months, then burned down in a fire that destroyed the entire block. Undeterred, Kyle tried again with the **Jazz Bar** on Chambers Street in 2005 and more than a decade later it's still in business.

There's more jazz at the **Queen's Hall** (*see p187*), which offers less regular but worth-the-wait bigger shows; and at July's **Edinburgh Jazz & Blues Festival** (*see p37*). It's also worth keeping an eye out for the **National Youth Jazz Orchestra of Scotland** (www.nyos.co.uk), and the **Scottish National Jazz Orchestra** (www.snjo.co.uk); its artistic director is the highly respected Edinburgh saxophonist Tommy Smith.

Venues

Jazz Bar

1A Chambers Street, Old Town (220 4298, www.the jazzbar.co.uk). Nicolson Street–North Bridge buses. **Open** 5pm-3am Mon-Fri; 2.30pm-3am Sat, Sun. **Admission** free-£5. **No credit cards. Map** p285 J8.
The hardest-working music venue in Edinburgh? With its non-stop programme of daily, multiple gigs, it could certainly claim to be. At the Jazz Bar, you can find everything from acoustic sessions at Sunday teatime to a 17-piece big band on Monday evenings, with a guest band filling the late Saturday slot into the small hours of Sunday. A fantastic place to drop by and listen to whatever's happening.

COMEDY

All praise the **Stand**. If it wasn't for the tireless efforts of this grungy basement comedy club in the New Town – and the **Beehive Inn** in the Old Town – Edinburgh's comedy scene would look decidedly threadbare. That is, of course,

excluding August, when the Fringe presents more comedy turns than anywhere else on the planet.

The two extremes, August on the one hand and the rest of the year on the other, make Edinburgh the feast-or-famine capital of LOL. This is particularly the case when it comes to top-of-the-range comedians, who flood here for the Fringe but otherwise limit their appearances to occasional forays into major venues such as the **Playhouse** (*see p190*). If comedy's your thing, choose the dates for your Edinburgh visit with care.

Venues

Monkey Barrel Comedy

Beehive Inn, Grassmarket, Old Town (box office 0845 500 1056, pub 225 7171, www.monkey barrelcomedy.com). Bus 2. **Open Bar** 9am-1am Mon-Sat; 12.30pm-1am Sun bar. *Food served* 10am-10pm Mon-Sat; 12.30-10pm Sun. **Admission** £10. **Map** p284 G8.
The Beehive Inn offers stand-up, generally at 8.30pm on Friday and Saturday, featuring up-and-coming Scottish comedians topped off with a visiting headliner. Stag and hen parties are welcome. The Facebook page (search for 'Monkey Barrel Comedy') often has more up-to-date listings than the website.

★ Stand

5 York Place, New Town (558 7272, www.thestand. co.uk). Bus 10, 11, 12, 16, 26, 41, 44, 44A. **Open** 7.30pm-1am Mon-Thur; 7pm-1am Fri, Sat; 12.30-3.30pm, 7.30pm-1am Sat. *Food served* 7-8.30pm Thur-Sat; 12.30-2.30pm Sun. **Admission** free-£13. **Map** p280 H5.
Housed in a basement on York Place, the Stand has endured as the shabbily appointed headquarters of capital comedy for a couple of decades, easily outliving the now-departed Edinburgh branch of Jongleurs. Here you can see complete unknowns on Mondays (the Red Raw showcase) as well as career comics such as Frankie Boyle, Josie Long, Richard Herring and Zoe Lyons. For gig listings, see website.

Performing Arts

At the Edinburgh Festival Theatre you could be watching a Scottish Ballet production one month, a Scottish Opera production the following month, and, a few weeks later, the Royal Scottish National Orchestra performing the score for *Back to the Future* as the movie is shown on the theatre's big screen. You're as likely to see the musical *Priscilla, Queen of the Desert* at the Playhouse as you are a contemporary piece by a leading European ballet company. The Queen's Hall can have a chamber quartet, a folk band and a soul legend on stage in the same week. If each of the city's leading venues kept itself strictly to a categorical silo, they would struggle financially; a catholic approach to programming is, therefore, the order of the day. Irrespective, Edinburgh still achieves an enviable calendar of classical music, ballet, contemporary dance, opera and theatre throughout those 11 months of the year when it's not August.

CLASSICAL MUSIC & OPERA

Although the Scottish capital was hardly a music-free zone in the pre-Georgian period, the city's first purpose-built concert space arrived in 1763 when **St Cecilia's Hall** (www.stcecilias.ed.ac.uk) was built on Niddry Street in the Old Town. Named in honour of the patron saint of music and musicians, it has survived for more than 250 years despite the catalogue of regeneration miscues, roistering and wrecklessness in the neighbourhood. Now owned by the University of Edinburgh, it serves as a concert venue and home for part of the university's collection of historic musical instruments. Lately, the university has been fully aware that St Cecilia's was looking a little unloved, and lacked 'streetscape appeal', so the building is currently undergoing a major refurbishment. This should restore some Georgian sparkle, although it's not expected to reopen until autumn 2016 at the earliest.

As for live music, rather than venue history, two of the most notable ensembles playing regularly in the city are the **Royal Scottish National Orchestra** (www.rsno.org.uk) and the **Scottish Chamber Orchestra** (www.sco.org.uk). The RSNO often appears at the Usher Hall, the SCO at both the Usher Hall and the smaller Queen's Hall.

A few other Scottish-based groups also play frequently in Edinburgh, chief among them the **BBC Scottish Symphony Orchestra** (www.bbc.co.uk/bbcsso). Watch out for **Mr McFall's Chamber** (www.mcfalls.co.uk), an ingenious experimental project that unites SCO players and other talented, broadminded musicians who are equally at home tackling

> ### IN THE KNOW ARTS VILLAGE
>
> With the **Traverse** (*see p191*), **Usher Hall** (*see p187*) and **Royal Lyceum** (*see p190*) all next to one another on the east side of Lothian Road, the **Filmhouse** (*see p174*) diagonally opposite, then the **Odeon Lothian Road** and the **Cameo Picturehouse** (*see p174*) on Home Street nearby, this part of South Edinburgh boasts quite a creative village, whether you want arthouse cinema, mainstream movies, music or theatre.

ARTS & ENTERTAINMENT

Frank Zappa as they are Franz Liszt. The energetic and distinctive **Scottish Ensemble** (www. scottishensemble.co.uk) is a string orchestra from Glasgow, while the **Edinburgh Quartet** (www. edinburghquartet.com) is a respected chamber group; both perform regularly in the capital.

Edinburgh has no dedicated opera house, but **Scottish Opera** (www.scottishopera.org.uk) often pops up at the city's Festival Theatre and opera productions are always a feature of the Edinburgh International Festival.

Major venues

Queen's Hall

85-89 Clerk Street, South Edinburgh (box office 668 2019, www.thequeenshall.net). Nicolson Street–North Bridge buses. **Box office** *In person* 10am-5.15pm Mon-Sat (or until 15mins after showtime). *By phone* 10am-5pm Mon-Sat. **Tickets** £4-£30. **Map** p285 K10.

Classical music makes up about half the programming at this 900-capacity venue, a former church that retains its old pews. The Scottish Chamber Orchestra, the Scottish Ensemble and the Edinburgh Quartet all perform regularly at the hall, which is also one of the main concert venues during August. Pop, rock and jazz are a major component too. *Photo p188.*

Reid Concert Hall

Bristo Square, Old Town (651 4336, www.eca. ed.ac.uk/reid school of music). Bus 2, 23, 27, 35, 41, 42, 45, 47, 60, 67. **Concerts** *Lunchtime*

Oct-May 1.10pm Tue, Fri. *Others* see website. **Tickets** *Lunchtime* free. *Others* free-£10. **Map** p284 H8.

Opened in 1859, and with a capacity of 300, this University of Edinburgh venue hosts a range of concerts and recitals during university term-time, although it's also used for lectures and rehearsals, so is very much an academic space. On selected days, you can wander in and enjoy a soul-enhancing, lunchtime concert from students or more established performers – for free. If you're having trouble finding the concert programme on the website, try a web search for 'Concerts at the University' and 'Edinburgh'.

Usher Hall

Lothian Road, South Edinburgh (box office 228 1155, www.usherhall.co.uk). Bus 1, 10, 11, 15, 15A, 16, 24, 34, 36, 475. **Box office** 10am-5.30pm Mon-Sat (or until 30mins after showtime); 1pm-30mins after showtime Sun. **Tickets** free-£50. **Map** p284 E8.

A great deal of time and money has been spent in recent years returning the Usher Hall to its former glory, from refurbishments of the auditorium to the restoration of the organ. First opened in 1914, it's in demand by everyone from the RSNO to rock acts such as the Flaming Lips. Many major classical music events during the Edinburgh International Festival are held here.

Churches

Some of Edinburgh's atmospheric churches are used as concert venues year-round. **Canongate Kirk** (153 Canongate, Old Town, 556 3515, www.canongatekirk.org.uk), the place of worship for Elizabeth II when she's at the Palace of Holyroodhouse, stages shows by military bands, chamber orchestras, choral groups and more. As well as holding free concerts at 6pm most Sundays (the St Giles at Six series), the **High Kirk of St Giles** (*see p57*) has informal lunchtime concerts during the week and celebrity organ recitals. **Greyfriars Kirk** (*see p73*) also holds regular musical events.

St Mary's Episcopal Cathedral

Palmerston Place, New Town (225 6293, www.cathedral.net). Bus 2, 3, 4, 12, 25, 26, 31, 33, 44, 44A. **Open & tickets** varies; see website. **Map** p283 C8.

St Mary's is unique in Scotland in maintaining daily sung services, but it also has a Steinway grand piano, a serious cathedral organ and choristers with an enviable reputation. There are all kinds of concerts and events throughout the year, especially in August. If you just want to drop by to hear choral evensong, however, it's held at 5.30pm Monday to Wednesday and Friday, and at 3.30pm on Sunday. Choral eucharist is sung at 5.30pm on Thursday.

Scottish Opera.

ARTS & ENTERTAINMENT

DANCE

The fan base for ballet and contemporary dance in Edinburgh is a devoted one, adventurous enough to enjoy visits from the likes of the Richard Alston Dance Company, Mark Morris Dance Group and Nederlands Dans Theater, as well as the Glasgow-based **Scottish Ballet** (www.scottishballet.co.uk). Throughout the year there are performances at the **Edinburgh Festival Theatre** (*see below*), although the **Playhouse** (*see p190*) and **Royal Lyceum** (*see p190*) also host shows during the Edinburgh International Festival. For smaller scale productions, it's worth checking the programme at the **Brunton** (*see p190*) in Musselburgh and the **Traverse** (*see p191*).

Venues

★ Dance Base

14-16 Grassmarket, Old Town (225 5525, www.dancebase.co.uk). Bus 2. **Open** 8am-10pm Mon; 9am-9.30pm Tue, Wed; 8am-9.30pm Fri; 9.30am-5.30pm Sat. **Map** p284 G8.
Not a venue for public performance as such, Scotland's national centre for dance could hardly be excluded from this section of the guide. A beautifully airy, purpose-built state-of-the-art danceteria, it has become the focal point for the capital's dance community. With four studios hosting an extensive programme of classes and workshops, it accommodates all levels and areas of interest.

Edinburgh Festival Theatre

13-29 Nicolson Street, South Edinburgh (529 6000, www.edtheatres.com). Nicolson Street–North Bridge buses. **Box office** *In person* 10am-6pm Mon-Sat (10am-8pm on performance days). *By phone* 11am-6pm Mon-Sat (11am-8pm performance days). **Tickets** £10-£80. **Map** p285 J8.
There has been a place of entertainment on this site for many years, but the current EFT began life as the Empire Palace Theatre in 1892, hosting the biggest old-time variety stars. By the mid 1980s, it was a run-down bingo hall, but a 1994 restoration turned it into the most opulent of receiving houses, with one of the biggest dance stages in the UK. Its programme is a mix of dance, opera and large-scale theatrical productions; it's a major venue for opera and dance during the Edinburgh International Festival.

THEATRE

You may get the chance to see big stars treading the Edinburgh boards during your stay – even outside the hallowed month of August – but the theatre scene in the city is not driven by celebrity or glamour. Audiences are just as likely to be attracted to the standard of the relevant ensemble or the promise of a new work by one of Scotland's many celebrated playwrights. In this regard, theatre is keenly attuned to the broader cultural life of Scotland; at its best, it can provide an illuminating snapshot of a nation.

The majority of venues used for theatre during the festival season revert to other functions for the rest of the year, but a core of full-time theatres and arts centres keep things ticking over. If you're looking for large-scale mainstream entertainment, it's most likely you'll end up at the **King's Theatre**, the **Edinburgh Playhouse** or the **Edinburgh**

Queen's Hall. *See p187.*

Festival Theatre (*see p188*); meanwhile, the city's two main producing houses, the **Traverse Theatre** (for the new and untried) and the **Royal Lyceum** (for the more established), sit back-to-back near the **Usher Hall** (*see p187*).

Many imaginative touring companies are based in the city, and often stage shows in less conventional spaces. The best include **Grid Iron** (www.gridiron.org.uk), renowned for site-specific performances held everywhere from the Edinburgh International Climbing Arena to the University of Edinburgh's Informatics Forum; **Wee Stories** (www.weestoriestheatre.org), which specialises in drama for children, both educational and entertaining; and **Stellar Quines** (www.stellarquines.com), offering a high-quality programme informed by a female perspective. And last but hardly least, standing apart from the crowd is the **National Theatre of Scotland** (www.nationaltheatrescotland. com), a roving company founded in 2006 that has already made its mark with a number of powerful and engaging new works.

INFORMATION & TICKETS

Time Out's Edinburgh website (www.time out.com/edinburgh) remains the best source for theatre listings in Edinburgh, but it's also worth taking a look at the extensive array of links to artists, companies and events provided by **theatreScotland** (www.theatrescotland.com). Tickets for many major venues can be booked through **Ticketline** (www.ticketline.co.uk) or **Ticketmaster** (www.ticketmaster.co.uk).

Venues

Assembly Roxy

2 Roxburgh Place, South Edinburgh (610 6611, www.assemblyfestival.com). Nicolson Street–North Bridge buses. **Tickets** £8-£14. **Map** p285 J8.
After coming into its own as a busy Fringe venue in recent times, the Roxy was taken over in 2012 by the Assembly organisation, best known for running temporary performance spaces in Edinburgh during August, especially those in and around George Square. The Roxy now serves as a multi-functional arts centre where you can see anything from stand-up comedy to live music and new theatre productions; it remains very busy during the Fringe.

Bedlam Theatre

Bristo Place, Old Town (box office 629 0430, www.bedlamtheatre.co.uk). Bus 2, 23, 27, 35, 41, 42, 45, 47, 60, 67. **Box office** from 45mins before performance. **Tickets** £3.50-£6. **Map** p284 H8.
This Victorian church is home to the Edinburgh University Theatre Company, which produces a rolling programme of student drama every week of the academic year. Relatively recent alumni include Mitch Benn (comic, musician, playwright),

Brunton. See p190.

ARTS & ENTERTAINMENT

Ella Hickson (director and playwright), Miles Jupp (actor and comedian) and Lucy Kirkwood (playwright). Go back far enough and both the iconic Ian Charleson (1949-90) and Alastair Sim (1900-76) appeared in Bedlam productions once upon a time.

Brunton

Ladywell Way, off Bridge Street, Musselburgh, East Lothian (665 2240, www.thebrunton.co.uk). Bus 15, 26, 30, 40, 44, 44A, 113. **Box office** 10am-6pm Mon-Fri; 10am-2.30pm, 3.30-6pm Sat. Until 7pm on performance days; no closure on Sat afternoons when there's a matinée. **Tickets** £6-£20.

Over the county line in East Lothian, the Brunton is around six miles east of Princes Street but only about half an hour away by bus. It's a comfortable civic space that's looking spruce after a £3.7 million refit in 2012; it hosts children's theatre, classical concerts, movies, musicals, pantomime, other populist shows, touring companies and tribute bands. Serious theatre heads might even be tempted to Musselburgh to see new work by Scottish writers. *Photo p189.*

Church Hill Theatre

Morningside Road, South Edinburgh (220 4348, www.assemblyroomsedinburgh.co.uk). Bus 5, 11, 15, 15A, 16, 23, 36.
A local authority-run venue under the same management as the Assembly Rooms (*see p85*) in George Street, the Church Hill is a small neighbourhood theatre holding 350, with additional studio space. The building itself was originally a late Victorian church. It's terribly underused, but does provide a venue for amateur groups such as Edinburgh People's Theatre (www.ept.org.uk) and Pit Pony Productions (search for 'Pit Pony Productions – Midlothian' on Facebook), plus it's busy with Fringe shows in August. Check the website for information about what's on.

Edinburgh Playhouse

18-22 Greenside Place, Calton Hill (0844 871 3014, www.playhousetheatre.com). Playhouse buses. **Box office** *In person* noon-6pm Mon-Sat (noon-8pm performance days). *By phone* 24hrs daily. **Tickets** £15-£38.50. **Map** p281 J5.
Owned by the Ambassador Theatre Group, the 3,059-seat Playhouse originally opened as a cinema in 1929, but is now the regular home for touring musicals such as *Dirty Dancing* or *Spamalot*. The Edinburgh International Festival regularly uses the auditorium, one of the largest of its kind in the British Isles, for bigger dance productions, while grown-up rock stars and headline comedians treat the Playhouse as a stopping-off point on UK tours.

King's Theatre

2 Leven Street, Tollcross, South Edinburgh (529 6000, www.edtheatres.com). Bus 10, 11, 15, 15A, 16, 23, 27, 36, 45. **Box office** *In person*

10am-6pm Mon-Sat. *By phone* 11am-6pm Mon-Sat. **Tickets** £14-£31. **Map** p284 F10.
Opened in 1906, this elegant old-time institution is managed, along with the Edinburgh Festival Theatre, by the Festival City Theatres Trust. The programme mixes musicals with star-studded serious drama, usually on pre- or post-West End tours, so you can see the likes of *A View from the Bridge* here as well as comedy, *The Woman in Black* and occasional polite live music. Every December, there's a popular pantomime.

North Edinburgh Arts

15A Pennywell Court, off Pennywell Road, Muirhouse (315 2151, www.northedinburgh arts.co.uk). Bus 27, 32, 37. **Tickets** 50p-£12.
Tucked away behind one of the least inviting shopping arcades in the city, this community arts centre has a programme aimed mainly at local children, with puppet theatre, comedy and creative play sessions to the fore. Children's theatre company Licketyspit (www.licketyspit.com) is based here, and there is also studio space, a garden and a café.

Out of the Blue Drill Hall

36 Dalmeny Street, Leith (555 7100, www.outof theblue.org.uk). Bus 7, 10, 12, 14, 16, 22, 25, 49. **Map** p281 L2.
An actual old drill hall built for the Royal Scots (Lothian Regiment), and dating from the early Edwardian era, this building has been the headquarters of Out of the Blue, an arts and educational trust, since 2004. The premises are generally used as studio space, for exhibitions, indoor markets and there is also a café. The hall also serves as a performance venue and has hosted shows during the Edinburgh International Festival, Fringe and Leith Festival. It's in tenement country between Leith Walk and Easter Road. Check the website for shows and prices.

Royal Lyceum

Grindlay Street, South Edinburgh (information 248 4800, box office 248 4848, www.lyceum. org.uk). Bus 1, 10, 11, 15, 15A, 16, 24, 34, 36, 47. **Box office** 10am-5pm Mon-Sat (10am-7pm performance days). **Tickets** £15-£29. **Map** p284 E8.

Royal Lyceum.

At its peak in the 1970s, the Lyceum was at the vanguard of a renaissance of local theatrical culture, and a breeding ground for leading directors such as Bill Bryden and Richard Eyre. The intervening decades were less radical but audiences remained loyal; the resident company remains one of the most popular in Scotland, with up to eight productions a year from farces to classics by Ibsen. The National Theatre of Scotland also stages productions here.

Scottish Storytelling Centre

43-45 High Street, Old Town (556 9579, www. tracscotland.org). Bus 35 or Nicolson Street–North Bridge buses. **Box office** 10am-6pm Mon-Sat (later on performance days). **Tickets** free-£36. **Map** p285 J7.

The former Netherbow Arts Centre now serves as a home for storytelling, arguably Scotland's one indigenous art form. The venue presents regular storytelling events pitched at all ages, with opportunities both to hear stories and to tell them. Other events include film, music and theatre performance.

▶ *For more on the centre, see p67.*

★ Traverse Theatre

10 Cambridge Street, South Edinburgh (228 1404, www.traverse.co.uk). Bus 1, 10, 11, 15, 15A, 16, 24, 34, 36, 47. **Box office** 10am-6pm Mon-Sat (later on performance days). **Tickets** £8-£16. **Map** p284 E8.

Founded in the 1960s, the Traverse was first housed in a former brothel on the High Street; a legend was born on its second night when actor Colette O'Neill was accidentally stabbed during a production of Sartre's *Huis Clos*. By the time the theatre moved to its second home in the Grassmarket in the late 1960s, the emphasis on European experimentalism had begun to shift towards home-grown fare from writers such as John Byrne and Tom McGrath. Then, in 1992, the theatre moved to this purpose-built expanse beneath a modern office complex. Two spaces (one housing up to 270, the other 115) showcase a lively array of new plays, as well as a rolling programme of touring shows. In August, it's always one of the hottest spots on the Fringe. There's a good café-bar, too – it's open long hours (*see p137*).

Escapes & Excursions

Glasgow

Marketeers sell Glasgow in various ways. Campaigns to brighten its image started in 1983 with 'Glasgow's Miles Better', then came 'Glasgow's Alive', 'Glasgow: Scotland with Style' and, since 2013, 'People Make Glasgow'. The gradual shift in emphasis is telling. 'Glasgow's Miles Better' was an attempt to reboot the self-respect of a city devastated by industrial decay. The most recent slogan doesn't shy away from continuing economic problems, but it does recognise the city's greatest resource.

While the Scottish capital is calmer, more genteel and prettier, Glasgow is brasher, livelier and tougher. Visitors and locals both have their preferences, but one city makes more sense when seen next to the other. Taken together, they're a fascinating pair.

One great thing about Glasgow is that it's an easy day trip from Edinburgh. Trains take around 50 minutes; they run from Edinburgh Waverley to Glasgow Queen Street every 15 minutes for most of the week, every half hour in the evenings and on Sundays. The last train back is at 11.30pm, daily, so you can have a full day out in Scotland's biggest city, take in dinner, a gig or a show and still make it back to the capital for a late bedtime.

The other great thing about Glasgow is the ease with which you can hop from area to area. Thanks to the Glasgow subway system (www. spt.co.uk) you can get from the grid-like city centre to the West End in less than ten minutes; the city's railway network (www.scotrail.co.uk) is also very useful. Although Glasgow sprawls off in all directions, the centre, riverside and the West End host more than enough sights, restaurants, bars and shops to keep the most experience-hungry day-tripper fully occupied for the course of their visit.

Glasgow has an origin myth dating to the Dark Ages, a medieval cathedral, bags of industrial heritage, one of Britain's major rivers, three of Britain's top dozen football stadiums, theatres, gig venues and a great deal more besides. The following pages are far from exhaustive, but they cover must-see attractions, great places to eat or drink, and details of the main shopping locations. For more about the city, the **Glasgow Information Centre** (170 Buchanan Street, 0845 859 1006, www.visitscotland.com) is the official tourist information office, and open daily –it should answer all your city queries. For the very latest events and things to do, head to www.timeout.com/glasgow.

IN THE KNOW DO SOME OLD

Glasgow grew at a remarkable rate in the 19th century, in the process eradicating the vast majority of the pre-industrial city. Originally, Old Glasgow ran from the Clyde, up Saltmarket and the High Street to the cathedral (*see p195*). The cathedral precinct today is the only place visitors can look around and get any real sense of the city's deeper history.

SIGHTS & MUSEUMS

FREE Gallery of Modern Art (GoMA)

Royal Exchange Square (0141 287 3050, www.glasgowlife.org.uk). Buchanan Street subway. **Open** 10am-5pm Mon-Thur, Sat; 11am-5pm Fri, Sun. **Admission** free. **Map** p197 E4 ❶

ESCAPES & EXCURSIONS

This grand 18th-century building has served as an art gallery since 1996. The permanent collection has works by major Scottish and international artists, and it also hosts first-class touring shows.

Glasgow Cathedral

Castle Street (0141 552 6891, www.historic-scotland.gov.uk). High Street rail. **Open** *Apr-Sept 9.30am-5pm Mon-Sat; 1-5pm Sun. Oct-Mar 9.30am-4.30pm Mon-Sat; 1-4.30pm Sun.* **Admission** varies; check website for details. **Map** p197 H3 ❷

St Mungo is said to have founded a place of worship at this site around the late sixth century AD. The earliest parts of the current building date back 800 years or more: it's an impressive example of medieval Scottish Gothic. In 2015, Historic Scotland planned to apply an admission charge for the first time. Around the cathedral precinct you will also find **Provand's Lordship**, Glasgow's oldest house, dating from the 15th century, and the **St Mungo Museum of Religious Life & Art** (admission to both is free; check www.glasgowlife.org.uk for opening times). Immediately east of the cathedral is the **Necropolis**, a handsome, atmospheric 19th-century cemetery inspired by Père Lachaise in Paris.

Glasgow Science Centre

50 Pacific Quay (0141 420 5000, www.glasgowsciencecentre.org). Exhibition Centre rail. **Open** varies; check website for details. **Admission** Science Mall £8.25; £6.25 reductions.

Extra fee for planetarium, IMAX documentaries, Glasgow Tower.

This futuristic titanium and glass structure (opened in 2001) boasts a Science Mall featuring three floors of hands-on science and technology. The well-run displays and excellent planetarium are popular with kids and armchair scientists. The centre also houses an IMAX cinema (www.cineworld.co.uk) and the Glasgow Tower, offering the best views in the city.

FREE Kelvingrove Art Gallery & Museum

Argyle Street (0141 276 9599, www.glasgowlife.org.uk). Kelvinhall subway. **Open** *10am-5pm Mon-Thur, Sat; 11am-5pm Fri, Sun.* **Admission** free. **Map** p196 A2 ❸

Kelvingrove (founded in 1901) has always been Glasgow's premier museum; after a 2006 refurbishment, it's in great shape. The impressive atrium sparkles, the ground floor exhibitions cover every subject under the sun, and the first floor has masterpieces by Botticelli, Dali, Rembrandt and Van Gogh.

FREE People's Palace & Winter Gardens

Glasgow Green (0141 276 0788, www.glasgowlife.org.uk). Bridgeton rail. **Open** *10am-5pm Tue-Thur, Sat; 11am-5pm Fri, Sun.* **Admission** free. **Map** p197 G6 ❹

Built in 1898, this venue houses a cherished exhibition that covers all aspects of Glaswegian life, but particularly the city's social and industrial history since the mid 18th century. The adjoining Winter Gardens are worth a look too.

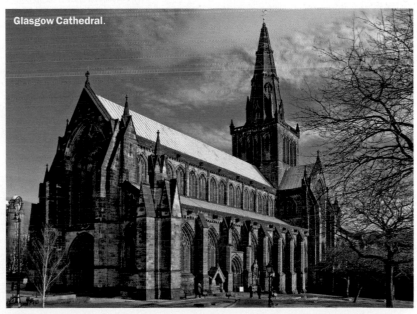

Glasgow Cathedral.

ESCAPES & EXCURSIONS

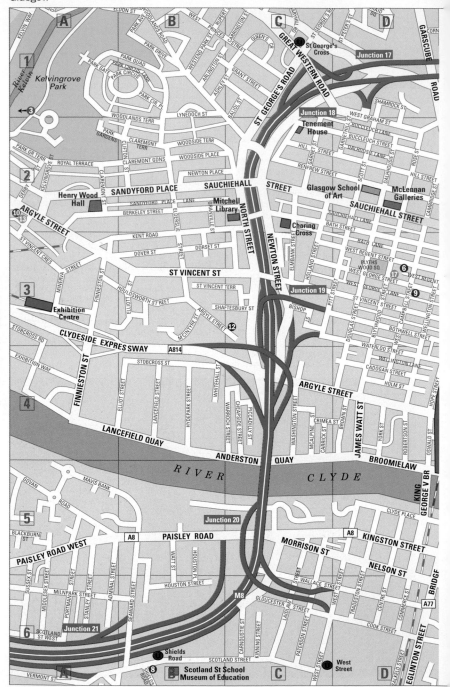

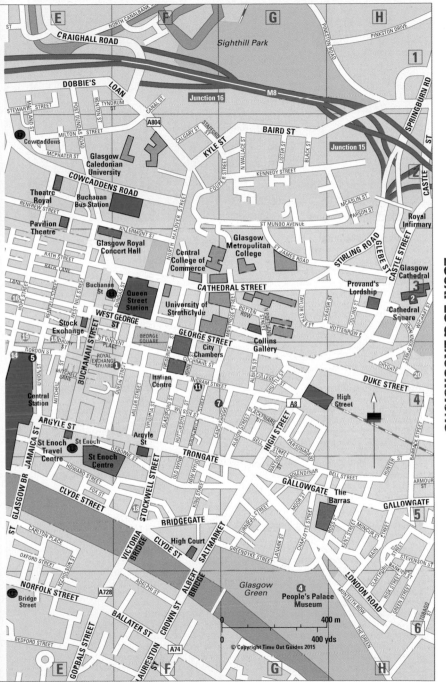

Kelvingrove Art Gallery & Museum. See p195.

FREE **Riverside Museum**
*100 Pointhouse Place (0141 287 2720, www.
glasgowlife.org.uk). Kelvinhall subway.* **Open**
10am-5pm Mon-Thur, Sat; 11am-5pm Fri, Sun.
Admission free.
An extraordinary building by Zaha Hadid, opened
in 2011, the award-winning Riverside houses the
city's transport collection. There are antique cars
and Porsches, steam engines, buses and trams,
but also enough social history to keep non-petrol-
heads happy. Berthed immediately outside is the
Tall Ship At Riverside, the *Glenlee*, a handsome,
late-Victorian barque (www.thetallship.com).

Scottish Football Museum
*National Stadium, Hampden (0141 616 6139,
www.scottishfootballmuseum.org.uk). King's Park
or Mount Florida rail.* **Open** 10am-5pm Mon-Sat;
11am-5pm Sun. **Admission** *Museum* £7; £3
reductions. *Stadium tour* £8; £3.50 reductions.
Joint ticket £11; £5 reductions.
This display of Scottish football history is comprehensive to the point of obsession, documenting all
the players, kits and trophies since 1867. An insight
into Glasgow and Scotland.

IN THE KNOW CREATIVE GENIUS

Charles Rennie Mackintosh, born in Glasgow
in 1868, was a talented architect and
designer responsible for some beautiful and
innovative buildings across the city. But his
career ran into trouble when he was in his
forties and he left Glasgow, never to return.
Spurned in his lifetime, he is now one of the
city's favourite sons; Mackintosh souvenirs
are everywhere. See the Charles Rennie
Mackintosh Society website (www.crm
society.com) for more details.

RESTAURANTS & CAFÉS

Alston Bar & Beef
*Glasgow Central Station, Gordon Street (0141 221
7627, www.alstonglasgow.co.uk). Buchanan Street
subway, or Glasgow Central rail.* **Open** noon-10pm
daily. **Main courses** £12-£33. **Map** p197 E4 ❺
Steakhouse
One of the city's most on-trend 2014 debuts, this
venue is downstairs within Glasgow Central Station
itself. Aside from the burgers, steak is the big selling
point, made from dry-aged beef hung for 35 days.

Babu Bombay Street Kitchen
*186 West Regent Street (0141 204 4042,
www.babu-kitchen.com). Buchanan Street
subway, or Charing Cross rail.* **Open** 8am-6pm
Mon, Tue; 8am-8pm Wed-Fri; 10am-6pm Sat.
Main courses £3.50-£8.50. **Map** p196 D3 ❻
Indian
An Indian street-food takeaway with a breezy,
Bollywood look and a few tables if people want to
sit in. Here you get breakfast chapati wraps, curries,
inventive snacks and more.

Dakhin
*89 Candleriggs (0141 553 2585, www.dakhin.
com). Argyle Street or High Street rail.* **Open**
noon-2pm, 5-11pm Mon-Fri; 1-11pm Sat, Sun.
Main courses £8-£27. **Map** p197 G4 ❼ **Indian**
Housed in a spacious, first-floor room, Dahkin is
accomplished South Indian cuisine; the nearby
Dhabba (44 Candleriggs, 0141 553 1249, www.the
dhabba.com) is the North Indian sister business.

Fish People Café
*350 Scotland Street (0141 429 8787, www.
thefishpeoplecafe.co.uk). Shields Road subway.*
Open noon-9pm Mon-Thur; noon-10pm Fri, Sat;
noon-4pm Sun. **Main courses** £9.50-£18.50.
Map p196 B6 ❽ **Fish & seafood**

An excellent reason to venture south of the Clyde, this caff has a surprisingly plush interior, is handily next door to a subway station and serves great seafood. Accessible and worthwhile.

Gamba

225A West George Street (0141 572 0899, www.gamba.co.uk). Buchanan Street subway. **Open** noon-2.15pm, 5-10pm Mon-Sat; 5-9pm Sun. **Main courses** £11.50-£28. **Map** p196 D3 **9** **Fish & seafood**
This distinguished basement venue has been one of the city's leading seafood restaurants for years. The approach is simple: good raw materials letting the quality shine through, with an international twist.

Gannet

1155 Argyle Street (0141 204 2081, www. thegannetgla.com). Exhibition Centre rail. **Open** noon-2.30pm, 5-9.45pm Tue-Sat; noon-3pm, 5-8.45pm Sun. **Main courses** £12-£28.50. **Map** p196 A2 **10** **Modern British**
New York-chic decor, a very talented kitchen and most of the main courses pitched at under £20. The Gannet has been a massive hit since it opened in 2013.

Hutchesons

158 Ingram Street (0141 552 4050, www.hutchesons glasgow.com). Buchanan Street subway, or Argyle Street or High Street rail. **Open** 9am-midnight Mon-Sat; 10am-midnight Sun. **Main courses** £15-£33. **Map** p197 F4 **11** **Brasserie**

Hutchesons

Gannet.

Housed in the gorgeous Hutchesons' Hall (1805), this grand brasserie opened in 2014. Set over three impressive floors, it specialises in seafood and great steaks, but you can also pop in for a drink.

Number 16

16 Byres Road (0141 339 2544, www.number16. co.uk). Kelvinhall subway. **Open** noon-2.30pm, 5.30-9pm Mon Sat; 5.30-8.30pm Sun. **Main courses** £13.50-£17. **Modern European**
A small space with a squeezed-in mezzanine, offering a Euro-bistro menu. It has been around since 1999 and remains a popular and dependable West End spot. Red walls and natural wood tones dominate the decor.

Two Fat Ladies at the Buttery

652 Argyle Street (0141 221 8188, www.two fatladiesrestaurant.com). Anderston rail. **Open** noon-3pm, 5.30-10.30pm Mon-Sat; noon-9pm Sun. **Main courses** £19-£28. **Map** p196 C3 **12** **Modern European**
The most illustrious restaurant in the Two Fat Ladies stable (see the website for the other four), this is an atmospheric, wood-panelled destination diner where the Scottish larder meets Modern European cooking.

Ubiquitous Chip

12 Ashton Lane (0141 334 5007, www.ubiquitous chip.co.uk). Hillhead subway **Open** 11am-1am daily. **Food served** noon-2.30pm, 5-11pm Mon-Sat; 12.30-3pm, 5-11pm Sun. **Main courses** £16-£27. **Modern European**
The Chip has evolved into a complex of venues since it opened in 1971. It now has a covered, cobbled courtyard restaurant, a more conventional dining room adjacent, a brasserie and three pubs. The signature dish is a venison haggis starter; mains offer Scottish ingredients in haute Modern European style.

PUBS & BARS

78

10-14 Kelvinhaugh Street (0141 576 5018, www.the78cafebar.com). Exhibition Centre rail. **Open** 12.30pm-midnight Mon-Wed; 12.30pm-1am Thur, Fri; noon-1am Sat, Sun. **Map** p196 A2 ④
A West End craft beer bar with a shabby-chic aesthetic, open fire and a vegan menu, the 78 is an amenable and relaxed place to eat and drink.

Champagne Central

Grand Central Hotel, 99 Gordon Street (0141 240 3700, www.thegrandcentralhotel.co.uk). Buchanan Street subway, or Glasgow Central rail. **Open** 10am-midnight Mon-Sat; 11am-midnight Sun. **Map** p197 E4 ④
Part of the Grand Central Hotel, this bar overlooks the concourse at Glasgow Central station. It's civilised, discreet and ornate; the cocktails are great and it offers afternoon tea, sandwiches and snacks.

Horseshoe

17-19 Drury Street (0141 204 4056, www. horseshoebar.co.uk). Buchanan Street subway, or Glasgow Central rail. **Open** 10am-midnight Mon-Sat; 12.30pm-midnight Sun. **Map** p197 E3 ④
Very near Central station, the Horseshoe is a classic Glasgow bar. There are few nods to the 20th century here, let alone the 21st; customers come for beer, chat, football and ludicrously cheap pub grub.

Inn Deep

445 Great Western Road (0141 357 1075, www.inndeep.com). Kelvinbridge subway. **Open** noon-midnight Mon-Sat; 12.30-11pm Sun.
By the River Kelvin, this bar (run by the craft brewing company the Williams Brothers) has a subterranean feel and a modern but robust pub grub menu. In summer, you can eat and drink alfresco. There's a dozen cask and craft keg beers on draught.

Òran Mór

Top of Byres Road (0141 357 6200, www.oranmor.co.uk). Hillhead subway. **Open** 9am-2am Mon-Wed; 9am-3am Thur-Sat; 10am-3am Sun.
A converted church at the north end of Byres Road plays host to this cavernous bar as well as some lively arts programming and a worthwhile restaurant. The Play, Pie and a Pint seasons – lunchtime theatre performances where admission includes a pie and a drink – are deservedly popular.

Pot Still

154 Hope Street (0141 333 0980, thepotstill.co.uk). Buchanan Street subway. **Open** 11am-midnight Mon-Sat; 12.30-11pm Sun. **Map** p197 E3 ④
You could walk past this fine, traditional bar and not give it a second glance, but it has good cask ales and an international reputation for its whisky range.

Salon

Blythswood Square, 11 Blythswood Square (0141 248 8888, www.townhousecompany.com). Buchanan Street subway, or Charing Cross rail. **Open** 11am-midnight Mon-Sat; 12.30pm-midnight Sun. **Map** p196 D3 ④
The Salon is the lounge bar at the upmarket hotel known simply as Blythswood Square in the eponymous city-centre square. You don't come here for a pint of craft IPA; you dress up and order a cocktail.

Scotia

112 Stockwell Street (0141 552 8681, www.scotia bar.net). Argyle Street rail. **Open** 11am-midnight Mon-Sat; 12.30pm-midnight Sun. **Map** p197 F5 ④
Glasgow's oldest functioning bar (1792) doesn't look much from the outside, but inside the establishment is gravid with history, offers good cask ale and bargain single malt whisky. In a word: authentic.

Stereo

22-28 Renfield Lane (0141 222 2254, www. stereocafebar.com). Buchanan Street subway. **Open** noon-1am Mon-Wed; noon-3am Thur-Sun. **Map** p197 E3 ④
With regular DJs, a basement used as a club and gig venue, and a vegan café menu on offer, Stereo is clearly more than just a city-centre bar. Then again, you can just pop in for a vodka and coke or a beer.

Vintage at Drygate

85 Drygate (0141 212 8815, www.drygate.com). High Street rail. **Open** 11am-midnight daily. **Map** p197 H4 ④
A craft beer venture with its very own brewery, these premises feel like a robust, airy canteen for the 21st century. The beer can be innovative and the food is interesting. Drygate is adjacent to the Wellpark Brewery, home of Tennent's, Scotland's best-selling lager brand.

SHOPS & SERVICES

Glasgow is second only to London among UK cities when it comes to shopping. Virtually every one of the country's favourite high-street chains

---

IN THE KNOW
CONNECT WITH YOUR INNER CELT

January is not the most appealing month anywhere, but in Glasgow it signals the start of the annual Celtic Connections music festival (www.celticconnections. com), a massive success since launching in 1994. Its remit goes way beyond the folk music of Britain and Ireland, with more than 2,000 artists taking part in 300 events across 20 city venues.

and big department stores is represented, but there are also plenty of independent retailers, hawking all manner of specialist goods to the shopaholic hordes.

In the city centre, three streets dominate the shopping landscape: **Argyle Street**, which crosses the bottom of **Buchanan Street**, which in turn connects with **Sauchiehall Street**. All three are pedestrianised for much of their extents; a good job, given the massive volume of foot traffic that descends upon them at weekends. The *primus inter pares* is Buchanan Street, home to malls, mainstream stores and a handful of chains.

Just east of here, the Merchant City has some of Glasgow's smarter clothing boutiques. There are individual brand stores such as Emporio Armani and Ralph Lauren, while others, such as Cruise, stock a wide and almost indiscriminate range of catwalk names.

Standing apart from the heady consumerist familiarity of the city centre, Glasgow's West End has its own commercial buzz, and is dotted with lively fashion boutiques and unusual gift shops. The main shopping streets are Byres Road and Great Western Road, which offer a jumble of cafés, clothing stores, grocers and music shops.

One-stop shopping

Glasgow's department stores generally think big, and none think bigger than **House of Fraser** (45 Buchanan Street, 0344 800 3728, www.house offraser.co.uk). The huge Victorian building is home to everything from LK Bennett boots for women to Ninja Ultima kitchen blenders and Sisley skincare kits for men at £160. Fraser's represents historical Glasgow as the business was founded here in 1849, making **John Lewis** (Buchanan Galleries, 0141 353 6677, www.john lewis.com), here only since 1999, a very fresh-faced incomer. **Debenhams** (97 Argyle Street, 0844 561 6161, www.debenhams.com) has most basics, while **Marks & Spencer** (2-12 Argyle Street, 0141 552 4546, www.marksandspencer.

com) is as popular for its above-average food hall as its inexpensive undies.

Several large shopping malls dot the city centre, each holding names familiar to shoppers from high streets across the UK. The most popular of the malls – and, not coincidentally, the most mainstream – is the shiny **Buchanan Galleries** (220 Buchanan Street, 0141 333 9898, www.buchanangalleries.co.uk), home to the likes of Hollister, Lego and the Whisky Shop. The **St Enoch Centre** (55 St Enoch Square, 0141 204 3900, www.st-enoch.co.uk), Europe's largest glass-covered structure when it opened in 1989, is similarly stocked with stores such as Arran Aromatics, Hotel Chocolat and Disney.

Some variation on the mall theme is provided by **Princes Square** (48 Buchanan Street, 0141 221 0324, www.princessquare.co.uk). You'll find Belstaff, French Connection and Ted Baker here, but also the less ubiquitous likes of jeweller Dower & Hall, and a Vivienne Westwood outlet. By far the town's most interesting mall isn't really a mall at all: the chain-free **De Courcy's Arcade** (5 Cresswell Lane, West End), home to a smile-inducing mix of antiques stores, cafés and galleries. Cresswell Lane runs parallel to the east side of Byres Road, just behind Hillhead subway.

Two regular markets merit mention. The **Barras** (Gallowgate, 0141 552 4601, www.glasgow-barrowland.com, open 10am-5pm Sat, Sun) comes alive with hundreds of traders and shopkeepers offering you – aye, you! – the bargain of a lifetime. Clothes, CDs, DVDs, sports socks, tea towels and just about everything else you can think of is on offer, although you might wonder where some of it came from. Even if you have no use for pre-owned videos or a polyester bra, it's worth the journey to hear the local dialect in full effect. For a much more containable experience, head to the **Weekend Craft & Design Fair** at the covered courtyard in Merchant Square (71 Albion Street, 0131 552 3038, www.merchant squareglasgow.com, open 11am-6pm Sat, noon-6pm Sun) for arts, gewgaws and gifts.

Around Edinburgh

The country immediately surrounding Edinburgh doesn't offer any of the capital's built-up urban charms – which is precisely its appeal. Things are different out here: just as marked by history, certainly, but altogether wilder, quieter and, in places, even hillier.

Due east of the city, along the coast, lie the handsome beaches, towns and villages of East Lothian. To the south lies Midlothian, home to the Pentland Hills and a succession of charming historic sights. West Lothian is where you'll find the 15th-century Linlithgow Palace. Slightly further afield, the Borders, Fife and Stirling areas each hold plenty of appeal.

ESCAPES & EXCURSIONS

EAST LOTHIAN

The coastal town of **Musselburgh** is only about five miles from the bustle of the Royal Mile, but it feels much further away. It was settled by the Romans, who established a port at the mouth of the River Esk as far back as AD 140, and it remains proud of its history. Since the early 14th century, it's been known as the 'honest toun', a sobriquet gained when the locals refused to claim a reward offered to them after they cared for Randolph, Earl of Moray, through illness. Post-war development hasn't been kind to the town's outer reaches, but its centre remains handsome; the chief attractions are run by the National Trust.

Towards Edinburgh sits **Newhailes** (Newhailes Road, 0844 493 2125, www.nts.org.uk, see website for seasonal opening times, £12.50, £9 reductions), at its core a 17th-century house with rococo interiors and an 18th-century garden (open all year round). And don't miss **Inveresk Lodge Garden** (24 Inveresk Village, 0844 493 2126, www.nts.org.uk, £3.50, £2.50 reductions): the building is private but the sublime gardens, sloping down to the Esk's peaceful banks, are open to all. Even so, most visitors tend to head to Musselburgh for sporting reasons, to play the famous **Musselburgh Links** golf course (www.musselburgholdlinks.co.uk) or for a day of horsey action at **Musselburgh Racecourse** (www.musselburgh-racecourse.co.uk).

After the racecourse, take the B1348 along the coast road, through the small towns of **Prestonpans**, **Cockenzie** and **Port Seton**. Once you get past the village of **Longniddry**, the scenic impact of the East Lothian coast really hits. Wildlife thrives here: the **Aberlady Bay Local Nature Reserve**, reached by crossing the footbridge from the car park just east of Aberlady village, is open all year, its sandy mudflats a big pull for bird-watchers. Golfers flock here too.

Virtually the entire coastline in this area is beach. It starts at Seton Sands, then goes round by Gosford Bay into Aberlady Bay. The most popular beach is **Gullane**, a shallow mile-long crescent behind the handsome town. Then comes another two miles of quiet, contemplative coastline to the north-east before the spacious **Yellowcraig Beach** near Dirleton. The weather may not allow sunbathing or swimming most of the year, but these East Lothian beaches have a beauty all their own.

Set a little way back from the coast, a mile or so past Gullane, the chocolate-box pretty village of **Dirleton** has the impressive remains of the 13th-century **Dirleton Castle** (01620 850330, www.historic-scotland.gov.uk, £5.50, £3.30-£4.40 reductions). The castle sits atop a hill surrounded by pleasant gardens: the north garden is associated with the Arts and Crafts movement of the 1920s, while the west garden was late Victorian and was reconstructed to its original plan in the early 1990s.

Tantallon Castle.

Two miles east sits the larger and busier town of **North Berwick**, a seaside settlement; Edinburghers head here during the summer to take advantage of the sands flanking the old harbour. Behind the town, **North Berwick Law** rises to a modest 613 feet, but it's a popular climb on a fine day and offers great views; geologically, it's the same as Arthur's Seat in Edinburgh, a basalt volcanic plug. But the main attraction is the **Scottish Seabird Centre** (01620 890202, www.seabird.org, £8.95, £4.95-£6.95 reductions); dolphins and otters have been spotted around here, but seabirds obviously predominate. **Seafari** (01620 890202, www.seafari.co.uk) runs various boat trips from the harbour, weather permitting, from Easter to October.

The coast past North Berwick is dominated by **Tantallon Castle** (01620 892727, www.historic-scotland.gov.uk, £5.50, £3.30-£4.40 reductions), a formidable cliff-edge fortification largely built in the 14th century. Past here, continue along the A198 until you reach the A1. Immediately east, by Dunbar, is the **John Muir Country Park**, which provides various habitats for wildlife. Born in Dunbar in 1838, he lived there until the age of 11, when his family emigrated; later, he was a hugely influential naturalist in the US.

Back along the A1, towards Edinburgh again, is the village of **East Linton**. Make a diversion into the minor roads south-west of the village to find the atmospheric remains of the 13th-century **Hailes Castle** (www.historic-scotland.gov.uk, free) overlooking the River Tyne. And just a short drive north of the A1, at a former World War II airfield, sits the popular **National Museum of Flight** (East Fortune Airfield, East Fortune, 0300 123 6789, www.nms.ac.uk, £10, £5-£8 reductions). The star attraction is a Concorde.

Keep heading west on the A1 and you soon come to the Royal Burgh of **Haddington**, a well-heeled market town. The main attraction is the 14th-century **St Mary's Church** (www.stmaryskirk.co.uk), also known as the Lamp of Lothian and keenly preserved by locals. An even more delightful settlement sits just four miles south along the B6369. Built by the Marquis of Tweeddale in the 18th century for his estate workers, **Gifford** is too charming for words. Continuing further south still and Longyester provides access to the **Lammermuir Hills**, which are popular with walkers.

Where to eat & drink

In Musselburgh, the original **S Luca** (32-38 High Street, 665 2237, www.s-luca.co.uk) sells the best ice-cream on the East Lothian coast; it has a branch in Edinburgh (*see p142*).

In Aberlady, **Ducks at Kilspindie House** (01875 870682, www.ducks.co.uk, three courses £49) has a kitchen operating to high standards; it's run by Malcolm Duck, who had a fine restaurant in Edinburgh for many years. The informal bistro on site is called Donald's.

For yet more upscale cooking, head to Gullane and **La Potinière** (01620 843214, www.la-potiniere.co.uk, set meals £20-£43), which takes a classic and polite French approach to its cuisine. At the edge of Gullane, **Chez Roux at Greywalls** (01620 842144, www.greywalls. co.uk, mains £16-£22.50) offers another classy French experience in an Edwardian-era hotel. The village also has amazing German cakes at **Falko Konditormeister** (1 Stanley Road, 01620 843168, www.falko.co.uk).

The seaside town of North Berwick has a good selection of eateries, while, in summer only, the **Lobster Shack** (www.lobstershack.co.uk)

IN THE KNOW LADIES DAY

Musselburgh Racecourse (*see p202*) stages around 25 meetings a year, but the event with the best party atmosphere is certainly Ladies Day, usually held in June. Aside from horse racing it features competitions, fashion shows, music and more. Posh frocks and fascinators optional.

---

at the harbour and the nearby **Rocketeer** (www.rocketeerrestaurant.co.uk) both offer great alfresco seafood. **Steampunk Coffee** (49A Kirk Ports, 01620 893030, www.steampunk coffee.co.uk) is your indoor port of call for coffee, brownies, breakfast and light meals.

Further east in Dunbar, the **Creel** (01368 863279, www.creelrestaurant.co.uk, set meals from £16.95) is a popular bistro with carefully sourced food.

Getting there

By bus Lothian Buses runs services from Edinburgh to Musselburgh, then on towards Port Seton. First Bus has more useful routes to North Berwick and Dunbar, and also to Haddington. For general bus details, *see p263*, or contact Traveline Scotland (0871 200 2233, www.travelinescotland.com).

By car For Haddington or Dunbar, take the A1 east out of the city. For all the towns and attractions along the coast, however, take the A198. From the A1, follow signs for Longniddry and pick up the A198 from there. Alternatively, just follow the B1348 coast road from Musselburgh going east.

By train A service runs from Waverley station to Musselburgh, Wallyford, Prestonpans, Longniddry, Drem and North Berwick. There are also trains from Waverley to Dunbar.

MIDLOTHIAN

The Pentland Hills rise south of Edinburgh and form the backbone of the **Pentland Hills Regional Park** (www.pentlandhills.org), which sprawls from the village of Carlops in the south and the area around Harperrig Reservoir in the south-west.

The park takes in a succession of summits, various lochs, 60 miles of signposted paths, two visitor centres and evidence of settlement dating from prehistoric times. Serious hillwalkers will be in heaven.

There's an entrance to the park just by the Flotterstone Inn (*see p205*), less than four miles south of the city bypass on the A702; this is also where you'll find one of the park's visitor centres. Less than a mile from the car park, the Glencorse Reservoir sits among heather-covered hills, a visual combination that forms one of those classic images of Scotland. If you're feeling energetic, you can keep walking around the reservoir and on to nearby Loganlea Reservoir, all on a good surface. Of course, the further you walk, the longer it will take to get back to the Flotterstone Inn for a drink.

Way down at the south end of the Pentlands, not far from the village of Dolphinton on the A702 (take the turning to Dunsyre), is **Little Sparta** (www.littlesparta.co.uk, £10), a fascinating sculpture garden created by the late artist Ian Hamilton Finlay. Mysterious, calming and full of subtle surprises, the garden is open only three afternoons a week from June to September, but it's well worth making the effort to see this quietly extraordinary place.

To the east side of the Pentlands is the village of **Roslin**, its fame boosted enormously by the appearance of the **Rosslyn Chapel** (440 2159, www.rosslynchapel.org.uk, £9, £7 reductions, accompanied under-18s free) in Dan Brown's thriller *The Da Vinci Code* and the subsequent movie. Sir William Sinclair (or St Clair), the Earl of Caithness and Orkney, decided to build the chapel in 1446, though the project may have been unfinished by the time he died in 1484. At any rate, the chapel has stood in its current form for well over 500 years. Its carvings, featuring hints

View from Scald Law, Pentland Hills.

IN THE KNOW THE PENTLANDS

Although not as lofty as hills and mountains elsewhere in Scotland, the **Pentland Hills** (see p204) immediately south of Edinburgh have a number of tops over 1,600 feet and offer an accessible day's walking from the city. The highest of the high is Scald Law at 1,900 feet.

of Celtic and Norse beliefs alongside Christian elements, are remarkable. One flourish has been interpreted as an ear of corn, which was unknown in Scotland in the 15th century; it's seen as evidence that one of the Sinclair family had visited North America before Columbus. However large a pinch of salt you take with such information, the chapel remains a strongly evocative place.

If you have the appetite for another ancient pile, try **Crichton Castle** (Crichton, 01875 320017, www.historic-scotland.gov.uk, £4.50, £2.70-£3.60 reductions). The residence of the Crichtons and then the Earls of Bothwell, the ruins contain a 14th-century tower house, a 15th-century great hall and an impressive diamond-dotted façade from the late 16th century. There's car parking nearby, but the final approach to the ruins is on foot along a rough track.

The Pentlands, Roslin and the likes of Crichton Castle are engaging, but Midlothian is better known for the aggregation of small towns and villages around Dalkeith that form the county's centre of gravity. These are not tourist magnets, but there are important, historical associations with coal mining around here. At Newtongrange, the **National Mining Museum Scotland** (Lady Victoria Colliery, 663 7519, www.national miningmuseum.com, £8.50, £6.50 reductions) explores the industry that once drove the local economy. The former colliery in which the museum is housed employed 2,000 people at its peak and produced 40 million tons of coal in its lifetime.

Where to eat & drink

South of the Pentlands on the A701 by Lamancha, **Whitmuir the Organic Place** (01968 661147, www.whitmuirtheorganicplace.co.uk, mains under £10) is an attractive café and farm shop where much of the produce is grown on site.

Closer to Edinburgh, on the A702, the **Flotterstone Inn** (01968 673717, www.flotterstoneinn.com, main courses £11-£16) is handy for Pentlands walks and offers bar and restaurant menus. On the A7 at Lothianbridge, the **Sun Inn** (663 2456, www.thesuninn edinburgh.co.uk, main courses £10-£22) does decent beer and good pub grub. Finally, the **Howgate** (01968 670000, www.howgate.com,

main courses £9-£27) is also out of the way: if you're driving to Penicuik on the A701, branch off on the B7026 before reaching the town and carry on for three miles. It looks like everyone's idealised notion of a farmhouse restaurant, but it also does good steaks.

Getting there

By bus Both Lothian Buses and First Bus have services from central Edinburgh to Roslin and Newtongrange; First Bus also stops by the Flotterstone Inn. For general bus details, see p263, or contact Traveline Scotland (0871 200 2233, www.travelinescotland.com). To get to Crichton Castle, you'll need a car.
By car The OS Landranger map 66 (Edinburgh) covers most of Midlothian.
By train A new line running from Edinburgh through Midlothian to the Borders is scheduled for completion late in 2015, with planned stations at Eskbank, Newtongrange and Gorebridge.

WEST LOTHIAN

Like its Midlothian neighbour, West Lothian has its share of old mining towns, but also a post-war new town in the shape of Livingston, built to take overspill from Glasgow. Driving through the county on either of its main routes, there's little of obvious interest to detain the visitor, but it's worth seeking out the main attractions.

Between Edinburgh and Queensferry, for example, is the village of **Dalmeny**, long identified with West Lothian but now technically part of the capital. The main attraction here is 12th-century **St Cuthbert's**, one of the finest Romanesque churches in Scotland. Nearby sits **Dalmeny House** (331 1888, www.dalmeny. co.uk, £9.50, £6.50-£8.50 reductions), a Gothic-revival mansion designed by William Wilkins in 1814 and home of the Earls of Rosebery. The property has a grand interior and, incongruously, an extensive collection of Napoleonic memorabilia. Opening hours are very limited: usually just four days a week during June and July – so check before setting out.

Just beyond the house is Queensferry, also known as South Queensferry to distinguish it from North Queensferry in Fife. This is the best place from which to see the enormous **Forth Rail and Road Bridges**. The wonderful rail bridge, a mile and a half long, is truly impressive and was regarded as the eighth wonder of the world when completed in 1890. The first road bridge was built in 1964, but traffic levels and safety concerns put a second bridge on the agenda from the 1990s. Work started west of the first bridge in 2011 and the second road bridge, dubbed the Queensferry Crossing, should be finished by the end of 2016.

ESCAPES & EXCURSIONS

Forth Rail Bridge. See p205.

The other main reason to visit Queensferry is to catch the *Maid of the Forth* ferry (331 5000, www.maidoftheforth.co.uk), which sails from South Queensferry to **Inchcolm Abbey** (01383 823332, www.historic-scotland.gov.uk, £5.50) on Inchcolm Island in the Firth of Forth. Founded in 1123, the abbey comprises a clutch of wonderfully preserved monastic buildings. Seal sightings are common during the boat trip.

West of Queensferry stands the astonishing **Hopetoun House** (331 2451, www.hopetoun house.com, £9.20, £4.90-£8 reductions), designed by William Bruce in 1699 and enlarged by William Adam in 1721. The elegant simplicity of the building belies the opulence within. Further west is **Blackness Castle** (01506 834807, www.historic-scotland.gov.uk, £5.50, £3.30-£4.40 reductions), used by Zeffirelli in his 1990 film version of *Hamlet*. Its walls have crumbled since they were built in the 1440s, but the view across the Forth from the castle promontory remains spectacular.

Just four miles south-west of Blackness is the attractive Royal Burgh of **Linlithgow**, most celebrated for the beautiful ruins of **Linlithgow Palace** (01506 842896. www.historic-scotland. gov.uk, £5.50, £3.30-£4.40 reductions) where Mary, Queen of Scots was born in 1542. The palace overlooks **Linlithgow Loch**, a scenic spot for a circular walk. Behind Linlithgow railway station lies the Union Canal Basin;

you'll find the **Linlithgow Canal Centre** (01506 671215, www.lucs.org.uk, free), which includes a museum and tearoom; it's also the departure point for canal cruises. For deeper local history, head to **Cairnpapple Hill** (01506 634622, www.historic-scotland.gov.uk, £4.50, £2.70-£3.60 reductions), off the minor roads between Linlithgow and Bathgate. This Neolithic burial site, dating back 5,500 years, sits on top of a hill with open views up and down the Forth.

Where to eat & drink

Driving out of Edinburgh to the first Forth Road Bridge on the A90, the black, space-age building on the right, just before the bridge, is the **Dakota Hotel** (319 3690, www.dakotahotels.co.uk, main courses £13-£30) with its contemporary Grill restaurant. In Queensferry itself, **Orocco Pier** (17 High Street, 0870 118 1664, www.oroccopier. co.uk, main courses £13-£35) has a restaurant, a café-bar and views of the bridges. A couple of doors along, the **Boat House** (22 High Street, 331 5429, www.theboathouse-sq.co.uk, main courses £18-£24) also has excellent views and acclaimed seafood. For a pint or pub grub with literary associations, try the **Hawes Inn** (7 Newhalls Road, 331 1990, www.vintageinn.co.uk, main courses £8-£17); Robert Louis Stevenson wrote part of *Kidnapped* here back in the 1880s.

The other main centre for food and drink in West Lothian is Linlithgow. **Livingston's** (52 High Street, 01506 846565, www.livingstons-restaurant.co.uk, set menus £19.50-£42.50) is a traditional Franco-Scottish affair; **Taste** (47 High Street, 01506 844445, www.taste-deli-cafe.co.uk) is a handy, café alternative. The best pint in the area can be had at the **Four Marys** (65 High Street, 01506 842171, www.thefourmarys.co.uk, main courses £9 or less), which also does bar food. But the best restaurant in West Lothian is just outside Linlithgow via the northbound A803: the **Champany Inn** (01506 834532, www.champany. com) is renowned for steak, shellfish and cheese.

IN THE KNOW
WHERE THE WILD BINGS ARE

Dotted around West Lothian you find huge old bings: spoil heaps from the shale oil industry that flourished here from the mid 19th century into the 20th. Once thought unsightly, the bings are now deemed ecologically important as primary succession habitats for fauna and flora.

Getting there

By bus Lothian Buses has services to Queensferry and Ratho; for anything else, it's First Bus and other providers. For general bus details, *see p263*, or contact Traveline Scotland (0871 200 2233, www.travelinescotland.com).

By car From the western edge of Edinburgh, the M9 will take you to Linlithgow, then on to Stirling; the M8 goes to Livingston, Bathgate and eventually Glasgow. For detailed navigation, the OS Landranger map 65 (Falkirk & Linlithgow) covers West Lothian.

By train Linlithgow and Dalmeny (by Queensferry) are on main lines and well served by trains from both Waverley and Haymarket stations. Suburban lines going west, also from Waverley and Haymarket, will take you to Livingston's two stations (North and South) and Bathgate.

STIRLING & AROUND

Given its associations with battles, military strategy, the Highlands and the Lowlands, plus control of the vital Forth Valley, **Stirling** can almost claim to be the historical heart of Scotland. A prominent castle on a rock, not unlike Edinburgh's, is the cherry on the cake.

Two of the country's most significant medieval victories against England took place near here: William Wallace's at Stirling Bridge in 1297 and Robert the Bruce's at Bannockburn in 1314, and both are commemorated in the town. Wallace is celebrated by the **National Wallace Monument** (Abbey Craig, Hillfoots Road, 01786 472140, www.nationalwallacemonument.com, £9.50, £5.90-£7.60 reductions), a Victorian tower that dominates the skyline. There's plenty of Wallace memorabilia on display, including his broadsword, and the views from the top are wonderful. The story of Bannockburn, meanwhile, is told at the **Battle of Bannockburn Visitor Centre** (Glasgow Road, 0844 493 2139, www.battleof bannockburn.com, £11, £8 reductions).

Stirling Castle Rock may have been first occupied as long ago as the Iron Age and the first documentary evidence of a castle comes from the 12th century, but today's **Stirling Castle** (01786 450000, www.stirlingcastle.gov.uk, £14.50, £8.70-£11.60 reductions, free under-5s) dates mainly from the 15th and 16th centuries. A magnificent sight, it's one of the finest castles in Scotland. Displays detail the history of the building: Alexander I died here in 1124; James II was born within its walls; Mary, Queen of Scots was crowned here; and Mary's son, James I & VI, was christened in the chapel. The Great Hall is impressive but the castle's exterior is its greatest asset: a dramatic structure that's particularly haunting at night.

Below, on Castle Wynd, is **Argyll's Lodging** (Castle Wynd, 01786 431319, www.historic-scotland.gov.uk, by guided tour only with a Stirling Castle admission ticket), a 17th-century townhouse with 16th-century elements that's named in honour of the ninth Earl of Argyll. Over the way is **Mar's Wark**, the impressive stone remains of what was once a grand, Renaissance-style house built by the Earl of Mar in the 16th century. Next door is the **Church of the Holy Rude** (www.holyrude.org), which has one of the few surviving medieval timber roofs in Scotland; the intended rebuild was halted by the Reformation.

Sightseeing in Stirling is dominated by its history. However, the town is also at the centre of old Stirlingshire, where you don't have to look too hard for evidence of 19th- and 20th-century industries; take, for instance, the gargantuan Grangemouth petrochemical refinery on the Forth. On the other side of the M9 from Grangemouth, the otherwise undistinguished town of **Falkirk** has one of Scotland's most amazing attractions. The **Falkirk Wheel** (Lime Road, Tamfourhill, 0870 050 0208, www. thefalkirkwheel.co.uk, boat trip £8.95, £4.95-£7.95 reductions) acts as a link between the Union Canal and the Forth & Clyde Canal, making it possible to travel from Edinburgh to Glasgow by boat. Although that may not sound too thrilling, there's a twist: the Union Canal is around 115 feet above the Forth & Clyde where they meet, and the wheel lifts canal boats from one to the other. It's a smooth marvel of modern engineering. You can watch it all happen from the sidelines, or take a boat ride and have a whirl on the wheel.

Just over four miles east of the Falkirk Wheel, along the Forth & Clyde Canal, you find Scotland's most extraordinary sculptures. The **Kelpies**, centrepiece of a major, green community space called the **Helix** (www.thehelix.co.uk), were completed in 2013 by Andy Scott and are in the shape of two enormous horse heads. People come to walk the canal, to picnic or just to marvel.

Further east of Falkirk, there's a different kind of transport by Bo'ness: the steam-powered trains of the **Bo'ness & Kinneil Railway** (01506 822298, www.srps.org.uk, day rover £10, £6-£9 discounts) regularly delight children and enthusiasts.

Where to eat & drink

One of the best bets in Stirling is **Hermann's** (58 Broad Street, 01786 450632, www.hermanns. co.uk, main courses £11.50-£23), a modern Austro-Caledonian crossover; the chef-proprietor is from the Tirol and does an accomplished schnitzel. Otherwise, **Brea** (5 Baker Street, 01786 446277, www.brea-stirling.co.uk, main courses £13-£15) is an informal bistro with burgers and pizza during the day, more ambitious cooking at

ESCAPES & EXCURSIONS

night. For a pint, pop into the **Portcullis Hotel** (Castle Wynd, 01786 472290, www.theportcullis hotel.com) near the castle or the **Settle Inn** (91 St Mary's Wynd, 01786 474609).

Getting there

By bus Scottish Citylink (0871 266 3333, www.citylink.co.uk) runs the main service from Edinburgh to Stirling; buses leave from St Andrew Square bus station once an hour. The journey, centre to centre, takes around 70 minutes.
By car Stirling is about 45 miles north-west of Edinburgh via the M9 motorway.
By train There are regular trains from Waverley and Haymarket stations to Stirling. The journey takes about 50 minutes.

THE BORDERS

The Borders covers the bottom right-hand corner of Scotland: everything between the Lothians and England, including the stretch of coast with the nature reserve at **St Abb's Head** and the fishing village of **Eyemouth**, towns such as **Galashiels, Hawick** and **Selkirk**, assorted abbey ruins, the **River Tweed** and some very handsome hill country. There are also recurring associations with Sir Walter Scott. The Borders uplands may lack the drama of the Scottish Highlands further north, but at beauty spots such as **Scott's View**, looking over the triple-peaked Eildon Hill (on the B6356 east of Melrose), the proportions of the landscape create a work of art.

It's here that the extent of the Borders' abbeys hits home. A series of institutions founded in the 12th century, with the most celebrated at **Kelso, Jedburgh** and **Dryburgh** (by the village of St Boswells), they lie in ruins today, but remain highly evocative; **Melrose Abbey** (Abbey Street, 01896 822562, www.historic-scotland. gov.uk, £5.50, £3.30-£4.40 reductions) is also where Robert the Bruce's heart was buried. (Historic Scotland also runs the other main abbeys; check its website for full details.)

Away from the ruins, a couple of gardens merit mention. **Dawyck Botanic Garden** (Stobo, on the B712, 01721 760254, www.rbge.org. uk, £6, £5 reductions) is an outpost of the Royal Botanic Garden Edinburgh. And just east of Peebles lies lovely **Kailzie Gardens** (Kailzie, 01721 720007, www.kailziegardens.com).

Around five miles south-east of Peebles is the tidy village of **Innerleithen**, where you'll find **Traquair House** (01896 830323, www.traquair. co.uk, £8.60, £4.30-£7.60 reductions). Reputedly the oldest inhabited dwelling in Scotland, it has served as a court for William the Lion, a hunting lodge for Scottish royalty, a refuge for Catholic priests and a centre for Jacobite sympathies. Every second summer, on odd-numbered years,

it hosts the two-day Traquair Fair, with craft stalls, food and drink, music and theatre. For literary buffs, Sir Walter Scott's former home at **Abbotsford** (near Melrose, 01896 752043, www.scottsabbotsford.co.uk, £8.75, £4.50-£7.50 reductions) on the banks of the Tweed, is unmissable: baronial, Gothic and romantic, just like the man himself.

Where to eat & drink

In Peebles, the **Sunflower** is well established and popular (4 Bridgegate, 01721 722420, www. sunflowerrestaurant.co.uk, main courses £10.50-£18.50), while the bistro-like **Osso** (1 Innerleithen Road, 01721 724477, www.ossorestaurant.com, main courses £12-£15) is very accomplished. For country house dining, try **Cringletie House Hotel** (01721 725750, www.cringletie.com, main courses £18-£24) just north of town.

In Melrose, **Burts Hotel** (Market Square, 01896 822285, www.burtshotel.co.uk, main courses £14-£25.50) has been a favourite for years, with polite decor and good Scottish ingredients. For something altogether more informal, the **Mainstreet Trading Company** (Main Street, 01835 824087, www.mainstreet books.co.uk) is a welcome stop in St Boswells; it has a bookshop, deli and café on the premises.

On the way towards Berwick-upon-Tweed, at Swinton, the **Wheatsheaf** (Main Street, 01890 860257, www.wheatsheaf-swinton.co.uk, main courses £16-£21) is dependable and award-winning. Otherwise, all the major Borders towns have options for eats and drinks.

Getting there

By bus Buses depart from St Andrew Square bus station to various Borders towns. For details, contact Traveline Scotland (0871 200 2233, www.travelinescotland.com).
By car The main route is the A68, but there are several other routes south from Edinburgh. You can drive via Fairmilehead and Penicuik to Peebles, for example; if you head down the A7, past Gorebridge and North Middleton, then take the B7007 off to the right, you'll be up into the bonny Moorfoot Hills. Alternatively, continue on the A7 and you reach Galashiels, with Selkirk and Melrose nearby.
By train A new line running from Edinburgh through Midlothian to the Borders is scheduled for completion late in 2015, with planned stations at Stow, Galashiels and Tweedbank.

FIFE

Given that you can see Fife across the Firth of Forth from many viewpoints in Edinburgh, and there are two major bridges that can get you there

– soon to be three – it's a surprise that more people don't take advantage of its attractions. One reason for this is that some of the choicest tend to be further away: the **Lomonds, Falkland Palace**, the **East Neuk** fishing villages and the historic town of **St Andrews**.

That said, the Royal Burgh of **Culross** on the north shore of the Forth is always worth a visit, an excellent example of how Scotland's past looked in the 17th and 18th centuries. Although you can just turn up and wander, the **Palace** and the **Town House** are particularly atmospheric, and run by the National Trust for Scotland (0844 493 2189, www.nts.org.uk, £10.50, £7.50 reductions). Meanwhile, in **Dunfermline**, the history of **Dunfermline Palace & Abbey** (01383 739026, www.historic-scotland.gov.uk, £4.50, £2.70-£3.60 reductions) goes back to the 11th century. It houses the remains of its founder St Margaret (once known as Queen Margaret) as well as other important Scottish monarchs such as David I and Robert the Bruce.

Head north into Fife and you reach the **Lomond Hills Regional Park** (www.fife coastandcountrysidetrust.co.uk) with its characteristic escarpment and upland bumps including West Lomond; at 1,712 feet it's the highest point in Fife. The historical interest in this part of the county is **Falkland Palace** (0844 493 2186, www.nts.org.uk, £12.50, £9 reductions) at the village of Falkland. Once a castle, then transformed into a palace in the 16th century, it was a country house favourite of the Stuart kings and queens.

In the eastern corner of Fife, jutting out into the North Sea, there's no single centre for visitors, but rather a succession of charming old fishing villages, from Elie to Crail, along a fine ten-mile strip of the East Neuk coast; driving or cycling through them on a sunny afternoon makes for a civilised jaunt.

The must-see town in Fife is **St Andrews** – around 50 miles by road from central Edinburgh. It boasts Scotland's oldest university, a legendary golf course, the ruins of a 12th-century cathedral and much else besides. For more information, see www.visitstandrews.com.

Where to eat & drink

Among the very best restaurants in Fife – and in Scotland – are **Sangster's** (51 High Street, 01333 331001, www.sangsters.co.uk, set meals £29.50-£47.50) in Elie, and the **Peat Inn** (01334 840206, www.thepeatinn.co.uk, main courses £23-£30), set in the countryside about seven miles south of St Andrews.

The fishing villages of the East Neuk cover all bases from fine dining to fish and chips. For the former, head for St Monans and **Craig Millar @ 16 West End** (01333 730327, www.16west end.com, set meals £18-£42); for the latter, try the **Anstruther Fish Bar** (42-44 Shore Street, Anstruther, 01333 310518, www.anstruther fishbar.co.uk).

St Andrews has no end of options, including the **Seafood Restaurant** (Bruce Embankment, 01334 479475, www.theseafoodrestaurant.com, set meals £22-£48.50), occupying a glass-walled pavilion in a striking setting. Over in Cupar, **Ostler's Close** (25 Bonnygate, 01334 6554574, www.ostlersclose.co.uk, main courses £19.50-£23.50) has been a shining presence on the dining-out scene for more than 30 years.

Getting there

By bus There are regular buses from St Andrew Square bus station in Edinburgh to Dunfermline and St Andrews, as well as various services that will take travellers around Fife. For details, contact Traveline Scotland (0871 200 2233, www.travelinescotland.com).
By car Edinburgh is linked to Fife by the Forth Road Bridge and, from 2016, by the Queensferry Crossing. OS Landranger map 59 (St Andrews, Kirkcaldy & Glenrothes) covers the entire eastern part of the county.
By train Much of Fife is peppered with railway stations and well served by trains from Waverley station. The two main exceptions are the East Neuk (no railway line) and St Andrews (nearest station Leuchars, six miles from town).

St Andrews' 12th-century cathedral.

In Context

History

*Edinburgh: important
British city or a capital
in its own right?*

**TEXT: WILL FULFORD-JONES
& KEITH DAVIDSON**

IN CONTEXT

Approach Edinburgh from any direction and the horizon will be dominated by Edinburgh Castle, prominent on its basalt outcrop. Catching the rays of the westering sun, visible from halfway to Glasgow and from the shores of Fife, here is a natural fortress whose occupants, secure on their dizzying heights, could survey the Firth of Forth and watch for raiders approaching in the distance.

The city of Edinburgh grew around this natural vantage point. The fortress on the rock was followed by the city itself, spreading down the castle's ridge to form the major roads and tiny passages of what would become the Old Town. Then, when it could no longer be contained, Edinburgh spilled northwards across the intervening valley into an elegant New Town, neatly ordered and regimented with its Georgian grids. Into the 21st century, the city continues to grow.

IN THE BEGINNING

Edinburgh's dramatic setting was shaped by fire and ice. Perhaps 350 million years ago, volcanoes spewed molten lava across desolate landscapes, helping to form the hills of the city. The landmass was much further south then, but it crept northwards as the continents played their slow, tectonic game of marriage and divorce. Relatively recently in geological terms, Scotland vanished under vast rivers of grinding, groaning frozen water, which carved out the country's mountainous landscape. With the disappearance of the final Ice Age glaciers around 15,000 years ago, the stage was set for the emergence of this most visually striking of cities.

Traces of human occupation here go back millennia. Rewind 6,000 years or more and you'd have seen hunter-gatherers foraging along the River Almond and the Water of Leith; go back three to four millennia and you would have witnessed farming, evidence of which can still be detected in the terraces on the flanks of Arthur's Seat. The hills of Edinburgh bear the signs of early fortification and hut settlements: when dredged in 1778, the waters of Duddingston Loch revealed caches of late Bronze Age weapons.

It's not known whether the Romans occupied the Castle Rock, although from their nearby forts at Cramond and Inveresk, they would have known of its strategic value. However, the Rock is known to have been a stronghold for Dark Age Celtic tribes. King Mynyddog of the Gododdin ruled from the Rock at the start of the seventh century when it was named Dun Eidyn, which probably meant, simply, 'fort on the rock'.

In AD 638, southern Scotland was conquered by King Oswald of Northumbria and the area stayed under Northumbrian rule for more than three centuries. Not until the Scottish kings pushed the Northumbrians back in the tenth century did the settlement become part of the northern polity once more.

THE FIRST CASTLE

Although there would have been a fortification on the Rock for centuries, the first known castle was established by Malcolm III, on the throne from 1058 to 1093. Also known as 'Canmore' (literally 'big head' but metaphorically 'great leader'), Malcolm is best remembered for his appearance in

Malcolm greeting Margaret on her arrival in Scotland.

Shakespeare's *Macbeth* and for his marriage to Margaret, the Hungarian-born, Anglo-Saxon princess who, fleeing the arrival of the Normans to England, arrived on the shores of the Forth in 1068 and wed the king in 1070. Margaret proved pious but energetic: mother to eight children, she played a central role in bringing the Scottish church closer to Rome, founded a priory at Dunfermline in Fife and established a ferry service for pilgrims across the Forth. The oldest extant building on the Rock, the 12th-century St Margaret's Chapel, is dedicated to her.

After a brief period when Malcolm's brother Donald sat on the throne in a joint rule with his nephew Edmund, three of Malcolm and Margaret's six sons went on to rule Scotland. Of these, it was David, the youngest, who had the greatest impact. During his three decades on the throne, he established a royal mint in Edinburgh, introduced feudalism to Scotland and established the first royal burghs (towns granted charters to hold markets and fairs).

Edinburgh folklore has it that David was hunting one day when he was knocked from his horse and attacked by a stag, only to be saved when a cross (or 'rood') appeared in his hand. His gratitude to God found a lasting monument in the shape of Holyrood Abbey, which he founded at the site in 1128 with the help of some construction-savvy Augustinian friars. The Gothic ruins of the Abbey can still be seen in the grounds of Holyroodhouse.

With the Abbey in place, Edinburgh grew along the spine of the volcanic ridge from the Castle Rock. The houses of the city's burgesses faced on to the High Street; behind them, long gardens rolled downwards, their lower walls used as part of the city defences. Parallel to the Canongate, the deep valley of the Cowgate developed into an entrance through which cattle were herded to market; the newly arrived Black Friars (Dominicans) established a friary at its eastern end in 1230. With a succession of religious orders arriving in the town, Edinburgh became an ecclesiastical centre of some importance.

1314 AND ALL THAT

While riding his horse along the Fife coast one stormy night in 1286, Scotland's King Alexander III fell to his death down an embankment at Kinghorn. Alexander left no surviving children to inherit his title; his young granddaughter, the 'Maid of Norway', then died before she could be brought to the throne. With Alexander's demise began a long, bloody chapter in Scotland's history of conflict with England: the Wars of Independence.

Repeatedly ravaged by English armies from 1296, Scotland nevertheless secured its independence after Robert the Bruce's victory at Bannockburn in 1314. The Treaty of Edinburgh was signed in 1328, formally ending hostilities between the two kingdoms. The following year, Bruce granted Edinburgh the status of royal burgh, giving its burgesses important fiscal privileges. However, Bruce died mere months later. David II, his son and successor, was only five years old when he ascended to the throne, a position of weakness that left the kingdom vulnerable to renewed strikes by the English. Edward III attacked first in 1333; and then, nine years after David's death in 1371, Richard II followed suit, besieging the castle and burning both the Canongate and St Giles.

David II had died without producing an heir and was succeeded by Robert II, the son of Walter the Steward and Robert the Bruce's daughter, Marjorie. Through this dynastic marriage, Robert II was able to start a dynasty of his own: the House of Stewart. During this period, Edinburgh emerged as Scotland's most populous burgh, a position it was to hold until Glasgow's spectacular 19th-century growth.

DEATH, DRAMA AND DYNASTY

The Stewarts proved to be dynastically tenacious, and went on to rule Scotland until 1603, then the whole of Britain and Ireland for much of the 17th century. As individuals, however, they were somewhat short-lived. Being king was a dangerous occupation: this was a period of lawlessness that saw successive monarchs murdered or killed in battle, the throne then passing to children who were too young to rule in their own right. This succession of Stewart child-kings in turn left Scotland vulnerable to the machinations of rival court factions. The family also, as we shall see, proved somewhat unimaginative in their choice of baby names.

James I (1406-37) tried to curb the power of the nobles, but himself became a victim of their machinations and was murdered at Blackfriars in Perth in 1437. His six-year-old son was hastily crowned James II by his mother at Holyrood Abbey, but his reign, during which Edinburgh's Old Town took shape and the city's first defensive wall was built, ended abruptly in 1460 when he was killed by an exploding cannon while besieging the English occupiers of Roxburgh Castle.

Robert the Bruce.

IN CONTEXT

UNION STREETS

Political branding in the New Town.

Scotland and England have been contractually bound in a constitutional marriage since 1707 and shared a monarchy for more than a century before that. The Union was not without its teething troubles, but as the 18th century progressed, Edinburgh saw value in the bigger markets and widened horizons afforded by the growing British Empire of which Scotland was now a core part. When the time came, in the 1760s, to name the streets of the New Town, the project could have been mistaken for a PR exercise for the relatively young British polity and the ruling royal house.

The initial plan was to have George Street as the main spine running east to west, named for King George III; it would link St Andrew Square and St George Square, named for the patron saints of Scotland and England. This changed when a different, unsaintly George Square was laid out in the south of Edinburgh; the New Town's St George Square was redubbed Charlotte Square to avoid confusion. It took its title from both Queen Charlotte and Princess Charlotte, the wife and eldest daughter of the king. Queen Charlotte, also known as Charlotte of Mecklenburg-Strelitz, was further honoured by another thoroughfare, the most northerly of the original New Town: Queen Street.

To emphasise the Union even more, streets were named after the official emblems of Scotland and England,

Thistle and Rose, while the most southerly of the New Town streets was to be named for the patron saint of Edinburgh, St Giles. At this point the king intervened, deciding that he wasn't going to allow anything in this prestigious development to have the same name as a London slum area. St Giles Street was hastily renamed for the royal princes, George and Frederick; it has been Princes Street ever since. The brothers later became George IV, and the Duke of York and Albany.

The three central streets running on a north–south axis were named Hanover for the ruling royal house, Frederick for the father of George III, and Castle, the most prosaic of all since it simply offered a pleasant view of Edinburgh Castle.

While the nomenclature had an obvious, political motivation around 250 years ago, modern Edinburgh associates this part of the New Town more with bars, hotels, restaurants and shops, although it has fine gardens too. Princes Street Gardens are always popular on sunny days, while St Andrew Square Garden was opened to the public in 2008 and has since been used as a Festival Fringe venue. As for Charlotte Square Gardens, the site is best known as the home of the annual Edinburgh International Book Festival; its associations with Charlotte of Mecklenburg-Strelitz and her princess daughter have long faded.

Charlotte Square Gardens.

'SHALL WE CALL HIM JAMES?'

Like his father and his grandfather, James III (1460-88) inherited the throne as a child, at a time when his kingdom was riven by warring factions and power struggles. Despite the turbulence, the Cowgate emerged as a fashionable quarter during James's reign, the French writer Froissart remarking on its fine aristocratic mansions, gardens and orchards. From around 1485, dwellings were also built in the Canongate. But the lack of a defensive wall left the area open to attack, and the Abbey itself was regularly sacked and looted.

Commerce was also flourishing. With Leith nearby, an important North Sea port, Edinburgh was ideally placed to capitalise on foreign trade opportunities. In 1469, when the town ceased to be ruled by the merchant burgesses, it became a self-electing corporation. Cloth sellers, beggars and fishwives plied their trade from booths around St Giles on the High Street; their stalls eventually became permanent fixtures. In 1477, James III chartered markets to be held in the Grassmarket, partly to alleviate the congestion caused by traders on the High Street. He granted the citizens of Edinburgh the Blue Blanket just five years later, a symbol of the independence of the municipality and the exclusive rights of the town's craftsmen.

A FLODDEN DISASTER

Holyrood witnessed an occasion of some grandeur on 8 August 1503, when the educated, cultured King James IV married Henry VII's 12-year-old daughter, Margaret Tudor. The events that followed were less splendid. As part of the marriage settlement, James had signed the Treaty of Perpetual Peace with England, but it failed to live up to its name. Only a decade after the accord, the French persuaded James to attack England, and the two countries went to war once more.

In 1513, James IV led his army into Northumberland. Despite the Scots' numerical superiority, the Battle of Flodden was a disaster for them: 10,000 were killed, including the king himself. In Edinburgh, disbelief at the defeat turned to panic when it was realised the English might press north and attack the city. Work on the Flodden Wall began, though the attack never materialised. Still visible in parts today, it had six entry points; when it was eventually

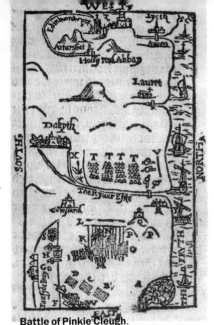

Battle of Pinkie Cleugh.

completed in 1560, it formed the town's boundary for a further two centuries.

The life of James's son, James V, who ascended to the throne in 1513 at the age of one, continued the dynastic turmoil but added a religious element. When he died in 1542, it was thought that stability could be ensured by marrying off his six-day-old daughter, the future Mary, Queen of Scots. But to whom? Opinion was split between those who wanted a French alliance (Cardinal Beaton and Mary's mother, Mary of Guise, among them) and those who preferred an English match.

Henry VIII of England sent the Earl of Hertford's army to Scotland to 'persuade' the Scots that a marriage to his son, Edward, was preferable. After landing at Leith in early summer 1544, Hertford looted both the Abbey and the Palace of Holyroodhouse in an episode known as the 'Rough Wooing'. Hertford's 10,000-strong forces then stormed the Netherbow, but were repulsed.

Three years later, in September 1547, the English returned. The Battle of Pinkie Cleugh was fought at Musselburgh, just outside the town, between the invading English and the defensive Scots. The Scots lost and were chased back to the gates, but the castle held; after French and Dutch reinforcements arrived in Leith the following year, the English were finally repelled and the port was reinforced to prevent further invasion. At the age of five, Mary was sent to live in France; by the time she returned 13 year later, Scottish politics was dominated by religious unrest and the Royal House of Stewart had changed its spelling to Stuart – easier to pronounce in French.

THERE'S SOMETHING ABOUT MARY

The Reformation Parliament declared Protestantism to be Scotland's official religion in 1560, and John Knox became leader of the Reformed Church. The faction that had previously been pro-French and pro-Mary now also became pro-Catholic, while Protestant forces rallied against them. Knox hated Mary; the period between her arrival at Leith from France in August 1561 and her abdication in 1567 saw much friction between the Catholic monarch and the Protestant Church.

Mary married Henry Darnley, a cousin, in 1565. Darnley was a grandchild of Margaret Tudor, which made Elizabeth I even more suspicious of the couple's claims on the English throne. Darnley was also a shameless manipulator: it has been suggested that the brutal assassination at Holyrood in 1566 of David Rizzio, Mary's favoured Italian secretary, was engineered by him in an attempt to cause Mary to miscarry, and perhaps even kill her, leaving the throne vacant for his assumption.

The couple's relationship became increasingly fractious until, in 1567, Darnley's house was blown up with him inside. While no proof of Mary's involvement emerged, suspicions fell on Lord Bothwell, one of her most loyal supporters. When the pair subsequently married, public opinion turned violently against the queen. Forced to abdicate in favour of her infant son, the future James VI, Mary escaped from imprisonment at Loch Leven Castle the following year, only to have her army defeated near Glasgow. She then took refuge in England, spending 19 years as the prisoner of Elizabeth I before her execution in 1587.

THE WISEST FOOL IN CHRISTENDOM

Born in a tiny room in Edinburgh Castle, James VI (1567-1625) became king when he was 13 months old; he was brought up by Protestant tutors, alienated from his mother, and subject to several kidnapping attempts as a child. By the time he took the reins of government, he had grown into a suspicious and wary man. His reign was a permanent headache, thanks to tussles with the economy, the nobles and, above all, the Church.

James VI's long-term ambition was to inherit the English crown from Elizabeth I. Elizabeth, meanwhile, was reluctant to formally recognise 'that false Scotch urchin' as her heir. However, on her death in 1603, Sir Robert Carey galloped from London to Holyrood in an incredible 36 hours to announce that James VI of Scotland was now also James I of England and Ireland. James, once described as 'the wisest fool in Christendom', left for London to be crowned, an event known as 'the Union of the Crowns'. He promised to return to Scotland every three years, but it was 16 years before Scotland saw him again.

The years following James's departure were characterised by social unrest, religious turmoil and a loss of national identity, as Scotland came to terms with absentee rule. Even so, the first years of the 17th century were not unprofitable for Edinburgh. Local merchants thrived, with goldsmiths, watchmakers and bookbinders all flourishing in Parliament Square, and the University of Edinburgh (which received its royal charter in 1582 as the Tounis College) continued to grow. The east wing of the Castle was rebuilt by Sir James Mason; Parliament House was begun in 1632 (the Scottish Parliament was, by then, resident in the city); and, a year later, Holyroodhouse was extended.

A DEAL WITH GOD

James died in 1625 and was succeeded by his son, Charles I, who wasn't crowned King of Scots until 1633. Charles made it his mission to impose uniformity throughout the British kingdom, but his attempt to force an Anglican liturgy on the Presbyterian Church met with strong resistance (not least from a market-trader named Jenny Geddes, who famously hurled her wooden stool at the pulpit in disgust during a service at the High Kirk of St Giles in 1637). As a result of this assault on their freedoms, a document called the National Covenant was drawn up, a contract with God no less, in order to assert the Scots' rights to spiritual and civil liberty. On the last day of February 1638, it was read from the

SHOT BY BOTH SIDES

The short and eventful life of Montrose.

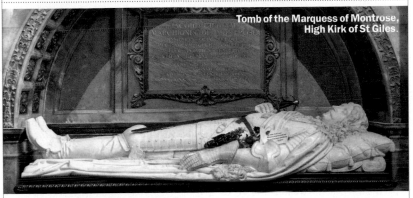

Tomb of the Marquess of Montrose, High Kirk of St Giles.

On 21 May 1650, the Marquess of Montrose – James Graham – was taken the short distance along the Royal Mile from the Old Tolbooth that served as Edinburgh's prison and hanged on a high gallows at the Mercat Cross. He was then dismembered, his head stuck on a spike at the Old Tolbooth where it remained for more than a decade, his torso buried on the Burgh Muir, and his arms and legs sent for display in Aberdeen, Glasgow, Perth and Stirling as a warning against impiety, murder and treason.

Visit the **High Kirk of St Giles** (see *p57*) today, however, and in one of the aisles you'll find a handsome memorial for the very same man, his reputation redeemed, his grand funeral at St Giles having taken place 11 years after his execution on the street outside.

Montrose was a Scottish nobleman who initially sided with the dissenting Covenanters after Charles I tried to impose an Episcopalian liturgy on Scotland in 1637. This was no obscure, ecclesiastical debate – it soon sparked the conflict known as the Bishops' Wars of 1639-40 and all the strife that followed. But in the 1640s, Montrose switched sides, plainly having decided his opposition to Episcopalianism was lower down his list of priorities than his support of the monarchy in general and Charles I in particular. With the approval of the king, the dashing Montrose then successfully led a Highland and Irish army against the Covenanters' forces in the years up to 1645, when he lost a decisive battle in the Borders. He subsequently fled abroad.

In England, meanwhile, the Civil War saw Charles I executed in 1649; his son, Charles II, reaffirmed Montrose's position as the king's man in Scotland. Montrose returned to the country in 1650 to rally pro-royalist forces in the far north, but failed, losing a battle in Ross-shire, and was soon captured by his enemies.

At this point, realpolitik rears its head. Exiled from England by the Civil War, Charles II needed the help of the Scottish Parliament to reclaim the English throne, which meant he had to give way on contentious religious matters. Montrose's incursion may have been used as a bargaining chip to pressure the Scots into a more amenable deal, but it wasn't to be. Charles II washed his hands of Montrose and the Marquess's fate was sealed.

The king did eventually regain his English throne in 1660, after which Montrose was disinterred and his parts reassembled; he was reburied with due pomp at St Giles in 1661, a popular decision at the time. Somewhat glossing over the historical complexities, Montrose has since been claimed as a bold cavalier, one of Scotland's most romantic figures. He may also have been one of its most expendable.

IN CONTEXT

FLIGHT CLUB

Edinburgh's century of air travel.

Edinburgh is Scotland's busiest airport, handling over ten million passengers a year. Although a great deal of the traffic is relatively local, around Britain and Ireland, there are also flights to major European cities and Mediterranean holiday destinations, and longer hauls to New York and elsewhere; it's an important hub. The terminal has the usual bars, cafés and shops, while the modern in-flight experience is not dissimilar to travelling by bus, albeit a bus more than six miles up in the air moving at three-quarters the speed of sound. A hundred years ago it was all rather different.

The history of powered flight starts with the Wright brothers in North Carolina, USA, in 1903 and moves quickly to the outbreak of the Great War in 1914. Early in the war, the UK government bought up sites across England and Scotland for use as airfields, including some farmland west of Edinburgh at Turnhouse. A simple airstrip was laid out in 1915 and a Royal Flying Corps squadron was stationed there from 1916; Turnhouse Aerodrome was born. Back then, flying to Edinburgh really was a gung-ho experience: biplanes with open cockpits, no radar, a maximum speed that could be exceeded by an average family car today and a flattened patch of grass for take-off and landing.

With the formation of the Royal Air Force (an amalgamation of the Royal Flying Corps and the Royal Naval Air Service) in 1918, the site became RAF Turnhouse, although the facilities remained rudimentary. In 1939, proper runways were laid, allowing Spitfires and other aircraft to operate here during World War II; in fact, Turnhouse was a key part of Britain's air defence during the conflict.

After the war, civil aviation took over. The first shuttle service linking Edinburgh and London started in 1947, courtesy of British European Airways. In 1960, the site was officially demilitarised, although the RAF maintained a presence as late as 1997.

From the 1950s to the 1970s, the modern Edinburgh Airport gradually took shape,

although it offered few scheduled flights outside the UK. Air travel across Europe was set to expand, however, thanks to the European Union's deregulation of the market from 1987 onwards, and the rise of budget airlines. In 1990, for example, the airport handled around 2.6 million passengers. That figure was 5.5 million by 2000, 8.6 million by 2010, and exceeded ten million four years later.

Touch down in Edinburgh today on a typical Airbus A320 and, once groundside, a tram will take you to the city centre in just over half an hour. The process is effective but it certainly lacks romance, a charge that could never be levelled at a Royal Flying Corps officer bumping down here on grass in his BE 12 biplane a century ago.

Air traffic control tower, Edinburgh Airport.

pulpit of Greyfriars Kirk; over the next two days, a host of lairds and burgesses came to sign it. Thus began the Covenanting Wars, which continued for decades. The city suffered: trade dropped off and, coincidentally, in 1644, plague killed a fifth of the population.

Charles I had been distracted from events in Edinburgh by the outbreak of civil war in England. Although Edinburgh Castle was held by forces loyal to the king, the rule of the Covenant held sway elsewhere in the town. By 1649, Oliver Cromwell had assumed power in England; on 30 January, Charles was executed. The Scots were outraged that their Parliament hadn't been consulted; but six days later, they proclaimed Charles II to be the King of Scotland, on the condition he accepted the Covenanters' demands.

Charles refused, and instead asked the Marquis of Montrose, who had been loyal to his father, to conquer Scotland for him. However, Montrose was defeated and captured by the Covenanters, brought to Edinburgh and, on 21 May 1650, executed (see p219 **Shot by Both Sides**). Cromwell's response was to invade, defeating the Scots under General Leslie at the Battle of Dunbar on 3 September 1650.

KILLINGS AND KINGS

Although Cromwell's government and the Scots' Presbyterian Church shared many of the same values, Cromwell's Scottish occupation proved increasingly unpopular. Charles II's return to the throne after Cromwell's death in 1658 – the Restoration – was greeted with relief.

It didn't last. When Charles reneged on acts made in favour of Covenanters, discontent simmered once again. After the Covenanters won a crucial victory at the Battle of Drumclog in 1679, Charles sent the Duke of Monmouth to crush them, which he did at the Battle of Bothwell Brig later that year. In a period since nicknamed 'the Killing Time', the survivors were marched to Greyfriars Kirkyard in Edinburgh and imprisoned for five months with little food, shelter or water. Many died or were executed; several hundred others were sent as slaves to Barbados.

James VII of Scotland (simultaneously James II of England and Ireland) ascended to the throne on the death of his brother Charles in 1685, but his Catholicism made him

unpopular. The Dukes of Argyll and Monmouth tried and failed to unseat him. But when James finally fathered a male heir in 1688, a group of English noblemen sought to replace him with William of Orange and his wife Mary, James's own Protestant daughter. James fled to France and the protection of Louis XIV.

ANE END OF ANE AULD SANG

Many in Edinburgh and the Scottish Parliament were delighted by William's ascension to the throne. However, supporters of the ousted James, named 'Jacobites' after the Latin for his name (Jacobus), mounted resistance. The misery of civil war was exacerbated by wretched living conditions. And when, in 1698, the so-called Darien Scheme, designed to enrich Scotland by establishing a trading colony in Panama, virtually bankrupted the nation, the citizens cracked, storming the Tolbooth and setting fire to the Cowgate.

The collapse of the Darien Scheme strengthened the hand of those who promoted a union of the Edinburgh and Westminster parliaments. After much discussion, argument and lavish bribery, the Act of Union became law in January 1707. When the Lord Chancellor of Scotland signed away Scotland's independence he said, 'There's ane end of ane auld sang.' It would be nearly 300 years before a Scottish Parliament sat again.

AN ENLIGHTENMENT IN THE NORTH

The 18th century is known in Edinburgh as the 'Age of Improvement'. The phrase refers to the massive building programme that was implemented in the 1760s, but also to the influence of the Enlightenment – the spirit of intellectual inquiry that flourished in the 18th century – among the lawyers, academics and churchmen of the city. Social and cultural improvements arrived apace.

However, at the start of the 18th century, the city's geography was still profoundly medieval in its nature: a cramped, towering settlement clinging grimly on to the hillside. By and large, the old city walls still formed the town boundaries, so as the population grew to well over 50,000 during the 18th century, the only way to build was up. This resulted in the 'lands'; six-, seven-, and even eight-storey buildings that were prone to collapse. Something needed to be done, but what?

'The contrast between New Town gentility and the low-life of the Old Town became a recurring theme that even modern Edinburgh has found hard to shift.'

THE STUARTS' LAST HURRAH

Gradually, Scottish cities such as Edinburgh and Glasgow came to realise that the union with England offered imperial opportunities for commerce and trade. Edinburgh was initially seduced by the glamour of Charles Edward Stuart, known as Bonnie Prince Charlie, who was the grandson of James VII: he 'took' the city in 1745 with the assistance of a number of Highland clans, before heading south in an abortive attempt to claim the British throne. However, the sympathies of the townspeople ultimately remained Hanoverian.

Shortly after Charlie's last stand at Culloden in 1746, finally ending the dynastic ambitions of the Stuarts, the city embarked on an ambitious building programme: the expansion of Edinburgh across the valley north of the Castle. The scheme was influenced by the appalling living conditions in the plague- and epidemic-prone Old Town, but it also had a political motivation: to show that Edinburgh was a civilised, cultured city. The competition to design the New Town was won in 1766 by a 22-year-old architect named James Craig, with a simple, elegant and harmonious plan. The two million cartloads of earth dug from the foundations of the New Town formed the basis for the Mound, begun in 1781 but not finished until 1830.

Around the same time, Edinburgh pushed southwards: George Square was laid out in 1766, and a new college for the university was built in 1789 at the old Kirk O'Fields. But it was the New Town that proved most popular. As the wealthier classes moved north, a social apartheid formed. The contrast between New Town gentility and the lowlife of the Old Town, with its cock-fighting dens and brothels, became a recurring theme that even modern Edinburgh has found hard to shift.

SCOTT CREATES SCOTLAND

The rise of the New Town virtually parallels the increased influence of Edinburgh-born Sir Walter Scott (1771-1832), a titan of the late 18th and early 19th centuries. Scott's character and interests were largely influenced by time spent as a child in the Borders, where he had been sent to recuperate from the polio that would leave him with a permanent limp. The romance of Borders legends and ballads enthralled the imaginative child, and the collection of similar stories from all parts of Scotland became a life-long passion.

After an Edinburgh education at the Royal High School and the Law Faculty of the university, Scott embarked on a starry legal career, while simultaneously finding great acclaim as a writer. But while his literary and legal works are rightly famed, his wider influence should not be underestimated: his tireless efforts almost single-handedly awakened the world to the romantic potential of Scotland, and paved the way for its rebranding after the Jacobite insurrections. It was thanks to Scott's enthusiasm and perseverance that the long-lost 'Honours of Scotland' – the crown, sceptre and sword of state – were uncovered from their hiding place in Edinburgh Castle and put on public display. He was also responsible for the first visit to Scotland of a British monarch since the reign of Charles II, when, in 1822, he persuaded George IV to visit the capital of 'North Britain'.

HERE COME THE VICTORIANS

Edinburgh added a number of notable features to its townscape in the first three decades of the 19th century, among them Waterloo Bridge in the east and Melville Street in the west. However, not every construction project ended in success; or, for that matter, ended at all. Begun in 1822, William Playfair's National Monument to honour the dead of the Napoleonic Wars was never completed: the Parthenon-like structure, dubbed 'Scotland's Disgrace', stands on Calton Hill to this day. Power was held by London and the intellectual

Edinburgh Castle from the Grassmarket, 1850s.

activity of the Enlightenment was declining. Edinburgh's pre-eminence was in the past.

Nevertheless, the city underwent a third period of expansion during the Victorian era, when suburbs such as Marchmont, Morningside and Bruntsfield were erected. The city that had become bipartite when the New Town was built found itself with still more facets, each with its own character. The solid Victorian suburbs were peopled by the growing middle classes; the grand New Town remained the area of choice for lawyers and judges; the teeming Old Town became a slum.

Through it all, the population of greater Edinburgh spiralled, growing from 100,000 at the start of the 19th century to 320,000 by 1881. One of the reasons for the dramatic increase was the influx of people from the Scottish Highlands, displaced by the Clearances, and from Ireland; they moved to the city in search of employment. Among them were William Burke and William Hare, a pair of Irish labourers who came to work on the Union Canal. They eventually abandoned digging ditches in favour of a more lucrative trade: supplying corpses to the university's Anatomy School. Burke and Hare infamously spurned the established – if somewhat unsavoury – practice of digging up recently buried bodies in favour of providing fresh ones, unfortunate by-products of a killing spree the pair carried out from their lodgings in the West Port off the Grassmarket. Convicted of 16 murders in 1829, Burke was hanged on the evidence

of his turncoat partner. A pocketbook made from his skin is still on display in the Pathology Museum at Surgeons' Hall on Nicolson Street.

With the increase in population came unemployment. Riots in 1812 and 1818 were both blamed on poor economic conditions; by the 1830s, outbreaks of cholera and typhoid had decimated the Old Town. The misery was compounded in 1824 when a fire destroyed much of the High Street, an early challenge for the world's first municipal fire service, formed just two months earlier.

A study conducted by Dr George Bell in the 1850s found that 159 of the Old Town's closes lacked drainage and fresh water; Bell also bemoaned the alcoholism. A decade or so earlier, a separate study undertaken by a young doctor named William Tait had found that the area contained an impressive 200 brothels. Attempts were made to restore the neighbourhood, but it was on a downward spiral that was to continue into the 20th century.

When the neglected Paisley Close on the High Street collapsed in 1861, killing 35 people, public outrage caused the city council to agree to the adoption of proper health and safety regulations. Dr Henry Littlejohn was appointed as the first Medical Officer of Health in Scotland. Littlejohn's report on sanitary conditions in Edinburgh coincided with the election of the philanthropic publisher William Chambers as Lord Provost, resulting in the improvement scheme of 1866 that cleared some of the slums of the Old Town and created new streets such as St Mary's and Blackfriars.

NOT AS BIG AS GLASGOW

The Victorian era (1837-1901) was also defined by technological advances, particularly in the fields of transport and medicine. In the early 17th century, London was 13 days away by coach; towards the end of the century, the journey could be done in a 'mere' four days. But with the advent of the age of steam, travel became far easier, to the benefit of all concerned.

During 1845 and 1846, rail tunnels were dug between Haymarket and Waverley stations, through the south flank of Calton Hill and under the Mound. Trains travelling through them brought tourists and travellers straight into the heart of the city, where they would

IN CONTEXT

emerge to face the Castle, the Gothic bulk of the Scott Monument (begun in 1840), the galleries at the foot of the Mound and the splendour of the Princes Street Gardens. In 1890, the completion of the Forth Bridge –hailed as the eighth wonder of the world – linked the city with towns north of the Forth, while a network of suburban lines facilitated expansion into outlying villages.

For all Edinburgh's expansion, Glasgow assumed increasing importance by the end of the Victorian era. The two international festivals held in Glasgow, in 1888 and 1901, far outshone the one staged at the Meadows in Edinburgh in 1886; at the same time, Glaswegians such as Charles Rennie Mackintosh were creating an artistic and architectural legacy still revered today. As the historian and journalist Allan Massie has pointed out, Edinburgh at the end of the 19th century was the biggest small town in Scotland.

THE WORLD WARS

The city's history during the first half of the 20th century was dominated by the two world wars. Scots made up ten per cent of British recruitment in the Great War, with 25 per cent of the Scottish male population marching off to fight. Long lists of the many who did not return can be seen in Sir Robert Lorimer's Scottish National War Memorial at the Castle.

Although a Zeppelin raid in 1916 killed 11 people and damaged buildings across the city (and also, with ironic accuracy, scored a direct hit on the German Church at Bellevue), Edinburgh generally suffered very few direct attacks in World War I. The city's lack of heavy industry meant it also escaped the worst ravages of the Depression in the 1930s. Although it was a peripheral victim of the first air raid of World War II, in October 1939, when the Luftwaffe attacked Royal Navy cruisers off Inchgarvie in the Forth, it was spared the ferocity of aerial attack experienced by other British cities. Still, Edinburgh already had plenty on its plate.

The suburbs went on sprawling outwards, but buildings were in general more likely to be pulled down than put up. At last, the city fathers (and private contractors) got to grips with the decaying Old Town, upgrading the city's infrastructure while also moving the population to outlying areas such as Niddrie

and Craigmillar. Even today, visitors to Edinburgh are often left unaware that the city is ringed by public-sector housing schemes, once innovative, then sorely neglected, now subject to regeneration.

TOWARDS THE MILLENNIUM

Many of Edinburgh's traditional industries, among them publishing, declined during the 20th century. However, the addition of three new universities (Heriot-Watt, Edinburgh Napier and Queen Margaret) helped to boost the city's already strong academic reputation, and the city has also become one of Europe's top financial centres, specialising in banking, fund management and insurance. As in the rest of Scotland, the tourist industry continues to be utterly vital to the local economy.

Easily the city's most notable cultural phenomenon of the 20th century was the establishment of the Edinburgh International Festival in 1947. The opening event featured music from the Vienna Philharmonic Orchestra, ballet from Sadler's Wells and theatre from the Old Vic. Just as crucial, though, were the eight theatre companies that, excluded from the official programme, staged shows in smaller venues. Others followed suit in subsequent years, and the ebullient, ever-expanding Edinburgh Festival Fringe began to make its own reputation.

A decade or two before the festivals were launched, Edinburgh saw a brief cultural flowering known as the Scottish Renaissance, centred around writers and artists such as James Bridie, Lewis Grassic Gibbon, Neil M Gunn, Hugh MacDiarmid and Edwin Muir. Something of that feel returned during the 1990s, thanks to the presence of a few vibrant publishing houses, such as Canongate, and the international success of Irvine Welsh's novel *Trainspotting*. The novel brought some realism to the city's image by throwing a spotlight on its ills, chiefly the heroin epidemic of the 1980s.

But the event that history may come to regard as the century's most significant happened right at its close. In 1997, Scots voted 74 per cent to 26 per cent for devolution and the re-establishment of a Scottish Parliament in Edinburgh. Whether that paves the way for full independence at some point in the 21st century, and the end of the union with England, remains to be seen.

KEY EVENTS

Edinburgh in brief.

AD 80 Romans invade Scotland, encountering a Celtic people, the Votadini.
AD 140 Roman forts established at Cramond on the Forth and Inveresk, East Lothian.
c AD 470 Emergence of the Gododdin kingdom in the Lothians in the post-Roman period, successors to the Votadini.
AD 638 Anglian Kingdom of Northumbria extended to the Forth, including Edinburgh.
1018 The Battle of Carham signals the Scots' final conquest of Lothian and the Borders.
1070 Malcolm III marries Margaret; around this time the first castle is built on Castle Rock.
1128 David I founds Holyrood Abbey.
1243 The High Kirk of St Giles is consecrated.
1286 Alexander III dies in a riding accident, precipitating the Wars of Independence.
1314 Thomas Randolph, Earl of Moray, retakes Edinburgh Castle from the English, three months before the Battle of Bannockburn.
1329 Robert the Bruce grants a royal charter to the city.
1349 The Black Death arrives in Edinburgh, with two or three further outbreaks before the century's end; huge numbers die.
1437 Edinburgh typically seen as the Scottish capital from this date.
1511 The Great Hall is built at Edinburgh Castle, during the reign of James IV.
1513 The Scots are defeated at Flodden; James IV is killed. Panic in the city.
1542 Mary, Queen of Scots inherits the throne from James V. She is six days old.
1544 Edinburgh is attacked by the English in the 'Rough Wooing'.
1560 The Reformation Parliament declares Protestantism to be Scotland's official religion.
1561 Mary returns from France to rule.
1582 University of Edinburgh receives its royal charter. Teaching starts the following year.
1603 James VI accedes to the English throne as James I; the Scottish court moves to London.
1638 The National Covenant is signed in Greyfriars Kirkyard.
1639 Parliament House is completed.
1681 James Dalrymple publishes Institutions of the Law of Scotland.
1695 Bank of Scotland founded.

1700 The Darien Scheme finally collapses.
1707 Following the Act of Union, the Scottish Parliament in Edinburgh is dissolved.
1727 The Royal Bank of Scotland is founded.
1736 The Porteous Riots take place.
1745 Bonnie Prince Charlie is based in the city during his campaign to restore the Stuarts.
1767 Work on the New Town begins.
1784 The last execution in the Grassmarket.
1820 The Royal Botanic Garden Edinburgh moves to its current Inverleith site.
1824 The Great Fire destroys much of the High Street, only months after the formation of the world's first municipal fire brigade.
1843 Church of Scotland split by the Disruption.
1854 Three railway stations at the east end of Princes Street acquire the collective name Waverley.
1865 The city's last public hanging.
1871 The first international rugby match, Scotland v England, is played at Raeburn Place.
1874 Heart of Midlothian FC founded.
1875 Hibernian FC founded.
1890 The Forth Rail Bridge opens.
1913 Edinburgh Zoo opens.
1925 Murrayfield Stadium opens.
1947 The first Edinburgh International Festival.
1966 Heriot-Watt College becomes a university.
1970 The city hosts the Commonwealth Games.
1986 The Commonwealth Games returns.
1996 The Stone of Destiny returns to Scotland and is housed in Edinburgh Castle.
1997 Scotland votes 74 per cent to 26 per cent in favour of a devolved parliament in Edinburgh.
1998 The Royal Yacht *Britannia* is berthed permanently at the Western Harbour, Leith.
1999 A Scottish Parliament convenes for the first time in 292 years.
2004 The new Scottish Parliament building at Holyrood is formally opened by Elizabeth II.
2008 Edinburgh's banking industry and property developers devastated by the credit crunch.
2011 Scotland elects its first majority SNP government at Holyrood.
2014 The tram system starts running. In the independence referendum Scotland votes 55 per cent to 45 per cent to remain in the UK.

Architecture

The past mixes easily with the present in the Scottish capital.

TEXT: KEITH DAVIDSON

Topographically, Edinburgh has been dealt a spectacular hand. The Pentland Hills lie to the south and a coastal plain stretches north and east to the Firth of Forth; Arthur's Seat and Castle Rock, along with Calton Hill and Salisbury Crags, lend geographic drama.

This magnificent setting has helped shape a city of two distinct characters: the architectural chaos of the Old Town looks across to the regularity of the New Town, a triumph of classical formality played out in a gridiron of well-disciplined streets. In 1995, UNESCO jointly designated the Old Town and the New Town a World Heritage Site, an honour that underlines the city's knack for seducing its visitors. For those interested in architecture, Edinburgh's building history, stretching back nearly 900 years, should prove endlessly rewarding.

EARLY DAYS

Under the ambitious rule of Malcolm III, Edinburgh's Castle Rock emerged as a fortified stronghold: what we know today as **Edinburgh Castle** (see p55). Named for Malcolm's wife, **St Margaret's Chapel** is the earliest architectural survivor from those times. Set within the Castle precincts, it's a small stone structure with a chevron-decorated chancel arch and was built at some point during the reign of David I (1124-53). The expanding settlement of Edinburgh was declared a royal burgh in 1125; three years later, **Holyrood Abbey** (see p64) was founded. As the medieval city grew, Holyrood became linked to the Castle Rock, defining today's **Royal Mile**.

Architecturally, little remains from Edinburgh's infant years. Few structures were made of stone or according to the most punctilious building standards; most houses were crudely constructed from wattle and post, covered in clay for insulation, thatched with straw, rushes or heather, and built to last decades at best. Of the handful of medieval stone structures still standing in Edinburgh, St Giles (today's **High Kirk of St Giles**; see p57) dates from the 12th century, but only fragments of the original building remain; the church was extensively remodelled in the late 14th century.

As its national stature grew during the reign of James III (1460-88), Edinburgh witnessed a surge in confidence. Apartments at Holyrood Abbey became a royal residence and the site soon saw a major addition: the **Palace of Holyroodhouse** (see p66). The castle was augmented by Crown Square and its baronial Great Hall, topped with a hammerbeam roof. Money was also pumped into houses of worship, Trinity College Church perhaps most notable among them. The building has long since been demolished, but its magnificent altarpiece, by Hugo van der Goes, was preserved, and is now on display in the Scottish National Gallery (see p88). Similar grand gestures resulted in a distinctive crown steeple being added to the central tower of St Giles towards the end of the 15th century. The resulting structure doubtless became a template for the crown steeples built in subsequent years on other Scottish churches.

John Knox House.

LIVING ARRANGEMENTS

Narrow lanes, known locally as wynds and closes, developed like ribs from the spine of the Royal Mile during medieval times. Simultaneously, many existing houses had extra storeys tacked on to them, often haphazardly. **John Knox House** (c1490; see p67) is one of the few remaining examples, but it's a relatively restrained one; some timber-framed structures protruded as far as seven feet into the street.

Eventually, building regulations began to rein in the speculators. But even as far back as the 15th century, local architects were experimenting with new ideas in a bid to solve a housing crisis brought on by a rising population. The rocky and uneven terrain of the Old Town, combined with the ancient 'feu' system of land tenure that granted leases in perpetuity, made horizontal development problematic. Expansion was thus forced upwards, leading to the birth of the tenement. Packed together along the Royal Mile these had more in common with northern Europe than England; their distinctive exteriors could be sandstone or harled (a protective covering of small stones and lime render). The five-storey **Gladstone's Land** (see p56) retains the once commonplace street arcade and an oak-panelled interior.

With the upper Royal Mile full of merchants, its lower reaches soon became the location of choice for the nobility. **Moray House** (c1628) with its pyramid-topped gate piers, and the vast, 17th-century **Queensberry House** (see p65), now part of the Scottish Parliament complex, are the grandest of the buildings that survive. Around this time, buildings regulators began introducing measures intended to reduce the risk of fire: a law in the 1620s stipulated that all buildings must have tile or slate roofs, and a 1674 edict forced developers to give their properties stone façades. The Palace of Holyroodhouse was rebuilt in the 1670s with a triumphant blend of Scottish and European influences, creating a thickset façade with turreted towers fronting an inner courtyard lined with classical arcades.

New wealth brought along in its well-heeled wake a new **Parliament House**; built next to St Giles in 1639, its presence added weight to Edinburgh's role as Scotland's capital. The building was given a classical overhaul in the early 19th century. Elsewhere, the city flaunted its internationalism, exemplified by its easy (if rather tardy) handling of the Renaissance 'palace' style in the grandly ornamented, 17th-century **George Heriot's School**, south of the Royal Mile. Churches were regularly built along the Royal Mile, among them John Mylne's handsome **Tron Kirk** (1647; see p54) and the Dutch-looking **Canongate Kirk** (1691; see p64), with its delicate, curving gables.

TIME FOR A MAKEOVER

The 1707 Act of Union with England provoked an identity crisis for Edinburgh, and some dubbed the city 'a widowed metropolis'. But the capital, not given to extended periods of mourning, soon came to see architecture as an essential way of asserting its character.

The collapse in 1751 of a Royal Mile tenement highlighted the run-down state of the Old Town and the need for 'modern' living quarters. George Drummond, the city's Lord Provost, drew up proposals the following year to expand Edinburgh, creating the grandiose Exchange (now the **City Chambers**; see p54) on the Royal Mile and, in the 1760s, the original **North Bridge** (see p54). The bridge, the first to cross Nor' Loch, offered easy access towards Leith and, importantly, to a swathe of redundant land north of the Old Town. This was to become the site of the New Town, created, stage by stage, from 1767.

Conceived as Edinburgh's 'civilised' face, the first part of the **New Town**, designed in 1766 by James Craig, was built to a regimented layout with proportion and grandeur as its hallmarks. One leading practitioner of this new style was Robert Adam, who designed **Charlotte Square** (see p86) as a grand full-stop to the west end of George Street. **General Register House** (1789; see p80), is another example of Adam's well-mannered classicism, its cupolas and pedimented portico a gracious retort to the haphazard gables of the Old Town.

IN CONTEXT

Queensberry House.

'Edinburgh liked the idea of being an intellectual "Athenian" metropolis, especially compared with the imperial "Roman" English capital of London.'

CLASSICAL REINVENTION

By the early 19th century, architecture had taken on an increasingly crucial role in expressing the city's newly cultivated identity. Edinburgh had been dubbed the 'Athens of the North' as early as 1762; though the nickname still raises eyebrows, the city's topography made the analogy plausible. What's more, Edinburgh liked the idea of being an intellectual 'Athenian' metropolis, especially compared with the imperial 'Roman' English capital of London. As the Scottish Enlightenment held

sway, architect William Playfair provided a stone and mortar representation of Calton Hill's status as Edinburgh's Acropolis with his **City Observatory** (1822; see p120), a cruciform mini-temple capped by a dome.

The Observatory stands next to another Playfair construction. Begun in 1826 to commemorate the Napoleonic Wars, the Parthenon-inspired **National Monument** (see p119) is built around 12 huge columns, set on a vast stepped plinth in an attempt at classical allusions. However, a funding crisis meant the structure was never completed; ever since, it's laboured under the nickname 'Edinburgh's Disgrace'. The monument later formed a visual link to Thomas Hamilton's **Royal High School** (1829) on the lower slopes of Calton Hill. Described as the 'noblest monument of the Scottish Greek revival', the structure was neoclassicism at its most authoritative.

Playfair's work can be seen elsewhere in the city. Even after the funding debacle that put paid to his plans for the National Monument, he remained a busy man, producing further classical expression in the forms of the **Royal Scottish Academy** (1826; see p81) and the **Scottish National Gallery** (1859; see p88), a monumental, temple-inspired duo on the Mound, parading an army of columns and classical trimmings.

SCOTTISH BARONIAL: TRUE LIES

George IV visited Edinburgh in 1822 at the behest of Sir Walter Scott, campaigner for the 'tartanisation' of Scotland. Scott's campaign bore fruit in brick and stone when the 1827 Improvement Act advised that new buildings and those in need of a facelift should adopt the 'Old Scot' style. Turrets, crenellations and crow-stepped gables elbowed their way back into the city's architecture; **Cockburn Street** (see p54) is a determined example of the style.

Elsewhere, several new public buildings masqueraded as rural piles airlifted from the Scottish Highlands. The old **Royal Infirmary** (1879; see p138), to the south of the Royal Mile, sports a central clock tower and an array of turrets. **Fettes College** (1870), north of the New Town, is an exuberant intermarriage of Highland baronial seat and French chateau. Its construction allegedly inspired JK Rowling when she was dreaming up Hogwarts.

This growing adventurousness included a degree of architectural promiscuity. The

Fettes College.

city's well-off institutions showed confident but sometimes florid excess, with a pick-and-mix approach to building style. The **Bank of Scotland** building, grandly posed on the precipice of the Mound, was given a neo-Baroque makeover in the years to 1878. The **British Linen Bank** on St Andrew Square, now owned by Bank of Scotland, instead opted for the Renaissance palazzo look, its Corinthian columns topped by six colossal statues.

The Gothic revival also made its mark. Augustus Pugin, the master of the decorated pinnacle and soaring spire, designed the Highland Tolbooth (1845) below Castle Esplanade; today, it's the **Hub** (see p58), headquarters of the Edinburgh International Festival. But the finest line in romantic Gothic came in the shape of George Meikle Kemp's **Scott Monument** (1844; see p89), a fitting memorial to the man who plundered Scotland's past to reinvent its future.

PRETTY VACANT

Little disturbed by industrialisation, late 19th-century Edinburgh saw no huge bursts of construction; in the 20th century, the impetus

---

WELCOME TO NEW WAVERLEY

The latest Old Town architectural controversy is hardly novel.

In January 2014, the City of Edinburgh Council's planning committee gave the green light to a major Old Town development, but only passed the proposal by eight votes to six. There was booing from the public gallery.

A matter of weeks later, a group of prominent Scottish authors including Candia McWilliam, Alexander McCall Smith and Irvine Welsh circulated a highly critical letter, describing the scheme as a 'massive, stale, sterile, modernist confection', while a conservation body decried the plan as a 'sad collection of concrete boxes with veneers of architectural motif'. Dissent made no difference; work went ahead.

A major part of the site, along New Street – between the Canongate and Calton Road – was once home to an enormous bus garage. In the development hubris of the pre-credit crunch years, property developers viewed it in much the same way that a raptor regards a baby vole. The initial plan, dubbed Caltongate, called for the garage to be demolished, along with two blocks of occupied tenements on the Canongate, to make way for new apartments, a conference centre, a hotel and more.

This caused a storm, not least among the Canongate residents who would be displaced. All the same, the garage was levelled in 2006 and most of the site cleared; then the credit crunch intervened. The developer went bust and everything stopped. In the years that followed, the site remained empty except for a few weeks every year in the festival month of August, when Snoozebox planted a temporary hotel there. Eventually, another developer came along with an even grander plan. It involves apartments, a civic square, three hotels, office space and shops around Calton Road, the Canongate, East Market Street and New Street. Distancing itself from its failed predecessor, the scheme is named New Waverley for its proximity to Waverley Station and the associations with Sir Walter Scott. This was the proposal simultaneously approved by the council and roundly condemned by many others in 2014.

The problem, as ever, is squaring the circle of the Old Town's medieval roots, its UNESCO World Heritage Site status and contemporary architecture. With large-scale commercial projects there is justifiable concern that budgets and timetables will produce functional, flat-topped buildings in an area with genuine medieval survivors, baronial tenements and a skyline of crow-stepped gables and turrets. Even the idea of a civic square is fairly alien to the Old Town, where side streets, narrow closes and pends are the order of the day.

Work started on New Waverley in late 2014 and will be completed in stages; New Waverley Square and the hotels should arrive by early 2016, the rest of the scheme thereafter. Watch this space for Edinburgh's response.

IN CONTEXT

Western Harbour.

to build was further anaesthetised by two world wars. Clean-cut 1930s modernism made scant impression, save for **St Andrew's House** (1939; see p118) on the lower reaches of Calton Hill. Designed by Thomas Tait, with an imposing, symmetrical façade, it's a true, authoritarian heavyweight.

With upwardly mobile residents siphoned off to the New Town, a large part of the Old Town had, by the Victorian era, developed into an overcrowded slum. As early as 1892, influential urban planner Sir Patrick Geddes, who inspired the revamp of Ramsey Gardens just below Castle Esplanade, had proposed seeding the area with members of the university as a means of adding to its intellectual weight, but his plan was not adopted. Instead, in the years before and after World War II, Edinburgh's residents were encouraged to decamp to a series of council-built satellite townships on the periphery of the city. This social engineering was a crude mirror of the earlier and socially exclusive New Town. Through it all, the fate of the crumbling Old Town remained in the balance.

In 1949, as part of a scheme drawn up by town planner Patrick Abercrombie, slum tenements and much of the grander George Square were demolished to create space for a new university campus. The sacrifice of George Square sent a rallying call to the preservation troops, and much of the Old Town was saved as a result. But other parts of the city suffered through explosions of 1960s brutalism. The subsequent backlash sent city planners retreating into an ultra-cautious approach. Accusations of architectural timidity reached their height in 1989, when a prime redundant site on the Royal Mile was given over to the Scandic Crown Hotel (now the **Radisson**

Blu Hotel; see p247), constructed in a Disneyfied, imitation-Old Town style.

A few modern buildings did eventually sneak past the planning department, most notably around the Exchange, the city's financial quarter to the west side of Lothian Road. Sir Terry Farrell's **Edinburgh International Conference Centre** (1995; see p146) on Morrison Street forms the nucleus of the area. The central parabolic sweep of the **Scottish Widows Building** (1998; see p146) on Morrison Street offers the only bold design gesture in the whole precinct.

HURRAH FOR HOLYROOD

The construction of the **Scottish Parliament** at Holyrood (see p67) was a textbook example of poor project management. The building was first discussed in 1997, although its original budget of £40 million seemed to have been plucked from thin air by a civil servant. Two years later, there was a site, a plan, an architect – celebrated Catalan Enric Miralles – and an apparently more realistic budget of £103 million. Partly inspired by the hulls of some small upturned boats he saw in Northumberland, Miralles's initial vision was far from conservative, but we'll never know how it would have developed under his watch: he died in 2000, aged just 44. The political driving force behind the project, Scottish First Minister Donald Dewar, himself passed away little more than three months later.

As the early years of the 21st century ticked by, the building became a national joke, and any potential merit was eclipsed by more pressing concerns: when would it open, and how much would it cost? The answers were autumn 2004 and £414.4 million, more than ten times the original budget.

Up close, the building's exterior can seem fussy and overly detailed, but it's impressive at a distance; take a walk up the Radical Road, under Salisbury Crags in Holyrood Park, and you'll get a much better sense of what Miralles was trying to achieve. Inside, themes of transparency and accountability are carried through in the design, while the main chamber combines the feel of a television studio with an achingly modern church and even has elements suggesting sailboat rigging. It's now more than a decade since the building opened, and outrage about its cost has subsided. More people have seen and been impressed by its interior, and – whisper it – some Scots are quite proud of the boldness and the philosophy behind the design.

Familiarity failing to breed contempt, as per the Parliament, also holds for Michael Hopkins's adjacent **Our Dynamic Earth** (see p66) , a huge tent-like structure completed in 1999. Despite the recent addition of student accommodation on Holyrood Road, the only other major local building of interest is the **Tun** (2002), Allan Murray Architects' reinvention of a former brewery building. That said, the once-neglected **Dumbiedykes** housing estate to the south side of the road did get a facelift between 2003 and 2008.

PERIPHERY AND CENTRE

Despite developments (and development failures) around the city centre, Edinburgh's main architectural focus – from the 1990s to the credit crunch – was its coastline. When Forth Ports was privatised in 1992, it started to consider new commercial uses for its derelict land; the result was residential development, the enormous **Scottish Government** building at Victoria Quay in Leith (1996) and the nearby **Ocean Terminal** shopping mall (2001; see p158). However, the latter pair were mere dots on the horizon compared to the grand plan once mooted for the waterfront, which called for thousands of homes and all kinds of commercial and retail properties between Granton in the west and Leith in the east. Although there's now no shortage of contemporary apartment complexes in the area, the recession saw that plan shelved. This is dramatised neatly on the reclaimed land to the seaward side of the Western Harbour between Leith and Newhaven. Home to a row of three plush,

contemporary developments (**Platinum Point**, **Western Harbour** and the **Element**), this stretch was supposed to form the centre of a whole new suburb rising from the sea. Since work stopped, it looks more like a trio of space arks that crash-landed on a mudbank.

Back in the city centre, the long-lived gap site created by the 2002 fire at South Bridge and the Cowgate was filled by the **SoCo** development (completed in 2014), featuring a supermarket, hotel, coffee shop and more. Given the time it took to get functional buildings back in situ, locals were simply grateful that the project didn't drag on for longer. Meanwhile, the controversy surrounding the **New Waverley** development is of a completely different order (see p231 **Welcome to New Waverley**); its first stage should be finished by 2016. **Quartermile** has also been a big name in recent years, an extensive, orthogonal complex containing 900 apartments and office space at the site of the old Royal Infirmary between Lauriston Place and the Meadows. Much of it has been let or sold, but it's not scheduled for completion until 2018.

Although new buildings with character can be found around the Old and New Towns, they tend to be discreet, as if the city retains strong memories of the planning decisions that blighted Edinburgh in the 1960s. The **Scottish Poetry Library** (1999; see p67) off the Canongate, and lofty apartments on **Old Fishmarket Close** (2004) are two subtle success stories, the **Chapel of St Albert the Great** (2012) off George Square Lane is simple and beautiful, while the **Scottish Storytelling Centre** (2006; see p67) on the Royal Mile is a neat conversion.

A current city-centre development that definitely has announced itself is **Edinburgh St James**, by the east end of Princes Street. Old tenements and other buildings were demolished from 1965 onwards to make way for government offices, a hotel and a shopping mall that were collectively regarded as a plug-ugly disaster for decades. The new scheme will spend £850 million, razing these buildings, then, over four levels, creating up to 250 homes, a six-screen cinema, a five-star hotel and lots and lots of shopping. On the grounds that nothing could be worse than what's already there, locals are quite looking forward to seeing how it all turns out. Barring problems, that should happen by 2020.

IN CONTEXT

Literary Edinburgh

The fiction spawned by the Scottish capital has traversed both the grandeur and the gutter with equal relish. Some of Edinburgh's scribes have hurled themselves into the city's netherworlds with rare and committed abandon; take the truculent patter of Irvine Welsh's *Trainspotting*, for instance. Others, though, have sought to cover up the malaise of poverty and marginalisation with the kind of quiet civility for which the city has long been renowned.

Alongside the idealised myth of the Scottish literary tradition has run a simultaneous eagerness to avoid discussing some of the city's harsher realities. For more than 150 years, Sir Walter Scott and the poet Robert Burns have remained the stalwart figureheads of the romantic and tourist-friendly tropes of Scottishness, shot through with lilac hues of heather and heroic scenery. The highbrow status of literature in Edinburgh rests upon the recollections of a decidedly selective memory.

Sir Walter Scott.

EARLY PIONEERS

During the late 18th century and into the 19th, Edinburgh's bookselling and publishing industries rivalled those of London, as the city busily preened itself in the wake of the success of its biggest national export, the Scottish Enlightenment. Yet the city responsible for publishing the first *Encyclopaedia Britannica* was by no means one big culture club. The smart drawing rooms of the New Town, favoured by those connected to Edinburgh's publishing industries, stood in stark contrast to the dank closes and squalid alleys of the working-class and still-shambolic Old Town.

The most influential literary figure of this period was **Sir Walter Scott**, who combined a soaraway legal career with a prolific sideline as a writer. Scott's blockbuster poems and novels introduced readers to the hitherto neglected landscape and heritage of the Scottish Borders and Highlands: epic poems such as *The Lay of the Last Minstrel* and *The Lady of the Lake*, and novels such as *The Heart of Midlothian* and *Rob Roy*, communicated the writer's deep knowledge of and love for his Scottish heritage to the literary world of the early 19th century.

One writer who adapted well to the two-faced nature of Edinburgh in the late 18th century was poet **Robert Burns**, a regular visitor. On the one hand, he enjoyed charming polite society in opulent Georgian villas following the success of his Kilmarnock poems in 1786; on the other, he wasn't averse to indulging in the licentiousness of the public inns. But it was left to **Robert Louis Stevenson** to pass comment on the city's curious duplicity in *The Strange Case of Doctor Jekyll and Mr Hyde*. Although the book is purportedly set in London, the topography it depicts is undeniably that of Edinburgh, with its veneer of bourgeois propriety casually bordering on impoverished slums.

Jekyll and Hyde scavenges its plot from the tale of Deacon Brodie, a wealthy Edinburgh cabinet-maker by day but a thief by night, who ended up hanged on gallows he himself had designed. However, the tale also recalls the infamous Edinburgh murderers William Burke and William Hare, who suffocated the incapacitated and elderly before selling their bodies as specimens for use on the university's dissecting tables (*see p223*). Against such a coarse backdrop of real-life violence, Jekyll and Hyde's Manichean tension between enlightened science and inexplicable savagery smacks of wry local satire, with a fitting taste for the Gothic that remains alive in the city's literature even today.

ARRIVALS AND DEPARTURES

Like many Edinburghers during the 19th century, Stevenson had a love-hate relationship with the city, finally abandoning its keen winds for the sunnier climes of

Western Samoa in the 1890s. In the 20th century, national treasure **Muriel Spark** also left the capital, although not before taking inspiration from James Gillespie's High School, her alma mater, while devising the setting for her 1961 novel *The Prime of Miss Jean Brodie*. Spark's old school still sits near the leafy surrounds of Morningside, but the book's shrewd take on Edinburgh's tempestuous religious history and schism between Calvinism and Catholicism takes its backdrop from the Tolbooth, St Giles Cathedral and, of course, John Knox House on the High Street. The book's title character even mentions her own lineage back to her cabinet-making namesake.

Edinburgh's uncommon literary talents and its even rarer topography have also pricked the interest of visiting writers. **Daniel Defoe**, **George Eliot** and **William Makepeace Thackeray** all spent time here, while **Mary Shelley** gave the city a rather unflattering cameo in *Frankenstein*. Although **William Wordsworth** dropped in on Sir Walter Scott, the *Edinburgh Review* of the time had little praise for the emerging poet. Wordsworth and his sister Dorothy spent a couple of nights at the Grassmarket's White Hart Inn; **Thomas de Quincey**, author of the celebrated *Confessions of an English Opium Eater*, went one better and moved here from London in 1826. Residing in a variety of addresses across town, partly in an effort to evade his creditors, the essayist died in the city in 1859.

NATIONAL PRIDE

The city's shape-shifting literary style is due in part to the problematic negotiations of its many identities, since it is a tradition unevenly built upon three languages: English, Gaelic and Scots. From the Union of the Crowns in 1603 and the subsequent shift in the dynamics of power between Scotland and London, through to the failure to secure devolution in 1979, the subsequent success

Ken Stott as Rebus.

IN CONTEXT

professor, Dr Joseph Bell, as he went about devising the character of Sherlock Holmes. More recently, **Alexander McCall Smith**, author of the *No.1 Ladies' Detective Agency* series, studied law at the university and went on to have a distinguished legal career. Even after shifting millions of books on both sides of the Atlantic, he continued teaching until 2005; he is now emeritus professor of medical law at the university.

THE MODERN WORLD

Despite McCall Smith's success, the city's two most notable contemporary novelists are probably **JK Rowling** and **Ian Rankin**, a pair of writers who embrace the city's taste for the Gothic in very different ways. Rowling's Harry Potter books have done much to colour the imaginations of youngsters attending Fettes College, on which Hogwarts may or may not have been partly modelled, but has also provided a not unwelcome fillip to the city's café culture. Rowling, famously, began sketching out her books in local coffeehouses; unsurprisingly, a host of Old Town establishments have since laid claim to being the birthplace of her first novel, *Harry Potter and the Philosopher's Stone*.

Rankin's series of Rebus books exposes the tartan noir of the town and its less touristy vistas, as his characters pass through Edinburgh's brothels, bars and banks with a world-weary demeanour. Rankin, born in Fife but resident in Edinburgh since his days at the university, has built a veritable industry from unassuming beginnings: his crime novels have been translated into 22 languages and adapted for TV, and form the basis for online quizzes, fan gatherings and tours of the city.

Rebus officially retired in *Exit Music* (2007), but has reappeared in a couple of novels since then, alongside Malcolm Fox who was introduced in *The Complaints* (2009). Fox is a detective who investigates other police officers, allowing the writer to look at ethics and tensions within the police force, even at the past and present behaviour of Rebus himself.

in 1997 then the vote against full independence in 2014, the literature of the city has moved with the often volatile rise and ebb of self-determination.

Scottish literature also has enduring links with the legal world, with a preponderance of lawyers and judges among its leading lights. **James Boswell**, **Sir Walter Scott**, **Robert Louis Stevenson**, **Henry Cockburn** and **Francis Jeffrey**, the first editor of the *Edinburgh Review*, were all scholars of law before they were writers. Even the *Review*'s motto carried judicial weight: '*judex damnatur ubi nocens absolvitur*', or 'the judge is condemned when the guilty is acquitted'. The *Review* was one of the most influential magazines of the 19th century before closing in 1929; more than 50 years after its demise, its name was recycled to launch a new publication, which continues to print the work of authors local, national and international.

The University of Edinburgh has educated many notable writers. Among them are **JM Barrie**, best known for his ever-young creation Peter Pan, and Edinburgh-born **Sir Arthur Conan Doyle**, who took inspiration from his

'This is a wonderful novel'
Melvyn Bragg, THE OBSERVER

J.K. ROWLING
THE CASUAL VACANCY

TIME MAGAZINE

SCOTLAND'S SCI-FI CAPITAL

Space is the place in Edinburgh.

Science fiction still has an image problem, but that's probably just how the science fiction community likes it. While the highbrow literary world gets on with its art, sci-fi is looked down upon as a wretched hive of *Star Wars* fans, fantasists and people who wear black T-shirts of some antiquity. It's interesting, then, that Edinburgh, with its rich heritage of mainstream novelists and poets, has become something of a sci-fi hub in recent years, largely thanks to the efforts of three writers.

The most prominent is certainly the late **Iain M Banks** who lived close to the city. His death in summer 2013 was a huge loss, not just to Scottish literature but also to Scotland. It prompted a heartfelt sadness among readers of both his conventional fiction (written under the name Iain Banks, with no middle initial) and his epic, imaginative sci-fi. His Culture series of novels, published between 1987 and 2012, were set in a tolerant, communitarian society – rooted in his own left-leaning politics – where he could be playful or devastating in turn, making wry jokes with the names of spaceships, then weaving stark issues of morality and choice into the narrative.

Another local sci-fi author, and a close friend of Banks, is **Ken MacLeod**, whose fiction is informed by his Western Isles heritage, former left-wing activism and background in the IT industry and biomechanics. He came to public attention with his Fall Revolution series, four books with a distinctly political voice, published from 1995 to 1999. Edinburgh and environs have featured as backdrops to his stories, particularly *The Execution Channel* (2007), *The Night Sessions* (2008) and *The Restoration Game* (2010).

The third in the triumvirate is **Charles Stross**, originally from Yorkshire but a long-time Edinburgh resident. A writer with a Stakhanovite output, he announced himself when six of his novels hit the shelves between 2003 and 2005 – and the pace has hardly let up since. Although he has a number of standalone titles to his name, and a densely comic collaboration with Cory Doctorow (*The Rapture of the Nerds*, 2012), Stross is perhaps best known for his Laundry Files series (featuring an occult branch of the British Secret Service and a hero called Bob) and the Merchant Princes series (universe-hopping fantasy). His books set in a near-future Edinburgh include *Halting State* (2007) and *Rule 34* (2011).

WHERE TO START
Iain M Banks *Consider Phlebas* (1987)
The first full-length Culture novel.
Ken MacLeod *The Night Sessions* (2008)
Artificial intelligence, ethics, murder and Presbyterianism in the Scottish capital a couple of decades from now.
Charles Stross: *Accelerando* (2003)
Humanity experiences the singularity – artificial intelligence accelerated to God-like status – and Stross teases out the implications.

IN CONTEXT

Iain M Banks.

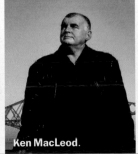

Ken MacLeod.

Charles Stross.

Drama aside, one delight shared by both Rebus and his maker is the Oxford Bar on Young Street in the New Town. A literary landmark since the days of Sydney Goodsir Smith, who was a regular during the 1950s, the bar has long been a favourite both with Rankin and his most famous invention.

The late **Iain Banks** lived just outside Edinburgh, but the city and its environs regularly crop up in his work or even inspire it: *The Bridge*, *Complicity* and *Whit* all carry with them shades of the Scottish capital. Banks may not have enjoyed the blockbuster sales of Rowling (or even, for that matter, Rankin), but his novels and epic space operas (*see p237* **Scotland's Sci-Fi Capital**) have carved him a special place in the local literary pantheon.

However, the novelist who most radically changed the outside world's view of contemporary Scots literature has been Leith-born **Irvine Welsh**, another in a long line of writers keen to expose an urban underbelly that the city fathers would rather conceal. In novels such as *The Acid House* and *Trainspotting*, Welsh writes of a town guilty of self-gentrification and of

pushing its predominantly working-class inhabitants into an outer ring of 'problem' housing schemes. A far cry from Muriel Spark's decorous aspersions, lovingly voiced in a clipped Morningside accent, Welsh's Edinburghers are afflicted by AIDS, drug addiction and privation, and bawl their complaints in the legendarily impenetrable local vernacular.

The prose of Rankin, Rowling, Scott, Spark and even Welsh earned Edinburgh the title of UNESCO's first City of Literature in 2004, while the annual presence of a major international book festival (*see p37*) on the city's cultural calendar has been of further benefit to the boom of literary tourism in the capital. With an abundance of endearingly haphazard second-hand bookshops, numerous bars bearing the names of famous local novels and authors, literary pub crawls around the Old Town and even a museum devoted to local scribes (Writers' Museum; *see p59*), the city has leapt at the chance to articulate and exploit its literary flair. It's to be hoped that the writers the city celebrates continue to expose the corners it doesn't want you to see.

Edinburgh International Book Festival.

The Brains Trust

Start with the statues. On the Royal Mile by the High Court sits David Hume, cast as a Roman senator. Opposite the City Chambers is Adam Smith, upright in his university gown. Down the hill, outside the Canongate Kirk, stands a youthful Robert Fergusson.

There are others. The corner of the National Portrait Gallery yields three more: artist Henry Raeburn, with paint palette; surgeon John Hunter, with skull; and geologist James Hutton, with lump of rock. On a high plinth in West Princes Street Gardens, Allan Ramsay wears a silk nightcap. And then there's a seated Sir Walter Scott (with dog) in the Scott Monument, the world's biggest memorial to a writer.

What connects these men is the Scottish Enlightenment. During the 18th century, Edinburgh set the cultural pace across Europe for cutting-edge artistic, philosophical and scientific thought. According to academic Arthur Herman, writing in *The Scottish Enlightenment* (2001), their intellectual adventurousness created the modern world. We still view our environment and ourselves through their framework.

SO WHY EDINBURGH?

The reasons for this flowering are hard to establish. Perhaps it was the less-than-equal union between Scotland and England in 1707 that led Edinburgh intellectuals to look to Europe, where Enlightenment ideas were taking hold. Perhaps it was the circulation of people passing through on their way to the Americas, bringing with them new ways of thinking. Perhaps it was the unusually high levels of literacy resulting from the free Scottish education system. Or perhaps it was a mere historical fluke that so many gifted people arrived here in quick succession.

Whatever the explanation, Edinburgh epitomises the dramatic shift from an era of religious doctrine, subservience and superstition to an age of individual will, rational thought and reason. This approach laid the ground rules for a secular society and governs the western mindset to this day. The significance of the men now enshrined in the city's statues is not only in the brilliance of their discoveries, novels, paintings, poetry and theories, but also in their embracing of a radical vision of human society.

Edinburgh cannot take all the credit. It was the German philosopher Immanuel Kant who coined the term 'Enlightenment', and the movement would have been nothing without the contributions of French intellectuals such as Denis Diderot, Jean-Jacques Rousseau and François Voltaire. It was Voltaire, however, who acknowledged Edinburgh's influence;

Allan Ramsay.

'We look to Scotland,' he said, 'for all our ideas of civilisation.'

Thanks to advances in science and the cool-headed logic of the scientific method, this was the era when people realised the world was best explained not by blind faith but empirical observation. Such a view encouraged a spirit of questioning that threatened the old orthodoxies, not least that of the Church. In place of faith came rational thought; in place of belief in the afterlife came the pursuit of happiness in this world. No wonder accusations of heresy abounded.

In this way, the movement affected all aspects of civilisation, from medicine to religion, from philosophy to literature, each development emboldening the other. As a result, by the following century, it had become possible for Charles Darwin to put forward an idea as iconoclastic as the theory of evolution by natural selection. It was, incidentally, the geological work of James Hutton laid out in *Theory of the Earth* (1785) and presented to the Royal Society of Edinburgh that helped set Darwin on his pioneering path. Hutton and the other figures who came to prominence in 18th-century Edinburgh represented the whole spectrum of human endeavour, their achievements embracing everything from economics to poetry.

THE LEADING FIGURES

Born in the city in 1711, **David Hume** was the pre-eminent figure in the Scottish Enlightenment, versed in philosophy, history, economics and diplomacy. His first published work, *A Treatise of Human Nature* (1739), had the subtitle 'an attempt to introduce the experimental method of reasoning into moral subjects', which gives a clear indication of how the different academic disciplines were influencing each other. His atheism stood in the way of a university career, and it was only posthumously that his *Dialogues Concerning Natural Religion* was published. However, his ideas were debated vigorously in France at the time, and his influence was still felt among empiricist philosophers in the 20th century.

It was Hume's *Essays Moral and Political* that set the agenda for the science of economics, of which his friend the Kirkcaldy-born **Adam Smith** was a master. Linked to the key Enlightenment thinkers of the day, Smith taught logic and moral philosophy at the University of Glasgow. In his book *An Inquiry into the Nature and Causes of the Wealth of Nations*, published in 1776, he set out the idea that economic growth was fuelled by the division of labour and made observations about the free market economy.

Perhaps the most fitting memorial to the Enlightenment... is not the collection of statues... but the chaotic cultural bonanza of the city's annual festivals.

In his monument to Smith on the Royal Mile (not far from his grave in the Canongate Kirkyard), sculptor Alexander Stoddart kept one of Smith's hands hidden, in reference to the economist's description of the 'invisible hand' that seemed to guide the economy in a way that made everyone better off. Although Smith was championed by free marketeers in the 1980s, he was also a moralist, one who was writing decades before the word 'capitalism' was even coined, and would have been shocked by the levels of rampant greed encouraged by the laissez-faire economists who claim his name.

His thoughts on ethics did not go unnoticed. According to the academic Robert Crawford, **Robert Burns** produced a 'straight versification of something in Adam Smith' when he wrote: 'O wad some Pow'r the giftie gie us/To see oursels as others see us.' Contrary to his image of an unschooled poet, Burns was a keen reader of Smith's *The Theory of Moral Sentiments*, published in 1759, the year of Burns's birth. Although more closely associated with Alloway and Dumfries, Burns spent time in Edinburgh in the 1780s, and was toasted by the literary establishment as a major Enlightenment figure.

The poet **Robert Fergusson** was another influence on Burns. He died tragically young, succumbing in 1784 at the age of 24, but not before making significant inroads in the writing of vernacular Scots. In turn, Fergusson had been influenced by **Allan Ramsay**, author of *The Gentle Shepherd* (1725) and a leading exponent of the Scots language. Emerging from the literary scene created by these writers, **Sir Walter Scott**, born in 1771, went on to popularise the historical novel, which he

effectively invented, and established an international career unparalleled in his day.

To these names can be added those of architect **Robert Adam**, whose works include General Register House; **James Boswell**, a noted diarist and the biographer of Samuel Johnson; philosophers **Francis Hutcheson** and **Thomas Reid**; scientists **John Playfair**, **William Cullen** and **Joseph Black**; **James Hutton**, who essentially invented modern geology; historian **Adam Ferguson**, the father of modern sociology; and economist **James Anderson**. Many were painted by the celebrated portraitist **Henry Raeburn**, himself an Enlightenment figure. And to underline the point, Edinburgh was where the first editions of the *Encyclopaedia Britannica* were published, between 1768 and 1771.

Perhaps the most fitting memorial to the Enlightenment, which ended two centuries ago, is not the collection of statues commemorating its leading figures, but the chaotic cultural bonanza of the city's annual festivals. In some wayward fashion, August in Edinburgh is a perfect tribute to the intellectual curiosity that defined the most vivacious intellectual shindig in history.

James Hutton.

Essential Information

Hotels

In Edinburgh, you can bunk with backpackers on the Royal Mile, share a modern apartment with friends, bed down in lush rooms attached to a couple of the city's finest restaurants and even go camping. But it's the hotels in Scotland's capital that really add a sense of occasion, with their character and their history. At the Balmoral, the grande dame with the clock tower at the east end of Princes Street, JK Rowling hid away to finish the last-ever Harry Potter novel. At the other end of Princes Street, the main dining room at the Waldorf Astoria Edinburgh – The Caledonian was created nearly a century ago and is a decorative fantasy, in late Baroque style, named for the mistress of Louis XV of France. Prestonfield in South Edinburgh is a refurbishment of an authentic 17th-century mansion, while the Edwardian-era building in the Old Town occupied by the Scotsman hotel was home to the daily newspaper of the same name for most of the 20th century. And if these prove to be out of your price range? There are now more budget hotels in the city than ever before.

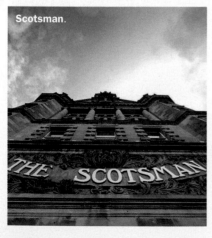

Scotsman.

OUR LISTINGS

Hotels in this guide are classified according to the advertised price of a basic double room. You can expect to pay over £210 a night for hotels in the **Deluxe** category, £140-£210 for **Expensive** hotels, £70-£140 a night for **Moderate** accommodation and under £70 for **Budget**.

It's important to emphasise that these prices are only a guideline, however: the variation in an individual hotel's tariff over the course of the year can be staggering. It can even change dramatically over the course of a week. Book a budget hotel sufficiently in advance and you could get a double room for the price of two hostel bunks, for example. But in August that same budget hotel can magically transform into a moderately priced property. Meanwhile, walk up to the reception of a deluxe property at that time of year and £600 for a double isn't unusual; opt for a suite and the bill for just one night could head into four figures.

Edinburgh is the most expensive place to stay in Scotland. and it's not just busy during the festivals frenzy. Average annual occupancy rates have crept towards the 80 per cent mark in recent years, thanks to near-constant business travel, the city's capital status and year-round tourism. This average does disguise an annual peak and trough, though. August sees more than 90 per cent of rooms taken by travellers, with rates spiralling upwards to match supply with demand. Other major events such as Hogmanay or big rugby matches also put pressure on availability and prices. Conversely, during the dog days of winter, before Christmas and after New Year, trade slackens off and well over a third of rooms are empty. At times like these, there are definitely bargains to be had.

If you have the ability to book ahead, do so. Several months in advance, discounts can be high, especially for the off-season period. It's also worth checking consolidator sites such as www.agoda.com, www.expedia.co.uk, www.hotels.com and www.lastminute.com, which often offer good rates, but be sure to check the small print. And there is always **Airbnb** (*see p260* **Stay With Me**).

If you do arrive with nothing pre-booked, the Edinburgh bureaux of the national tourist agency, **VisitScotland**, run an accommodation service and can usually find you something. It has a desk at Edinburgh Airport and an office above Princes Mall at the east end of Princes Street (for both, *see p270*).

G&V Royal Mile. *See p247.*

THE OLD TOWN
Deluxe

Scotsman
20 North Bridge, EH1 1YT (556 5565, www.thescotsmanhotel.co.uk). Nicolson Street–North Bridge buses. **Map** p284 H7.
Filling the lofty, baronial premises vacated by the *Scotsman* newspaper, this hotel leapt straight into the city's upper echelon of places to stay when it opened back in 2001, although it may be showing its age a little now. The ornate North Bridge Brasserie (*see p61*) is the hotel restaurant, while the Spa & Healthclub has Edinburgh's most space-age swimming pool. The rooms have contemporary elements but remain in keeping with the building's Edwardian-era origins.

★ The Witchery by the Castle
352 Castlehill, EH1 2NF (225 5613, www.thewitchery.com). Bus 23, 27, 41, 42, 67. **Map** p284 G7.
The Witchery is not a hotel but a collection of nine preposterously gorgeous suites, some with kitchens, hidden away in two buildings at the top of the Royal Mile. Each suite is dark, theatrically gothic and lavishly furnished with antiques, leather and velvet,

Victorian baths and grand beds, complementing the magical atmosphere at the associated Witchery restaurant (*see p61*). Impure indulgence.

Expensive

★ G&V Royal Mile

1 George IV Bridge, EH1 1AD (220 6666, www. quorvuscollection.com). Bus 23, 27, 41, 42, 67. **Map** p284 H7.

These new-build premises opened in 2009 as Hotel Missoni, transforming into G&V Royal Mile in 2014. Design remains a big feature: expect bright colours and bold fabrics. Its flagship restaurant, Cucina (*see p59*), is noted for Italian eats, the location couldn't be more central and some rooms come with views across the Old Town rooftops to St Giles. *Photo p245.*

Radisson Blu

80 High Street, EH1 1TH (557 9797, www. radissonblu.co.uk/hotel-edinburgh). Bus 35 or Nicolson Street–North Bridge buses. **Map** p285 J7.

Sympathetic in situ or just historical pastiche? Originally the Scandic Crown, the faux-medieval exterior of this hotel divided opinion when it opened in 1989. With a clean, modern interior it was substantially refurbished in 2005 while another revamp for the guest bedrooms is scheduled for winter 2015/16.

Moderate

Apex International

31-35 Grassmarket, EH1 2HS (0845 365 0000, www.apexhotels.co.uk). Bus 2. **Map** p284 G8.

The Apex chain now operates eight properties in Edinburgh, Dundee and London, but this big Old Town hotel is where it all began back in 1996. Its 169 rooms offer contemporary-style accommodation; the north-facing rooms on the upper floors (some with balconies) offer great views across the Grassmarket to Edinburgh Castle. There's a small pool and gym.

Other locations Apex City, 61 Grassmarket, Old Town, EH11 2JF; Apex Haymarket, 90 Haymarket Terrace, New Town, EH12 5LQ; Apex Waterloo Place (*see p256*).

Carlton

19 North Bridge, EH1 1SD (472 3000, www. thehotelcollection.co.uk). Nicolson Street–North Bridge buses. **Map** p285 J7.

The beast on the Bridges, this large old city centre property has gone through a number of identities in the last decade or so, because of changes in branding and ownership, but has now reverted to the nice, simple Carlton Hotel. It's not in the same class as the nearby Balmoral (*see p249*), but the styling of the public spaces is grand (in a 1980s kind of way), rooms have expansive views and there's a small pool with a sauna and steam room. Good location too; the Royal Mile is round the corner.

Fraser Suites

12-26 St Giles Street, EH1 1PT (221 7200, http:// edinburgh.frasershospitality.com). Bus 23, 27, 41, 42, 67. **Map** p284 H7.

This international company, with properties from Barcelona to Wuxi, opened just off the Royal Mile by

Fraser Suites.

Stay Central.

the High Kirk of St Giles, in 2009. It's sort of a hotel, but not, given that all rooms and suites – done out in classic style – have at least some self-catering facilities. You can choose from a basic room with microwave and fridge, to a suite with a full kitchen and a stunning outlook. The adjacent Broadsheet Bistro restaurant (www.thebroadsheetbistro.co.uk) opens at 7am (8.30am weekends), so you don't even have to cook your own breakfast.

Hotel du Vin
11 Bristo Place, EH1 1EZ (08447 364255, www. hotelduvin.com). Bus 2, 23, 35, 41, 42, 47, 60, 67. **Map** p284 H8.
Edinburgh joined the posse of British towns and cities with a Hotel du Vin in late 2008, and its sheer, well-executed reliability has made it a discreet hit ever since. Jammed ergonomically into what seems like an impossible urban space between Bristo Place and Forrest Road, it has a busy, traditional bistro, a bar, a whisky snug, a courtyard with 'cigar shack', a wine-tasting room and characteristically lush, modern decor in all the bedrooms.

Macdonald Holyrood
81 Holyrood Road, EH8 8AU (0844 879 9028, www. macdonaldhotels.co.uk). Bus 6, 60. **Map** p285 L7.
Opened in 1999 to capitalise on the other developments around Holyrood, chief among them the

Scottish Parliament, the Macdonald has always had a business feel. There's a gym, a small swimming pool and spa treatment rooms. The restaurant, Rocca@ Holyrood, arrived in 2013 and is better than you might expect, while all the rooms were refurbished in 2014.

Stay Central
139 Cowgate, EH1 1JS (622 6801, www.stay central.co.uk). Bus 2 or Nicolson Street–North Bridge buses. **Map** p284 H8.
Once Tailors Hall, then simply the Central, this hotel has long had a reputation as a good-times crash pad adjacent to one of the city's biggest and liveliest pubs, the Three Sisters (622 6802, www. thethreesistersbar.co.uk). The accommodation was completely refurbished and restyled in 2013-14: from the hostel-like reception area to attractive modern rooms ranging in size from small doubles to quads and even up to your own private dormitory (resembling a self-contained cabin with games area). The theme is fun and quirky, from room design to bike and skateboard hire, and that bustling bar next door.

Hostels

Brodies Hostels *93 High Street, EH1 1SG (556 6770, www.brodieshostels.co.uk). Bus 35 or Nicolson Street–North Bridge buses.* **Map** p285 J6.
Shared dormitories; private double rooms.

Budget Backpackers *37-39 Cowgate, EH1 1JR (226 6351, www.budgetbackpackers.com). Bus 2, 23, 27, 35, 41, 42, 45, 47, 60, 67.* **Map** p284 H8. Shared dormitories; female-only dormitories; private twin rooms.

Castle Rock Hostel *15 Johnston Terrace, EH1 2PW (225 9666, www.castlerockedinburgh.com). Bus 2.* **Map** p284 F8. Shared dormitories; private double and quad rooms.

Cowgate Tourist Hostel *96-112 Cowgate, EH1 1JN (226 2153, www.cowgatehostel.com). Bus 2.* **Map** p284 H7. Shared dormitories; private twin and double rooms.

Edinburgh Backpackers Hostel *65 Cockburn Street, EH1 1BU (220 2200, www.hoppo.com/edinburgh). Nicolson Street–North Bridge buses.* **Map** p284 H7. Shared dormitories; private twin, double, triple and quad rooms.

Edinburgh Metro SYHA Hostel *11 Robertson's Close, Cowgate, EH1 1LY (524 2090, www.syha.org.uk). Bus 35 or Nicolson Street–North Bridge buses.* **Map** p285 J7. Single rooms and private apartments of 4-6 rooms. Open July, Aug only.

High Street Hostel *8 Blackfriars Street, EH1 1NE (557 3984, www.highstreethostel.com). Bus 35 or Nicolson Street–North Bridge buses.* **Map** p285 J7. Shared dormitories; private twin and quad rooms.

Royal Mile Backpackers *105 High Street, EH1 1SG (557 6120, www.royalmilebackpackers.com). Bus 35 or Nicolson Street–North Bridge buses.* **Map** p285 J7. Shared and female-only dormitories.

St Christopher's Inns *9-13 Market Street, EH1 1DE (226 1446, www.st-christophers.co.uk). Bus 6, 41, 42, 67.* **Map** p284 H7. Shared and female-only dormitories; private twin rooms.

Smart City Hostels *50 Blackfriars Street, EH1 1NE (524 1989, www.smartcityhostels.com). Bus 35 or Nicolson Street–North Bridge buses.* **Map** p285 J7. Shared and female-only dormitories; private twin, triple, quad and six-person rooms.

THE NEW TOWN

Deluxe

★ Balmoral

1 Princes Street, EH2 2EQ (556 2414, www.roccofortehotels.com). Nicolson Street–North Bridge buses or Princes Street buses. **Map** p284 H6.

With a Michelin-starred restaurant (Number One; *see p93*), an acclaimed spa, the elevated service standards you would expect from a Rocco Forte hotel and a clock tower that has been a defining feature of the Edinburgh skyline for more than a century, the Balmoral pretty much has the lot. The rooms and suites, some of which have classic castle views, factor modern lines and design elements into an Edwardian fabric for an overall package that easily merits five-star status. Other amenities include the Palm Court (great for tea in the afternoon or champagne in the early evening; *see p93*), a brasserie, a chic bar and another bar dedicated to Scotch whisky.

Balmoral.

ESSENTIAL INFORMATION

★ Chester Residence

9 Rothesay Place, EH3 7SL (226 2075, www. chester-residence.com). Bus 19, 37, 41, 43, 47, 113. **Map** p283 C7.

The award-winning Chester Residence isn't exactly a hotel, but it's certainly close enough to merit inclusion here. It's a collection of 23 fully serviced luxury suites, set within four West End Georgian townhouses. It's like staying in your own apartment, but with housekeeping and a 24-hour concierge to take care of you. The suites range in size and price, but all are designer-plush, rather beautiful and less than a ten-minute stroll from Princes Street.

Tigerlily

125 George Street, EH2 4JN (225 5005, www. tigerlilyedinburgh.co.uk). Bus 10, 11, 12, 16, 24, 29, 41, 42, 61 or Princes Street buses. **Map** p284 E6.

Tigerlily really did break the mould when it arrived on the scene in 2006, representing the apogee of demotic chic near the height of an economic bubble when anyone with a credit card could have the kind of experience usually reserved for a reality televison star on the upswing. The bedrooms are designed to the hilt, but somehow the effect isn't overwhelming; this remains one of the best-loved hotels in the city, to say nothing of its sparkly, opulent bar and restaurant. The basement is home to a nightclub called Lulu. The place is best appreciated by a certain demographic, of course; your gran probably wouldn't like it, nor that curmudgeonly uncle.

★ Waldorf Astoria Edinburgh – The Caledonian

Princes Street, EH1 2AB (222 8888, www. waldorfastoria3.hilton.com). Princes Street buses. **Map** p284 E7.

The Caledonian has been around since 1903; it was originally a railway hotel, although the adjacent station was knocked down many years ago. Part of the Hilton group for some time, it was expensively refurbished, rebranded and given its current name in 2012. All the same, there is genuine Edwardian heritage here, appealing and understated room design plus two dining options from celebrated restaurateurs the Galvin brothers (for the Pompadour by Galvin, *see p101*). The Caledonian is big, far from brash and provides the west end of Princes Street with a suitably grand full stop.

Expensive

Le Monde

16 George Street, EH2 2PF (270 3900, www. lemondehotel.co.uk). Bus 6, 10, 11, 12, 16, 23, 26, 27, 41, 42, 44, 44A, 61, 67. **Map** p284 G6.

A themed designer hotel that arrived not long before the financial crisis, Le Monde contains only 18 rooms and junior suites, but all are individual, all reflect the city that gives them their name. The New York, for example, smacks of chic, Manhattan loft living; the Atlantis has a large piece of arty aquamarine glass on one wall for that sub-sea feel. It has a central location by the George Street shops, and

Waldorf Astoria Edinburgh – The Caledonian.

is fun if you're up for it. It's not the greatest place to eat, but there are lots of alternatives nearby.

Nira Caledonia

10 Gloucester Place, EH3 6EF (225 2720, www.nira caledonia.com). Bus 24, 29, 36, 42. **Map** p280 E5.
One hotel occupying two townhouses at 6 and 10 Gloucester Place, this venue was previously known as Christopher North House and Number 10 before transforming into Nira Caledonia in 2012. It has a classic New Town frontage outside, while the interior is tasteful and boutiquey with mod cons; some rooms overlook Gloucester Place, and there are garden suites to the rear. Staff are engaging and highly professional. The restaurant, Blackwood's Bar & Grill, specialises in very good steaks. Part of a select little group of hotels, Nira Caledonia's three sister establishments are in Italy, Mauritius and Switzerland. *Photo p252.*

Rutland Hotel

1-3 Rutland Street, EH1 2AE (229 3402, www.therutlandhotel.com). Princes Street buses. **Map** p284 E7.
This building at the west end of Princes Street is quite the pleasure palace these days. In the basement is Heads & Tales (a gin bar with two stills in situ), on the ground floor is the Huxley (a more workaday, modern bar-café) and upstairs is steak restaurant Kyloe. The Rutland functions as a hotel too, with a dozen boutique rooms, some with views towards the castle, St Cuthbert's, St John's and Princes Street.

Moderate

Bonham

35 Drumsheugh Gardens, EH3 7RN (274 7400, www.thebonham.com). Bus 19, 37, 41, 43, 47, 113. **Map** p283 C7.
Originally three Victorian townhouses, the Bonham has at various points been used for private housing, a medical practice and a hall of residence for female medical students. It opened as a hotel in 1998, with contemporary art, stylish furniture, attractive bedrooms and a decent, spacious restaurant. Nearly two decades on, it has lasted very well indeed. Its local partner hotel is the Edinburgh Residence (*see below*).

Edinburgh Residence

7 Rothesay Terrace, EH3 7RY (274 7403, www. theedinburghresidence.com). Bus 19, 37, 41, 43, 47, 113. **Map** p283 C7.
One of the city's most popular places to stay, the Edinburgh Residence is a collection of large suites stuffed with antique furniture, all with hotel services attached, not a hotel as such. Breakfast is served in your room, as are snacks and sandwiches 24 hours a day, and more formal meals from 6pm to 9pm daily. Aficionados wouldn't stay anywhere else. The Edinburgh Residence is owned by Hapimag, the members-only holiday organisation based in Switzerland, but it's open to all.

George

19-21 George Street, EH2 2PB (225 1251, www.thegeorgehoteledinburgh.co.uk). Bus 6, 10, 11, 12, 16, 23, 26, 27, 41, 42, 44, 44A, 61, 67. **Map** p284 G6.
The core of the George, one of Edinburgh's old stagers, dates from the late 18th century, although the building has only been a hotel since 1881. Owned by Principal Hayley Hotels for just over a decade, the venue benefited from a major refurbishment under its stewardship, completed in 2009. Rooms are all quietly contemporary; some have fantastic views north to the Forth and Fife. The Tempus bar restaurant is modern and slick, but this remains an Edinburgh hotel of some tradition.

Nira Caledonia. *See p251.*

Hotel Indigo
*51-59 York Place, EH1 3JD (556 5577, www.
hiedinburgh.co.uk). Bus 10, 11, 12, 16, 26, 41,
44, 44A.* **Map** p280 F4.
By the York Place tram terminus, just round the
corner from Broughton Street with all its eating and
drinking options, Hotel Indigo is a deceptively large
affair with 60 rooms behind its terraced exterior.
Not all the rooms are big, but they are stylish, modern and fun – with a complimentary minibar (soft
drinks and snacks only). It sets a good standard for
a moderately priced modern hotel in the city. A second branch opened on Princes Street in spring 2015;
check the website for details.

Howard
*34 Great King Street, EH3 6QH (557 3500, www.
thehoward.com). Bus 23, 27, 61.* **Map** p280 F4.
A Georgian New Town fantasy of a hotel, the
Howard occupies three adjacent buildings that
date from the 1820s. You can almost imagine King
George IV fiddling with the television remote while
Mrs Fitzherbert dallies in a voluminous, free-standing
bath. There are butlers to look after you, modern
comforts alongside period fixtures and furniture,

and a discreet dining room, the Atholl. It's as romantic as anything and popular with honeymooners. If
you're not staying, do pop in for afternoon tea.

Old Waverley
*43 Princes Street, EH2 2BY (556 4648,
www.oldwaverley.co.uk). Princes Street buses.*
Map p280 H6.
Dating from 1848, the Old Waverley is one of the
capital's oldest hotels. Although it has undergone a
refurbishment in recent years, it still retains a genteel, old-fashioned look. Some rooms face south, giving rather splendid views: the castle, Old Town and
Scott Monument. If yours doesn't, you could always
try for a window seat in the bar as recompense. It's
not cutting-edge, but you'll step out of the front door
straight on to Princes Street, so the location couldn't
be more central.

Rick's
*55A Frederick Street, EH2 1HL (622 7800, www.
ricksedinburgh.co.uk). Bus 10, 11, 12, 16, 24, 29,
41, 42, 61.* **Map** p284 F6.
Quite the trendsetter when it opened back in 2000,
Rick's has stayed the course with a formula that

sees a bar-restaurant specialising in cocktails occupying the basement space, with ten boutique rooms above. Plenty of hotel interiors look like this now, so Rick's no longer has a unique selling point, but it's always worth checking the website to see what offers are available.

Royal Scots Club

29-30 Abercromby Place, EH3 6QE (556 4270, www.royalscotsclub.com). Bus 23, 27, 61. **Map** p280 G5.

The Royal Scots Club was founded in 1919 as a tribute to those who died in World War I. Nearly a century later, it still exudes a sense of history and tranquillity; the atmosphere may be more akin to an old-fashioned private gentlemen's club than a hotel, but it's open to all. The rooms have been furnished in traditional style, some with four-poster beds, but aren't averse to modern comforts; a few rooms offer views north over the rooftops to the Firth of Forth.

Roxburghe

38 Charlotte Square, EH2 4HQ (527 4678, www.theroxburghe.com). Princes Street buses. **Map** p284 E7.

The Roxburghe is a large operation with an elegant address, no distance at all from the west end of Princes Street. Lurking away behind the classical New Town façade (the building is the work of 18th-century architect Robert Adam) is a small pool, a gym and 199 bedrooms and suites done up in a style not uncommon in the city: the 19th century meets the 21st. It changed hands in 2014 and is now under the same ownership as the George (*see p251*).

Budget

Cityroomz

25A Shandwick Place, EH2 4RG (229 6871, www.cityroomz.com). Princes Street buses. **Map** p283 D7.

Opened as a Travelodge in 2009, this place became Cityroomz in 2013, part of the Sleeperz group that has hotels in Cardiff and Newcastle. It's budget, it's basic, but rooms have all the necessary facilities (safe, hairdryer, iron, television, wi-fi). Effectively, it's Travelodge-style accommodation with slightly funkier decor, given some soul by the brilliant reception staff. It's dead central too, just yards from Princes Street.

Frederick House

42 Frederick Street, EH2 1EX (226 1999, www.frederickhousehotel.com). Bus 10, 11, 12, 16, 24, 29, 41, 42, 61. **Map** p280 H6.

Frederick House is a former office space in an old listed building, less than five minutes' walk from Princes Street. Its conversion to a hotel saw traditional decor throughout; it may be too aesthetically middlebrow for some, but there's little quibbling with the value. The prices, already very decent, get even cheaper out of season; check the website for special offers. There's no dining room or bar, but there are lots of options nearby. Breakfast is served at Rick's (*see p252*) across the street.

Motel One

18-21 Market Street, EH1 1BL (220 0730, www.motel-one.com). Princes Street buses. **Map** p284 H7.

Cityroomz.

ESSENTIAL INFORMATION

A relatively young German budget chain that only appeared in Edinburgh at the end of 2012, this hotel couldn't be handier. Waverley Station is across the street one way, Princes Street Gardens the other. The breakfast buffet is decent, there's a 24-hour residents' bar, the rooms have appealing cool blue touches in the decor and the televisions are all from German manufacturer Loewe. A second property opened on Princes Street itself in 2014.
Other location 10-15 Princes Street, EH2 2AN (550 9220).

Tune

7 Clifton Terrace, EH12 5DR (347 9700, www. tunehotels.com). Bus 2, 3, 4, 12, 25, 26, 31, 33, 44, 44A. **Map** p283 B8.
This 2013 arrival is from a Malaysian company that mostly operates in Indonesia, Malaysia and the Philippines, although it has been making inroads to the British market lately. Opposite Haymarket Station, Tune is on the tram line from the airport and only ten minutes' walk from Princes Street. It's also enormous, with 179 rooms, done up in contemporary style with a red, white and black colour scheme. The catch is the pay-as-you-go pricing structure. Rooms are cheap, but there are extra daily charges for the hairdryer (£2), safe (£3), wi-fi (£4), TV (£4) and even towels and toiletries (£2) – or you can have the lot for £10, the 'comfort package' option. If you find this budget airline-style approach offputting, then clearly Tune isn't for you. But even if you fork out £6 a day for wi-fi and a towel on top of the room rate, a night here still won't break the bank. *Photo p256.*

Hostels

Baxter Hostel *5 West Register Street, EH2 2AA (556 8609, www.thebaxter.eu). Princes Street buses.* **Map** p284 H6. Shared and female-only dormitories.
Caledonian Backpackers *3 Queensferry Street, EH2 4PA (226 2272, www.caledonianbackpackers. com). Princes Street buses.* **Map** p283 D7. Shared and female-only dormitories.
Code Hostel *50 Rose Street North Lane, EH2 2NP (659 9883, www.codehostel.com). Princes Street buses.* **Map** p284 F6. Shared dormitories with 'pod' beds.
Edinburgh Belford Hostel *6-8 Douglas Gardens, EH4 3DA (220 2200, www.hoppo.com/belford). Princes Street buses.* **Map** p283 B7. Shared dormitories; private twin and double rooms.

STOCKBRIDGE
Moderate

Channings

12-16 South Learmonth Gardens, EH4 1EZ (315 2226, www.channings.co.uk). Bus 19, 36, 37, 41, 43, 47, 113. **Map** p279 B4.
Noted polar explorer Sir Ernest Shackleton once lived in these premises, which is why Channings has three suites and two rooms on a Shackleton theme. Guests can now lie in an enormous bathtub and soak up the narrative from a huge print of his ship, the *Endurance*, on the wall adjacent. The other rooms

Motel One. *See p253.*

THE CHAIN GANG

Familiarity need not breed contempt.

Premier Inn Edinburgh
Leith Waterfront.

At the last count, there were 11 **Premier Inns** in and around Edinburgh, another 11 **Travelodges**, four hotels with the **Holiday Inn Express** branding, four from the **Ibis** chain, two plain **Holiday Inns**, two **Novotels**, an **easyHotel** and a **Jurys Inn**. As far as chain hotels go, this is a far from definitive list, but you get the idea.

If all you want is a familiar name – and a bargain bed in most cases – then you can book on the web and know exactly what you're going to get. The critic might say that the whole point of these hotels is their uniformity; that it doesn't matter where you are, they all look the same on the inside. With the curtains drawn, a Novotel on the outskirts of Edinburgh is really no different to one in Leeds, Liverpool or Luton. That's largely true, of course, but what you find outside the front door makes an enormous difference.

The **Premier Inn Edinburgh Leith Waterfront** (51-53 Newhaven Place, Newhaven, 0871 527 8360, www.premierinn. com) has a great location, for example. It's actually at Newhaven Harbour, immediately west of Leith – despite the name – although it could hardly be more 'waterfront'. There are several good options for eats and drinks nearby (including C-Shack, Loch Fyne Seafood & Grill, and Porto & Fi; for all, see p164), and a short coast walk before breakfast is a definite option.

Back in the city centre, the **Travelodge Edinburgh Central Princes Street** (Mouse Lane, off South St Andrew Street, New Town, 0871 559 1855, www.travelodge.co.uk) has what is possibly the worst-looking entrance of any city hotel. Meuse Lane is a bin-strewn blemish behind the Topman and Topshop store on Princes Street. But the location is bang central, the interior is absolutely fine and you could get off the train at Waverley Station and be checked in within minutes.

Ignoring the chains also means ignoring brand-new hotels. As part of the SoCo development on South Bridge, the **Hotel Ibis Edinburgh Centre South Bridge** (77 South Bridge, Old Town, 292 000, www.ibis.com) only opened for business in early 2014. It has an attractive 'urban courtyard' space at the back – between Chambers Street and the Cowgate – and Old Town views from its upper floors.

The chains may not have much character in themselves, unless reception staff inject it, which is always welcome, but a well-chosen chain hotel can still be a good option when visiting Edinburgh.

Tune. *See p254.*

are equally crisp and handsome, with the north-facing accommodation giving views to the Firth of Forth and Fife. It's a discreet hotel that feels far away from central Edinburgh, but isn't; you could walk to the east end of Princes Street in around 15 minutes.

Budget

Inverleith Hotel

5 Inverleith Terrace, EH3 5NS (556 2745, www. inverleithhotel.co.uk). Bus 8, 23, 27. **Map** p280 E2.
Sitting cheek by jowl with the beautiful Royal Botanic Garden Edinburgh (*see p105*), the Inverleith is a small and pleasantly furnished townhouse dating from the early Victorian period. There's no restaurant (although room service brings light meals and sandwiches), but guests are unremittingly positive about the service, friendliness and comfort of the ten rooms (especially the suite with the four-poster bed). Staying here can be quite a bargain in the off-season. The owners also have self-catering apartments; check the website for details.

CALTON HILL & BROUGHTON

Deluxe

No.11 Brunswick Street

11 Brunswick Street, EH7 5JB (557 6910, www.11brunswickst.co.uk). Playhouse buses. **Map** p281 K4.
Quietly making its debut on the Edinburgh hotel scene in 2013, No.11 only has ten rooms, but they include a couple of four-poster suites at the top end and a couple of very well-appointed singles at the other end of the size scale. The interior design is in keeping with the neo-classical terrace that the hotel calls home; the brasserie is worth a look, whether or not you're staying overnight.

★ Paul Kitching 21212

3 Royal Terrace, EH7 5AB (0845 222 1212, www.21212restaurant.co.uk). Playhouse buses. **Map** p281 K5.
The four gorgeous guestrooms above Paul Kitching's restaurant (*see p121*) have the same visual panache as his celebrated cooking. Each of the rooms is slightly different, but all have a contemporary boutique feel, are spacious, super-plush and clearly cut from the same aesthetic fabric. Two have great views north over Royal Terrace Gardens to the Forth; the other pair overlook the enclosed gardens behind.

Expensive

★ Glasshouse

2 Greenside Place, EH1 3AA (525 8200, www. theglasshousehotel.co.uk). Playhouse buses. **Map** p281 J5.
Behind the façade of the former Lady's Glenorchy Church (1846), between the Playhouse Theatre and the Omni Centre, the Glasshouse has been delighting guests with its contemporary design, roof garden (with views to Calton Hill), comfortable beds and amenable service since 2003. The clean, modern lines of the interior design make for a cultured and classy atmosphere; the name is echoed in the floor-to-ceiling windows that offer impressive views over the city. The most desirable rooms either give access to the roof garden or have balconies looking over Edinburgh. *Photo p258.*

Moderate

Apex Waterloo Place

23-27 Waterloo Place, EH1 3BH (523 1819, www.apexhotels.co.uk). Princes Street or Nicolson Street–North Bridge buses. **Map** p281 J6.

Established as a hotel in 1819, this building has been used for a variety of different purposes during its life, most recently as local authority offices. After a refurbishment, it reopened in 2009 and became the Edinburgh flagship for the Apex chain. Modern, central, comfortable and handsome in the 21st-century style, the rooms at the front of the building overlook Waterloo Place and the historic Old Calton Burial Ground. Some premium rooms in a 1970s-era extension at the back have expansive views over the north of the city; others don't – ask when booking.
Other locations Apex City, 61 Grassmarket, Old Town, EH11 2JF; Apex Haymarket, 90 Haymarket Terrace, New Town, EH12 5LQ (for both, 0845 365 0000); Apex International (*see p247*).

Budget

Parliament House
15 Calton Hill, EH1 3BJ (478 4000, www. parliamenthouse-hotel.co.uk). Princes Street or Playhouse buses. **Map** p281 J6.
Parliament House opened back in the days when people thought the new Scottish Parliament would be at the former Royal High School building on nearby Regent Road, which explains the slightly off-kilter name. Still, the location is pretty decent, tucked away down a lane at Calton Hill. Georgian and Edwardian buildings were connected to create the hotel premises, and rooms vary in outlook and size as a result. With a clean, crisp design, some of them are cavernous, while others are cosy; a number are blessed with views of Arthur's Seat or the Old Town.

Hostels

Edinburgh Central SYHA Hostel *9 Haddington Place, EH7 4AL (524 2090, www.syha.org.uk). Bus 7, 10, 11, 12, 14, 16, 22, 25, 49.* **Map** p281 J4.
Shared male-only or female-only dormitories; private single, double, triple and quad rooms.

SOUTH EDINBURGH
Deluxe

Sheraton Grand Hotel & Spa
1 Festival Square, Lothian Road, EH3 9SR (229 9131, www.sheraton.com). Bus 1, 2, 10, 11, 15, 15A, 16, 22, 24, 30, 34, 36, 47, 61. **Map** p284 E8.
Yes, it's a Sheraton and yes, it's often the preserve of expenses-girded business travellers, but it is very, very accomplished. Some of the rooms are ultra-slick with rich wooden floors, huge beds and Edinburgh Castle views; the One Square restaurant makes a serious effort with the food, and the One Spa – with a clientele of its very own – is arguably the best facility of its kind in Scotland.

Expensive

★ Prestonfield
Priestfield Road, Prestonfield, EH16 5UT (225 7800, www.prestonfield.com). Bus 2, 14, 30.
Set in parkland just south of Arthur's Seat, the main building here dates from 1687. Much of its original character remains, from the approach

ESSENTIAL INFORMATION

Paul Kitching 21212.

Glasshouse. See p256.

along a tree-lined drive to the ornate fixtures and fittings inside. Opulent antique furniture and lush upholstery complement the look of the hotel and its well-known restaurant, Rhubarb (*see p140*); the guestrooms, both in the old house or the modern extension out back, are similarly sumptuous, and contemporary comforts abound (plasma-screen TVs, wi-fi). Peacocks and Highland cattle amble about in the grounds. Genuinely impressive.

Moderate

DoubleTree by Hilton
34 Bread Street, EH3 9AF (221 5555, www. doubletree3.hilton.com). Bus 1, 2, 10, 11, 15, 15A, 16, 24, 34, 35, 36, 47. **Map** p284 E8.
Launched as the Point Hotel in 1995, sold and rebranded as the Mercure Point from 2008, sold again and re-rebranded as the DoubleTree by Hilton in 2014, the signage firms of Edinburgh love this place. Whatever name it has this week, it inhabits a former department store just off Lothian Road and is credited with being the city's first boutique hotel given how radical its interior looked in the mid 1990s. The accommodation was freshened up again in 2014; the castle-view suites remain a pretty tempting option.

Edinburgh City Hotel
79 Lauriston Place, EH3 9HA (622 7979, www. edinburghcityhotel.com). Bus 2, 23, 27, 35, 45, 47. **Map** p284 F9.

Two minutes' walk from Tollcross, less than ten minutes from the Grassmarket, this hotel doesn't exactly announce itself to passers-by, and has that vaguely Gothic look shared by many Edinburgh buildings – it was originally a Victorian maternity hospital. Inside, however, achingly modern beds, furniture and facilities have been transplanted into the 19th-century fabric. Friendly, helpful staff; 24-hour residents' bar; happy guests.

★ Salisbury Hotel
43-45 Salisbury Road, EH16 5AA (667 1264, www.the-salisbury.co.uk). Nicolson Street–North Bridge buses. **Map** p285 L11.
Discreetly tucked away on a side road in Newington, the Salisbury merits praise for the standard of accommodation ('now' meets 'then' decor with all mod cons), the friendly approach from its staff, the breakfasts, the food in its basement bistro (Café Cassis) and even the mellow space of its garden. You could splash out on a four-poster if you wanted, but all the rooms are handsome and very good value. It's about a mile south of the Royal Mile.

Salisbury Green
Pollock Halls, 18 Holyrood Park Road, EH16 5AY (651 2007, www.edinburghfirst.co.uk). Bus 2, 14, 30, 33. **Map** p285 M11.
Not to be confused with the relatively nearby Salisbury Boutique Hotel (*see above*), Salisbury Green opened for business in 2006. It's an impressive mid 18th-century mansion, which has been

ESSENTIAL INFORMATION

The Hotel Ceilidh-Donia is in the suburbs, around a mile and a half south of the Royal Mile. But that's the only downside with this fine property, a much-loved, family-run bargain in an attractive Victorian terrace. The rooms are comfortable, the rates are appealing and there is an evening bar menu for residents (5-7pm Mon-Thur), plus the entire property has a fair amount of character. The proprietors are dog owners; take note if you're allergic.

WEST EDINBURGH
Moderate

Dunstane
4 West Coates, EH12 5JQ (337 6169, www. dunstane-hotel-edinburgh.co.uk). Bus 12, 26, 31. Map p283 A8.
A mid 19th-century villa a short distance west of Haymarket Station, Dunstane House has been a hotel for decades. The slightly Gothic frontage gives way to a handsome interior, with many period features (including some four-poster beds). The owners hail from Orkney, which explains the Orcadian influence in the hotel restaurant, Skerries, and perhaps the whisky focus in its Stane Bar. It's also bigger than you think; there's a similarly well-appointed annexe over the road. With 38 rooms in total, the Dunstane is in the first rank of smaller local hotels. *Photo p261.*

Edinburgh Marriott Hotel
111 Glasgow Road, EH12 8NF (334 9191, www.marriott.co.uk). Bus 31, 100, X12.
Way out west, over four miles from Princes Street en route to the airport, the Marriott is one of those hotels that many visitors pass by without noticing. That's a pity. It's a business hotel and not exactly characterful, but the location and market position of the property can mean decent special offers and weekend deals. The rooms are comfortable, modern and well equipped, and amenities include a small pool.

Budget

Brooks Hotel
70-72 Grove Street, EH3 8AP (228 2323, www. brooksedinburgh.com). Bus 1, 34, 35. Map p283 D9.
Just off Fountainbridge, a few minutes from Lothian Road, Brooks offers the delights of a rear courtyard and 46 rooms refurbished in 2013. They say it's a boutique B&B, but it's really too big for that; more of a hotel with no restaurant but an honesty bar and a room service menu available noon-10pm daily.

LEITH & THE COAST
Moderate

Malmaison
1 Tower Place, EH6 7DB (0844 693 0652, www. malmaison.com). Bus 16, 22, 35, 36. Map p288 Y2.

converted into a hotel but retains many of its original fixtures and fittings in public areas. Part of the University of Edinburgh's Edinburgh First events management and accommodation business, it sits within the grounds of an extensive halls of residence, so there are lots of students around in term time. Rooms are in the modern style; most people love the location by Holyrood Park.

Ten Hill Place
10 Hill Place, EH8 9DS (662 2080, www.tenhill place.com). Nicolson Street–North Bridge buses. Map p285 J8.
Home to the Royal College of Surgeons of Edinburgh for aeons, Surgeons' Hall is a grand classical building. Behind the frontage, however, it's an extensive complex, with a modern extension to the original Georgian premises that includes a conference venue and this hotel (access is around the corner on Hill Place). The rooms are slick and modern, and the central location is a selling point – the quartet of premium fourth-floor Skyline rooms have brilliant views across the city. There's a restaurant on site, but also lots of options in the immediate vicinity. For those with a strong constitution, the pathology exhibition at Surgeons' Hall Museums (*see p138*) is a must-see.

Budget

★ Hotel Ceilidh-Donia
14-16 Marchhall Crescent, EH16 5HL (667 2743, www.hotelceilidh-donia.co.uk). Bus 2, 14, 30, 33.

STAY WITH ME

Airbnb has changed the rules.

ESSENTIAL INFORMATION

The internet has a great deal to answer for. **Airbnb** (www.airbnb.co.uk) started in San Francisco in 2008, but now, just a few years later, operates virtually everywhere with accommodation from Anchorage via Kinshasa to Zhangzhou.

For those who don't know, it's a web-based way of bringing travellers together with private hosts. The travellers pay as they would in a hotel or B&B; the hosts rent out a range of accommodation from a spare room to an entire apartment or house.

For a solo traveller, it can be a cheap way to find a bed for the night and someone to chat with. For a couple going on holiday, it could be a way of renting a one-bedroom apartment for a weekend city break. Families also use Airbnb to rent entire houses during the school summer holidays. Payment is via the website – Airbnb takes a cut – while travellers can review accommodation and hosts can review travellers. In theory, this weeds out cynical hosts and careless, heedless guests.

In Edinburgh there are literally hundreds of places available via the site. Backpacker hostels have a presence, but it's mostly private individuals offering anything from a sofa bed in their living room for £20 a night to entire two-bedroom apartments at over £400 a night.

Given that the Airbnb phenomenon has gone from nowhere to near-ubiquity in just a few years, it has taken the questions time to catch up with the answers. Are hosts running unregulated and potentially dangerous apartment lets? Are they insured – buildings, contents, liability – for having strangers come to stay? If someone in Edinburgh can make several thousand pounds simply by renting out their flat for a few weeks in August, are they declaring it as taxable income? Are they paying the same local taxes that hotels or guesthouses would pay?

In autumn 2014, the New York attorney general said that nearly three-quarters of Airbnb rentals in that city were illegal. Into 2015 the number of Airbnb properties in New York was still in four figures.

A compromise will doubtless be reached, eventually. Asking Airbnb hosts to take on the same regulatory burden as big hotel chains is clearly unworkable, but city authorities and national tax agencies won't tolerate significant commercial activity that goes untaxed. Meanwhile, Airbnb is surely now too successful to simply shut down.

There are now 13 Malmaisons across the UK, from Aberdeen to London via Belfast, but this property was the very first. Launched in 1994, it's housed in a former seamen's mission in a grand building where the Water of Leith meets the docks. When it opened, it brought a sense of boutique style to the waterfront; more than 20 years later, it may be showing its age but the suites and rooms are still furnished with plush intent (check the website for special offers). If you want to go out for something to eat, Leith has everything from Michelin-starred wonders to naughty cake.

LONGER STAY

If you're here for more than a few nights, as many visitors are during the festival period, it may pay to investigate a short-stay apartment, or a suite in an apart-hotel. Some are basic and cheap; others are much more luxurious and expensive. During the various August festivals and in the run-up to Hogmanay, the short lettings market is busy and prices inflate. As with hotels, book as early as possible for the best deals. Always bear in mind that many of these properties will demand a minimum stay, which could be anywhere from a few nights to a fortnight.

Historic properties

Although many short-stay apartments are in modern blocks, or buildings of Georgian or Victorian origin, a number have even more historical resonance. Operated by the National Trust for Scotland, **Gladstone's Flat** is housed in a 17th-century building off the Royal Mile. **Pilrig House** is another 17th-century property, located in a park halfway between Leith and the city centre. And on the south side of the city, at the foot of the Pentland Hills, sit **Swanston Cottages**, traditional farm cottages renovated to modern standards.

Gladstone's Flat *477B Lawnmarket, Old Town, EH1 2NT (458 0305, www.nts.org.uk). Bus 23, 27, 41, 42, 67.* **Map** p284 G7.
Pilrig House *30 Pilrig House Close, Leith, EH6 5RF (554 4794, www.pilrighouse.com). Bus 11, 36.*
Swanston Cottages *111 Swanston Road, South Edinburgh, EH10 7DS (445 2239, www.swanston. co.uk). Bus 4, 18 to Oxgangs Road/Swanston Road then 0.5 mile walk.*

Apartments

Apartments and apart-hotels vary in size, quality and price. Whether you're a couple looking for a two-week stay or whether you need to house a largeish party over a long weekend, you should find something to suit.

Dunstane. *See p259.*

Calton Apartments

44 Annandale Street, Calton Hill & Broughton, EH7 4AZ (556 3221, www.townhousehotels. co.uk). Bus 7, 10, 11, 12, 14, 16, 22, 25, 49. **Map** p281 J3.

Eight apartments in a modern residential building, these are simply furnished, fairly central and cheap at certain times of year. Parking is free. The owners of Calton Apartments also run Frederick House Hotel (*see p253*).

Canon Court Apartments

20 Canonmills, Stockbridge, EH3 5LH (0800 027 1035, www.canoncourt.co.uk). Bus 8, 23, 27, 36. **Map** p280 F3.

Simple, modern self-catering accommodation not far from the Royal Botanic Garden Edinburgh (*see p105*). There are a couple of good cafés nearby too: Bluebird and Water of Leith Café-Bistro (for both, *see p108*).

Fountain Court Apartments

622 6677, www.fountaincourtapartments.com.
Fountain Court rents out a variety of modern serviced apartments in several locations: three lots in Grove Street (West Edinburgh), one lot by the Union Canal basin (West Edinburgh), one lot on Young Street (New Town) and the Royal Garden Apartments on Queen Street (New Town). Check the website for offers and short lets.

Holyrood Aparthotel

1 Nether Bakehouse, Holyrood, Old Town, EH8 8PE (524 3200, www.holyroodaparthotel.com). Bus 35, 60. **Map** p285 L7.

Modern serviced apartments, tucked away just a few minutes from the Royal Mile, the Scottish Parliament and the Palace of Holyroodhouse.

Kew House & Balbirnie Apartment

1 Kew Terrace, West Edinburgh, EH12 5JE (313 0700, www.kewhouse.com). Bus 12, 26, 31.
Kew House is effectively a B&B, but with a separate, serviced apartment to let nearby in a modern terrace. There's not much character, but it's roughly halfway between Murrayfield Stadium and Haymarket and so handy for both.

Numberfive Self-Catering Apartments

5 Abercorn Terrace, Portobello, EH15 2DD (669 1044, www.numberfive.com). Bus 15, 26, 40, 45, 113.
This seaside special sits just a couple of minutes' walk from the beach. The accommodation is basic but decent; there is a good discount for stays of more than six weeks.

West End Apartments

2 Learmonth Terrace, Stockbridge, EH4 1PQ (332 0717, www.holidaysinedinburgh.com). Bus 19, 36, 37, 41, 43, 47, 113. **Map** p279 B4.

Housed in a Victorian terrace, these apartments have a traditional feel, and some come with great views. They're located just on the other side of Dean Bridge from the West End.

Apartment agencies

The agencies and websites listed below all offer a variety of self-catering accommodation around the city, ranging from luxurious family houses to more basic and affordable apartments. Some properties are let out for as little as a week at a time; others require minimum terms of a month or longer.

Festival Beds differs from the others in that it deals in B&B accommodation, not self-catered apartments, and rents properties only during August (the others operate year-round).

Apartments in Edinburgh *556 8309, www.apartmentsinedinburgh.com.*
Clouds *550 3808, www.clouds.co.uk.*
Edinburgh Apartments *www.festival apartments.com.*
Edinburgh-Flats *513 9969, www.edinburgh-flats.com.*
Factotum *0845 119 6000, www.factotum.co.uk.*
Festival Beds *225 1101, www.festivalbeds.co.uk.*
Festival Flats *01620 810620, www.festival flats.net.*

UNIVERSITY ACCOMMODATION

The city's universities not only rent rooms in their halls of residence during academic vacations, but offer much more besides. Their business diversification in recent years has seen events management added to the list of activities, with add-ons such as hotel ownership and self-catering apartments for let all year round.

Some of the more basic accommodation is relatively far-flung at campuses on the edge of the city; some of the options are very comfortable and only about 15 minutes' walk from Princes Street. Whether you just want something simple for a quick overnight stay or you're coming to the capital for a couple of weeks, the universities are worth checking out.

Edinburgh Conference Centre (Heriot-Watt University) *451 3115, www.edinburgh-conference.com.*
Edinburgh First (University of Edinburgh) *651 2189, www.edinburghfirst.co.uk.*
Edinburgh Napier University Conferences & Lettings *455 3722, www.napier.ac.uk.*
Queen Margaret University *474 0000, www. www.qmuc.ac.uk.*

Getting Around

ARRIVING & LEAVING

By air

Edinburgh Airport *0844 448 8833, www.edinburghairport.com.*
Edinburgh Airport is about seven miles west of the city centre. The airport is served by all major UK and European airlines, while direct USA flights are run by United (to Chicago O'Hare and New York Newark; www.united.com).

The smoothest way to and from the airport is on the newly opened **Edinburgh Trams** system (*see p264*), which stops at various locations in the west of the city before reaching the centre. The service operates from early morning until late at night – check the website for the full timetable. Tickets cost £5 single/£8 return.

Despite the introduction of the tram system in 2014, the well-established **Airlink 100 bus** kept going, also shuttling travellers to and from the airport. Its main advantage over the trams: it's slightly cheaper, at £4 single/£7 return. Cheaper still, although it takes an hour or more to reach the centre, is the **Lothian Buses 35 bus**, which runs to and from the airport via the west of the city, the centre, then to Leith; a single fare is £1.50. In the wee, small hours when neither trams nor the usual buses are running, the **N22 night bus** goes from the airport to Leith via the city centre and back; a ticket costs £3. See www.lothianbuses.com for details of the Airlink 100, 35 and N22 services.

A **taxi** journey to central Edinburgh usually takes around 25 minutes and costs £20-£25.

By road

National Express (0871 781 8181, www.nationalexpress.com) operates coach services between Edinburgh and destinations in England and Wales. Coaches run by **Scottish Citylink** (0871 266 3333, www.citylink.co.uk) serve a variety of towns around Scotland, while **Megabus** (www.megabus.com) runs budget bus services from Edinburgh to a number of Scottish destinations (including Glasgow) and some in England. All buses and coaches arrive and depart from St Andrew Square bus station.

Edinburgh Bus Station *St Andrew Square, New Town. Princes Street buses.* **Open** 4.30am-midnight Mon-Thur, Sun; 4.30am-12.30am Fri, Sat. *Ticket hall* 6.30am-7.45pm Mon-Sat; 8am-8pm Sun. **Map** p280 H5.

By train

Waverley Station *Waverley Bridge, New Town (0845 748 4950, www.nationalrail.co.uk). Princes Street buses.* **Open** 4am-12.45am Mon-Sat; 6am-12.45am Sun. **Map** p280 H6.
From here, you can catch trains to London (on the East Coast main line), Glasgow (every 15mins during the day Mon-Sat, every 30mins evenings Mon-Sat and all day Sun) and all other Scottish destinations. Check www.nationalrail.co.uk and www.scotrail.co.uk for details. Edinburgh's other city-centre railway station, Haymarket, is just over a mile west of Waverley.

PUBLIC TRANSPORT

Public transport in Edinburgh is reasonably fast and reliable, and certainly a better option than driving. For lost property, *see p267*.

Buses

The city and its suburbs are well served by a comprehensive bus network. **Lothian Buses** (554 4494, www.lothianbuses.com) runs the majority of services throughout Edinburgh and into East Lothian and Midlothian; it's these services that are listed throughout the guide.

Several parts of town are served by a large number of buses. In these cases, rather than list each bus on every occasion, we've broken them into groupings. These groupings are listed below, detailing the bus routes that serve the respective streets or areas:

Nicolson Street–North Bridge buses 3, 5, 7, 8, 14, 29, 30, 31, 33, 37, 47, 49.
Playhouse buses 1, 4, 5, 7, 10, 11, 12, 14, 16, 19, 22, 25, 26, 34, 44, 45, 49.
Princes Street buses 1, 3, 4, 10, 11, 12, 15, 15A, 16, 19, 22, 24, 25, 26, 29, 30, 31, 33, 34, 37, 41, 43, 44, 44A, 47, 61, Airlink 100, 104, 113.
Night Buses, operated by Lothian Buses, run nightly on 11 routes

around the city. Most routes run hourly from around midnight, but a few are more frequent (including the N22 between Leith and the airport; *see left*). For more, see www.lothianbuses.com.

Single journeys in Edinburgh cost £1.50 for adults or 70p for children aged 5-15; under-5s travel free, up to a maximum of two kids per adult passenger. There are no single fares on the city's Night Bus network; unlimited travel costs £3 a night. Exact change is required for all single fares.

Daytickets allow for unlimited travel on Lothian Buses (excluding the Airlink 100 bus, tour services and Night Buses). Daytickets cost £3.50 (£2 for 5-15s), and are available when you board your first bus of the day.

For longer periods, the **Ridacard** affords the holder unlimited travel on the network. The card costs £17 for one week (£13 for accredited students, £9 for 5-15s) or £51 for four weeks (£40 students, £29 5-15s). Ridacards are also valid on trams, the Airlink 100 bus and Night Buses, but not tour buses. You can buy a Ridacard from Lothian Buses Travelshops (*see below*). There is also a Transport for Edinburgh app that allows for mobile ticketing on Lothian Buses and trams, and the citysmart card that you can preload with cash then use on buses and trams; more details from the Lothian Buses website.

In addition to Lothian Buses, **First Group** (01324 602200, www.firstgroup.co.uk) runs some services between Edinburgh and outlying areas. Tickets are not transferable between operators.

Lothian Buses Travelshops
27 Hanover Street, New Town. Princes Street buses. **Open** 9am-6pm Mon-Fri; 9am-5.30pm Sat. **Map** p280 G6. *Waverley Bridge, New Town. Princes Street buses.* **Open** 9am-7pm Mon, Thur; 9am-6pm Tue, Wed, Fri; 9am-5.30pm Sat; 10am-5.30pm Sun. **Map** p280 H6.

Trains

Rail services within Scotland are run by ScotRail, although the franchise changed hands in 2015 with Dutch company Abellio taking up the reins. Service details are available from the ScotRail website (www.scotrail.co.uk)

or on 0330 303 0112. The information desk at Waverley station (*see p263*) can also help.

As well as Waverley and Haymarket stations, Edinburgh has several suburban stations; check with ScotRail for full details.

Trams

After a highly controversial and eye-poppingly expensive gestation period, **Edinburgh Trams** finally started rolling in 2014, with just one line rather than the network originally planned. The route goes from York Place in the city centre, via St Andrew Square, Princes Street, Haymarket and Murrayfield out to the west of Edinburgh, then on to the airport. It's efficient and cheap for any journey other than to the airport itself; tickets are integrated with those of the Lothian Buses network.

For more information, see www.edinburghtrams.com or call 555 6363.

TAXIS

Black cabs

Most of Edinburgh's taxis are black cabs carrying up to five passengers. When a taxi's yellow 'For Hire' light is on, you can hail it in the street. The fare structure is complicated, varying according to time of day, number of passengers, distance travelled and waiting time, but it can be downloaded from the City of Edinburgh Council website (www.edinburgh.gov.uk).

Phoning for a taxi is advisable at night or if you're outside the city centre. To book, contact **Central Taxis** (229 2468, www.taxis-edinburgh.co.uk), **City Cabs** (228 1211, www.citycabs.co.uk) or **Computer Cabs** (272 8000, www.comcab-edinburgh.co.uk).

More taxis now accept payment by credit or debit card, but always check if it's 'cash only' as you get in. For lost property, *see p267.*

Private hire cars

Minicabs (saloon cars) are generally cheaper than black cabs and may be able to carry more passengers (specify when booking). Cars must be booked in advance. Reputable firms include **Bluebird** (603 1682) and **Persevere** (555 2323).

Complaints

Complaints about a taxicab or private hire company journey should be made in writing to the City of

Edinburgh Council Licensing Section, 249 High Street, Edinburgh EH1 1YJ. Be sure to make a note of the date and time of the journey and the licence number of the vehicle.

DRIVING

If you're planning on staying within Edinburgh during your visit, driving isn't recommended. For one thing, the city is reasonably small and very accessible either on foot or with public transport. For another, the number of pedestrianised and one-way streets, plus the Old Town's medieval street plan, means the traffic can be terrible. Also, parking is difficult and pricey.

If you're a member of a motoring organisation in your home country, check to see if it has a reciprocal agreement with a UK equivalent.

AA (Automobile Association) *0870 600 0371, www.theaa.com.*
RAC (Royal Automobile Club) *0870 572 2722, www.rac.co.uk.*

Car hire

All firms below have branches at or near the airport; several also have offices in the city centre. Shop around for the best rate and always check the level of insurance included in the price.

Arnold Clark *333 0124, www.arnoldclarkrental.com.*
Avis *0844 544 6004, www.avis.co.uk.*
Budget *0844 544 4605, www.budget.co.uk.*
Enterprise *348 4000, www.enterprise.co.uk.*
Europcar *0371 384 3406, www.europcar.co.uk.*
Hertz *0843 309 3025, www.hertz.co.uk.*
Thrifty *335 3900, www.thrifty.co.uk.*

Parking

Edinburgh is divided into parking zones and in the heavily controlled central zones charges apply 8.30am-6.30pm Mon-Sat. In the peripheral and extended zones immediately surrounding these, charges apply 8.30am-5.30pm Mon-Fri. Charges vary from £1/hr to £3.20/hr depending how close to the centre you are; there are also maximum time limits on parking. You can pay at a pay-and-display machine or by mobile with RingGo (instructions on the pay-and-display machine). Look out for 'resident only' parking bays; you need the relevant permit to park at these during the restricted periods indicated.

Information on parking in Edinburgh, including a map of parking locations, can be found at www.edinburgh.gov.uk/parking.

If you park illegally, you may get a ticket (the fine is £60, reduced to £30 if paid within 14 days) or even towed (a release fee of £150 plus the fine applies). If you think your car has been towed, call the police on 101 – they have a record of all towed cars.

If you don't fancy taking your chances with street parking, there are commercial car parks around the city. This is more expensive than street parking but less hassle, and there's no limit on the amount of time you can park your car. All the car parks listed below are open 24 hours. Rates vary; call for details. There's a full list of city-centre car parks, complete with a map, at www.edinburgh.gov.uk/parking.

Castle Terrace *Old Town (0345 050 7080, www.ncp.co.uk).* **Rates** £7.10/2hrs-£24.10/24hrs. **Map** p280 E8.
Q-Park Omni *Greenside Row, Calton Hill & Broughton (0870 442 0104, www.omniedinburgh.co.uk).* **Rates** £3/1hr-£17/24hrs. **Map** p281 J5.
Quartermile *South Edinburgh (0345 050 7080, www.ncp.co.uk).* **Rates** £1.50/1hr-£16/24hrs. **Map** p284 G9.
Waverley *New Street, Old Town (0345 222 4224, www.apcoa.co.uk).* **Rates** £3/1hr-£17/24hrs. **Map** p281 J6.

CYCLING

With repurposed old railway lines, road-edge cycle lanes and shared lanes for buses, taxis and bikes, Edinburgh can be a decent city to get around by bicycle. The City of Edinburgh Council even has a series of maps titled 'Explore Edinburgh By Bike', downloadable from its website (www.edinburgh.gov.uk). That said, embedded tram lines in the city centre, steep, cobbled streets in the Old Town and the ever-present threat of motor traffic present safety hazards to the cyclist. Mind how you go. Also be careful when tethering a bike and leaving it for any period of time: a nice bike in an obscure corner plus breakable lock equals theft risk.

WALKING

Edinburgh is best explored on foot. The usual caution should be exercised at night, but generally walking around the city is safe, rewarding and – for the able-bodied – a lot more sensible than trying to drive and park.

Resources A-Z

AGE RESTRICTIONS

You have to be 18 to drink in Scotland, although some bars and clubs admit only over-21s. The legal age for driving is 17 but some car rental firms won't hire cars to people under 25; ask when booking. The age of consent is 16.

ATTITUDE & ETIQUETTE

Edinburgh is, on the whole, an informal city. A handful of high-end restaurants may insist on a jacket or jacket and tie, but this is generally a pretty relaxed place, whether you're here doing business or on holiday.

CONSUMER

If you pay with a credit card, you can cancel payment or get reimbursed if there's a problem. The Citizens Advice Bureau (see p266) can also help.

CUSTOMS

Citizens entering the UK from outside the EU must adhere to duty-free import limits:

● 200 cigarettes or 100 cigarillos or 50 cigars or 250g of tobacco
● 4 litres still table wine plus either 1 litre spirits or strong liqueurs (above 22% abv) or 2 litres fortified wine (under 22% abv), sparkling wine or other liqueurs
● cash worth €10,000 (or equivalent sum in other currency)
● other goods (including electronic devices and perfume) to the value of no more than £390.

The import of meat, poultry, fruit, plants, flowers and protected animals is restricted or forbidden.

In theory, people arriving from an EU country can bring in unlimited amounts of most goods although very large amounts of alcohol or tobacco products will be more likely to prompt a spot-check by customs officials. For more details, see www.gov.uk.

DISABLED

It's forbidden to widen the entrances of, or add ramps to, listed buildings; and parts of the Old Town have wheelchair-unfriendly narrow pavements. However, new buildings are required by equal-opportunity legislation to be fully disabled-accessible.

Lothian Buses' (see p263) new fleet of vehicles is accessible to passengers in wheelchairs; call 555 6363 for details. Newer black taxis (see p264) are wheelchair-accessible; always specify when booking.

Most theatres and cinemas are fitted with induction loops for the hard of hearing. Ask when booking.

For more on disabled living in Edinburgh, contact **Grapevine**, part of the Lothian Centre for Integrated Living.

Grapevine *Norton Park, 57 Albion Road, EH7 5QY (475 2370, www. lothiancil.org.uk).* **Open** *By phone* 10am-4pm Mon-Thur.

DRUGS

Both hard and soft drugs are illegal in Scotland, as they are in the rest of the UK. For the Know the Score drugs helpline, see p266.

ELECTRICITY

The UK electricity supply is 230V, 50Hz rather than the 120V, 60Hz used in the US; standard UK plugs are three-pin. Foreign visitors will need to run appliances via an adaptor.

CONSULATES

For a list of local consular offices, see www.visitscotland.com/travel/information. The majority of consulates do not accept personal callers without an appointment.

Honorary Consul of Canada
07702 359916 mobile, www. canadainternational.gc.ca. There is no office.
Irish Consulate General
16 Randolph Crescent, EH3 7TT (226 7711, www.irishconsulate scotland.co.uk)
US Consulate General
3 Regent Terrace, EH7 5BW (enquiries 556 8315, www. usembassy.org.uk/scotland).

EMERGENCIES

In the event of a serious accident, fire or incident, call 999 and specify whether you require an ambulance, the fire service or the police. The standard European number 112 also works. *See also* p266 **Helplines**.

GAY & LESBIAN

Several campaigning groups maintain offices in Edinburgh, such as **Stonewall Scotland** (474 8019, www.stonewallscotland. org.uk) and the **Equality Network** (467 6039, www.equality-network. org). Young people should check the **LGBT Youth Scotland** website at www.lgbtyouth.org.uk. There is a comprehensive list of gay community organisations on the **LGBT Health & Wellbeing Centre** website, *see p266*.

For more on the local gay scene, including local bars and cafés, *see* pp176-178.

LGBT Health & Wellbeing Centre
*9 Howe Street, New Town (523 1100,
LGBT Helpline 0300 123 2523,
www.lgbthealth.org.uk). Bus 24,
29, 42.* **Open** *Centre varies. LGBT
Helpline* noon-9pm Tue, Wed. **Map**
p280 F5. Health advice and support,
along with social events.

HEALTH

National Health Service (NHS)
treatment is free to EU nationals,
UK residents, workers from overseas
in most cases and those studying
here. All can register with a doctor
(commonly known as a general
practitioner, or GP). There are
no NHS charges for accident and
emergency treatment, treatment
of some communicable diseases
(including STDs), compulsory
psychiatric treatment, treatment
imposed by court order or family
planning. If you aren't eligible
to see an NHS doctor, you will be
charged the cost price for medicines
prescribed by a private doctor.

If you don't fit into any of the
above categories but want to find out
if you still qualify for free treatment,
see www.nhsinform.co.uk.

Accident & emergency

Royal Hospital for Sick Children
*9 Sciennes Road, South Edinburgh
(536 1000). Bus 5, 41, 42, 67.*
24-hour casualty department for
children aged under 13.
Royal Infirmary of Edinburgh *51
Little France Crescent, Old Dalkeith
Road, South Edinburgh (536 1000).
Bus 7, 8, 18, 21, 24, 33, 38, 49.*
24-hour casualty department for
adults and children aged 13 and over.

Abortion, contraception, HIV & sexual health

Lothian Sexual Health *2A
Chalmers Street, South Edinburgh
(536 1070, www.lothiansexualhealth.
scot.nhs.uk). Bus 23, 27, 35, 45, 47.*
Open *By appointment* 9am-4pm
Mon-Fri. *Drop-in clinic* 8.30-10am
Mon-Fri. **Map** p284 G9. Confidential
advice and information about
contraception, pregnancy, sexually
transmitted diseases and sexual
problems. For people aged 18 and
under there is a dedicated service
on the premises called Healthy
Respect (07831 527976, www.
healthyrespect.co.uk).

Waverley Care *3 Mansfield
Place, Broughton (558 1425,
www.waverleycare.org). Bus 8.*
Open 10am-4pm Mon-Thur;

10am-1pm Fri. **Map** p280 H4. Care
and support for people living with
HIV or hepatitis C.

Dentists

Chalmers Dental Centre
*3 Chalmers Street, South Edinburgh
(www.nhslothian.scot.nhs.uk).
Bus 23, 27, 35, 45, 47.* **Open**
9am-5pm Mon-Fri. **Map** p284 G9.
A walk-in clinic offering emergency
care – but arrive before 3pm if you
want to be seen that day. Out of
hours, phone the Lothian Dental
Advice Line on 536 4800; staff can
arrange for you to receive emergency
care if necessary.

Hospitals

See below **Accident & emergency**.

Opticians

The Alexandra Pavilion offers a free
walk-in service for emergency eye
complaints.

Princess Alexandra Eye Pavilion
*45 Chalmers Street, South Edinburgh
(536 1000). Bus 23, 27, 35, 45, 47.*
Open phone for times. **Map** p284 G9.

Pharmacies

Edinburgh doesn't have a 24-hour
pharmacy, but some of the pharmacies
attached to larger supermarkets stay
open into the evening.

HELPLINES

Alcoholics Anonymous *0845 769
7555, www.alcoholics-anonymous.
org.uk.* **Open** 24hrs daily. Help with
drinking problems and alcohol
addiction.

Childline *0800 1111, www.childline.
org.uk.* **Open** 24hrs daily. Counselling
for children and young people aged
up to 18.

Know the Score Drugs Helpline
*0800 587 5879, www.knowthescore.
info.* **Open** 8am-11pm daily.
Straightforward and confidential
advice about drugs.

NHS 24 *111, www.nhs24.com.*
Open 24hrs daily. Health helpline
provided by the NHS.

Rape Crisis Scotland *08088
010302, www.rapecrisisscotland.
org.uk.* **Open** 6pm-midnight daily.
Crisis support for anyone affected
by sexual violence: female, male
or child.

Samaritans *08457 909090,
www.samaritans.org.* **Open** 24hrs
daily. Emotional support for anyone
in distress.

Terence Higgins Trust (THT) Direct
0808 802 1221. **Open** 10am-8pm
Mon-Fri. Information, support and
advice about sexual health and HIV.
Victim Support *0808 168 9111,
www.victimsupport.org.uk.* **Open**
8am-8pm Mon-Fri; 9am-7pm Sat,
Sun. Specialist support for victims
of crime and witnesses.

Drop-in centres

Advice Shop *249 High Street, Old
Town (200 2360, www.edinburgh.
gov.uk). Nicolson Street–North
Bridge buses.* **Open** 8.30am-4pm
Mon, Wed, Thur; 10am-4.30pm
Tue; 8.30am-3pm Fri. **Map** p285 J7.
Advice on debt problems and
welfare benefits.
Citizens Advice Bureau *58 Dundas
Street, New Town (appointments 557
3681, www.citizensadviceedinburgh.
org.uk). Bus 23, 27.* **Open** *General
appointments* 9.10am-4pm Mon, Tue,
Thur; 9.10am-1pm Wed; 1-4pm Fri.
Legal Clinic 6-8pm Wed. *Employment
Matters Clinic* 5.45-7.30pm Thur (call
603 7714 to book). *Drop-in sessions*
1.30-4pm Wed; 9.10am-1pm Fri.
Map p284 F4.
Free advice on legal, financial and
personal matters. Aside from this
city-centre branch, there are four
other offices around Edinburgh;
see website for details.

ID

ID is not widely required in the
UK, but you will need a passport or
driver's licence (with photograph)
for changing money, cashing
travellers' cheques and so on.

INSURANCE

Non-nationals should arrange
baggage, medical and trip-
cancellation insurance before
departures. Medical centres will
ask for your insurance company
and policy number; keep the details
with you at all times.

LEFT LUGGAGE

Edinburgh Airport *344 3486,
www.edinburghairport.com.*
Open 4am-10pm daily. Left luggage
facilities are run by Luggage-Point
(www.luggage-point.co.uk).
St Andrew Square Bus Station
Open 6am-midnight daily. There
are lockers of various sizes in the
station costing £3-£10.
Waverley Station *558 3829,
www.left-baggage.com.* **Open**
7am-11pm daily. The left luggage
facilities are run by Excess Baggage.

LEGAL HELP

If a legal problem arises, contact your embassy, consulate or high commission (see p265). You can get advice from any **Citizens Advice Bureau** (see p266) or one of the organisations listed below. If you need financial assistance, be sure to ask about legal aid eligibility. For leaflets explaining how the system works, write to the Scottish Legal Aid Board. Advice on problems concerning visas and immigration can be obtained from the Immigration Advisory Service.

Edinburgh & Lothians Regional Equality Council *14 Forth Street, FH1 3LII (556 0411, www.elrec. org.uk)*. Experts in discrimination, human rights and race relations.
Immigration Advisory Service *69 Buchanan Street, Glasgow, G1 3HL (enquiries & appointments 0141 314 3581, helpline 0844 887 0111; www.iasservices.org.uk)*. Immigration law specialists.
Law Society of Scotland *26 Drumsheugh Gardens, EH3 7YR (226 7411, www.lawscot.org.uk)*. Professional body that can help the public find a suitable solicitor.
Scottish Legal Aid Board *44 Drumsheugh Gardens, EH3 7SW (office 226 7061, Legal Aid information line 0845 122 8686, www.slab.org.uk)*. Legal aid information for the public.

LIBRARIES

The **Central Library** (see p55) stocks an enormous range of publications, and has a large reference section. You don't need to live locally to use the library for reference, but only local residents are permitted to join the lending library.

The **National Library of Scotland** (see p58) is a deposit library. The Reading Rooms are open for reference and research; admission is by ticket only to approved applicants.

University of Edinburgh Main Library *George Square, South Edinburgh (650 3409, www.ed. ac.uk)*. Bus 2, 23, 27, 35, 41, 42, 45, 47, 60, 67. **Open** *Term-time* 7.30am-2.30am daily. *Helpdesk services* 9am-7.50pm Mon-Thur; 9am-6.50pm Fri; noon-4.50pm Sat; noon-6.50pm Sun. For vacation opening times, call or check website. **Map** p284 H9. The library is open to university staff, students and academic visitors, university alumni, staff and students from other universities with a reciprocal membership agreement, NHS Lothian staff, Office of Lifelong Learning students and members of some other institutions. The general public can also become members: reference access is free but charges apply for external borrowing.

LOST PROPERTY

Always inform the police if you lose anything, if only to validate insurance claims; the non-emergency police number is 101. A lost passport should also be reported at once to your embassy or consulate (see p265), if relevant. Below are the details of the lost property offices for items left on public transport.

Edinburgh Airport *344 3486, www.edinburghairport.com*. **Open** 4am-10pm daily. The lost property office is run by Luggage-Point (www.luggage-point.com).
Lothian Buses *27 Hanover Street, New Town (475 0652, www.lothianbuses.com)*. *Princes Street buses*. **Open** 10am-1.30pm, 2-5.30pm Mon-Fri. **Map** p280 G6.
Taxis & general *Police Scotland, Fettes Avenue, Stockbridge (311 3141)*. *Bus 19, 24, 29, 37, 38, 42, 113*. **Open** 8am-noon, 1-3.45pm Mon, Tue, Thur; 8am-noon, 1-3pm Wed, 8am-noon, 1-3.30pm Fri. **Map** p279 A4. All property that has been left in black cabs, as well as in the street or shops, gets sent here.
Waverley Station *550 2333, www. excess-baggage.com*. **Open** 9am-5pm Mon-Fri. Waverley's lost property facilities are operated by Excess Baggage. For items lost in other stations or on trains, contact the individual station or train operator.

MEDIA

Scotland's print media has had a torrid time in recent years. Broadsheet and tabloid newspapers alike have seen sales drop as new generations grow up taking alternative media for granted, advertising has migrated elsewhere and it's generally accepted that some venerable old names of the Scottish press are on borrowed time. Meanwhile, the BBC, Scottish Television and ITV Border provide broadcast television coverage, and the BBC and various independents run radio stations.

Newspapers

Much of the London-based press prints Scottish editions. Nationals such as the *Guardian* and Sunday sibling the *Observer* (left-leaning, London-centric), the *Times* and the *Sunday Times* (right-slanted, business-heavy), and the *Sun* (gossip, soft porn) are all widely sold; the *Metro* freesheet also publishes a Scottish edition with an entertaining letters page. Alongside the imports sit a number of exclusively Scottish papers, as below.

The Scotsman & Scotland on Sunday
www.scotsman.com
Based in Edinburgh, the *Scotsman* was once a broadsheet newspaper of record but it has changed hands twice since 1995. It now has the attitudes of a right-leaning, mid-market tabloid and its circulation is in freefall. *Scotland on Sunday* is its sister paper.
The Herald & Sunday Herald
www.heraldscotland.com
This Glasgow-based broadsheet is the longstanding rival to the *Scotsman* but has adapted more ably to the contemporary media environment. Its sister, the *Sunday Herald*, was the only Scottish newspaper to back a yes vote during the 2014 independence referendum.
The National
www.thenational.scot
Published by Newsquest (owner of the *Herald* and *Sunday Herald*), the *National* was launched in the wake of the 2014 referendum to fill an obvious market gap. It is the only daily newspaper in Scotland to support Scottish independence.
Daily Record & Sunday Mail
www.dailyrecord.co.uk, www. sundaymail.co.uk
Published by the Mirror Group in Glasgow, the *Record* is effectively the house journal for the Scottish Labour Party and for Glasgow's two biggest football teams. More of the same on Sundays with the *Sunday Mail*.
Evening News
www.edinburghnews.scotsman.com
From the same publisher as the *Scotsman*, this is Edinburgh's daily evening tabloid. It still does a decent job of being a local paper relevant to the citizens of the capital.

Magazines

The city's most high-profile magazine is the **List** (free, www.list.co.uk). Issued every two months, it contains full cultural listings for Edinburgh and Glasgow, and also publishes an annual Eating & Drinking Guide to both cities (£5.95). The *List* has competition from the **Skinny** (free, www.theskinny.co.uk).

ESSENTIAL INFORMATION

Tradition still has a place on Scots' magazine shelves. In print since 1739, the monthly **Scots Magazine** (£3, www.scotsmagazine.com) is a kind of McReader's Digest, while the fresher **Scottish Field** (£3.95, www.scottishfield.co.uk) offers lifestyle features. The literary scene is covered by journals such as the **Edinburgh Review** (£7.99, www. edinburgh-review.com) while the fortnightly **Holyrood** (subscription only, www.holyrood.com) covers current affairs and politics.

Radio

UK national stations are accessible in Edinburgh, chief among them the five main BBC stations (www.bbc. co.uk/radio): **Radio 1** (97-99 FM, youth-slanted pop), **Radio 2** (88-91 FM, adult pop and dadrock), **Radio 3** (90-93 FM, classical), **Radio 4** (92-95 FM, current affairs and culture) and **Radio 5 Live** (693 & 909 MW, news and sport). There are also a number of other national stations available solely on digital.

Several stations are unique to Scotland. **BBC Radio Scotland** (92-95 FM) commands respect for its mix of talk and music-based programming. **Forth 1** (97.3 FM, www.forth1.com) is a sort of East Central Scotland Radio 1 with commercials, while **Forth 2** (1548 AM, www.forth2.com) plays older music. To get properly local, check out **Shore Radio** (www.shoreradio. org), a Leith-based, online station that has everything from drivetime shows to studio guests playing live and late-night dance music sessions.

Television

Both the BBC and ITV in Scotland opt in and out of the UK-wide output, with BBC Scotland (www.bbc.co.uk) and Scottish Television (www.stv.tv) contributing regularly to their respective networks. ITV Border (www.itv.com) straddles England and Scotland.

MONEY

The UK currency is the pound sterling (£). One pound equals 100 pence (p). 1p and 2p coins are copper; 5p, 10p, 20p and 50p coins are silver; the £1 coin is gold; the £2 coin is silver with a gold edge. The euro may be accepted in some shops in tourist areas.

Three banks (Bank of Scotland, Clydesdale Bank, Royal Bank of Scotland) issue their own Scottish paper notes. The colours of the notes

varies between the three, but they're not far from the following: blue £5; brown £10; purple/pink £20; green £50; red/maroon £100.

Banks & ATMs

Scottish banking did not have a good credit crunch. Bank of Scotland is now owned by Lloyds Banking Group, with the UK state holding a significant stake in the latter. Royal Bank of Scotland (RBS), meanwhile, was effectively nationalised by the UK state in 2008 to stop it going bust; the Clydesdale has been owned by the National Australia Bank since 1987. Consequently, the independence of all three 'Scottish' banks is moot, although these brands dominate the market in Scotland. In terms of operation and regulation they're no different to London-domiciled banks such as Barclays or HSBC.

Branches of the big three are generally open 9am/9.15am to 4.30pm/5pm Mon-Fri, but some remain open later, and open on Sat mornings. ATMs, usually situated outside banks, give 24-hour access to cash; most will also allow you to draw money on a card tied to an international network such as Visa.

Other major banking names in Edinburgh include the following.

Barclays *Unit 2, 10-15 Princes Street, New Town (03457 345345, www.barclays.co.uk). Princes Street buses.* **Open** 9am-5pm Mon-Fri; 9am-3pm Sat. **Map** p280 H6.

HSBC *76 Hanover Street, New Town (existing customers 03457 404404; new customers 03456 040626; www. hsbc.co.uk). Bus 6, 23, 27 or Princes Street buses.* **Open** 9am-5pm Mon-Fri; 9am-2pm Sat. **Map** p280 G6.

NatWest *8 George Street, New Town (0845 366 1965, www.natwest.com). Princes Street buses.* **Open** 9am-5pm Mon-Fri; 10am-3pm Sat. **Map** p280 G6.

TSB *28 Hanover Street, New Town (517 9998, www.tsb.co.uk). Bus 6, 23, 27 or Princes Street buses.* **Open** 9am-5pm Mon, Tue, Fri; 10am-5pm Wed; 9am-5.30pm Thur; 10am-4pm Sat. **Map** p280 G6.

Bureaux de change

Bureaux de change charge fees for cashing travellers' cheques or exchanging currency. Commission rates vary greatly; it pays to shop around. There are bureaux de change at the airport and Waverley station; others are scattered around in areas popular with tourists.

Most banks offer currency exchange; rates are usually better than at bureaux de change. Commission is often charged for cashing travellers' cheques in foreign currencies, but not for sterling travellers' cheques, provided you cash them at a bank affiliated to the issuing bank. Get a list when you buy your cheques. When changing currency or travellers' cheques, you'll need photo ID, such as a passport or driver's licence.

Lost/stolen credit cards

Report lost or stolen credit cards both to the police and the 24-hour phone lines listed below. Inform your bank by phone and in writing.

American Express *01273 696933, www.americanexpress.com.*
Diners Club *0845 862 2935, www.dinersclub.co.uk.*
MasterCard *0800 964767, www.mastercard.co.uk.*
Visa *0800 891725, www.visaeurope.com.*

Tax

By law, sales tax (also known as Value Added Tax or VAT) is included in all prices advertised in UK shops. However, a handful of hotels may quote you rates that exclude tax. Always check when booking.

OPENING HOURS

In general, business hours are 9.30am-5.30pm Mon-Fri. Most shops are open 9am-5.30pm Mon-Sat and 11am-5pm on Sun. Many restaurants are open all day; some stay open well beyond 11pm. Officially, closing time for pubs is 11pm, but most pubs have licences to sell alcohol until 1am. Many shops, restaurants, pubs and clubs in Edinburgh operate longer hours during the August festivals. For bank opening hours, *see left*.

POLICE

In an emergency, call 999. For non-emergency matters, call 101. The main police station for the city centre is at 2 Gayfield Square, Broughton.

POSTAL SERVICES

The UK has a fairly reliable postal service. If you have a query, contact customer services on 03457 740740.

Post office hours vary, although most branches will be open by 9am and many closed by 5pm weekdays, 9am-12.30pm Sat. Some larger post offices have more extensive hours, while smaller branches shut for lunch, even for one or more afternoons each week. Three central post offices are listed below; for others, check online at www.royalmail.com.

You can buy individual stamps at post offices, and books of first- or second-class stamps at newsagents and supermarkets that display the appropriate sign. A first-class stamp for a regular letter costs 62p; second-class stamps are 53p. For details of other rates, see www.royalmail.com.

Edinburgh City *5-6 Princes Mall, Princes Street, New Town. Princes Street buses.* **Open** 8.30am-6pm Mon, Wed-Fri; 9am-6pm Tue; 9am-5.30pm Sat. **Map** p280 H6.
Frederick Street *40 Frederick Street, New Town. Bus 41, 42, 61 or Princes Street buses.* **Open** 9am-5.30pm Mon, Wed-Fri; 9.30am-5.30pm Tue; 9am-12.30pm Sat. **Map** p280 F6.
St Mary's Street *46 St Mary's Street, Old Town. Nicolson Street–North Bridge buses.* **Open** 9am-5.30pm Mon-Fri; 9am-12.30pm Sat. **Map** p285 J7.

Poste restante

If you intend to stay in Edinburgh for a while, friends and family in your home country can write to you care of a local post office, where mail will be kept at the enquiries desk for up to one month. For details, see www.postoffice.co.uk/poste-restante or call 03457 740740 to check which Edinburgh post offices offer the service. Take photo ID (ideally a passport) when you collect your mail.

RELIGION

BAPTIST: Charlotte Baptist Chapel *204 Rose Street, New Town (225 4812, www.charlottechapel.org). Princes Street buses.* **Services** 11am, 6.30pm Sun. **Map** p284 E7.
BUDDHIST: The Portobello Buddhist Priory (27 Brighton Place, Portobello, 669 9622, www.portobellobuddhist.org.uk) offers daily meditation. The **Edinburgh Buddhist Centre** (35-37 Bread Street, South Edinburgh, 07599 718556, www.edinburghbuddhist centre.org.uk) offers classes and meditation sessions.
CATHOLIC: St Mary's Cathedral *61 York Place, New Town (556 1798, www.stmaryscathedral.co.uk). Playhouse buses.* **Services** 10am, 12.45pm Mon-Fri; 10am, 6pm (vigil) Sat; 9.30am, 11.30am, 6pm (vigil), 7.30pm Sun. **Map** p280 H5.

EPISCOPALIAN: St Mary's Cathedral *Palmerston Place, New Town (225 6293, www.cathedral. net). Bus 2, 3, 4, 12, 25, 26, 31, 33, 44, 44A.* **Services** 7.30am, 1.05pm, 5.30pm Mon-Wed, Fri; 7.30am, 11.30am, 1.05pm, 5.30pm Thur; 7.30am Sat; 8am, 10.30am, 3.30pm Sun. **Map** p283 C8.
HINDU: Edinburgh Hindu Mandir & Cultural Centre *St Andrew Place, Leith (07890 726117, www. edinburghhindumandir.org.uk). Bus 12, 16, 35.* **Meetings** see website. **Map** p288 Y4.
JEWISH: Edinburgh Hebrew Congregation *4A Salisbury Road, South Edinburgh (07734 291836, www.ehcong.com). Bus 2, 3, 5, 7, 8, 14, 29, 30, 31, 33, 37, 47, 49.* **Services** phone for details. **Map** p285 L11.
METHODIST: City of Edinburgh Methodist Church *25 Nicolson Square, South Edinburgh (662 8635, www.edinburghmethodist.com). Nicolson Street–North Bridge buses.* **Services** 11am, 6.30pm Sun. **Map** p286 J8.
MUSLIM: Edinburgh Central Mosque *50 Potterrow, South Edinburgh (667 1777, www. edmosque.com). Bus 2, 41, 42.* **Prayer times** Jumu'a 1.10pm Fri. **Map** p285 J8.
PRESBYTERIAN: High Kirk of St Giles *High Street, Old Town (225 4363, www.stgilescathedral.org.uk). Bus 23, 27, 28, 35, 41, 42.* **Services** noon Mon-Sat; 8am, 10am, 11.30am, 6pm, 8pm Sun. **Map** p284 E4.
QUAKER: Quaker Meeting House *7 Victoria Terrace, Victoria Street, Old Town (225 4825, www.quaker scotland.org). Bus 2, 23, 27, 41, 42, 45.* **Main meetings** 12.30pm Wed; 11am Sun. **Map** p284 D4.
SIKH: Guru Nanak Sikh Gurdwara *1 Sheriff Brae, Leith (553 7207, www.edinburghsikhs.com). Bus 7, 10, 14, 16, 21, 22, 34, 35, 36.* **Services** phone for details. **Map** p288 X3.

SAFETY & SECURITY

Violent crime is relatively rare in central Edinburgh, but it still pays to use common sense. Keep your wallet and other valuables out of sight; and never leave bags, coats or purses unattended.

Edinburgh's city centre is a generally safe and civilised place, but the lairy pub culture on the Cowgate and Lothian Road can be a little unpleasant at closing time. Ill-lit parks such as the Meadows have been the scene of (infrequent) assaults down

the years. Women should also avoid the Leith backstreets, traditionally the city's main red-light district. Away from the centre, in Edinburgh's peripheral housing schemes, the environment is rougher and poverty more apparent. Best avoided.

SMOKING

Smoking is now banned in enclosed public spaces across Scotland, including all restaurants and pubs. The law is strictly enforced, as fines for establishments that break it are punitively high.

STUDY

A good deal of Edinburgh's character is defined by its big student population. Most study at one of four universities, of which the most prestigious is the **University of Edinburgh** (Old College, Old Town, South Edinburgh, 650 1000, www.ed.ac.uk). Granted a royal charter in 1582, it's since been joined by **Heriot-Watt University** in 1966 (Riccarton Campus, Currie, 449 5111, www.hw.ac.uk), **Edinburgh Napier University** in 1992 (Craighouse Road, South Edinburgh, 0845 260 6040, www.napier.ac.uk) and **Queen Margaret University** in 2007 (Queen Margaret Drive, Musselburgh, 474 0000, www.qmuc. ac.uk). There's also the **Edinburgh College of Art** (Lauriston Place, South Edinburgh, 651 5800, www.eca.ed.ac.uk) although, formally, it has been part of the University of Edinburgh since 2011.

TELEPHONES
Dialling & codes

The area code for Edinburgh is 0131; Glasgow's area code is 0141. To reach a UK number from abroad, dial the international access code (011 if you're in the USA, for example) or the '+' symbol on a mobile phone; then 44 for the UK; then the area code, omitting the first 0; then the number. So to call Edinburgh Castle (*see p55*) from the USA, for example, dial 011 44 131 225 9846.

Numbers beginning 075, 077, 078 and 079 are mobile phones (cellphones). From landlines, a call to phone numbers beginning 0800 or 0808 are toll-free; those prefixed 084 and 087 are charged at up to 15p a minute; and numbers beginning 09 are billed at very high rates indeed. The cost of calls from mobiles to these numbers vary: check with

ESSENTIAL INFORMATION

ESSENTIAL INFORMATION

LOCAL CLIMATE

Average temperatures and monthly rainfall in Edinburgh.
The warmest months are June-Sept, the coldest Dec-Mar;
the drier months Feb-May, the wettest Jan, July, Sept and Oct.

	High (°C/°F)	Low (°C/°F)	Rainfall (mm/in)
Jan	7/45	1.5/35	67/2.7
Feb	7.5/45.5	1.5/35	47/1.9
Mar	9.5/49	3/37	52/2
Apr	12/53	4.5/40	40/1.6
May	15/59	7/45	49/1.9
June	17/63	10/50	61/2.4
July	19/66	11.5/53	65/2.6
Aug	19/66	11.5/53	60/2.4
Sept	16.5/62	9.5/49	64/2.5
Oct	13/55.5	6.5/44	76/3
Nov	9.5/49	4/39	62/2.4
Dec	7/45	1.5/35	61/2.4

your mobile provider. To call abroad from Edinburgh, dial the international access code (00) or the '+' symbol, then the country code (86 for China, 91 for India, 1 for the USA), then the local number.

Mobile phones

Almost everyone has at least one, most people have a smartphone these days and one of the biggest growth sectors in the industry in recent years has been the switch to 4G services. The website of the UK's independent regulator for the communications industry, Ofcom, has useful information on getting the best from your mobile phone: http://consumers.ofcom.org.uk/phone/mobile-phones.

Operator services

Operator *100*.
International operator *155*.
Directory enquiries *118 500*.

TIME

Edinburgh operates on Greenwich Mean Time (GMT). Clocks move an hour forward to run on British Summer Time (BST) at 1am on the last Sunday in March, and return to GMT on the last Sunday in October.

TIPPING

Tipping around ten per cent in barbers, some bars (but not pubs), cafés, hairdressers, restaurants and taxis is normal. In Edinburgh, and Scotland more generally, a minority of bars and restaurants add service automatically to bills; always check to avoid paying twice.

TOILETS

Shopping malls and major department stores have public toilets, while the City of Edinburgh Council (www.edinburgh.gov.uk) maintains a decent number of public toilets around town. The most central are on Castlehill and Hunter Square in the Old Town, and in Princes Street Gardens in the New Town. See website for full details.

TOURIST INFORMATION

The **Edinburgh Information Centre** at the east end of Princes Street is the main tourist office in the city. As well as distributing a wealth of information on tours and attractions, staff can book hotels and event tickets, car hire and coach trips. There's also internet access and a bureau de change. The information point at the airport has a smaller range of services, but can help with tours and hotels. There are other centres around the Lothians; see www.visitscotland.com for details.

Edinburgh Information Centre
Above Princes Mall, 3 Princes Street, New Town (0845 225 5121). Princes Street buses. **Open** *9am-5pm Mon-Sat; 10am-5pm Sun; later in July & Aug.* **Map** p280 H6.
Edinburgh Airport Tourist Information Desk *Edinburgh Airport (473 3690).* **Open** *Apr-Oct 7.30am-9.30pm daily. Nov-Mar 7.30am-7.30pm Mon-Fri; 7.30am-7pm Sat, Sun.*

VISAS & IMMIGRATION

EU citizens do not require a visa to visit the UK; citizens of Australia, Canada, New Zealand and the USA

can enter with only a passport for tourist visits of up to six months as long as they can show they can support themselves during their visit and plan to return. Go online to www.ukvisas.gov.uk to check your visa status for visiting, studying or working well before you travel, or contact the British embassy, consulate or high commission in your own country.

WEIGHTS & MEASURES

In Edinburgh, as throughout the British Isles, the traditional imperial system of measurement is used for some things, metric for others. Yes, it's confusing. Road distances are in miles, for example, while grammes and kilogrammes are used for foodstuffs; beer in a pub comes in pints; in the supermarket it's in 500ml bottles. Try to view this as an endearing quirk.

WHEN TO GO

There isn't really a best or a worst time to visit Edinburgh: it all depends on what you want from your visit. The cultural festivals mean August is the liveliest and most interesting month of the year. However, it's also the busiest: the pavements are awash with tourists and street performers, and there are queues everywhere. Similarly, Hogmanay draws hordes each year, but is not to everyone's taste. For more on August's festivals, *see pp32-37*; for other special events, *see pp28-31*.

The changeable Scottish weather further complicates matters. Winters can be wet and chilly, summers are pleasant although far from hot, but rain is a threat all year round. There's no guarantee of good weather at any time, but between the months of June and September, when the days are long and you can finally take off your anorak, the city is probably at its most amenable. For a chart detailing the local climate, *see above*.

WOMEN

Women travelling on their own face the usual hassles, but Edinburgh is generally a safe place. Take the same precautions you'd take in any big city. Many of the city's black cab firms now give priority to lone women, whether booked by phone (recommended after midnight) or flagged in the street. *See p269* for general safety and security tips.

Further Reference

BOOKS

Fiction

Kate Atkinson *One Good Turn*
Crime novel set in Edinburgh
during the August festivals.
Iain Banks *Complicity*
A visceral, body-littered thriller,
with spot-on characterisation of
both the city and the protagonists.
Iain M Banks *The Culture series*
Soaring, imaginative, utopian space
opera from the writer of *Complicity*,
who used a middle initial for his sci fi.
Ron Butlin *Night Visits*
'Edinburgh at its grandest, coldest
and hardest,' as the *Times Literary
Supplement* put it.
Laura Hird *Born Free*
Family life on a local housing estate.
James Hogg *The Private Memoirs
and Confessions of a Justified Sinner*
A jibe against religious bigotry
in the 17th and 18th centuries.
Paul Johnston *The Quint
Dalrymple series*
Futuristic detective fiction set in
a dystopian Edinburgh city state
with a year-round festival.
Ken MacLeod *The Night Sessions*
Artificial intelligence, faith, religion
and terrorism in the Edinburgh of
the 2030s.
Alexander McCall Smith
The Sunday Philosophy Club;
44 Scotland Street
The prolific prof's output includes
a detective series and a collection
of whimsical stories.
Ian Rankin *Resurrection Men*
One of Rankin's best novels in the
celebrated Inspector Rebus series.
JK Rowling *Harry Potter and
the Philosopher's Stone*, et al
Rowling actually did write the
first Potter books in cafés around
Edinburgh; the last was written
in the Balmoral Hotel.
Sir Walter Scott *The Heart of
Midlothian*
Scott's 1818 novel contains, among
many other tales, an account of the
Porteous lynching of 1736.
Muriel Spark *The Prime of
Miss Jean Brodie*
Schoolteacher Jean Brodie makes a
stand against the city's intransigence
and convention.
Robert Louis Stevenson
Edinburgh: Picturesque Notes; *The
Strange Case of Dr Jekyll & Mr Hyde*
A perceptive study; fiction inspired
by local criminal Deacon Brodie.

Charles Stross *Halting State*
Near future sci fi with gaming,
spies, police and thieves.
Irvine Welsh *Trainspotting*;
Porno; *Filth*; *The Acid House*
The city's underbelly uncovered.

Non-fiction

Neil Ascherson *Stone Voices*
An unusual, insightful and tender
meditation on Scotland.
Donald Campbell *Edinburgh:
A Cultural and Literary History*
A digressionary wander through
Edinburgh's cultural past.
David Daiches *Edinburgh*
A highly readable and academically
sound history of the city.
Jan-Andrew Henderson
The Town Below the Ground
The Old Town's underground slums.
James U Thomson *Edinburgh
Curiosities*
A look at the city's history that
reveals its dark and quirky nature.
AJ Youngson *The Making of
Classical Edinburgh*
An exhaustive account of the city's
Georgian townscape.

Poetry

Robert Burns *The Complete
Poems & Songs*
Scotland's national poet.
William Dunbar *Selected Poems*
Vibrant, bawdy poems by the local
poet, priest and member of James
IV's court.
Robert Fergusson *Selected Poems*
Born in Edinburgh in 1750,
Fergusson died in poverty in the
city's Bedlam just 24 years later.
William Topaz McGonagall
Poetic Gems
Was it all a prank? Decide for
yourself in this collection by
Edinburgh-born McGonagall, one of
the worst poets in literary history.

FILMS

The 39 Steps *dir Alfred Hitchcock*
(1935)
Hitchcock's adaptation of John
Buchan's novel; Ralph Thomas's
1959 remake featured the city
more prominently.
Chariots of Fire *dir David
Puttnam* (1981)
Based on the 1924 Olympics,
Puttnam's film has spectacular
shots of Salisbury Crags.

The Da Vinci Code *dir Ron
Howard* (2006)
Rosslyn Chapel plays a key role in
Howard's blockbuster adaptation
of Dan Brown's novel.
The Illusionist *dir Sylvain
Chomet* (2010)
Animated feature about a French
illusionist in Scotland, and illusions.
The Prime of Miss Jean Brodie
dir Ronald Neame (1969)
The brilliant Maggie Smith
misconceives la crème de la crème.
Sunshine on Leith *dir Dexter
Fletcher* (2013)
Two soldiers come home from
Afghanistan and try to build a life;
music by the Proclaimers.
Trainspotting *dir Danny Boyle*
(1996)
Edinburgh's heroin underclass,
largely shot in Glasgow.

WEBSITES

Caledonian Mercury
www.caledonianmercury.com
Online newspaper with analysis
and discussion.
City of Edinburgh Council
www.edinburgh.gov.uk
Local government services.
Edinburgh Festivals
www.edinburgh-festivals.com
Umbrella site for the various festivals
from WOW247 and the *Scotsman*.
Edinburgh's Hogmanay
www.edinburghshogmanay.com
Details of New Year festivities.
Historic Scotland
www.historic-scotland.gov.uk
Government body in charge of
historic monuments.
The List
www.list.co.uk
Local online resource for city events.
National Trust for Scotland
www.nts.org.uk
Information on all NTS properties
around the country.
Scotland's People
www.scotlandspeople.gov.uk
Trace your family history.
Scottish Government
www.scotland.gov.uk
The official government website.
Time Out Edinburgh
www.timeout.com/edinburgh
Your online complement to this book.
VisitScotland
www.visitscotland.com
Scotland's official tourist agency;
Edinburgh has its own
comprehensive section.

ESSENTIAL INFORMATION

Index

INDEX

INDEX

INDEX

Maps

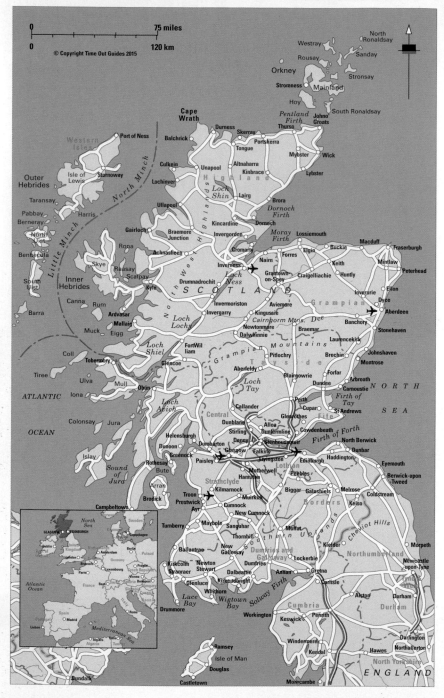

MAPS

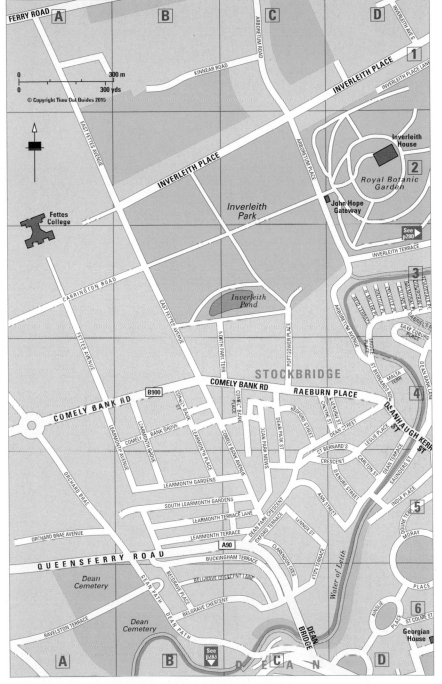

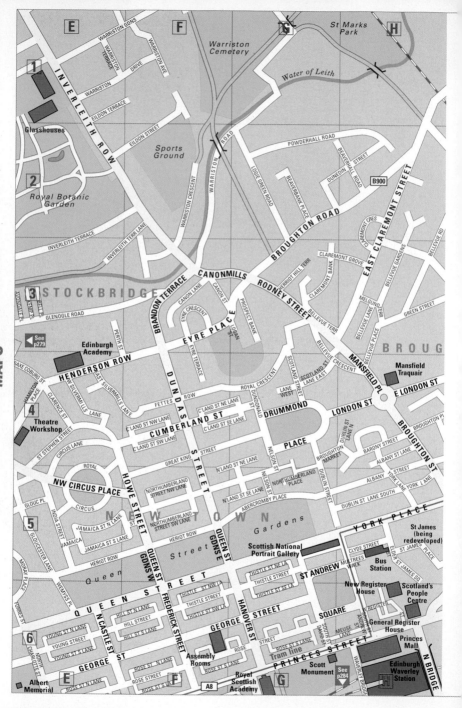

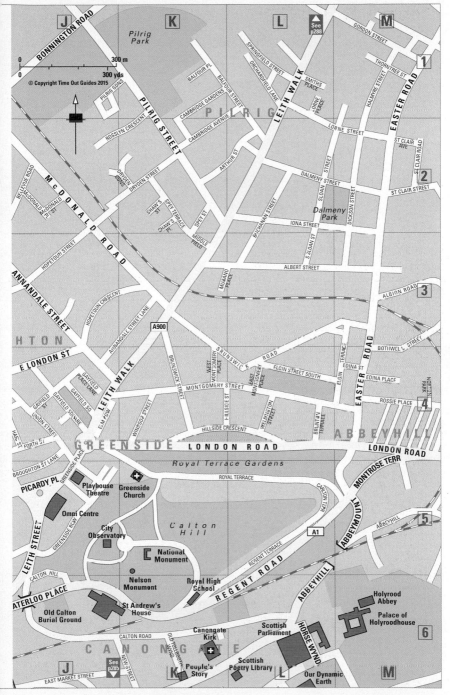

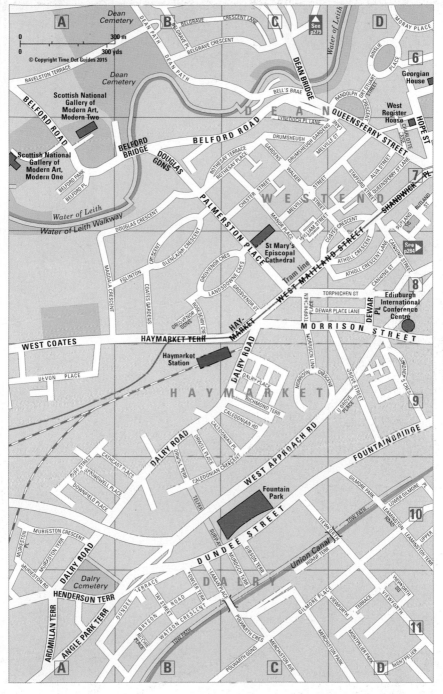

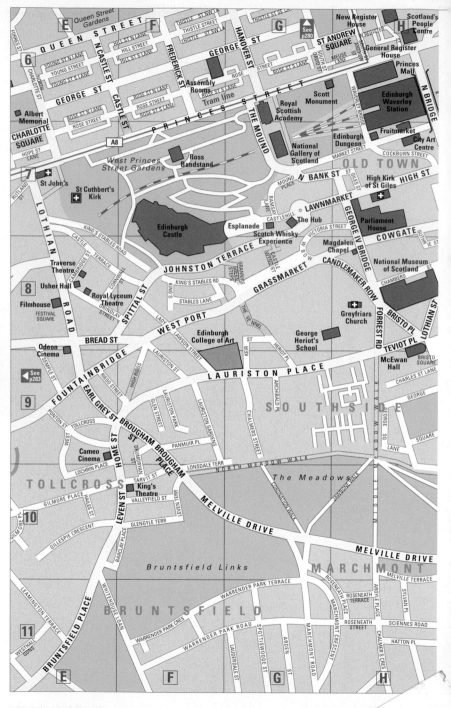

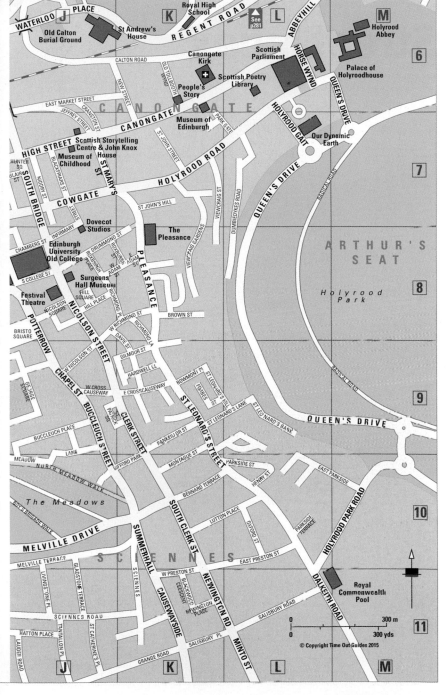

Street Index

STREET INDEX

STREET INDEX

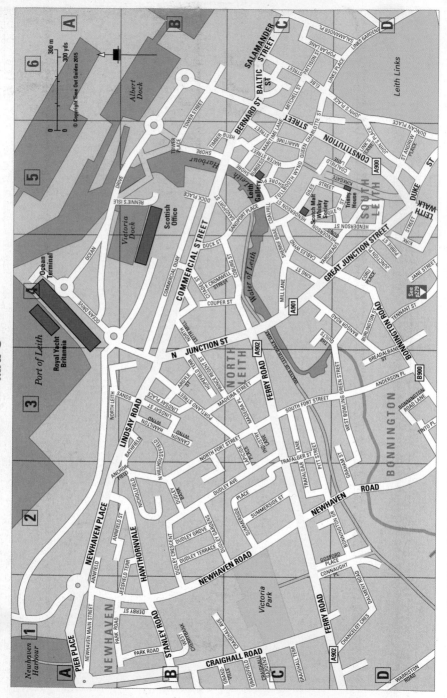

MAPS